MITCHELL'S
ELECTRONIC FUEL INJECTION
TROUBLESHOOTING GUIDE

DOMESTIC VEHICLES

FISHER BOOKS

Publishers: Bill Fisher
Helen Fisher
Howard Fisher
Tom Monroe, P.E., S.A.E.

Cover Design: David Fischer

Cover Photo: Bill Keller

Fisher Books
P.O. Box 38040
Tucson, AZ 85740-8040
(602) 292-9080

Library of Congress Cataloging-in-Publication Data
Mitchell's electronic fuel injection troubleshooting guide. Domestic vehicles.
    p.   cm.
  ISBN 1-55561-032-3
  1. Automobiles—Motors—Electronic fuel injection systems-
-Maintenance and repair—Handbooks, manuals, etc. I. Mitchell International. II. Title: Electronic fuel injection troubleshooting guide. Domestic vehicles.
TL214.F78M57   1989
629.25'3—dc20   89-25780
                                                                CIP

Prepared Expressly for
FISHER BOOKS
by
MITCHELL INTERNATIONAL, INC.

© 1989 MITCHELL INTERNATIONAL

CONTENTS

APPLICATION	FUEL INJECTION SYSTEM	
AMC (Jeep/Eagle for 1988)		
1983-84 Alliance, Encore - 1.4L California	AMC AFC	Pg. 8
1983-84 Alliance, Encore - 1.4L & 1.7L Federal	AMC TBI	Pg. 27
1985-89 Alliance, Cherokee, Comanche, Encore, Wagoneer, Wrangler - 1.4L, 1.7L & 2.5L	AMC/Renix	Pg. 36
1987 GTA - 2.0L	Renix II	Pg. 64
1987-89 Cherokee, Comanche, Wagoneer - 4.0L	AMC Multi-Point	Pg. 16
CHRYSLER MOTORS		

1983-84 E-Class
1983-88 New Yorker, 600 - 2.2L, 2.5L
1984-89 Acclaim, Daytona, Dynasty, Laser,
   New Yorker, Spirit - 2.5L
1985-89 Aries, Caravan, Caravelle, Horizon,
   Lancer, LeBaron, LeBaron GTS, Omni,
   Reliant, Shadow, Sundance, Voyager - 2.2L, 2.5L
1988-89 Dakota, "D" & "W" Series Pickup,
   Ramcharger, Van - 2.5L, 3.9L, 5.2L
	Chrysler TBI	Pg. 159

1984 E-Class
1984-89 Daytona Turbo, Laser XE, New Yorker,
   600, Caraville, Shadow, Sundance - 2.2L Turbo
1985-88 Lancer, LeBaron, LeBaron GTS,
   Omni GL, Shelby Charger, Limousine - 2.2L
1987 Caravelle, Shadow, Sundance
1988-89 Caravan, Dynasty, New Yorker, Voyager - 3.0L
1989 Acclaim, Daytona, Lancer, LeBaron,
   Shadow, Spirit, Sundance - 2.5L Turbo
	Chrysler Multi-Point	Pg. 128

1983-88 Colt 1983-89 Conquest	Mitsubishi ECI	Pg. 93
1987-89 Colt Vista 1989 Colt, Colt Wagon, Raider	Mitsubishi MPFI	Pg. 109
EAGLE		
1989 Medallion	Bosch AFC	Pg. 77
1989 Premier - 3.0L	Eagle Multi-Point	Pg. 197
1989 Premier - 2.5L	Eagle TBI	Pg. 204
FORD MOTOR CO.		

1980 Versailles - 5.0L
1981 LTD, Marquis - 5.0L California
1981-84 Continental, Mark VI, Town Car - 3.8L & 5.0L
1983-84 Country Squire, Crown Victoria,
   Grand Marquis, Mark VII - 5.0L
1983-84 LTD, Marquis - 3.8L
1985 Colony Park, Continental, Crown Victoria, Grand Marquis,
   LTD, Mark VII, Marquis, Town Car - 5.0L
1985-86 Capri, Cougar, LTD, Marquis,
   Mustang, Thunderbird - 3.8L
1987 Cougar, Thunderbird - 3.8L
	Ford High Pressure CFI	Pg. 210

APPLICATION	FUEL INJECTION SYSTEM	
FORD MOTOR CO. (Cont.) 1985-87 Tempo, Topaz - 2.3L 1986-88 Taurus - 2.5L 1987-89 Escort, EXP, Lynx - 1.9L	Ford Low Pressure CFI	Pg. 225
1983 Escort, EXP, Lynx, LN7 - 1.6L 1984 Capri, Cougar, Mustang, Thunderbird - 2.3L Turbo 1984-85 Escort, EXP, Lynx - 1.6L & 1.6L Turbo 1985-87 Aerostar, Ranger - 2.3L 1985-89 Bronco, "E" Series, "F" Series - 5.0L	Ford Multi-Point	Pg. 241
1987-89 Festiva, Merkur XR4Ti, Scorpio, Tracer	Bosch AFC	Pg. 77

1985-88 Capri, Cougar, Mustang,
      Thunderbird - 2.3L Turbo
1986-89 Bronco II, Ranger - 2.3L & 2.9L
1986-87 Capri, Colony Park, Continental, Cougar, Country Squire,
      Crown Victoria, Grand Marquis, LTD, Mark VII, Marquis,
      Mustang, Thunderbird, Town Car - 5.0L
1986-89 Escort, EXP, Lynx - 1.9L H.O.
1986-89 Aerostar, Sable, Taurus - 3.0L
1987-89 Bronco, "E" Series, "F" Series - 4.9L, 5.8L, 7.5L
1987 Colony Park, Continental, Cougar, Country Squire,
      Mark VII, Thunderbird, Town Car - 5.0L
1987-89 Mustang, Tempo, Topaz - 2.3L
1987-89 Mark VII, Mustang - 5.0L
1988 Colony Park, Cougar, Country Squire,
      Grand Marquis, Thunderbird, Town Car,
      LTD Crown Victoria - 5.0L H.O
1989 Probe - 2.2L & 2.2L Turbo
	Ford Multi-Point	Pg. 241

GENERAL MOTORS
Cadillac
  1980 Eldorado, Seville - 6.0L Federal
  1981-85 Brougham, DeVille,
      Eldorado, Seville - 6.0L
  1984-87 DeVille, Eldorado,
      Fleetwood, Seville - 4.1L
  1982-86 Limousine - 6.0L
  1988-89 Deville, Eldorado,
      Fleetwood, Seville - 4.5L
	General Motors DFI	Pg. 328

Buick
  1982-89 Century, Skylark, Somerset - 2.5L
  1988-89 Century & Skyhawk - 2.5L
  1982-89 Skyhawk - 1.8L & 2.0L, 2.0L H.O.
  1989 Electra Wagon, LeSabre Wagon - 4.3L
Cadillac
  1983-87 Cimarron - 2.0L
	GM EFI - Single Unit	Pg. 396

APPLICATION	FUEL INJECTION SYSTEM
GENERAL MOTORS (Cont.)	
Chevrolet	
1982-86 Camaro, Celebrity, Citation - 2.5L	
1988-89 Camaro - 5.0L	
1983-88 Cavalier - 2.0L, 2.0L H.O.	
1986 Astro, S10 - 2.5L & 2.8L	
1987-88 S10, T10 - 2.5L & 2.8L	
1986-89 Astro, El Camino, Caprice,	
Malibu, Monte Carlo - 4.3L, 5.0L & 5.7L	
1987-89 Beretta, Corsica - 2.0L H.O.	
1987-89 Celebrity - 2.5L	
1987-89 Blazer, "G" Series,	
"R" Series, "V" Series - 4.3L	
1987-89 Blazer, "G" Series, "P" Series, "R" Series,	
Suburban, "V" Series - 5.0L & 5.7L	
1987-89 "P" Series, "R" Series,	
Suburban, "V" Series - 7.4L	
1988-89 "G" Series - 7.4L	
GMC	
1986-87 Safari, S15 - 2.5L & 2.8L	
1988-89 S15, T15 - 2.5L & 2.8L	
1986-89 Caballero, Safari - 4.3L	
1987-89 "G" Series, Jimmy,	
"R" Series, "V" Series - 4.3L	
1987-89 "G" Series, Jimmy, "P" Series,	
"R" Series, Suburban, "V" Series - 5.0L & 5.7L	
1987-89 "P" Series, "P" Series, "R" Series,	
Suburban, "V" Series - 7.4L	
1988-89 "G" Series - 7.4L	
1989 LeMans - 1.5	
Oldsmobile	
1982-86 Calais, Cutlass Ciera, Omega - 2.5L	
1983-87 Firenza - 1.8L & 2.0L	
1988 Firenza - 2.0L, 2.0L High Output	
1987-89 Calais, Cutlass Ciera - 2.5L	
1989 Cutlass Cruiser - 4.3L	
Pontiac	
1982-86 Firebird, Phoenix, 6000 - 2.5L	
1988-89 Firebird - 5.0L	
1983-87 Sunbird (2000) - 1.8L & 2.0L	
1988-89 Sunbird - 2.0L	
1984-89 Fiero, Grand Am - 2.5L	
1986 Bonneville, Grand Prix,	
Parisienne - 4.3L	
1987-88 Grand Prix - 4.3L	
1987-88 6000 - 2.5L	
1989 Safari Wagon - 4.3L	GM EFI - Single Unit Pg. 396
Chevrolet	
1982-84 Camaro - 5.0L	
1982-84 Corvette - 5.7L	
Pontiac	
1982-84 Firebird - 5.0L	GM EFI - Dual Unit Pg. 386
Chevrolet	
1988 Nova - 1.6L	GM EFI Pg. 449

4

APPLICATION	FUEL INJECTION SYSTEM
GENERAL MOTORS (Cont.)	
Chevrolet	
1988-89 Spectrum - 1.5L	Isuzu EFI Pg. 459
Chevrolet	
Sprint Turbo	Bosch AFC Pg. 77
Buick	
1984-88 Century, Regal - 3.8L	
1985-87 Electra, LeSabre, Somerset, Skylark - 3.0L & 3.8L	
1986 Skyhawk - 1.8L Turbo	
1987-89 Century - 2.8L & 3.3L	
1987-89 Electra, LeSabre - 3.8L	
1987 Skyhawk Turbo - 2.0L	
1987 Skylark, Somerset - 3.0L	
1988-89 Skylark - 2.3L, 3.0L & 3.3L	
1988-89 Regal - 2.8L & 3.1L	
1989 Reatta, Riviera - 3.8L	
Cadillac	
1985-88 Cimarron - 2.8L	
1989 DeVille, Eldorado, Fleetwood - 3.8L	
Chevrolet	
1985-86 Camaro, Cavalier, Celebrity, Citation II, Corvette - 2.8L, 5.0L & 5.7L	
1987-89 Beretta, Camaro, Cavalier, Celebrity, Corsica - 2.8L	
1987-89 Camaro, Corvette - 5.0L, 5.7L	
Oldsmobile	
1984-86 Calais, Cutlass Ciera, Firenza, Ninety-Eight, Regency, Toronado - 2.8L & 3.8L	
1987-88 Calais - 3.0L	
1987 Firenza - 2.8L	
1987-89 Cutlass Ciera, Cutlass Cruiser, Delta 88, Ninety-Eight - 3.3L & 3.8L	
1988-89 Cutlass Supreme - 2.8L & 3.1L	
1988 Cutlass Calais - 2.3L	
Pontiac	
1984-86 Sunbird (2000) Turbo - 1.8L	
1985-86 Fiero, Firebird, Grand Am, 6000 - 2.8L, 3.0L & 5.0L	
1987-89 Sunbird Turbo - 2.0L	
1987-89 Fiero, Firebird, 6000 - 2.8L	
1987 Grand Am - 3.0L	
1987-89 Bonneville - 3.8L	
1987-89 Firebird - 5.0L & 5.7L	
1988-89 Grand Am - 2.0L Turbo & 2.3L	
1988-89 Grand Prix - 2.8L	
1989 Grand Prix - 3.1L Turbo	GM Port Fuel Injection Pg. 467
Buick	
1986-89 Riviera - 3.8L	
1988-89 Reatta - 3.8L	
Cadillac	
1987-89 Allanté - 4.1L, 4.5L	
Oldsmobile	
1986-89 Toronado - 3.8L	GM Digital Port Fuel Injection Pg. 350

HISTORY & DEVELOPMENT

Fuel injection is not new. The Robert Bosch Company began experimentation with gasoline injection as early as 1912. Twenty years later, this work resulted in the first fuel injected aircraft engines. However, it was not until 1952 that Bosch installed a fuel injection system in a production automobile.

American development of fuel injection systems began in the 1940's. Both Enderle and Hilborn were working on fuel injection systems for use in race cars at that time. However, it wasn't until 1957 that fuel injection was used on a domestic production vehicle. In that year, Chevrolet introduced a Rochester produced fuel injection system on Corvette and other models. During the same year, Pontiac produced a limited number of vehicles with fuel injection. All of these were mechanical systems.

Chrysler offered the first electronic fuel injection system in 1958. Manufactured by Bendix and dubbed the "Electrojecter" system, its high cost ($400-$500) was a major factor in its lack of popularity, as less than 100 Electrojector equipped vehicles were sold.

Shortly thereafter, Bendix decided to sell the manufacturing rights to this system. The Bosch company, believing that there might be a future for electronic fuel injection, bought those rights and continued with development. In 1968 the first Bosch electronic fuel injection system was offered on a Volkswagen.

In the early 1970's, the government became concerned for the state of the environment and our increasing dependence on foreign oil supplies. Legislation established minimum fuel efficiency standards and maximum pollution levels for vehicles sold in the United States. Automobile manufacturers realized that fuel systems would be hard pressed to meet future standards. The complexity and cost of carburetor-based fuel systems began to rise.

Meanwhile, work in the electronics industry was resulting in the availability of reliable and inexpensive solid state components. These advances were applied to computer control of electronic fuel injection systems. Fuel injection became more dependable and cost competitive. Finally in 1975, an electronic fuel injection system reappeared in a domestic built Cadillac Seville.

ADVANTAGES

Fuel injection, particularly when computer controlled, has several advantages over conventional carburetion. These include improved driveability under all conditions, improved fuel control, an increase in engine efficiency and power, and a reduction in exhaust pollutant levels. In fact, fuel injection advantages have resulted in several models being offered ONLY with fuel injection in California. This state has had more stringent pollution standards than the rest of the nation, and only fuel injection systems could meet those standards.

FUEL INJECTION THEORY

Most electronic fuel injection systems only inject fuel during part of an engine combustion cycle. The engine fuel requirements are measured by airflow past a sensor, or by intake manifold pressure (vacuum). These sensors convert airflow or pressure to electrical signals which are sent to a micro-computer. This computer processes the signals to determine the fuel requirements of the engine. Once these requirements are determined, the computer sends an electrical signal to the injector(s). The signal controls the length of time that the injector stays open. This interval is known as the injector "pulse width".

On port-type injection systems, fuel injectors are mounted in the intake manifold runners. One injector is directed at the back of each intake valve. The other system in use has a "throttle body" assembly, mounted on the intake manifold in the position usually occupied by a carburetor. This assembly usually contains one or two fuel injectors. Aside from the difference in injector location, operation of the different systems is quite similar.

CAUTION
- *When diagnosing or repairing fuel system related components, there is always the possibility of the presence of highly flammable fuel or fuel vapors which may be ignited causing dangerous flames or even explosion resulting in possible personal injury.*

SAFETY TIPS

Before starting to test, diagnose or repair any electronic fuel injection systems and/or electronic ignition systems, ensure you perform the following:

- – Always maintain an appropriate type fire extinguisher within easy reach of the work area.
- – Do not smoke when working on or near any fuel related component.
- – Never allow open flames from matches, lighters, torches or other devices around the work area and avoid spark generation from grinders, gas or electric welding equipment, electrical wiring, flints, etc.
- – Avoid the presence of incandescent materials including smoking materials, soldering irons, and the breakage of burning light bulbs.
- – Clean up excessive fuel spills immediately and properly dispose of fuel soaked materials.
- – Always use and wear safety equipment.
- – Always test and diagnose fuel systems in a well ventilated area.
- – Always refer to each article in the manual for precautions and notes before testing and diagnosing any fuel injection system or electronic ignition system.
- – Electronic fuel systems operate at higher than normal pressures. If any line is loosen without properly depressurizing fuel system, pressurized fuel may squirt out, thus possibly causing a fire or personal injury.
- – Never place your fingers under a fuel injector to check spray pattern and pressure. Fuel in injectors is under pressure and may penetrate the skin, without breaking the skin. This may cause blood poisoning.
- – Do not apply voltage to fuel injector with fuel line connected, unless proper procedures and precautions are used. Failure to do so may cause fire or personal injury.
- – When working around electronic ignition systems, always use care when testing. Remember, some systems operate at extremely high voltages. Failure to use caution may lead to component damage, personal injury and/or fire.

Alliance & Encore (California)
NOTE
 • *For detailed applications, refer to FUEL INJECTION SYSTEMS APPLICATION CHARTS at the front of this publication.*

DESCRIPTION

All Alliance/Encore models manufactured for sale in California up to 1984 are equipped with an airflow controlled electronic fuel injection system manufactured by Bosch. *See Fig. 1.* The system determines engine fuel requirements by measuring intake airflow. This information is combined with information from various engine sensors to determine specific fuel requirements for any engine operating condition.

The system consists primarily of the fuel system and the control system. The fuel system includes the fuel pump, fuel filter, pressure regulator, fuel injectors (one per cylinder) and the cold start injector.

The control system includes the control relay, airflow meter, auxiliary air valve, Electronic Control Unit (ECU), coolant temperature thermo time switch, Throttle Position Sensor (TPS), coolant temperature sensor, ignition control module, and oxygen sensor (O_2).

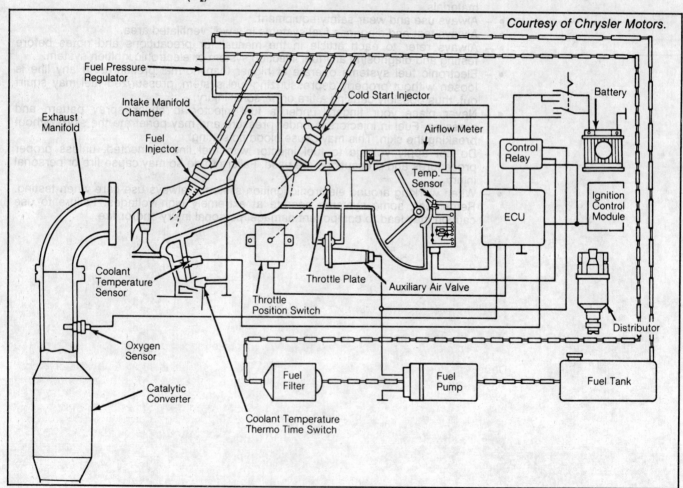

Courtesy of Chrysler Motors.

Fig. 1: AMC Airflow Controlled Fuel Injection System Schematic

OPERATION

FUEL DELIVERY

Fuel is supplied to the fuel rail assembly, mounted on the intake manifold, by an in-tank electric fuel pump. A constant system pressure of about 36 psi (2.5 kg/cm²) is maintained by the pressure regulator. The regulator is mounted on the intake manifold chamber in line with the fuel return line. It contains a spring controlled diaphragm which is exposed to fuel pressure on one side and intake manifold pressure on the other. Fuel delivered in excess of that required to maintain system pressure is by-passed by the regulator and returned to the fuel tank via the fuel return line.

Fuel is supplied to the intake manifold from the fuel rail assembly, through the fuel injectors. *See Fig. 2.* The injectors are electro-magnetic solenoid valves. A needle valve in the injector is held against a seat by a coiled spring. An electrical armature at the back of the valve reacts to electrical signals from the ECU by pulling the injector needle off its seat. This allows fuel to be injected into the intake manifold. Since fuel pressure is maintained at a constant level, the amount of fuel injected is dependent only upon the length of time that the injector is held open (injector "on" time). All injectors are fired simultaneously, twice per engine revolution.

Courtesy of Chrysler Motors.

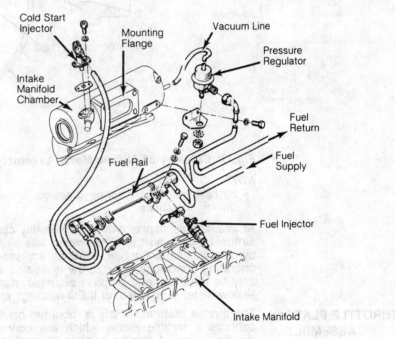

Fig. 2: Fuel Delivery System
NOTE
- *Throttle body assembly mounts to flange on intake manifold chamber.*

During starter engagement, with cold engine, additional fuel is supplied by the cold start injector. Power to the injector is routed through the coolant temperature thermo time switch, mounted in the water jacket of the cylinder head. The switch contains an electrical contact on the end of a bi-metallic strip. When the engine is cold, the contact is closed and power to the injector is supplied. The switch maintains power to the injector for a maximum of 8 seconds at a coolant temperature of -4°F (-20°C). At 95°F (35°C), power to the injector is cut off.

AIRFLOW METER

The airflow meter is located in line with the intake air duct, between the air cleaner and the intake manifold. All engine air is drawn through the airflow meter, which contains a tunnel with a measuring flap and dampening flap (offset 90 degrees on the same casting). *See Fig. 3.* The measuring flap swings in the air stream against the pressure of a calibrated spring. A potentiometer connected to the flap supplies the ECU with a voltage signal. This signal is directly proportional to the degree of flap opening (air flow).

A temperature sensor is fitted in the airflow meter to measure the temperature of incoming air. Resistance value of the sensor varies with air temperature. A voltage signal from the sensor combines with airflow meter output voltage, resulting in a voltage signal to the ECU which indicates air density as well as volume. This information is used to determine engine fuel requirements under various conditions.

Courtesy of Chrysler Motors.

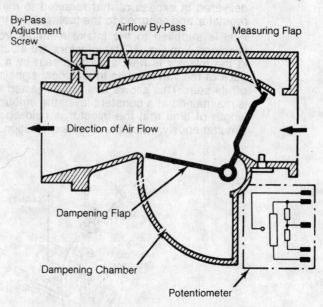

Fig. 3: Cutaway of Airflow Meter Assembly

NOTE
- *By-pass adjustment screw is sealed. No adjustment should be required under normal driving conditions.*

At idle, the air flap is almost completely closed. Idle air requirements are met by an airflow by-pass built into the meter. Idle air flow rate can be adjusted with the airflow by-pass adjustment screw, although adjustment should not be necessary under normal operating conditions. The screw is sealed with a tamper resistant cap which should only be removed, and by-pass adjusted, during a major engine overhaul, when a new airflow meter is installed, or if CO readings are excessively high.

THROTTLE PLATE ASSEMBLY

The throttle plate assembly is mounted on the intake manifold chamber. *See Fig. 6.* It contains 2 throttle plates which are connected to the accelerator by conventional throttle linkage. A throttle position switch is built into the assembly. This switch informs the ECU, by electrical signal, of wide open throttle and idle conditions. The ECU uses this information to adjust air/fuel mixture to meet engine demands.

On vehicles equipped with air conditioning or power steering, a fast idle valve is built into the assembly to compensate for decreased idle speed during AC operation, during an extreme turn, or engine starting condition. Additional idle air is supplied through an auxiliary air circuit in the throttle assembly.

ELECTRONIC CONTROL UNIT (ECU)

The ECU is the "brain" of the fuel injection system. It is a pre-programmed, solid state computer which receives and interprets data from various engine sensors and switches. This data is used to determine the amount of fuel required by the engine to maintain efficiency and minimize exhaust emissions under varying operating conditions. Control signal outputs from the ECU control fuel injector operation.

AUXILIARY AIR VALVE

The auxiliary air valve supplies extra air to the engine to increase idle speed during cold engine operation. Air is supplied to the valve from in front of the throttle valve assembly. An air passage in the valve reacts to engine heat (valve is mounted to cylinder head) and an integral heater, opening or closing in response to temperature changes. Air from the valve is delivered to the intake manifold chamber, by-passing the throttle valve assembly.

OXYGEN SENSOR

The oxygen sensor is located in the exhaust manifold. The outer surface of the sensor is in contact with exhaust gases, while the inner surface is exposed to outside air. A voltage signal, created by the difference in oxygen contents, is transmitted to the ECU.

This signal is a measurement of the unburned oxygen in the exhaust gas, which is directly related to the intake air/fuel mixture. In this way, the ECU is kept up to date on air/fuel ratio so that necessary adjustments to mixture can be made.

COOLANT TEMPERATURE SENSOR

The coolant temperature sensor is screwed into the cylinder head water jacket, adjacent to the thermo time switch. It detects engine coolant temperature and provides a voltage signal to the ECU. Air/fuel ratio corrections for cold engine operation are determined from this signal.

CONTROL RELAY

The control relay is mounted on the front of the right shock tower. It controls power input to the fuel pump, injectors, ECU, auxiliary air valve, and the throttle position switch.

TROUBLE SHOOTING

PRELIMINARY CHECKS

Several driveability problems may result from faulty or poor wiring, loose and/or leaking hose connections, or basic engine systems malfunctions. To avoid unnecessary component testing, check the following areas before beginning trouble shooting of the system.
- Intake air system leaks.
- Electrical connections at all components.
- Vacuum lines (secure and leak-free).
- Battery charge and water level.
- Ignition components (distributor, plugs, coil).
- Ignition timing.
- Engine compression.
- Valves properly adjusted.
- Correct oil pressure.
- Correct fuel pressure.

Most of the engine symptoms listed may also be caused by either a leaking cold start injector or a defective airflow meter. Check these components first when diagnosing engine malfunctions.

ENGINE TURNS OVER BUT WILL NOT START

1) Check fuel tank, filter, and lines for blockage or other restriction. Test control relay wiring harness and control relay. Check fuel pump for proper operation.

2) Check cold start injector operation and wiring. Check for defective thermo time switch. Check auxiliary air valve operation. Check temperature sensor resistance values. Check control relay, thermo time switch, and airflow meter.

3) If malfunction remains after checking and/or repairing noted systems, replace ECU with a known good unit. If malfunction ceases, original ECU is defective and must be replaced.

ENGINE STARTS, THEN DIES

1) Check fuel tank, filter and lines for blockage or other restriction. Check auxiliary air valve operation. Check temperature sensor resistance values. Check throttle position switch.

2) Check and adjust idle speed as needed. Check for proper adjustment of air by-pass screw and vane on airflow meter. Check cold start injector and operational injectors.

3) If malfunction remains after checking and/or repairing noted systems, replace ECU with a known good unit. If malfunction ceases, original ECU is defective and must be replaced.

ROUGH IDLE

1) Check fuel tank, filter, and lines for blockage or other restriction. Check auxiliary air valve operation. Check temperature sensor resistance values.

2) Check throttle position sensor. Adjust or replace as needed. Check and adjust idle speed. Check individual injector operation. Ensure that air by-pass screw on airflow meter is properly adjusted. Check coolant temperature sensor.

3) If malfunction remains after checking and/or repairing noted systems, replace ECU with a known good unit. If malfunction ceases, original ECU is defective and must be replaced.

IDLE SPEED INCORRECT

Check basic idle speed adjustment, cold start injector, airflow meter vane, air by-pass adjustment, and auxiliary air valve operation.

INCORRECT CO VALUE

1) Check temperature sensor resistance values. Check and adjust idle speed as needed.

2) Check individual fuel injectors for correct operation. Check and adjust air by-pass and vane on airflow meter.

ERRATIC ENGINE RPM Check individual fuel injector operation. Check and adjust air by-pass and vane on airflow meter.

ENGINE MISS WHILE DRIVING Check injector operation. Replace ECU with a known good unit and test drive vehicle. If miss is gone, original ECU is defective and should be replaced.

POOR FUEL ECONOMY Check temperature sensor resistance values. Check cold start injector. Check and adjust air by-pass screw and vane on airflow meter.

LACK OF POWER 1) Check fuel tank, filter, and lines for blockage or other restriction. Check and adjust or replace throttle position switch as needed. Check individual fuel injectors for correct operation.

2) If malfunction remains after checking and/or repairing noted systems, replace ECU with a known good unit. If malfunction ceases, original ECU is defective and must be replaced.

DIAGNOSIS & TESTING

NOTE
- *When testing resistance or voltage values, a high impedence (digital) volt-ohm meter must be used.*

INTAKE AIR SYSTEM 1) Air leaks in the intake system can cause various engine difficulties. To check for leaks, disconnect the air line from the auxiliary air valve at the intake manifold chamber. Apply a soapy water solution to any connections or joints where leaks are likely to occur.

2) Plug the exhaust pipe with a shop towel or rag. Apply compressed air (15 psi maximum) at the hose connection on the intake manifold chamber and open the throttle plate. The presence of bubbles or foam at any point in the system indicates the presence of air leaks. Repair as needed.

FUEL SYSTEM **Fuel Pump Pressure & Regulator**

1) Install a "T" fitting in the fuel line between the pressure regulator and the fuel rail assembly, then connect a pressure gauge. Disconnect the vacuum line from the pressure regulator and attach a vacuum pump in its place.

2) Locate the system diagnostic connectors on the inner fender well, just in front of the right side shock tower. The connectors are covered by plastic caps. Flip off the caps and install a jumper wire between terminals 5 and 6 of the small connector (D1). See Fig. 4. Fuel pressure should be about 36 psi (2.5 kg/cm²). If not, check fuel pump, fuel filter, and electrical circuits.

3) Apply about 16 in. Hg vacuum to the regulator. Fuel pressure should drop to 23 psi (2 kg/cm²). If not, replace the pressure regulator.

Courtesy of Chrysler Motors.

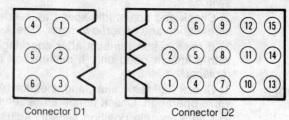

Connector D1   Connector D2

Fig. 4: Diagnostic Connector Identification

NOTE
- *Connectors are mounted on right side inner fender well.*

Fuel Pump Output

1) Disconnect the fuel return line from the pressure regulator. Attach a length of fuel hose to the regulator with the other end in a graduated container.

2) Attach a jumper wire between terminals 5 and 6 of diagnostic connector D1 to activate pump. The system should pump at least one pint of fuel in 30 seconds. If not, ensure that the fuel filter is not clogged. Check pump and regulator operation and electrical circuits.

Fuel Injector Leakage

1) Disconnect all injector wire connectors. Remove the fuel rail assembly, with injectors, and cold start injector, from intake manifold.

2) Attach a jumper wire between terminals 5 and 6 of diagnostic connector D1 and observe injectors. No fuel should leak from any injector.

Injector Operation

1) Remove fuel rail assembly, with injectors, from intake manifold. Attach jumper wire between terminals 5 and 6 of diagnostic connector D1.

2) Briefly apply 12 volts power and ground to one injector at a time. Injector should spray fuel and cut off cleanly when power is removed. If not, replace faulty injector.

AUXILIARY AIR VALVE

Air Valve

1) With air valve assembly at about 70°F (20°C), disconnect electrical connector and air hoses, air passage in valve should be slightly open. This may be verified by looking into the valve from the inlet or outlet sides.

2) Use jumper wires to connect valve terminals directly to battery voltage. Observe the air passage. Passage should be completely closed in about 10 minutes. If not, replace the valve.

Wiring Harness

Disconnect wiring connector from valve. With the engine cold, connect a test light to the connector terminals and start engine. The light should come on. If not, replace or repair wiring harness.

COOLANT TEMPERATURE THERMO TIME SWITCH

1) Check resistance values between switch terminals and from each terminal to ground. Checks should be made with coolant temperature below 85°F (30°C) and again with coolant temperature above 105°F (40°C).

2) Compare results to values given in THERMO TIME SWITCH RESISTANCE VALUES (OHMS) table. If resistance is incorrect, replace switch.

THERMO TIME SWITCH RESISTANCE VALUES (OHMS)

Measured Between Terminals	Below 85°F (30°C)	Above 105°F (40°C)
G and W	25-40	50-80
G and Ground	25-40	50-80
W and Ground	0	100-160

CONTROL RELAY & ECU GROUND

With the ignition switch in the "ON" position, engine coolant temperature below 77°F (25°C) and engine off, check resistance value between terminal 3 of diagnostic connector D1 and terminal 7 of connector D2. If value is anything other than zero, check control relay and ECU ground circuits.

COOLANT TEMPERATURE SENSOR

1) Remove the sensor from the engine (DO NOT remove when engine is hot). Measure resistance in relation to temperature after a temperature stabilization time of at least 10 minutes. With the temperature about 70°F (20°C), the resistance should be around 25,000 ohms. With the temperature of 175°F (80°C), the resistance should be around 320 ohms.

2) If readings are incorrect, check wiring circuit and repair as needed. If circuit is good, replace temperature sensor.

THROTTLE POSITION SWITCH

1) With the ignition switch in the "ON" position, engine coolant temperature below 77°F (25°C) and engine off, test resistance between terminals 13 and 6 of diagnostic connector D2. Reading should be infinite. Check between terminals 4 and 13 of connector D2 with throttle closed. Resistance should be zero.

2) With meter between terminals 4 and 6 of D2 and throttle at wide open position, resistance should be zero. If any of the readings are incorrect, check wiring circuit and repair as needed. If circuit is good, replace TPS.

ADJUSTMENTS

IDLE SPEED

1) Warm engine to operating temperature. Connect tachometer positive lead to diagnostic connector D1-1. Connect tachometer negative lead to diagnostic connector D1-3.

2) Turn off all accessories. Wait for electric cooling fan to cycle on and off. Turn adjustment screw on throttle plate to adjust idle speed. See Fig. 5.

Courtesy of Chrysler Motors.

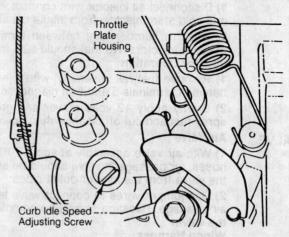

Fig. 5: Adjusting Idle Speed

THROTTLE POSITION SWITCH

With the throttle plate assembly removed, loosen the 2 throttle position switch (TPS) attaching screws. *See Fig. 6.* Hold the throttle plates against the idle stop and slowly rotate the TPS in the direction of the throttle plate opening until the inner stop is felt. Tighten the attaching screws. If properly adjusted, the clicking of the microswitch in the TPS should be heard just before the stop is reached.

Courtesy of Chrysler Motors.

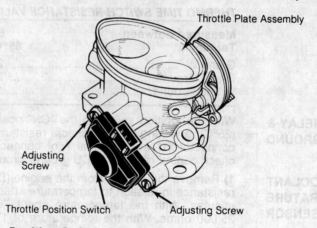

Fig. 6: Throttle Position Switch Adjustment

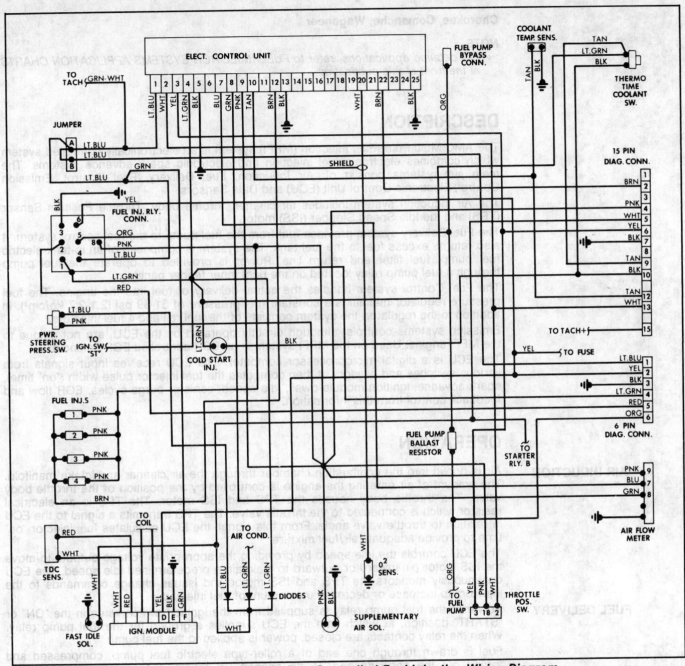

Fig. 7: AMC Airflow Controlled Fuel Injection Wiring Diagram

Cherokee, Comanche, Wagoneer

NOTE
- *For detailed applications, refer to FUEL INJECTION SYSTEMS APPLICATION CHARTS at the front of this publication.*

DESCRIPTION

The AMC Multi-Point Fuel Injection (MPFI) system is an electronically controlled system which combines electronic fuel injection and electronic spark advance systems. The main sub-systems consist of: Air Induction, Fuel Delivery, Fuel Control, Emission Control, Electronic Control Unit (ECU) and Data Sensors.

The Air Induction system includes air cleaner, throttle body, Throttle Position Sensor (TPS) and the Idle Speed Stepper (ISS) motor.

The Fuel Delivery system provides fuel from the fuel pump to the fuel control system. It also returns excess fuel to the fuel tank. The system is composed of an in-tank electric fuel pump, fuel filter and return line. Power is provided to operate the fuel pump through a fuel pump relay located on the right inner fender panel.

The Fuel Control system handles the actual delivery of fuel into the engine. The fuel pressure regulator maintains a constant fuel pressure of 31-39 psi (2.1-2.7 kg/cm²). In addition to the regulator, the system consists of the fuel rail and 4 fuel injectors.

Emission systems controls, although directly operated by the ECU, are not unique to the MPFI engine. Used on the MPFI system is an ECU controlled EGR solenoid.

The ECU is a digital microprocessor computer. The ECU receives input signals from various switches and sensors. It then computes the fuel injector pulse width ("on" time), spark advance, ignition module dwell, idle speed, canister purge cycles, EGR flow and feedback control from this information.

OPERATION

AIR INDUCTION

Air is drawn into the combustion chamber through the air cleaner and intake manifold. The amount of air entering the engine is controlled by the position of the throttle body valve. The throttle body houses the TPS and ISS motor. The TPS is an electrical resistor which is connected to the throttle valve. The TPS transmits a signal to the ECU in relation to throttle valve angle. From this signal, the ECU calculates fuel injector "on" time to provide adequate air/fuel mixture.

The ECU controls the idle speed by providing the appropriate voltage outputs to move the ISS motor pin inward or outward to maintain a predetermined idle speed. The ECU continuously monitors the TPS and ISS motor and issues change commands to the injectors to increase or decrease the amount of fuel injected.

FUEL DELIVERY

Power to the fuel pump relay is supplied from the ignition switch when in the "ON" or "START" position, at which time the ECU supplies a ground for the fuel pump relay. When the relay contacts are closed, power is applied to the fuel pump.

Fuel is drawn through one end of a roller-type electric fuel pump, compressed and forced out the opposite end. Pump capacity is greater than the maximum engine consumption so that the pressure in the fuel system is always maintained.

FUEL CONTROL

The fuel control system handles the actual delivery of fuel into the engine. *See Fig. 1.* Fuel from the pump enters the fuel rail, injectors and pressure regulator. Based upon a manifold vacuum signal, the pressure regulator maintains a constant fuel pressure in the system of approximately 31-39 psi (2.1-2.7 kg/cm²) by allowing excess fuel to return to the fuel tank.

Fuel injectors are electrically operated solenoid valves which are powered by the ECU. The ECU determines injector pulse width ("on" time) based upon input from the various sensors.

Courtesy of Chrysler Motors.

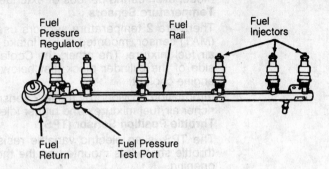

Fig. 1: Fuel Control Components

EMISSION CONTROL

The ECU controls the EGR valve operation. By energizing the EGR solenoid, vacuum is shut off, making this system non-operative. When the engine reaches normal operating temperatures, the ECU de-energizes the solenoid. When de-energized, the solenoid allows vacuum to flow to the EGR valve. The ECU will energize the solenoid whenever EGR action is undesirable, during idle, cold engine operation, wide open throttle and rapid acceleration or deceleration.

ECU

The ECU is a digital microprocessor computer. Data sensors provide the ECU with engine operating information in the form of varying electrical signals. The computer analyzes this information and corrects air/fuel ratio, ignition timing, and emission control as needed to maintain efficient engine operation. Other ECU output signals control the upshift indicator light (manual transmission only), ignition module dwell and A/C clutch operation.

UPSHIFT INDICATOR

On vehicles equipped with a manual transmission, the ECU controls the upshift indicator light. The indicator light is normally illuminated when the ignition switch is turned on without the engine running. The light is turned off when the engine is started.

The indicator light will be illuminated during engine operation in response to engine load and speed. If the gears are not shifted, the ECU will turn the light off after 3 to 5 seconds. A switch located on the transmission prevents the light from being illuminated when the transmission is shifted to the highest gear.

IDLE SPEED STEPPER (ISS) MOTOR

The Idle Speed Stepper (ISS) motor, located on the throttle body, is controlled by the ECU. The motor controls the airflow inside the throttle body by-pass passage to control idle speed.

B+ LATCH RELAY

The B+ latch relay, located on right front inner fender panel, is energized by the ECU for 3-5 seconds after ignition is turned off. This allows the ECU to reposition the ISS motor for the next start.

DATA SENSORS & SWITCHES

Manifold Absolute Pressure (MAP) Sensor

The MAP sensor is located in the engine compartment on the firewall behind the engine. The MAP sensor monitors manifold vacuum via a vacuum line from the intake manifold to the sensor.

The sensor supplies an electrical signal which keeps the ECU informed of manifold vacuum and barometric pressure conditions. This information is combined with data supplied by other sensors to determine correct air/fuel ratio.

Oxygen Sensor

The oxygen (O_2) sensor is mounted in the exhaust manifold where it is exposed to exhaust gas flow. Its function is to monitor oxygen content of exhaust gases and to supply the ECU with a voltage signal directly proportional to this content.

If the oxygen content of exhaust gases is high (lean air/fuel mixture), the voltage signal to the ECU is low. As oxygen content decreases (mixture becomes richer), signal voltage increases.

In this way, the ECU is kept constantly informed of air/fuel ratio. It can then alter fuel injector "on" time, in response to these signals, to obtain the best air/fuel ratio of 14.7:1 under any given operating conditions.

The O_2 sensor is equipped with a heating element that keeps the sensor at proper operating temperatures. Maintaining correct sensor temperatures at all times guarantees a more accurate signal to the ECU. By using an O_2 heater, the fuel control system may also enter the "closed loop" operating mode sooner and maintain this mode, even during periods of extended idle.

Temperature Sensors

There are 2 temperature sensors used on this system. The Manifold Air Temperature (MAT) sensor, mounted in the intake manifold, measures the temperature of incoming air/fuel mixture. The other, the Coolant Temperature Sensor (CTS), located on the left side of the cylinder block just below the exhaust manifold, measures temperature of engine coolant.

Information provided by these 2 sensors to the ECU allows the ECU to demand slightly richer air/fuel mixtures and higher idle speeds during cold engine operation.

Throttle Position Sensor (TPS)

The TPS is an electric variable resistor which is regulated by the movement of the throttle shaft. It is mounted on the throttle body and senses the angle of throttle blade opening.

A voltage signal of up to 5 volts at wide open throttle is produced by the sensor. Voltage varies with throttle angle changes. This signal is transmitted to the ECU where it is used to adjust air/fuel ratio during acceleration, deceleration, idle, and wide open throttle conditions.

A dual TPS is used on models with automatic transmissions. This dual TPS not only provides the ECU with input voltages but also supplies the automatic transmission with input signals relative to throttle position.

Knock Sensor

The knock sensor (detonation sensor) is located on the lower left side of the cylinder block just above the oil pan. The knock sensor picks up detonation vibration from the engine and converts it to an electrical signal for use by the ECU.

The ECU uses this information to determine when a change in ignition timing is required. The knock sensor allows for engine operation on either "premium" unleaded or "regular" unleaded fuel.

When knock occurs, the ECU retards the ignition timing in one or more cylinders until detonation is eliminated.

Speed Sensor

The speed sensor is secured by special shouldering bolts to the flywheel/drive plate housing. The speed sensor senses TDC and engine speed by detecting the flywheel teeth as they pass the pick-up coil during engine operation. *See Fig. 2.*

The flywheel has a large trigger tooth and notch located 12 small teeth before each TDC position. When a small tooth and notch pass the magnetic core in the sensor, the concentration and collapse of the magnetic field created induces a small voltage spike into the sensor pick-up coil windings. These small voltage spikes are sent to the ECU, allowing the ECU to count the teeth as they pass the sensor.

Courtesy of Chrysler Motors.

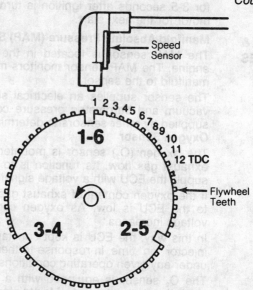

Fig. 2: Speed Sensor Operation

When a large tooth and notch pass the magnetic core in the sensor, the increased concentration and collapse of the magnetic field induces a higher voltage spike than the smaller teeth. The higher spike indicates to the ECU that a piston will soon be at TDC position, 12 teeth later.

Ignition timing for the cylinder is either advanced or retarded as necessary by the ECU based upon the inputs from all sensors.

Engine Switches

Several switches provide operating information to the ECU. These include the Park/Neutral switch (automatic transmission only), air conditioning clutch, and Sync Pulse switch. When the A/C or Park/Neutral switches supply the ECU with an "on" signal, the module signals the ISS motor to change idle speed to a specific RPM.

With the A/C on and the throttle blade above a specific angle, the ECU de-energizes the A/C relay, preventing the air conditioning clutch from engaging until throttle blade angle is reduced.

The Sync Pulse switch, located within the distributor, generates a signal to the ECU, helping to properly synchronize injector opening with intake valve opening.

TESTING & DIAGNOSIS

PRELIMINARY CHECKS

The following systems and components must be in good condition and operating properly before assuming a fuel injection system malfunction.
- Air filter.
- All support systems and wiring.
- Battery connections and specific gravity.
- Compression pressure.
- Electrical connections on components and sensors.
- Emission control devices.
- Ignition system.
- All vacuum line, fuel hose and pipe connections.

NOTE
- *The ECU is an extremely reliable part and must be the final component replaced.*

GENERAL PRECAUTIONS

In order to prevent damage to components and injury to operator, use the following precautions:
- Turn ignition off before unplugging or connecting any connector.
- NEVER apply more than 12 DC volts to any component when testing.
- Disconnect battery cables before charging.
- Remove ECU from vehicle if ambient temperature could exceed 176°F (80°C).
- NEVER modify or circumvent any system functions.

SYSTEM TESTING

Fuel System Test

Remove the cap from the pressure test port located in the fuel rail. *See Fig. 3.* Connect Fuel Pressure Gauge (J-34730-1) to the pressure fitting. Start vehicle. Pressure should be approximately 31 psi (2.1 kg/cm²) with the vacuum hose connected to the pressure regulator and 39 psi (2.6 kg/cm²) with the vacuum hose removed from the pressure regulator.

Check the fuel pump flow rate. A good fuel pump will deliver at least one liter of fuel per minute with the fuel return line pinched off. If the fuel pump does not pump adequately, inspect the fuel system for a plugged fuel filter or filter sock.

Fuel pump flow rate can be checked by connecting one end of an old A/C gauge hose to the fuel test port on the fuel rail and inserting the other end of the hose into a container of at least one liter or more capacity. Run the fuel pump by installing a jumper wire into diagnostic connector terminals D1-5 and D1-6. Be sure to pinch off the fuel return line or most of the fuel will be returned to the fuel tank.

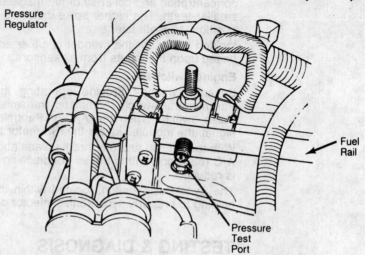

Courtesy of Chrysler Motors.

Pressure Regulator

Fuel Rail

Pressure Test Port

Fig. 3: Fuel System Pressure Test Components

EGR Solenoid Test

Verify that vacuum is present at vacuum fitting "C" of the EGR solenoid. *See Fig. 4.* Remove vacuum connector from "A" and "B". Connect a vacuum gauge to "B". Start and idle the engine. There should be no vacuum at "B". Disconnect electrical connector "D" from the solenoid. There should now be vacuum at "B".

Courtesy of Chrysler Motors.

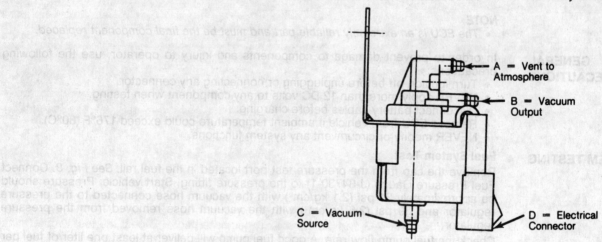

A = Vent to Atmosphere

B = Vacuum Output

C = Vacuum Source

D = Electrical Connector

Fig. 4: EGR Solenoid Test Points

MAP Sensor Test

Inspect the MAP sensor hoses and connections. Repair as necessary. With the ignition on and the engine off, test the MAP sensor output voltage at the MAP sensor connector terminal "B". (Marked on sensor body). *See Fig. 5.* The output voltage should be 4 to 5 volts.

To verify the wiring harness condition, test ECU terminal C-6 for the same voltage described.

Test MAP sensor supply voltage at the sensor connector terminal "C" with the ignition on. The voltage should be 4.5-5.5 volts. The same voltage should also be at terminal C-14 of the ECU wire harness connector.

Test the MAP sensor ground circuit at the ECU connector between terminal D-3 and terminal B-11 with an ohmmeter. If the ohmmeter indicates an open circuit, inspect for a defective sensor ground connection, located on the right side of the cylinder block. If the ground connection is good the ECU may need to be replaced.

Courtesy of Chrysler Motors.

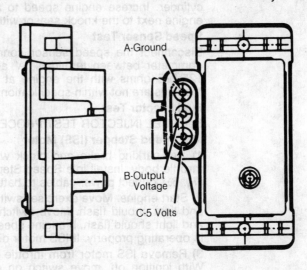

A-Ground

B-Output Voltage

C-5 Volts

Fig. 5: MAP Sensor Test Points

O₂ Sensor Test

Disconnect the O₂ sensor connector. Connect an ohmmeter to terminals "A" and "B" (marked on the connector) of the O_2 sensor connector. Resistance should be between 5 and 7 ohms. Replace the sensor if the ohmmeter indicates an infinity reading.

CTS Test

Disconnect the wire harness connector from the CTS switch. Test the resistance of the sensor. The resistance should be below 1000 ohms with the engine warm. See CTS & MAT SENSOR TEMPERATURE-TO-RESISTANCE VALUES table. Test the resistance of the wire harness between ECU terminal D-3 and the sensor connector terminal. Repeat the test at terminal C-10 of the ECU and the sensor connector terminal. Repair the wire harness if an open circuit is indicated.

MAT Sensor Test

Disconnect the wire harness connector from the MAT sensor. Test the resistance of the sensor with a high impedance digital ohmmeter. The resistance should be less than 1000 ohms with the engine warm. See CTS & MAT SENSOR TEMPERATURE-TO-RESISTANCE VALUES table. Replace the sensor if the resistance is not within the specified range. Test the resistance of the wire harness between the ECU wire harness connector terminal D-3 and the sensor connector terminal. Repeat the test with terminal C-8 at the ECU and the sensor connector terminal. Repair the wire harness if the resistance is greater than one ohm.

CTS & MAT SENSOR TEMPERATURE-TO-RESISTANCE VALUES

°F	°C (Approximate)	Ohms
212	100	185
160	70	450
100	38	1,600
70	20	3,400
40	4	7,500
20	-7	13,500
0	-18	25,000
-40	-40	100,700

TPS Test

See THROTTLE POSITION SENSOR TEST PROCEDURE chart in this article.

Knock Sensor Test

With engine at normal operating temperature, check engine timing to verify proper adjustment. Set to specification if necessary. Connect a timing light to the No. 1 cylinder. Increse engine speed to approximately 1600 RPM and hold. Knock on the engine next to the knock sensor with a metallic object. The timing should retard.

Speed Sensor Test

Disconnect the speed sensor connector from the ignition control module. Place an ohmmeter between terminals "A" and "B" (marked on connector). Reading should be 125-275 ohms with the engine at normal operating temperature. Replace sensor if readings are not within specifications.

Fuel Injector Test

See FUEL INJECTOR TEST PROCEDURE chart in this article.

Idle Speed Stepper (ISS) Motor

1) Set parking brake and block wheels. With ignition off, unplug ISS connector, at throttle body. Install Idle Speed Stepper Motor Exerciser (Ele. CT01) to ISS motor. *See Fig. 6.* Connect power cables to battery. Red light on exerciser should lite.

2) Start engine. Move exerciser switch to "High" position. Engine speed should increase and light should flash. Move switch to "Low" position. Engine speed should decrease and light should flash. If engine speed increased and decreased as specified, ISS motor is operating properly. If ISS motor did not chgange idle speed, turn ignition off.

3) Remove ISS motor from throttle body, but do not unplug exerciser from ISS motor. With ignition off, move switch on exerciser to "High" and "Low" position and watch pintle inside ISS motor for movement. If pintle does not move, replace ISS motor and retest. If pintle does move, check components for damage and recheck.

Courtesy of Chrysler Motors.

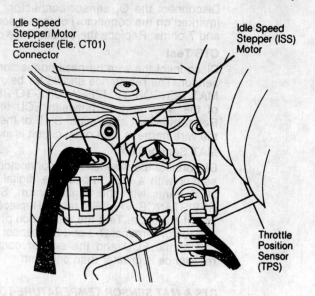

Idle Speed Stepper Motor Exerciser (Ele. CT01) Connector

Idle Speed Stepper (ISS) Motor

Throttle Position Sensor (TPS)

Fig. 6: Testing Idle Speed Steeper (ISS) Motor

ADJUSTMENTS

NOTE

- *Idle speed and air/fuel mixture are controlled by the ECU and are non-adjustable. On-car adjustment procedures for other components should not be necessary during normal vehicle operation or maintenance. Adjustments of components should only be required when a faulty component is replaced with a new one.*

Courtesy of Chrysler Motors.

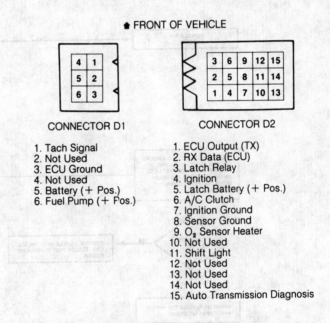

CONNECTOR D1

1. Tach Signal
2. Not Used
3. ECU Ground
4. Not Used
5. Battery (+ Pos.)
6. Fuel Pump (+ Pos.)

CONNECTOR D2

1. ECU Output (TX)
2. RX Data (ECU)
3. Latch Relay
4. Ignition
5. Latch Battery (+ Pos.)
6. A/C Clutch
7. Ignition Ground
8. Sensor Ground
9. O_2 Sensor Heater
10. Not Used
11. Shift Light
12. Not Used
13. Not Used
14. Not Used
15. Auto Transmission Diagnosis

Fig. 7: Diagnostic Connector Identification

Courtesy of Chrysler Motors.

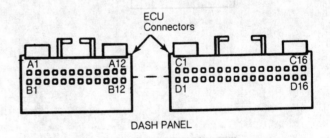

Fig. 8: ECU Connector Identification

Courtesy of Chrysler Motors.

FUEL INJECTOR TEST PROCEDURE CHART

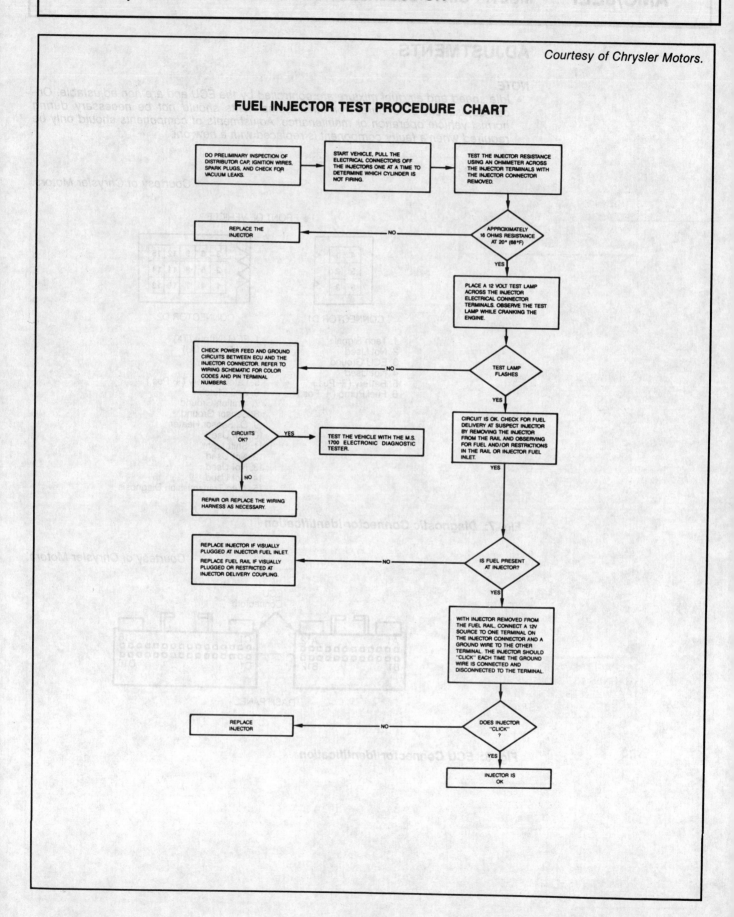

DO PRELIMINARY INSPECTION OF DISTRIBUTOR CAP, IGNITION WIRES, SPARK PLUGS, AND CHECK FOR VACUUM LEAKS.

START VEHICLE, PULL THE ELECTRICAL CONNECTORS OFF THE INJECTORS ONE AT A TIME TO DETERMINE WHICH CYLINDER IS NOT FIRING.

TEST THE INJECTOR RESISTANCE USING AN OHMMETER ACROSS THE INJECTOR TERMINALS WITH THE INJECTOR CONNECTOR REMOVED.

APPROXIMATELY 16 OHMS RESISTANCE AT 20° (68°F)

REPLACE THE INJECTOR — NO

YES

PLACE A 12 VOLT TEST LAMP ACROSS THE INJECTOR ELECTRICAL CONNECTOR TERMINALS. OBSERVE THE TEST LAMP WHILE CRANKING THE ENGINE.

CHECK POWER FEED AND GROUND CIRCUITS BETWEEN ECU AND THE INJECTOR CONNECTOR. REFER TO WIRING SCHEMATIC FOR COLOR CODES AND PIN TERMINAL NUMBERS. — NO — TEST LAMP FLASHES

YES

CIRCUITS OK? — YES — TEST THE VEHICLE WITH THE M.S. 1700 ELECTRONIC DIAGNOSTIC TESTER.

CIRCUIT IS OK. CHECK FOR FUEL DELIVERY AT SUSPECT INJECTOR BY REMOVING THE INJECTOR FROM THE RAIL AND OBSERVING FOR FUEL AND/OR RESTRICTIONS IN THE RAIL OR INJECTOR FUEL INLET.

NO

REPAIR OR REPLACE THE WIRING HARNESS AS NECESSARY.

YES

REPLACE INJECTOR IF VISUALLY PLUGGED AT INJECTOR FUEL INLET.

REPLACE FUEL RAIL IF VISUALLY PLUGGED OR RESTRICTED AT INJECTOR DELIVERY COUPLING. — NO — IS FUEL PRESENT AT INJECTOR?

YES

WITH INJECTOR REMOVED FROM THE FUEL RAIL, CONNECT A 12V SOURCE TO ONE TERMINAL ON THE INJECTOR CONNECTOR AND A GROUND WIRE TO THE OTHER TERMINAL. THE INJECTOR SHOULD "CLICK" EACH TIME THE GROUND WIRE IS CONNECTED AND DISCONNECTED TO THE TERMINAL.

REPLACE INJECTOR — NO — DOES INJECTOR "CLICK"?

YES

INJECTOR IS OK

Courtesy of Chrysler Motors.

THROTTLE POSITION SENSOR TEST PROCEDURE CHART

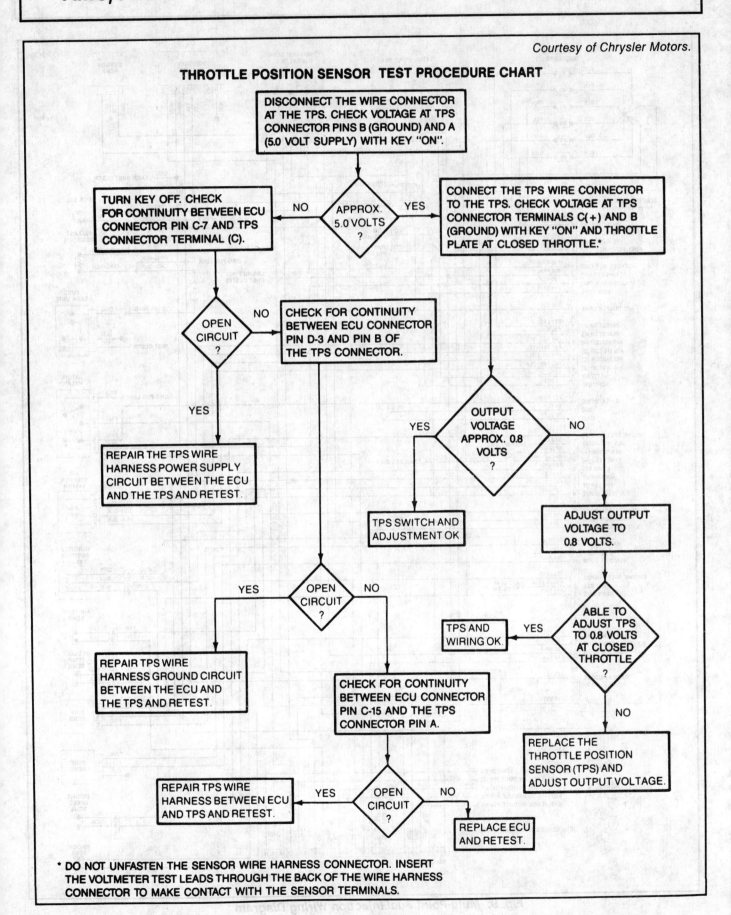

* DO NOT UNFASTEN THE SENSOR WIRE HARNESS CONNECTOR. INSERT THE VOLTMETER TEST LEADS THROUGH THE BACK OF THE WIRE HARNESS CONNECTOR TO MAKE CONTACT WITH THE SENSOR TERMINALS.

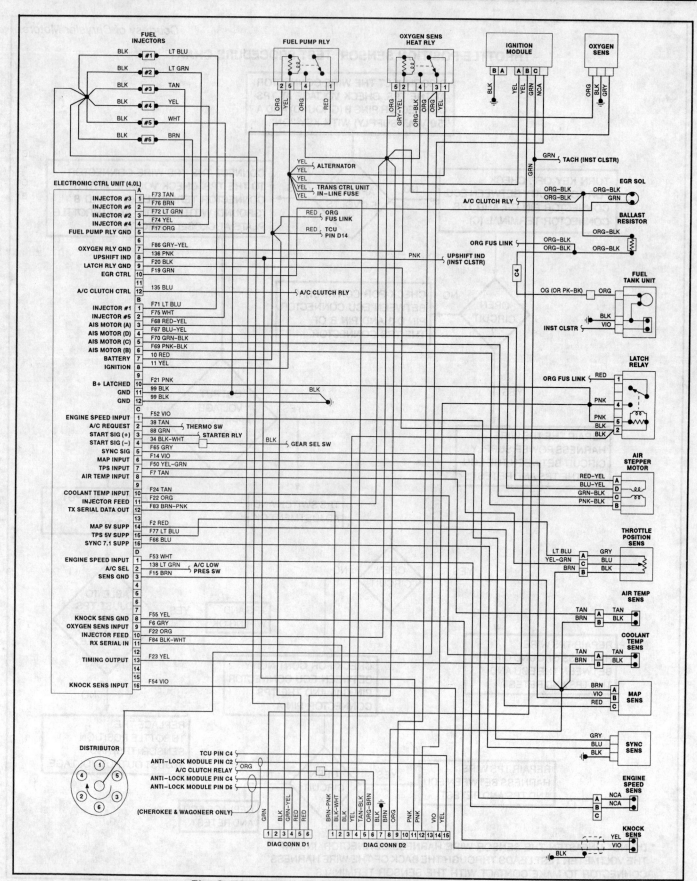

Fig. 9: Multi-Point Fuel Injection Wiring Diagram

Alliance & Encore (Federal)

NOTE
- *For detailed applications, refer to FUEL INJECTION SYSTEMS APPLICATION CHARTS at the front of this publication.*

DESCRIPTION

The AMC throttle body fuel injection system (TBI) uses an electronically controlled fuel injector to inject a metered spray of fuel above the throttle blade in the throttle body.

The TBI system is comprised of 2 sub-systems: the Fuel System and the Control System. Major components of the fuel system include an in-tank electric fuel pump, fuel filter, pressure regulator and a fuel injector.

Major components of the control system include the manifold air/fuel mixture temperature sensor, coolant temperature sensor, manifold absolute pressure sensor, wide open throttle switch, closed throttle switch, oxygen sensor, electronic control unit, throttle position sensor and an idle speed control motor.

NOTE
- *The closed throttle and wide open throttle switches, throttle position sensor, idle speed control motor and fuel pressure regulator are all located on the throttle body assembly.*

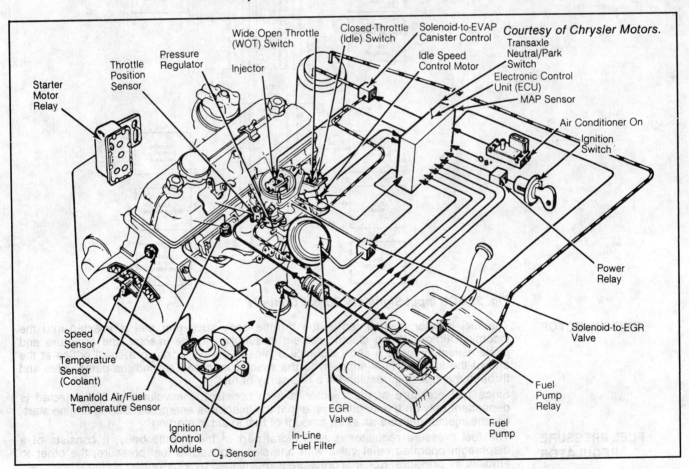

Fig. 1: AMC Throttle Body Fuel Injection System Components

OPERATION

ELECTRONIC CONTROL UNIT (ECU)

The ECU is located below the glove box near the fuse panel. It receives information from the various engine sensors to determine engine operating conditions at any particular moment. The ECU responds to these signals by sending a control signal to the fuel injector to determine the amount of time that the injector will be left open (injector "on" time). *See Fig. 2.*

Courtesy of Chrysler Motors.

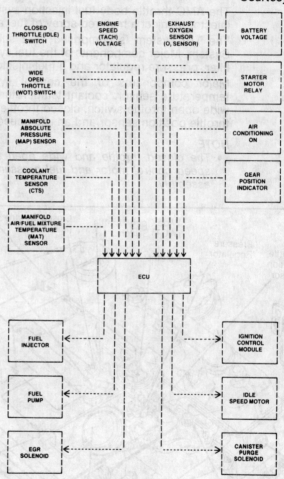

Fig. 2: ECU Input Signals & Output Controls

FUEL INJECTOR

The fuel injector is mounted in the throttle body such that fuel is injected into the incoming airflow. When electric current is supplied to the injector, the armature and pintle assembly move a short distance against a spring, opening a small orifice at the end of the injector. Fuel supplied to the injector is forced around the pintle valve and through this opening, resulting in a fine spray of fuel.

Since fuel pressure at the injector is kept constant, the volume of fuel injected is dependent only on the length of time that the injector is energized. During engine start-up, the injector delivers an extra amount of fuel to aid in starting.

FUEL PRESSURE REGULATOR

The fuel pressure regulator is an integral part of the throttle body. It consists of a diaphragm operated relief valve with one side exposed to fuel pressure, the other to ambient air pressure. Nominal pressure is maintained by a calibrated spring.

Fuel delivered by the pump in excess of that required for engine operation is returned to the fuel tank via the fuel return line.

FUEL PUMP

The fuel pump is an electrically operated roller type pump which is located in the fuel tank. It contains an integral check valve which is designed to maintain fuel pressure in the system after the engine has stopped. Pump operation is controlled by the ECU.

IDLE SPEED CONTROL (ISC) MOTOR

The ISC motor is an electrically driven actuator that changes the throttle stop angle by acting as a moveable idle stop. It controls engine idle speed and maintains a smooth idle during sudden engine deceleration.

Throttle stop angle is determined by input information from the air conditioner compressor (on or off), Park/Neutral switch, and throttle position sensor (wide open or closed).

For cold engine starting, the throttle is held open for a longer period to provide adequate engine warm-up prior to normal operation. When starting a hot engine, throttle open time is shorter.

Under normal engine operating conditions, engine idle is maintained at a pre-programmed RPM which may vary slightly due to engine operating conditions. Under certain engine deceleration conditions, the throttle is held slightly open.

DATA SENSORS & SWITCHES

Oxygen (O₂) Sensor

The amount of oxygen in exhaust gases varies according to the air/fuel ratio of the intake charge. The oxygen sensor detects this content and transmits a low voltage signal to the ECU.

The oxygen sensor is located in the exhaust pipe adapter. The outer surface of the sensor is exposed to exhaust gases, the inner surface to outside air. The difference in the amount of oxygen contacting the inner and outer surfaces of the sensor creates a pressure, which results in a small voltage signal. This signal, which is a measure of the unburned oxygen in the exhaust gas, is transmitted to the ECU.

If the amount of oxygen in the exhaust system is low (rich mixture), the sensor voltage signal will be high. If the mixture is lean, the oxygen sensor will generate a low voltage signal.

Manifold Air Temperature (MAT) Sensor

The MAT sensor is installed in the intake manifold in front of an intake port. This sensor provides a voltage signal to the ECU representing the temperature of the air/fuel mixture in the intake manifold.

Coolant Temperature Sensor

The coolant temperature sensor is installed in the engine water jacket and provides a voltage signal to the ECU. The ECU determines cold engine operation from this signal and responds by enriching the fuel mixture.

Manifold Absolute Pressure (MAP) Sensor

The MAP sensor detects absolute pressure in the intake manifold as well as ambient atmospheric pressure. This information is supplied to the ECU as an indication of engine load. The sensor is mounted in the passenger compartment under the middle of the dash. A vacuum line from the throttle body supplies the sensor with manifold pressure information.

Wide Open Throttle (WOT) Switch

The WOT switch is mounted on the side of the throttle body. The switch provides a voltage signal to the ECU under wide open throttle conditions. The ECU responds to this signal by increasing the amount of fuel delivered by the injector.

Closed Throttle (Idle) Switch

This switch is integral with the idle speed control motor and provides a voltage signal to the ECU which increases or decreases the throttle stop angle in response to engine operating conditions.

Throttle Position Sensor (TPS)

The TPS is a variable resistor mounted on the throttle body and connected to the throttle shaft. It is found on vehicles with automatic transaxles only. Movement of the throttle cable causes the throttle shaft to rotate (opening or closing the throttle). The sensor detects this movement and provides the ECU with an appropriate voltage signal. The ECU uses this signal to determine engine operating conditions for the automatic transmission control system.

TROUBLE SHOOTING

PRELIMINARY CHECKS

The following systems and components must be in good condition and operating properly before assuming a fuel injection system malfunction.
- Air filter.
- All support systems and wiring.
- Battery connections and specific gravity.
- Compression pressure.

- Electrical connections on components and sensors.
- Emission control devices.
- Fuel system pressure and flow.
- Ignition system.
- Vacuum hose to MAP sensor.
- Vacuum line, fuel hose and pipe connections.

SYSTEM DIAGNOSIS

The AMC TBI fuel injection system is equipped with a self-diagnostic capacity. When a failure occurs within the system, a trouble code is stored in the ECU.

To recall trouble codes, install a test light between pins 2 and 4 of diagnostic connector D2. Push the WOT switch lever on the throttle body to wide open position and close the ISC motor plunger (activates closed throttle switch). Turn the ignition switch on and observe the test bulb.

The ECU test bulb will light for a moment, then go out. This should always occur, regardless of whether or not any trouble codes are stored, to indicate that the ECU is functional.

The bulb will then indicate all stored trouble codes as a series of brief flashes (flash, flash for code 2, flash, flash, flash, for code 3, etc.). If multiple trouble codes are stored in the ECU, the first code stored is indicated first, followed by a short pause, and any remaining codes. After a longer pause, the cycle repeats. Each trouble code indicates the malfunction of a specific sensor or sensors. Note codes.

Some injection system difficulties or abnormalities may occur without setting a trouble code. If this happens, refer to specific problem descriptions listed after trouble code references.

TROUBLE SHOOTING

Code 1

Check/test MAT sensor resistance.

Code 2

Check/test coolant temperature sensor resistance and MAT sensor, replace if needed.

Code 3

Check/test WOT switch, closed throttle switch, and associated wiring harness.

Code 4

Check/test closed throttle switch and MAP sensor, also associated hoses and wiring harness. Replace as needed.

Code 5

Check/test WOT switch and MAP sensor, also associated hoses and wiring harness. Replace as needed.

Code 6

Check/test oxygen sensor operation, fuel system pressure, fuel pressure, EGR solenoid control, canister purge, secondary ignition circuit, or PCV circuit.

No Test Bulb Flash

Check/test battery voltage, ECU, WOT, coolant temperature switch, defective test bulb, relays, and wiring harness.

Continuous Fuel Pump Operation

Check/test fuel pump, fuel pump relay, and wiring harness for short to ground.

No Fuel Pump Operation with Starter Motor Engaged

Check/test fuel pump, starter motor relay, and wiring harness.

Poor Idle when Started Cold

Check/test ISC motor, ECU, or wiring harness.

Erratic Idle Speed

Check/test ISC motor, ECU, or wiring harness.

Battery Loses Charge with Key Off

Check/test throttle position switch, MAP sensor, coolant temperature sensor, starter motor relay, ECU, ignition switch, wiring harness, or associated hoses.

Poor Idle, Oxygen Sensor Inactive

Check/test EGR solenoid valve, ECU, or wiring harness.

Canister Purge Erratic

Check/test EGR solenoid valve.

Low Fuel Pressure

Check/test fuel pressure, fuel pump, fuel filter, fuel pump ballast resistor, fuel pressure regulator, fuel injector, throttle body, hoses, or associated wiring.

Excessive Fuel Pressure

Check/test fuel return line and fittings for restrictions.

Engine Will Not Start

Check/test WOT switch, low fuel supply or pressure, fuel filter, fuel pump, ECU, secondary ignition circuit, battery voltage, fuel injector, ISC motor, starter motor relay, coolant temperature sensor, spark plugs, fuel lines, or associated wiring harness.

Engine Starts But Will Not Idle

Check/test fuel pump, MAP sensor, ECU, or ignition system.

Poor Fuel Economy/Driveability

Check/test fuel pressure, ECU, oxygen sensor, fuel injector, WOT switch, air filter, EGR valve and hoses, canister purge, or associated hoses and wiring.

DIAGNOSIS & TESTING

NOTE
* The ECU is extremely reliable and should be the last component to be replaced if a doubt exists concerning the cause of a fuel injection system failure. When test calls for volt-ohmmeter, use ONLY a digital high impedence volt-ohmmeter.

MANIFOLD AIR/FUEL TEMPERATURE SENSOR

1) Disconnect the wiring harness connector from the MAT sensor. Test resistance of the sensor with an ohmmeter. Resistance ranges from 300 ohms to 300,000 ohms (10,000 ohms at room temperature). Replace sensor if outside of specified range.

2) Test resistance of the wiring harness between pin 13 of ECU harness connector J2 and the sensor connector, and between pin 11 of connector J2 and sensor connector. See Fig. 9. Repair harness if resistance is greater than one ohm.

COOLANT TEMPERATURE SENSOR

1) Disconnect wiring harness from sensor. Disconnect the wiring harness connector. Test resistance of sensor. If resistance is not 300-300,000 ohms (10,000 ohms at room temperature), replace the sensor.

2) Test resistance of the wiring harness between pin 14 of ECU connector J2 and the sensor connector. Test resistance between pin 11 of connector J2 and sensor connector. Repair wiring harness if any open circuit is found.

WIDE OPEN THROTTLE (WOT) SWITCH

1) Disconnect the wiring harness from the WOT switch and test resistance while opening and closing switch manually. When switch is closed, resistance should be infinite. A low resistance should be indicated at the wide open position. Test switch operation several times. Replace switch if defective. Reconnect wiring harness.

2) With ignition switch on, test for WOT switch voltage between pin 6 and pin 7 (ground) of diagnostic connector D2. Voltage should be zero with switch in wide open position and greater than 2 volts in any other position.

3) If voltage is always zero, test for short circuit to ground in the wiring harness or switch. Check for open circuit between pin 19 of ECU connector J2 and the switch connector. Repair or replace as needed.

4) If voltage is always greater than 2 volts, test for an open wire or connector between the switch and ground. Repair as needed.

CLOSED THROTTLE SWITCH

NOTE
* It is important that all testing be done with the idle speed control motor plunger in the fully extended position, as it would be after a normal engine shut down. If it is necessary to extend the motor plunger to test the switch, an ISC motor failure can be suspected. Refer to ISC motor adjustment.

1) With ignition on, test switch voltage at diagnostic connector D2 between pin 13 and pin 7 (ground). Voltage should be close to zero at closed throttle and greater than 2 volts off closed throttle position.

2) If the voltage is always zero, test for a short circuit to ground in the wiring harness or switch. Test for an open circuit between pin 20 of ECU connector J2 and throttle switch.

3) If voltage is always more than 2 volts, test for an open circuit in the wiring harness between the ECU and switch connector. Check for open circuit between the switch connector and ground. Repair or replace wiring harness as needed.

AMC — THROTTLE BODY FUEL INJECTION

MANIFOLD ABSOLUTE PRESSURE SENSOR

1) Check MAP sensor vacuum hose connections at throttle body and sensor. Test MAP sensor output voltage at MAP sensor connector pin B (as marked on sensor body) with the ignition switch on and the engine off. Output voltage should be 4.0-5.0 volts.

NOTE
- *Voltage should drop 0.5-1.5 volts with hot engine, at idle.*

2) Test pin 12 of ECU connector J2 for 4.0-5.0 volts to verify wiring harness condition.

3) Check for MAP sensor supply voltage of 4.5-5.5 volts at sensor connector, pin C, with ignition on. Similar voltage should be present at pin 2 of ECU connector J2. Repair or replace wiring harness if required. Test for sensor ground between pin 13 of ECU connector J2 and pin A of sensor connector.

4) Check for ground from pin 13 of ECU connector J2 to pin F of connector J1. If an open circuit is indicated, check for a good sensor ground on the flywheel housing near the starter motor.

5) If ground is good, the ECU must be replaced. Before replacing ECU, check to see if pin 13 of ECU connector J2 is shorted to 12 volts. If so, correct the condition before replacing ECU.

OXYGEN (O₂) SENSOR

1) Test continuity of harness between O₂ sensor connector and pin 9 of ECU harness connector J2. Ensure that the wiring harness is not shorted to ground. Repair or replace as necessary.

2) Test continuity between sensor ground (exhaust manifold) and pin 13 of ECU connector J2. Repair harness if needed.

3) Check sensor operation by driving vehicle with a test light (No. 158 bulb) connected between pins 2 and 4 of diagnostic connector D2.

4) Bulb lighted at start is normal operation for test circuit. If the bulb does not light after warm up, the O₂ sensor is functioning normally. If the bulb stays lit or lights after the engine warms up, replace the O₂ sensor.

5) Before installing new sensor, check for system failures which may have caused O₂ sensor malfunction. System failures which can affect O₂ sensor are: EGR solenoid control, canister purge control, PCV system, secondary ignition circuit and fuel delivery system.

ELECTRONIC CONTROL UNIT

If all components have been checked and/or repaired, but a system failure or problem still exists, ECU may be at fault. However, it is extremely important to note that ECU is a very reliable unit and must always be the final component replaced, if a doubt exists concerning cause of an injection system failure.

NOTE
- *To avoid replacing a good ECU, take it to an AMC/Renault dealer and have it tested.*

ADJUSTMENTS

NOTE
- *On-car adjustment procedures should not be necessary during normal vehicle operation or maintenance. Adjustment of the listed components should only be required when a faulty component is replaced with a new one.*

IDLE SPEED CONTROL MOTOR

1) With air cleaner removed, air conditioner off (if equipped) and engine at normal operating temperature, connect a tachometer to terminals 1 (pos) and 3 (neg) of the small diagnostic connecter (D1). *See Fig. 3.* Turn ignition off and observe ISC motor plunger. The plunger should move to fully extended position.

NOTE
- *Diagnostic connectors and relays are mounted on a common bracket, just in front of the shock tower.*

32

Courtesy of Chrysler Motors.

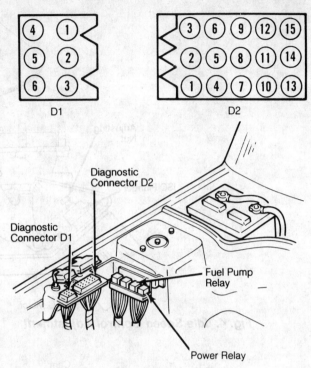

Fig. 3: TBI System Diagnostic Connector Location With Terminal Identification

2) Disconnect the ISC motor wire connector and start the engine. Idle speed should be 3300-3700 RPM. If not, turn adjusting nut on plunger until correct idle is obtained. *See Fig. 4.*

3) Hold the closed throttle switch plunger all the way in while opening the throttle. Release the throttle. The throttle lever should not make contact with the plunger. If contact is made, inspect throttle linkage and/or cable for binding or damage. Repair as needed.

4) Reconnect the ISC motor wire connector and turn ignition off for 10 seconds. Motor should move to fully extended position. Start the engine. Engine should idle at 3300-3700 RPM for a short time and then fall to normal idle. Turn off engine and remove tachometer.

5) When final adjustments to the ISC motor have been made, apply a thread sealer to adjustment screw threads to prevent movement. Install air cleaner. Since step **3)** may set a trouble code, remove the negative battery cable for 10 seconds to clear ECU memory.

WIDE OPEN THROTTLE SWITCH

1) Remove the throttle body assembly from the engine and loosen the WOT switch retaining screws (2). Hold throttle in wide open position and attach a throttle angle gauge to the flat surface of the lever. *See Fig. 4.*

2) Rotate scale to align the 15 degree mark with the pointer. Level the gauge. Rotate scale to align zero with the pointer and close the throttle enough to center the bubble. This positions the throttle at 15 degrees before wide open.

3) Adjust the WOT switch lever on the throttle cam so that the plunger is just closed. Tighten the retaining screws and remove the gauge.

FUEL PRESSURE REGULATOR

1) Remove the air cleaner and connect a tachometer to terminals 1 (pos) and 3 (neg) of diagnostic connector D1. With a fuel pressure gauge attached to the throttle body test fitting, start engine and hold at about 2000 RPM.

2) Adjust fuel pressure to obtain 14.5 psi (1.0 kg/cm²) by turning the adjusting screw in to increase pressure, out to decrease pressure. When proper system pressure is obtained, seal the screw with a lead plug. Turn off ignition, disconnect tachometer and cap test fitting. Install air cleaner.

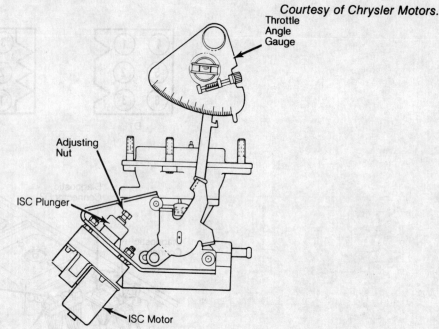

Courtesy of Chrysler Motors.

Fig. 4: Idle Speed Control Adjustment

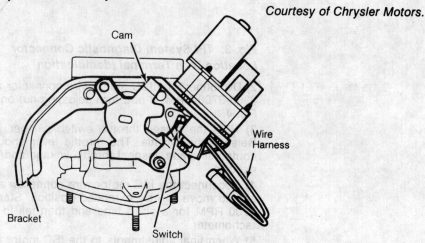

Courtesy of Chrysler Motors.

Fig. 5: Wide Open Throttle Switch

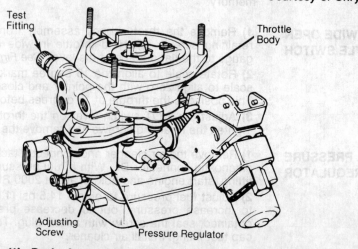

Courtesy of Chrysler Motors.

Fig. 6: Throttle Body Assembly

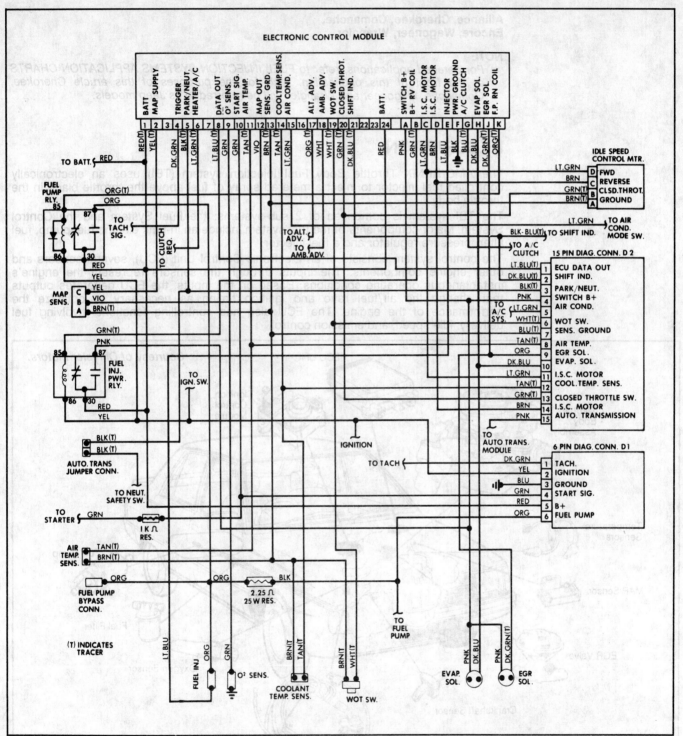

Fig. 7: AMC Throttle Body Fuel Injection System Wiring Diagram

**Alliance, Cherokee, Comanche,
Encore, Wagoneer, Wrangler**

NOTE
- *For detailed applications, refer to FUEL INJECTION SYSTEMS APPLICATION CHARTS at the front of this publication. Also for the purposes of this article Cherokee, Comanche, Wagoneer and Wrangler will be referred to as Jeep models.*

DESCRIPTION

The AMC/RENIX Throttle Body Fuel Injection system (TBI) uses an electronically controlled fuel injector to inject a metered spray of fuel above the throttle blade in the throttle body.

The TBI system is comprised of 2 sub-systems: the Fuel System and the Control System. Major components of the fuel system include an in-tank electric fuel pump, fuel filter, pressure regulator and a fuel injector.

The control system consists of an Electronic Control Unit (ECU), several sensors and other engine components. The inputs through the sensors represent the engine's instantaneous operating conditions. Based on the inputs, the ECU generates outputs that change the air/fuel ratio and ignition timing as necessary to optimize the performance of the engine. The ECU also has controlling functions involving fuel delivery, idle speed, and emission control.

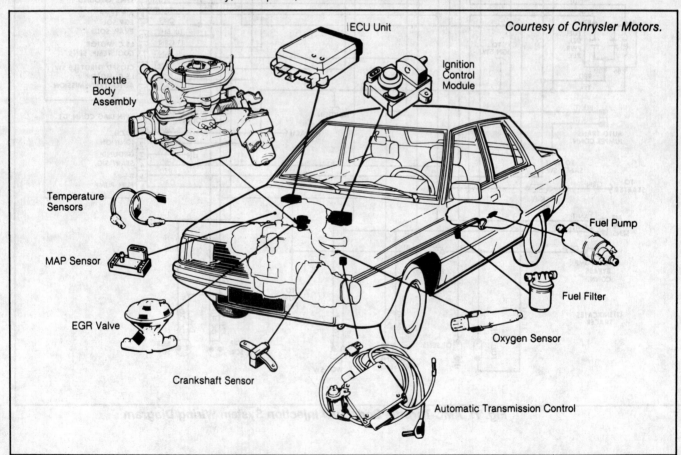

Courtesy of Chrysler Motors.

**Fig. 1: AMC/RENIX Throttle Body Fuel Injection System Components
(2.5L Jeep Models Similar)**

Input components of the control system include the air conditioning select switch, Manifold Air/Fuel Temperature (MAT) sensor, Coolant Temperature Sensor (CTS), Manifold Absolute Pressure (MAP) sensor, Wide Open Throttle (WOT) switch, closed throttle (idle) switch, Oxygen Sensor (O_2), ECU, crankshaft position/speed sensor, Throttle Position Potentiometer (TPP) or Throttle Position Sensor (TPS), gear position switch (auto. trans.), battery voltage, starter motor relay, A/C temperature control and a knock sensor. Two additional inputs are used on Jeep models; the power steering pressure switch and the load swap relay.

Outputs from the ECU include control of the fuel pump, fuel injector, upshift indicator (man. trans. only), ignition control module, A/C compressor clutch, and an Idle Speed Actuator (ISA) and motor assembly.

NOTE
- *Not all vehicles use all of the listed sensors, switches and relays.*

OPERATION

ELECTRONIC CONTROL UNIT (ECU)

The ECU is located below the glove box adjacent to the fuse panel on Alliance and Encore models, or under the instrument panel above the accelerator pedal on all Jeep models. It receives information from the various engine sensors to determine engine operating conditions at any particular moment. The ECU responds to these signals by sending a control signal (pulse) to the fuel injector to determine the amount of time that the injector will be left open (injector "on" time).

FUEL INJECTOR

The fuel injector is mounted in the throttle body such that fuel is injected into the incoming airflow. When electric current is supplied to the injector, the armature and pintle assembly move a short distance against a spring, opening a small orifice at the end of the injector. Fuel supplied to the injector is forced around the pintle valve and through this opening, resulting in a fine spray of fuel. Since fuel pressure at the injector is kept constant, the volume of fuel injected is dependent only on the length of time that the injector is energized by the ECU. *See Fig. 2.* During cold engine start-up, the injector delivers an extra amount of fuel to aid in starting.

Courtesy of Chrysler Motors.

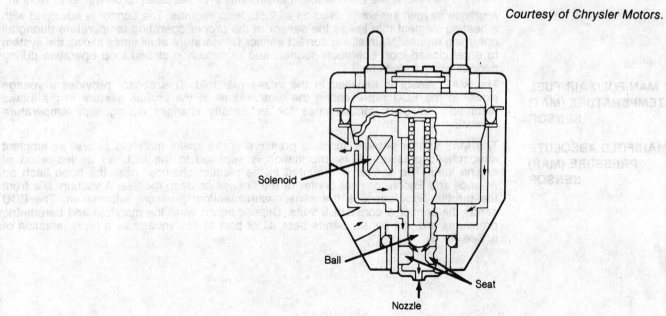

Fig. 2: Sectional View of Fuel Injector Assembly

FUEL PRESSURE REGULATOR

The fuel pressure regulator is an integral part of the throttle body. The pressure regulator has a spring chamber that is vented to the same pressure as the tip of the injector.

Because the differential pressure between the injector nozzle and the spring chamber is the same, the volume of fuel injected is dependent only on the length of time the injector is energized.

The fuel pump delivers fuel in excess of the maximum required by the engine and the excess fuel flows back to the fuel tank from the pressure regulator via the fuel return line. The fuel pressure regulator is not controlled by the ECU.

FUEL PUMP

The fuel pump is an electrically operated roller type pump which is located in the fuel tank. It contains an integral check valve which is designed to maintain fuel pressure in the system after the engine has stopped. Pump operation is controlled by the ECU.

FUEL PUMP BALLAST RESISTOR

The fuel pump ballast resistor (1983-86 Alliance and Encore) is attached to the right side of the plenum chamber. The purpose of the resistor is to reduce the fuel pump speed after the engine is started-up (ignition switch returned to the "ON" position). The resistor is by-passed when the ignition switch is in the "START" position. The fuel pump ballast resistor is not used on some 1986 systems.

IDLE SPEED ACTUATOR (ISA) MOTOR

The ISA motor is an electrically driven actuator that changes the throttle stop angle by acting as a movable idle stop. It controls engine idle speed and maintains a smooth idle during sudden engine deceleration.

The engine idle speed and engine deceleration throttle stop angle are controlled by the electric motor driven actuator.

The ECU controls the ISA motor by providing the appropriate voltage outputs to produce the idle speed or throttle stop angle required for the particular engine operating condition.

Throttle stop angle is determined by input information from the air conditioner compressor (on or off), transaxle (Park or Neutral), and throttle position (wide open or closed).

For cold engine starting, the throttle is held open for a longer period to provide adequate engine warm-up prior to normal operation.

Under normal engine operating conditions, engine idle is maintained at a pre-programmed RPM which may vary slightly due to engine operating conditions. Under certain engine deceleration conditions, the throttle is held slightly open.

OXYGEN SENSOR (O₂)

The O_2 sensor used on Alliance/Encore models is located in the exhaust pipe. The voltage output from this sensor, which varies with the oxygen content in the exhaust gas, is supplied to the ECU.

The ECU utilizes it as a reference voltage to vary the air/fuel ratio to obtain a ratio of 14.7:1. This ratio is the most efficient environment for the catalytic converter to work in.

A unique oxygen sensor is used on all 2.5L Jeep engines. The sensor is equipped with a heating element that keeps the sensor at the proper operating temperature during all operating modes. Maintaining correct sensor temperature at all times allows the system to enter closed loop operation sooner, and to remain in closed loop operation during periods of extended idle.

MANIFOLD AIR/FUEL TEMPERATURE (MAT) SENSOR

The MAT sensor is installed in the intake manifold. This sensor provides a voltage signal to the ECU representing the temperature of the air/fuel mixture in the intake manifold. The ECU compensates for air density changes during high temperature operation.

MANIFOLD ABSOLUTE PRESSURE (MAP) SENSOR

The MAP sensor detects absolute pressure in the intake manifold as well as ambient atmospheric pressure. This information is supplied to the ECU as an indication of engine load. The sensor is mounted in the plenum chamber near the hood latch on Alliance and Encore or near center of the firewall on Jeep models. A vacuum line from the throttle body supplies the sensor with manifold pressure information. The ECU sends the sensor a constant 5 volts. Depending on what the manifold and barometric pressures are, the sensor sends back all or part of this voltage as a representation of it. *See Fig. 3.*

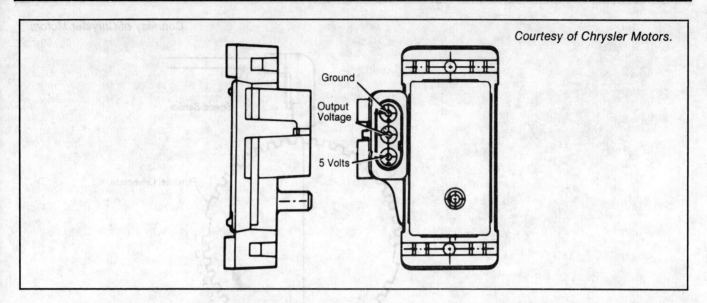

Courtesy of Chrysler Motors.

Ground

Output Voltage

5 Volts

Fig. 3: MAP Sensor

COOLANT TEMPERATURE (CTS) SENSOR	The CTS is installed in the engine water jacket and provides a voltage signal to the ECU. The ECU determines engine operating temperature from this signal and responds by enriching or leaning the fuel mixture.

This sensor also compensates for fuel condensation in the intake manifold, controls engine warm-up idle speed, increases the ignition advance and inhibits the EGR when the coolant is cold.

WIDE OPEN THROTTLE (WOT) SWITCH

The WOT switch is mounted on the side of the throttle body. The switch provides a voltage signal to the ECU under wide open throttle conditions. The ECU responds to this signal by increasing the amount of fuel delivered by the injector.

NOTE
- *If WOT switch is closed (throttle wide open) during engine cranking, fuel injector will be deactivated by ECU to clear flooded engine.*

CLOSED THROTTLE (IDLE) SWITCH

This switch is integral with the ISA motor and provides a voltage signal to the ECU which increases or decreases the throttle stop angle in response to engine operating conditions.

UPSHIFT INDICATOR

Manual Transmission Only

On vehicles equipped with a manual transmission, the ECU controls the upshift indicator light. The indicator light is normally illuminated when the ignition switch is turned on without the engine running. The light is turned off when the engine is started.

The indicator light will be illuminated during engine operation in response to engine load and speed. If the gears are not shifted, the ECU will turn the lamp off after 3 to 5 seconds. A switch located on the transmission prevents the light from being illuminated when the transmission is shifted to the highest gear.

SPEED SENSOR

The speed sensor is attached to the transmission drive plate housing. This sensor detects the flywheel drive plate teeth as they pass during engine operation and provides engine speed and crankshaft angle information to the ECU. *See Fig. 4.*

The flywheel drive plate has a large trigger tooth and notch located 90 degrees and 12 small teeth before each top dead center (TDC) position. When a small tooth and notch pass the magnet core in the sensor, the concentration and then collapse of the magnetic field induces a small voltage (spike) into the sensor pick-up coil winding. These small voltage spikes enable the ECU to count the teeth as they pass the sensor. When a large trigger tooth and notch pass the magnet core in the sensor, the increased concentration and then collapse of the magnetic field induces a higher voltage (spike) into the sensor pick-up coil winding. The higher voltage (spike) indicates to the ECU that a piston will be at the TDC position 12 teeth later. The ignition timing for the cylinder is either advanced or retarded as necessary by the ECU according to the sensor inputs.

Courtesy of Chrysler Motors.

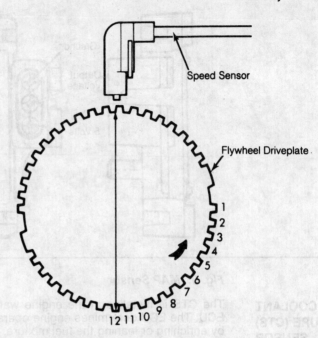

Fig. 4: Speed Sensor & Flywheel Driveplate

A/C CONTROLS	The A/C inputs indicate to the ECU that the A/C switch is in the "ON" position and when the compressor clutch must engage to lower the temperature. The ECU changes the engine idle speed according to A/C compressor operation.
KNOCK SENSOR	The knock sensor (1985-87 1.7L engine) is located in the cylinder head. This sensor provides an input to the ECU that indicates detonation (knock) during engine operation. When detonation (knock) occurs, the ECU retards the ignition advance to eliminate the detonation (knock) at the applicable cylinder.
POWER STEERING PRESSURE SWITCH	This switch is used only on 2.5L Jeep models. A pressure sensing switch is included in the power steering system. The switch increases the idle speed during periods of high pump load and low engine RPM; such as during parking maneuvers.
	Input signals from the pressure switch to the ECU are routed through the A/C request and A/C select input circuits. When pump pressure exceeds 275 psi (19 kg/cm²), the switch contacts close transmitting an input signal to the ECU. The ECU raises engine idle speed immediately after receiving the pressure switch input signal.
LOAD SWAP RELAY	This relay is only used on 1986-88 2.5L Jeep models with A/C and power steering. The relay works in conjunction with the power steering pressure switch to disengage the A/C compressor clutch.
	If the compressor clutch is engaged when the power steering pressure switch contacts close, the input signal from the switch to the ECU also activates the load swap relay. The relay contacts open cutting off electrical feed to the compressor clutch. The clutch remains disengaged until the pressure switch contacts reopen and engine idle returns to normal.
THROTTLE POSITION SENSOR (TPS)	The TPS (2.5L) is mounted on the throttle body assembly. The sensor provides the ECU with an input signal of up to 5 volts to indicate throttle position. At minimum throttle opening (idle speed), a signal input of approximately one volt is transmitted to the ECU. As throttle opening increases, signal input voltage increases to a maximum of 5 volts at higher throttle openings.
	The ECU uses the TPS voltage to determine current engine operations conditions. On 1989 models, TPS used on manual transmission models is not the same as that used on automatic models. A second 3-pin connector used on automatic transmission models provides throttle position information for transmission control unit.
THROTTLE POSITION POTENTIOMETER	The throttle position potentiometer (1.4 and 1.7L) is a simple potentiometer which provides variable voltage signals to the ECU based upon throttle positions between wide open throttle and closed throttle.

RELAYS & SOLENOIDS

Starter Motor Relay

The engine starter motor relay provides an input to the ECU that indicates the starter motor is engaged.

System Power Relay

The system power relay is located on the right strut tower. This relay is initially energized during engine start-up and remains energized until 3 to 5 seconds after the engine is stopped.

B+ Power Latch Relay

The B+ power latch relay is located on right front inner fender panel on 1986-89 models. This relay supplies current to the ISA motor enabling the ECU to extend the ISA during engine start.

Fuel Pump Control Relay

The fuel pump control relay is located on the front of the right strut tower. Battery voltage is applied to the relay via the ignition switch and it is energized when a ground is provided by the ECU. When energized, voltage is applied to the fuel pump.

Cold Start Relay

This relay enables the system power relay to have maximum voltage when the starter motor is engaged for a cole engine start-up. The cold start relay is used only on the 1986-87 1.7L engine.

A/C Clutch Relay

The ECU controls the A/C compressor clutch via this relay.

EGR Valve/Evaporative Canister Purge Solenoid

When solenoid is energized by ECU, it prevents vacuum action on the EGR valve and canister. Solenoid is energized during engine warm up, closed throttle, wide open throttle and rapid acceleration/deceleration. If solenoid wire connector is disconnected, EGR and canister will operate at all times.

TROUBLE SHOOTING (1983-88)

PRELIMINARY CHECKS

The following systems and components must be in good condition and operating properly before assuming a fuel injection system malfunction.
- Air filter.
- All support systems and wiring.
- Battery connections and specific gravity.
- Compression pressure.
- Electrical connections on components and sensors.
- Emission control devices.
- Fuel system pressure and flow.
- Ignition system.
- Vacuum hose to MAP sensor.
- All vacuum line, fuel hose and pipe connections.

TROUBLE SHOOTING

NOTE
- *The ECU is an extremely reliable part and must be the final component replaced if doubt exists.*

Fuel Pump Operates Continuously

1) Check fuel pump relay for possible short; check pin 6 of ECU connector D2 for short to ground with key off and fuel pump relay disconnected.

2) Disconnect the ECU from the wire harness. If pump stops, check the ECU with ECU Tester (MS-1700).

Fuel Pump Will Not Operate When Starter Is Engaged

1) Check fuel pump operation, repair or replace as necessary.

2) Check fuel pump relay. Check wiring harness from relay to the fuel pump.

Idle Speed Erratic

1) Check ISA motor operation, verify that motor is not jammed in either the extended or retracted position. All motor tests should be done with motor disconnected from wiring harness.

2) Check the closed throttle switch and associated wiring. Check ECU for output signals to the ISA motor.

Fuel Pressure Low

1) Check fuel pump voltage, nominally it should be 7.5 volts.

2) Check fuel filter for obstructions. Replace the fuel pump ballast resistor, and check resistor wiring harness.

3) Check fuel pressure regulator for proper operation. With engine idling gently pinch off fuel return hose, the fuel pressure should rise greatly.

4) Check fuel pressure out of the pump; several cycles of the key "on" then "off" should produce a pressure well over 20.0 psi (1.4 kg/cm²).

TESTING & DIAGNOSIS (1983-88)

DIAGNOSTIC TOOLS

NOTE
- *To diagnose fuel system problems, no specialized service equipment is needed. If ECU Tester (MS-1700) is used, however, the following diagnostic procedures are not applicable.*

To test fuel system the service technician must have available the following equipment:
- A digital volt-ohmmeter (or a meter with a minimum input impedance of one megaohm).
- A 12-volt test lamp (type 1892 or equivalent) and an assortment of jumper wires and probes.
- A hand vacuum pump with gauge.
- An ignition timing light.

DIAGNOSTIC TESTS

There are 7 different test flow charts that are used to thoroughly evaluate the fuel injection system:

Ignition Switch "OFF" Test

This test checks system power for the ECU memory keep-alive power.

Ignition Switch "ON" Power Test

This test checks system power function and fuel pump power function.

Ignition Switch "ON" Input Test

This test checks closed throttle (idle) switch, Wide Open Throttle (WOT) switch, MAP sensor, gear position (P/N) switch (auto. trans. only), coolant temperature switch (cold), and the MAT sensor (cold). This test procedure checks all of the related wiring circuits as well.

System Operational Test

This test checks engine start-up circuit, fuel injector and circuit, closed loop air/fuel mixture function, coolant temperature sensor function, MAT sensor function, knock sensor and closed loop ignition retard/advance function, EGR and canister purge solenoid function, idle speed control function, and air conditioning control function.

Basic Engine Test

This test indicates probable failures in related engine components that are not part of the fuel injection system.

Man. Trans. Upshift Test

This test checks the up-shift indicator lamp function on vehicles with manual transmissions.

Throttle Position Sensor Test

Tests the operation of the throttle position sensor. *See Fig. 7.*

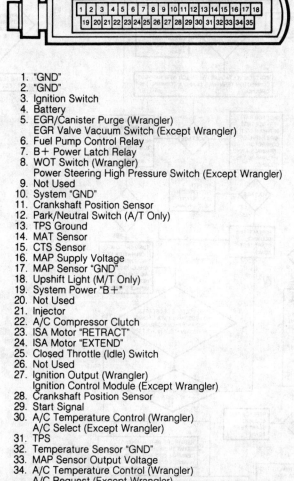

Courtesy of Chrysler Motors.

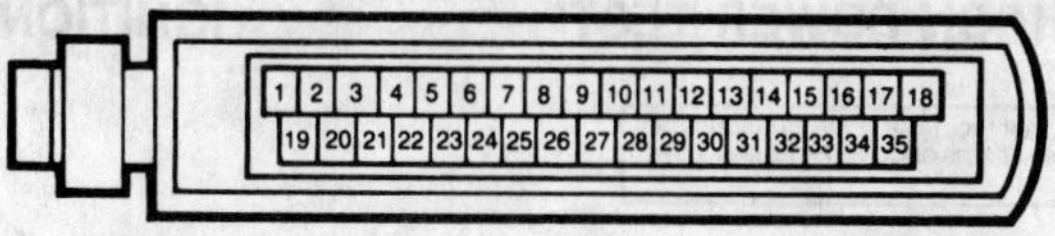

1. Ground
2. Ground
3. Ignition Switch
4. Battery
5. EGR Valve/
   Canister Purge
6. Fuel Pump Relay
7. System Power Relay
8. WOT Switch
9. Not Used
10. System Ground
11. Speed Sensor
12. Park/Neutral Switch
    (Auto. Trans.)
13. Knock Sensor
14. MAT Sensor
15. Coolant Temp. Sensor
16. MAP Sensor
    (Supply Voltage

17. MAP Sensor (Ground)
18. ECU Data (Output)
19. System Power (Batt. Pos.)
20. Not Used
21. Injector
22. A/C Compressor Clutch
23. ISA Motor Retract (Reverse)
24. ISA Motor Extend (Forward)
25. Closed Throttle (Idle) Switch
26. Not Used
27. Ignition
28. Speed Sensor
29. Start
30. A/C Clutch Control
31. Sensor Ground
32. Knock Sensor
33. MAP Sensor (Output Voltage)
34. A/C Temperature Control
35. Oxygen Sensor

Fig. 5: 1983-88 ECU Connector Terminal Identification

Courtesy of Chrysler Motors.

1. "GND"
2. "GND"
3. Ignition Switch
4. Battery
5. EGR/Canister Purge (Wrangler)
   EGR Valve Vacuum Switch (Except Wrangler)
6. Fuel Pump Control Relay
7. B+ Power Latch Relay
8. WOT Switch (Wrangler)
   Power Steering High Pressure Switch (Except Wrangler)
9. Not Used
10. System "GND"
11. Crankshaft Position Sensor
12. Park/Neutral Switch (A/T Only)
13. TPS Ground
14. MAT Sensor
15. CTS Sensor
16. MAP Supply Voltage
17. MAP Sensor "GND"
18. Upshift Light (M/T Only)
19. System Power "B+"
20. Not Used
21. Injector
22. A/C Compressor Clutch
23. ISA Motor "RETRACT"
24. ISA Motor "EXTEND"
25. Closed Throttle (Idle) Switch
26. Not Used
27. Ignition Output (Wrangler)
    Ignition Control Module (Except Wrangler)
28. Crankshaft Position Sensor
29. Start Signal
30. A/C Temperature Control (Wrangler)
    A/C Select (Except Wrangler)
31. TPS
32. Temperature Sensor "GND"
33. MAP Sensor Output Voltage
34. A/C Temperature Control (Wrangler)
    A/C Request (Except Wrangler)
35. O_2 Sensor Input

Fig. 6: 1989 ECU Connector Terminal Identification

IGNITION SWITCH "ON POWER TEST"

IGNITION SWITCH "ON POWER TEST" (Cont.)

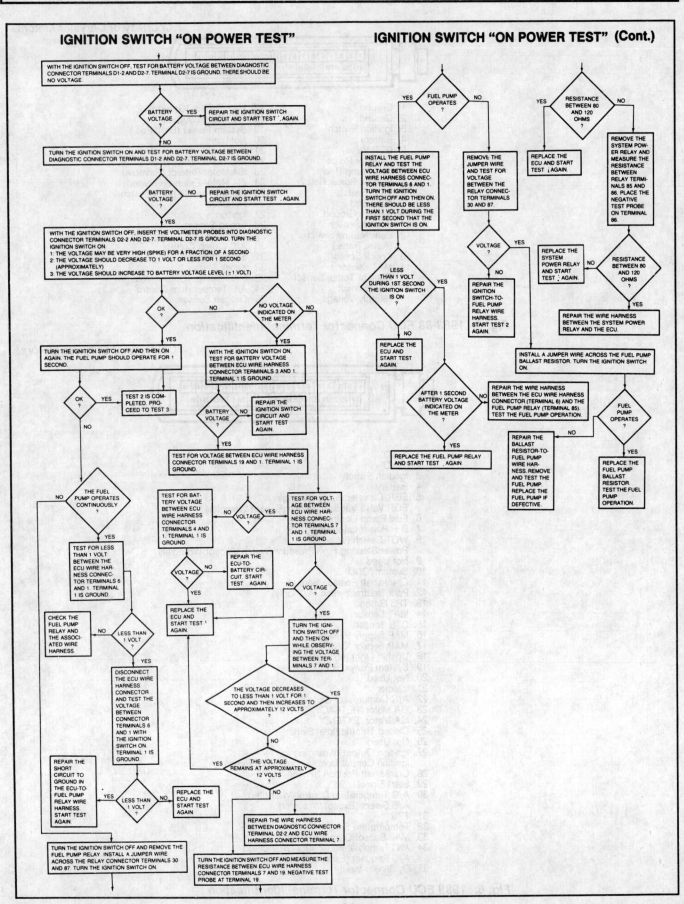

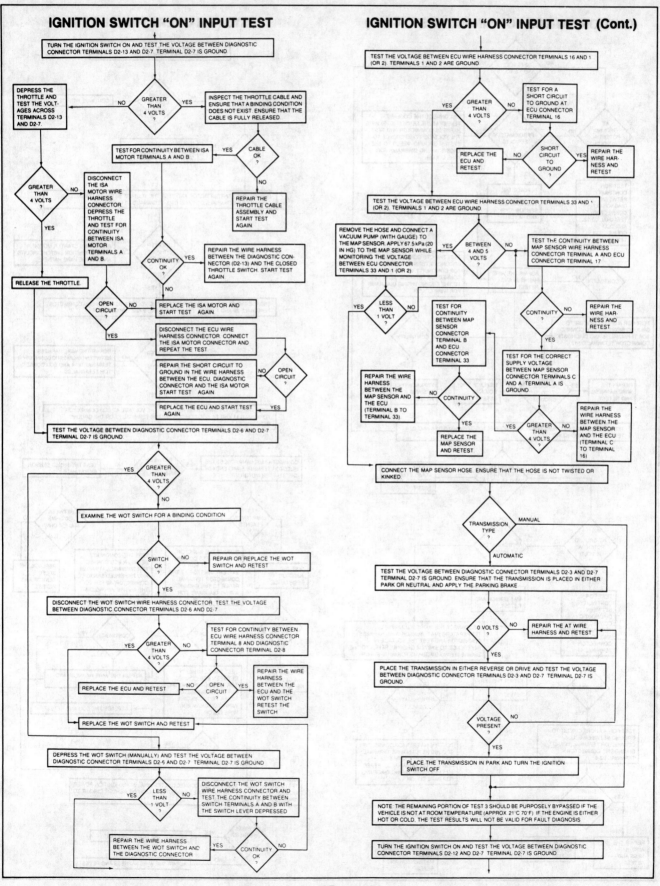

IGNITION SWITCH "ON" INPUT TEST

IGNITION SWITCH "ON" INPUT TEST (Cont.)

IGNITION SWITCH "ON" INPUT TEST (Cont.)

SYSTEM OPERATIONAL TEST

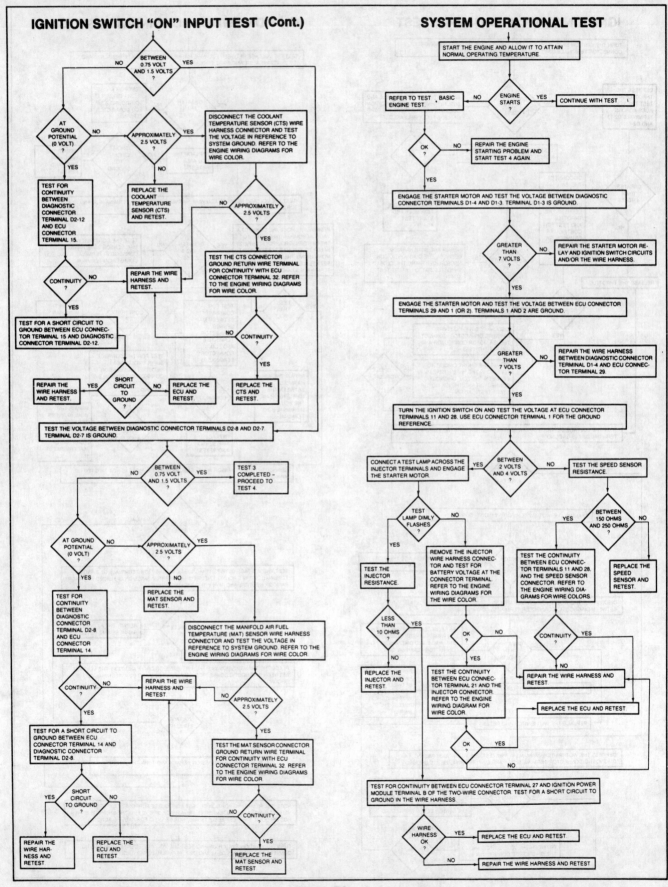

SYSTEM OPERATIONAL TEST (Cont.)

SYSTEM OPERATIONAL TEST (Cont.)

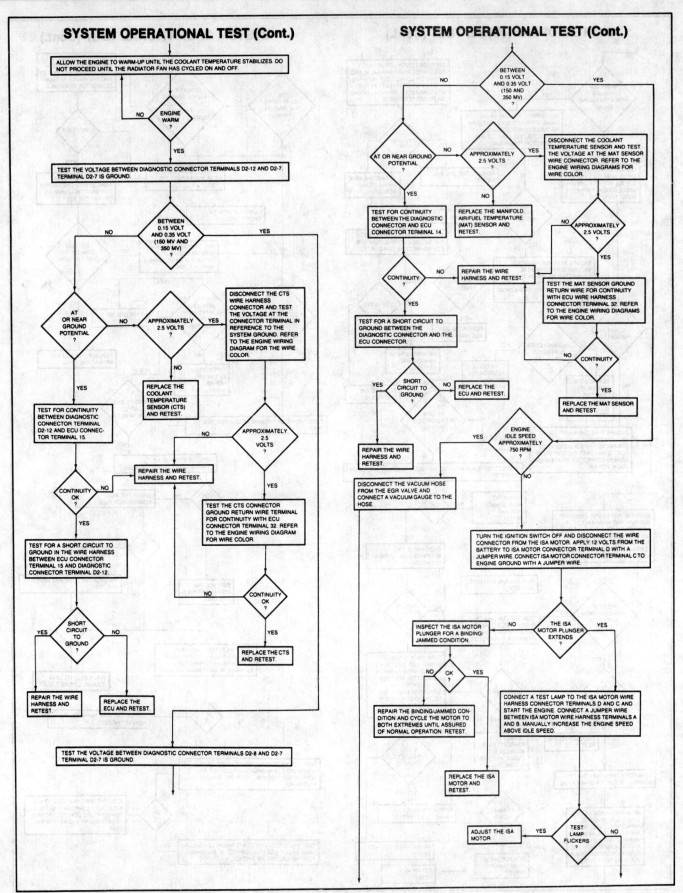

SYSTEM OPERATIONAL TEST (Cont.)

SYSTEM OPERATIONAL TEST (Cont.)

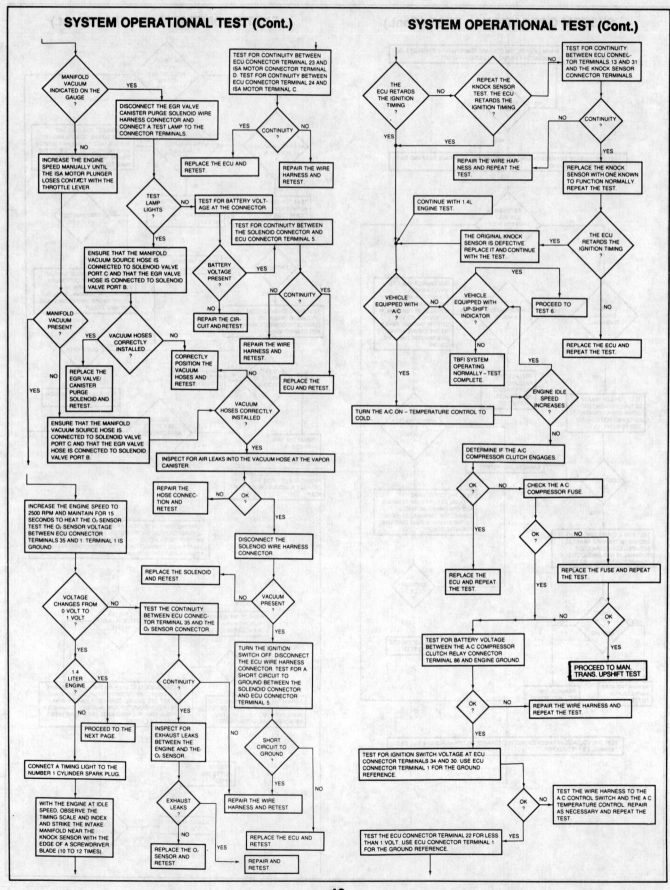

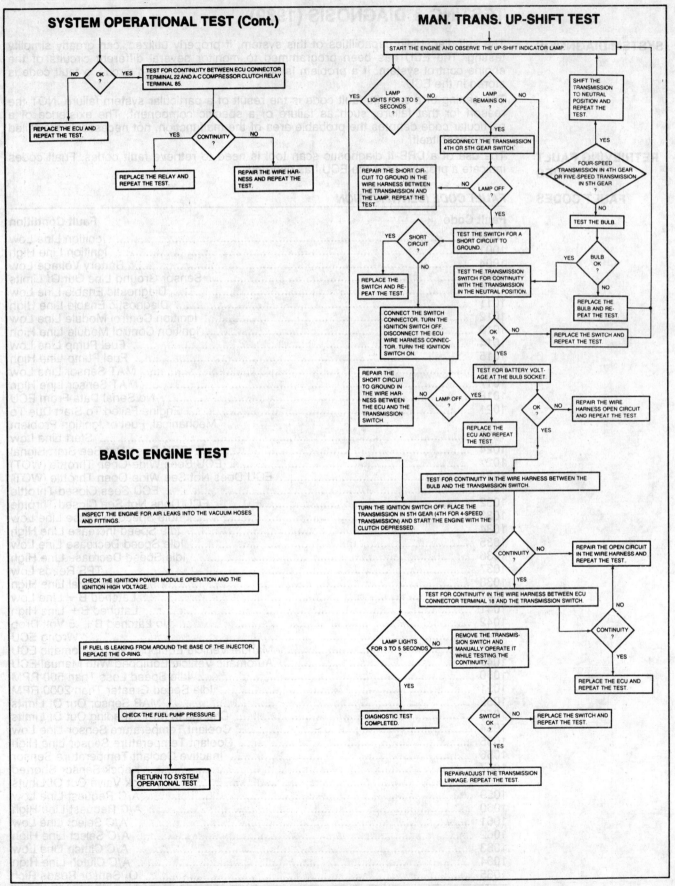

SYSTEM OPERATIONAL TEST (Cont.)

OK? — NO → REPLACE THE ECU AND REPEAT THE TEST.

OK? — YES → TEST FOR CONTINUITY BETWEEN ECU CONNECTOR TERMINAL 22 AND A/C COMPRESSOR CLUTCH RELAY TERMINAL 85.

CONTINUITY? — YES → REPLACE THE RELAY AND REPEAT THE TEST.

CONTINUITY? — NO → REPAIR THE WIRE HARNESS AND REPEAT THE TEST.

BASIC ENGINE TEST

INSPECT THE ENGINE FOR AIR LEAKS INTO THE VACUUM HOSES AND FITTINGS.

↓

CHECK THE IGNITION POWER MODULE OPERATION AND THE IGNITION HIGH VOLTAGE.

↓

IF FUEL IS LEAKING FROM AROUND THE BASE OF THE INJECTOR, REPLACE THE O-RING.

↓

CHECK THE FUEL PUMP PRESSURE.

↓

RETURN TO SYSTEM OPERATIONAL TEST

MAN. TRANS. UP-SHIFT TEST

START THE ENGINE AND OBSERVE THE UP-SHIFT INDICATOR LAMP.

LAMP LIGHTS FOR 3 TO 5 SECONDS? — NO → **LAMP REMAINS ON?** — NO → SHIFT THE TRANSMISSION TO NEUTRAL POSITION AND REPEAT THE TEST.

LAMP REMAINS ON? — YES → DISCONNECT THE TRANSMISSION 4TH OR 5TH GEAR SWITCH.

↓

LAMP OFF? — NO → REPAIR THE SHORT CIRCUIT TO GROUND IN THE WIRE HARNESS BETWEEN THE TRANSMISSION AND THE LAMP. REPEAT THE TEST.

LAMP OFF? — YES → TEST THE SWITCH FOR A SHORT CIRCUIT TO GROUND.

↓

SHORT CIRCUIT? — YES → REPLACE THE SWITCH AND REPEAT THE TEST.

SHORT CIRCUIT? — NO → TEST THE TRANSMISSION SWITCH FOR CONTINUITY WITH THE TRANSMISSION IN THE NEUTRAL POSITION.

↓

CONNECT THE SWITCH CONNECTOR. TURN THE IGNITION SWITCH OFF. DISCONNECT THE ECU WIRE HARNESS CONNECTOR. TURN THE IGNITION SWITCH ON.

OK? — NO → REPLACE THE SWITCH AND REPEAT THE TEST.

OK? — YES → TEST FOR BATTERY VOLTAGE AT THE BULB SOCKET.

LAMP OFF? — NO → REPAIR THE SHORT CIRCUIT TO GROUND IN THE WIRE HARNESS BETWEEN THE ECU AND THE TRANSMISSION SWITCH.

LAMP OFF? — YES → REPLACE THE ECU AND REPEAT THE TEST.

OK? — NO → REPAIR THE WIRE HARNESS OPEN CIRCUIT AND REPEAT THE TEST.

OK? — YES → TEST FOR CONTINUITY IN THE WIRE HARNESS BETWEEN THE BULB AND THE TRANSMISSION SWITCH.

FOUR-SPEED TRANSMISSION IN 4TH GEAR OR FIVE-SPEED TRANSMISSION IN 5TH GEAR — YES → SHIFT THE TRANSMISSION TO NEUTRAL POSITION AND REPEAT THE TEST.

FOUR-SPEED TRANSMISSION IN 4TH GEAR OR FIVE-SPEED TRANSMISSION IN 5TH GEAR — NO → TEST THE BULB.

BULB OK? — YES → TEST THE TRANSMISSION SWITCH FOR CONTINUITY WITH THE TRANSMISSION IN THE NEUTRAL POSITION.

BULB OK? — NO → REPLACE THE BULB AND REPEAT THE TEST.

TURN THE IGNITION SWITCH OFF. PLACE THE TRANSMISSION IN 5TH GEAR (4TH FOR 4-SPEED TRANSMISSION) AND START THE ENGINE WITH THE CLUTCH DEPRESSED.

↓

CONTINUITY? — NO → REPAIR THE OPEN CIRCUIT IN THE WIRE HARNESS AND REPEAT THE TEST.

CONTINUITY? — YES → TEST FOR CONTINUITY IN THE WIRE HARNESS BETWEEN ECU CONNECTOR TERMINAL 18 AND THE TRANSMISSION SWITCH.

CONTINUITY? — NO → ... YES → REPLACE THE ECU AND REPEAT THE TEST.

LAMP LIGHTS FOR 3 TO 5 SECONDS? — YES → DIAGNOSTIC TEST COMPLETED.

LAMP LIGHTS FOR 3 TO 5 SECONDS? — NO → REMOVE THE TRANSMISSION SWITCH AND MANUALLY OPERATE IT WHILE TESTING THE CONTINUITY.

↓

SWITCH OK? — NO → REPLACE THE SWITCH AND REPEAT THE TEST.

SWITCH OK? — YES → REPAIR/ADJUST THE TRANSMISSION LINKAGE. REPEAT THE TEST.

TESTING & DIAGNOSIS (1989)

SYSTEM DIAGNOSIS

The self-diagnostic capabilities of this system, if properly utilized, can greatly simplify testing. The ECU has been programmed to monitor several different circuits of the engine control system. If a problem is sensed with a monitored circuit, a fault code is stored in the ECU.

The setting of a specific fault code is the result of a particular system failure, NOT the reason for that failure, such as failure of a specific component. The existence of a particular code denotes the probable area of the malfunction, not necessarily the failed component itself.

RETRIEVING FAULT CODES

The use of a DRB-II diagnostic scan tool is need to retrieve fault codes. Fualt codes indicate a problem area the ECU has denoted.

FAULT CODES

FAULT CODE IDENTIFICATION

Fault Code	Fault Condition
1000	Ignition Line Low
1001	Ignition Line High
1004	Battery Voltage Low
1005	Sensor Ground Line Out Of Limits
1010	Diagnostic Enable Line Low
1011	Diagnostic Enable Line High
1012	Ignition Control Module Line Low
1013	Ignition Control Module Line High
1014	Fuel Pump Line Low
1015	Fuel Pump Line High
1016	MAT Sensor Line Low
1017	MAT Sensor Line High
1018	No Serial Data From ECU
1021	Engine Failed To Start Due To Mechanical, Fuel or Ignition Problem
1022	Start Line Low
1024	ECU Does Not See Start Signal
1027	ECU Sees Wide-Open Throttle (WOT)
1028	ECU Does Not See Wide-Open Throttle (WOT)
1031	ECU Sees Closed Throttle
1032	ECU Does Not See Closed Throttle
1033	Idle Speed Increase Line Low
1034	Idle Speed Increase Line High
1035	Idle Speed Decrease Line Low
1036	Idle Speed Decrease Line High
1037	TPS Reads Low
1038	Park/Neutral Line High
1040	Latched B+ Line Low
1041	Latched B+ Line High
1042	No Latched B+ .5 Volt Drop
1047	Wrong ECU
1048	Manual Vehicle Equipped With Automatic ECU
1049	Automatic Vehicle Equipped With Manual ECU
1050	Idle Speed Less Than 500 RPM
1051	Idle Speed Greater Than 2000 RPM
1052	MAP Sensor Out Of Limits
1053	Change In MAP Reading Out Of Limits
1054	Coolant Temperature Sensor Line Low
1055	Coolant Temperature Sensor Line High
1056	Inactive Coolant Temperature Sensor
1057	Knock Sensor Shorted
1058	Knock Valve Out Of Limits
1059	A/C Request Line Low
1060	A/C Request Line High
1061	A/C Select Line Low
1062	A/C Select Line High
1063	A/C Clutch Line Low
1064	A/C Clutch Line High
1065	O_2 Sensor Reads Rich

FAULT CODE IDENTIFICATION (Cont.)

Fault Code	Fault Condition
1066	O$_2$ Sensor Reads Lean
1067	Latch Relay Line Low
1068	Latch Relay Line High
1069	No Tach
1070	A/C Cut-Out Line Low
1071	A/C Cut-Out Line High
1073	ECU Does Not See Crankshaft Position/Engine Speed Sensor Signal
1200	ECU Defective
1202	Injector Shorted To Ground
1209	Injector Open
1218	No Voltage At ECU From Power Latch Relay
1219	No Voltage At ECU From Shift Light
1220	No Voltage At ECU From EGR Solenoid
1221	No Injector Voltage
1222	Ignition Control Module Not Grounded
1223	No ECU Tests Run

PRELIMINARY CHECKS & PRECAUTIONS

Subsystem Checks

Before testing fuel injection system for cause of malfunction, check that following subsystems and components are in good condition:

- Battery and charging system.
- Engine state of tune.
- Emission control devices.
- Fuel system pressure and delivery volume.
- Wiring connectors and components.

General Precautions

In order to prevent injury to operator or damage to system or components, use the following techniques:

- Turn ignition off before connecting or disconnecting any components
- DO NOT apply DC voltage greater than 12 volts or any AC voltage to system.
- Disconnect battery before charging.
- Remove ECU from vehicle if ambient temperature could exceed 176°F (80°C).
- Never modify any system function.

COMPONENT TESTING

NOTE

- Only use a high impedance digital type meter, when testing.

Coolant Temperature Sensor (CTS)

Unplug CTS connector. Measure resistance between CTS terminals. See CTS/MAT SENSOR RESISTANCE VALUES table. If resistance is not as specified, replace sensor. If resistance is okay, check wiring between CTS and ECU. If resistance is greater than one ohm, repair wiring.

CTS/MAT SENSOR RESISTANCE VALUES [1]

°F (°C)	Ohms
212 (100)	185
160 (70)	450
100 (38)	1600
70 (20)	3400
40 (4)	7500
20 (-7)	13,500
0 (-18)	25,500
-40 (-40)	100,700

[1] – Values given are approximate.

Manifold Air Temperature (MAT) Sensor

Unplug MAT sensor connector. Measure resistance between MAT sensor terminals. See CTS/MAT SENSOR RESISTANCE VALUES table. If resistance is not as specified, replace sensor. If resistance is okay, check wiring between CTS and ECU. If resistance is greater than one ohm, repair wiring.

MAP Sensor

1) Ensure MAP sensor vacuum hose connections are secure. Connect a voltmeter between MAP sensor terminal "B" (marked on sensor connector) and ground. See Fig.

4. Turn ignition on. Voltmeter reading should be 4-5 volts. Start engine and warm to normal operating temperature. Voltmeter reading should be 1.5-2.1 volts.

2) Check ECU terminal No. 33 for same voltmeter reading. If voltage readings vary between MAP sensor and ECU, check wiring. Turn ignition off. Connect a voltmeter between MAP sensor terminal "C" and ground. Turn ignition on. Voltage should be 4.5-5.5 volts. Also check voltage at ECU terminal No. 16. Turn ignition off.

3) Using an ohmmeter, check for continuity between MAP sensor terminal "A" and ECU terminal No. 17. If there is not continuity, repair wiring as necessary. Ensure there is continuity between ECU terminals No. 17 and No. 2. If there is no continuity, check for defective sensor ground in flywheel housing, near starter. If ground is okay, check for voltage at ECU terminal No. 17. If there is voltage, repair and retest. If there is no voltage an terminal No. 17, replace ECU.

Throttle Position Sensor

For testing procedure for TPS, *see Fig. 7.*

Courtesy of Chrysler Motors.

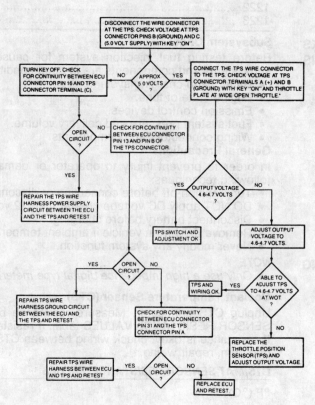

Fig. 7: Throttle Position Sensor Testing

ADJUSTMENTS

NOTE
- *On-car adjustment procedures should not be necessary during normal vehicle operation or maintenance. Adjustment of the listed components should only be required when a faulty component is replaced with a new one.*

IDLE SPEED ACTUATOR MOTOR

1) With air cleaner removed, air conditioner off (if equipped) and engine at normal operating temperature, connect a tachometer to terminals D-1 (pos) and D-3 (neg) of the small diagnostic connector located on right fenderwell, behind relay block on Cherokee, Comanche and Wagoneer. On Wrangler, connector is located on bracket, fender side of battery and on the heater plenum on all other models. *See Figs. 8 and 9.* Turn ignition off and observe ISA motor plunger. The plunger should move to fully extended position.

2) Disconnect the ISA motor wire connector and start the engine. Idle speed should be 3300-3700 RPM. If not, turn adjusting screw on plunger until correct idle is obtained. *See Fig. 10.*

3) Hold the closed throttle switch plunger all the way in while opening the throttle. Release the throttle. The throttle lever should not make contact with the plunger. If contact is made, inspect throttle linkage and/or cable for binding or damage. Repair as necessary.

4) Reconnect the ISA motor wire connector and turn ignition off for 10 seconds. Motor should move to fully extended position. Start the engine. Engine should idle at approximately 3500 RPM for a short time and then fall to normal idle. Turn off engine and remove tachometer.

5) When final adjustments to the ISA motor have been made, apply a thread sealer to adjustment screw threads to prevent movement. Install air cleaner. Since step **3)** may set a trouble code, remove the negative battery cable for 10 seconds to clear ECU memory.

Courtesy of Chrysler Motors.

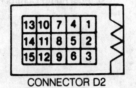
CONNECTOR D2

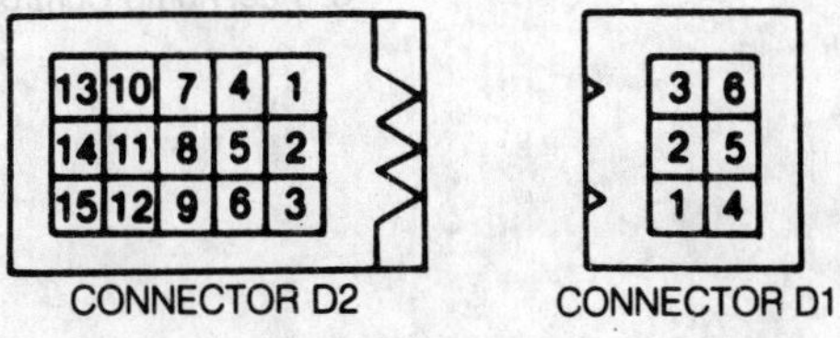

CONNECTOR D1

FENDER

Connector D2	Connector D1
1. ECU Data Output	1. Tach (RPM) Voltage (Input)
2. System Power Relay	2. Ignition
3. Park/Neutral Switch	3. Ground
4. System Power (Batt. Pos.)	4. Starter Motor Relay
5. A/C Clutch	5. Battery
6. WOT Switch	6. Fuel Pump
7. Ground	
8. MAT Sensor	
9. Ignition Power Module	
10. EGR Valve/Canister Purge Solenoid	
11. ISA Motor Forward	
12. Coolant Temp. Switch	
13. Closed Throttle Switch	
14. ISA Motor Reverse	
15. Auto. Trans. Diagnosis	

Fig. 8: 1983-88 Fuel Injection Diagnostic Connectors

Courtesy of Chrysler Motors.

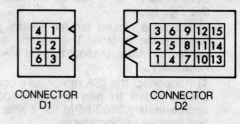

CONNECTOR
D1

CONNECTOR
D2

ALL MODELS (EXCEPT WRANGLER)

CONNECTOR "D1"	CONNECTOR "D2"
1. Tachometer Input	1. Upshift Light (M/T Only)
2. Ignition	2. "B+" Power Latch Relay
3. "GND"	3. Park/Neutral Switch
4. Start Solenoid	4. "B+" Power Latch Relay
5. Battery	5. A/C Clutch Relay
6. Fuel Pump Control Relay	6. Power Steering Switch
	7. "GND"
	8. MAT Sensor
	9. Ignition Control Module
	10. EGR Valve Solenoid
	11. ISA Motor "EXTEND"
	12. CTS
	13. Closed Throttle Switch
	14. ISA Motor "RETRACT"
	15. Not Used

WRANGLER

CONNECTOR "D1"	CONNECTOR "D2"
1. Tachometer Input	1. Upshift Light (M/T Only)
2. Ignition	2. "B+" Power Latch Relay
3. "GND"	3. Park/Neutral Switch
4. Start Solenoid	4. "B+" Power Latch Relay
5. Battery	5. Not Used
6. Fuel Pump Control Relay	6. WOT Switch
	7. Not Used
	8. MAT Sensor
	9. Ignition Control Module
	10. EGR Valve Solenoid
	11. ISA Motor "EXTEND"
	12. CTS
	13. Closed Throttle Switch
	14. ISA Motor "RETRACT"
	15. Not Used

Fig. 9: 1989 Fuel Injection Diagnostic Connectors

Courtesy of Chrysler Motors.

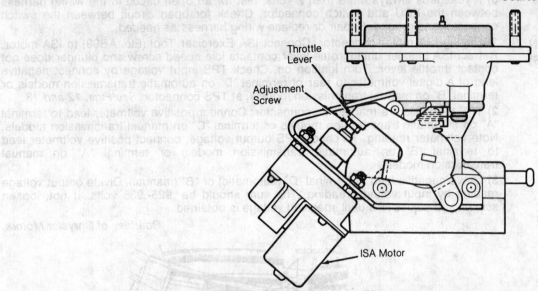

Throttle
Lever

Adjustment
Screw

ISA Motor

Fig. 10: Idle Speed Actuator (ISA) Motor Adjustment

FUEL PRESSURE REGULATOR

1) Replacement fuel pressure regulator must be adjusted to establish correct pressure. Remove air filter elbow and hose. Connect tachometer leads to diagnostic connector "D1", attaching positive lead to terminal "D1-1" and negative lead to terminal "D1-3". Remove screw plug and install pressure gauge and test fitting. *See Fig. 11.*

2) Start engine and allow to idle. Turn screw at bottom of regulator to set correct pressure. Turning screw inward increases pressure, outward to decrease pressure. Fuel pressure should be 14-15 psi (.97-1.03 kg/cm²). Install lead sealing ball to cover regulator adjustment screw after adjustment. Turn engine off and remove equipment.

Courtesy of Chrysler Motors.

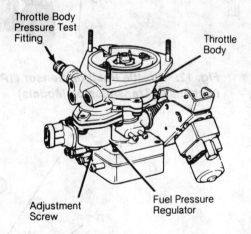

Throttle Body
Pressure Test
Fitting

Throttle
Body

Adjustment
Screw

Fuel Pressure
Regulator

Fig. 11: Adjusting Fuel Pressure Regulator

CLOSED THROTTLE SWITCH

NOTE
- It is important that all testing be done with the idle speed control motor plunger in the fully extended position, as it would be after a normal engine shut down. If it is necessary to extend the motor plunger to test the switch, an ISA motor failure can be suspected. Refer to ISA motor adjustment.

1) With ignition "ON", test switch voltage at small diagnostic connector D2 between pin 13 and pin 7 (ground). Voltage should be close to zero at closed throttle and greater than 2 volts off closed throttle position.

2) If the voltage is always zero, test for a short circuit to ground in the wiring harness or switch. Test for an open circuit between pin 20 of ECU connector J2 and throttle switch.

3) If voltage is always more than 2 volts, test for an open circuit in the wiring harness between the ECU and switch connector. Check for open circuit between the switch connector and ground. Repair or replace wiring harness as needed.

THROTTLE POSITION SENSOR (TPS)

1) Unplug ISA motor connector. Connect ISA Exerciser Tool (Ele. AB99) to ISA motor. Retract ISA plunger until throttle lever contacts idle speed screw and plunger does not contact throttle lever. Turn ignition on. Check TPS input voltage by connect negative lead of a digital voltmeter to rear of terminal "D" on automatic transmission models or terminal "B" on manual transmission models, at TPS connector. See Figs. 12 and 13.

2) TPS terminals are marked on connector. Connect positive voltmeter lead to terminal "A" on automatic transmission models or terminal "C" on manual transmission models. Note voltmeter reading. To check TPS output voltage, connect positive voltmeter lead to terminal "B" on automatic transmission models or terminal "A" on manual transmission models.

3) Leave negative lead at terminal "D" (automatic) or "B" (manual). Divide output voltage reading by input voltage reading. The sum should be .925-.935 volts. If not, loosen screws and adjust TPS until specified voltage is obtained.

Courtesy of Chrysler Motors.

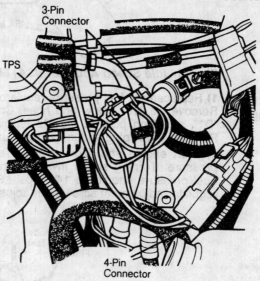

Fig. 12: Throttle Position Sensor (TPS) Connector Location (Automatic Transmission Models)

Courtesy of Chrysler Motors.

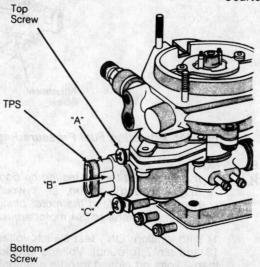

Fig. 13: Throttle Position Sensor (TPS) Connector Location (Manual Transmission Models)

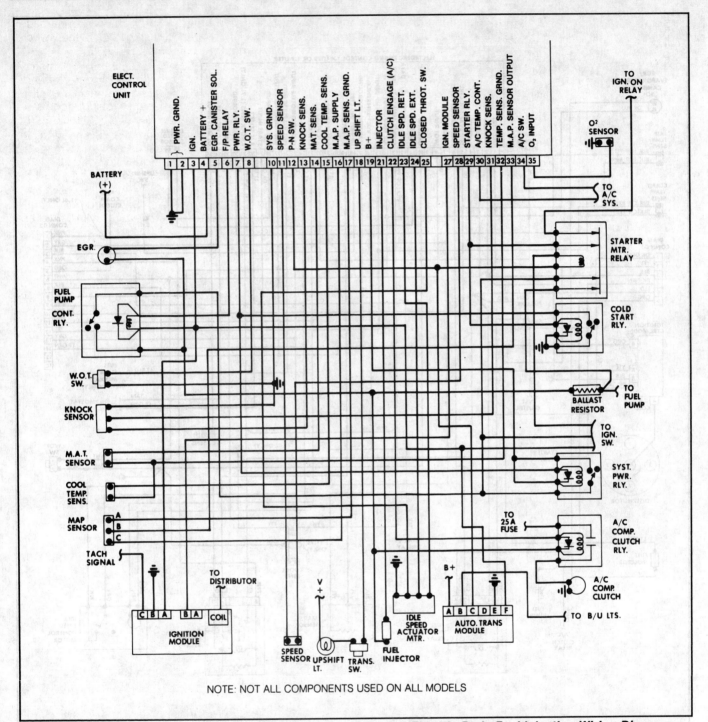

Fig. 14: 1983-85 Alliance & Encore Throttle Body Fuel Injection Wiring Diagram

NOTE: NOT ALL COMPONENTS USED ON ALL MODELS

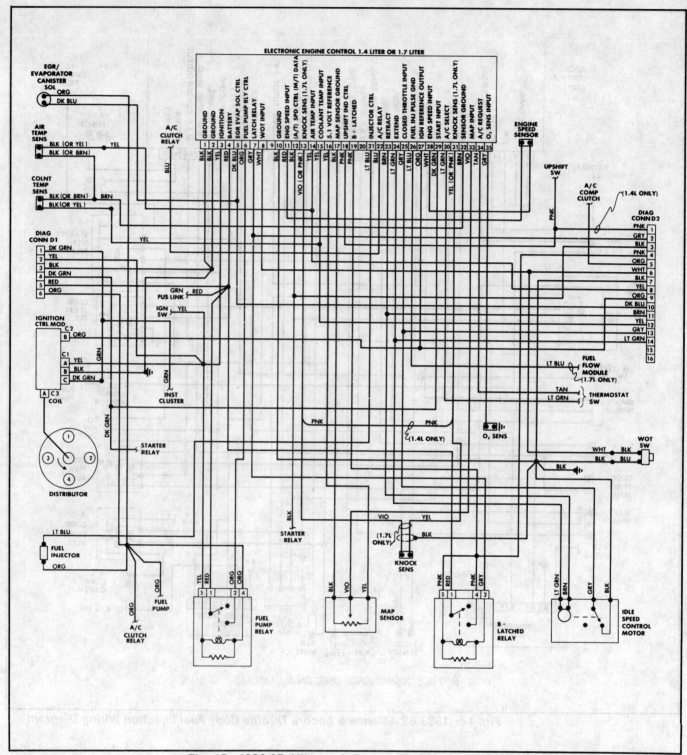

Fig. 15: 1986-87 Alliance & Encore Throttle Body Fuel Injection Wiring Diagram

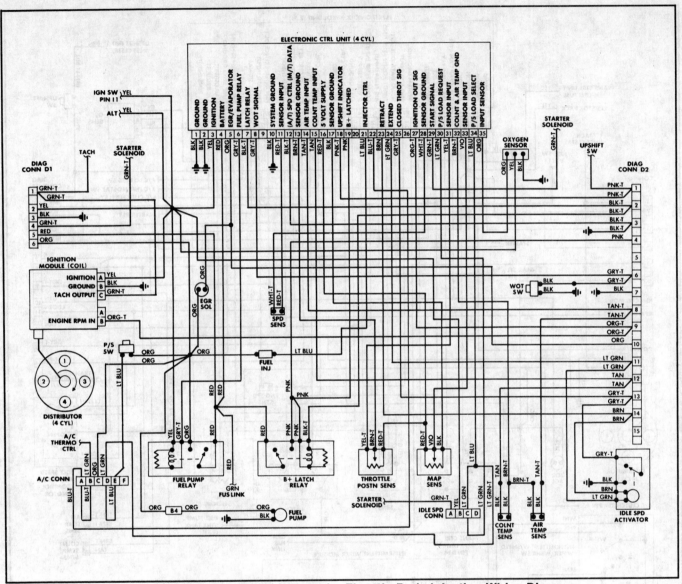

Fig. 16: 1987-88 Wrangler Throttle Body Injection Wiring Diagram

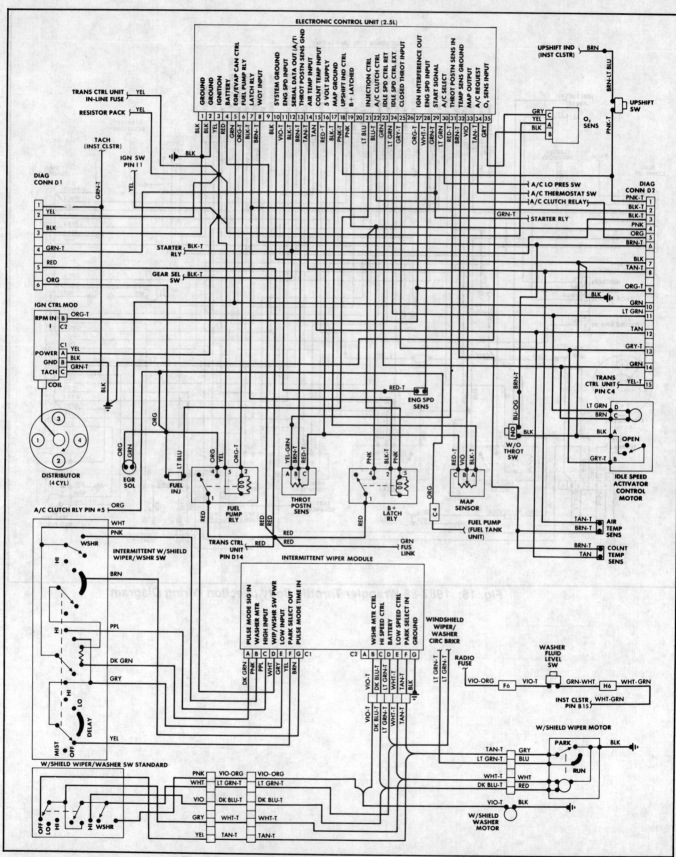

**Fig. 17: 1987-88 Comanche Throttle Body Injection Wiring Diagram
(1986 Wiring Diagram Unavailable)**

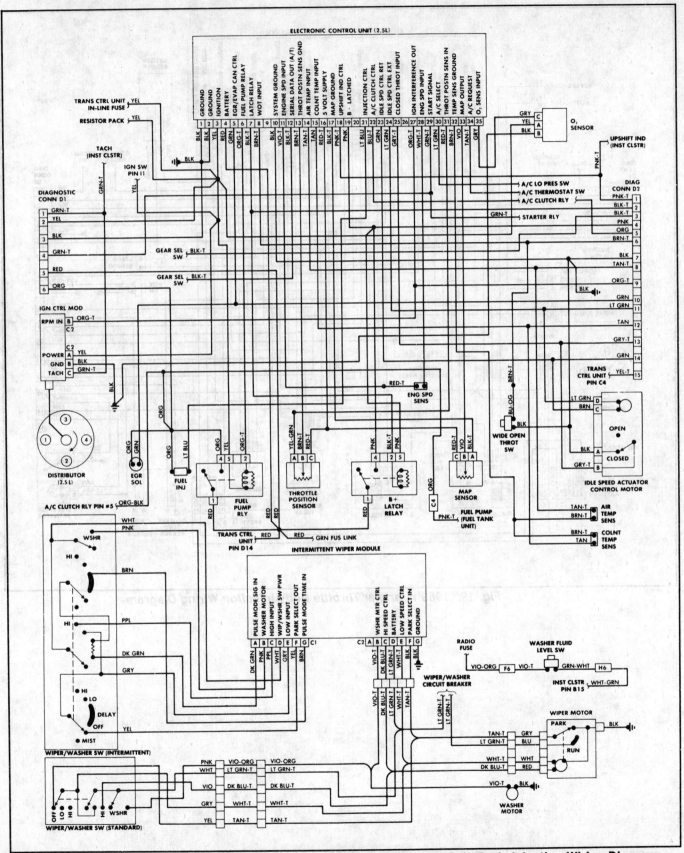

Fig. 18: 1987-88 Cherokee & Wagoneer Throttle Body Injection Wiring Diagram

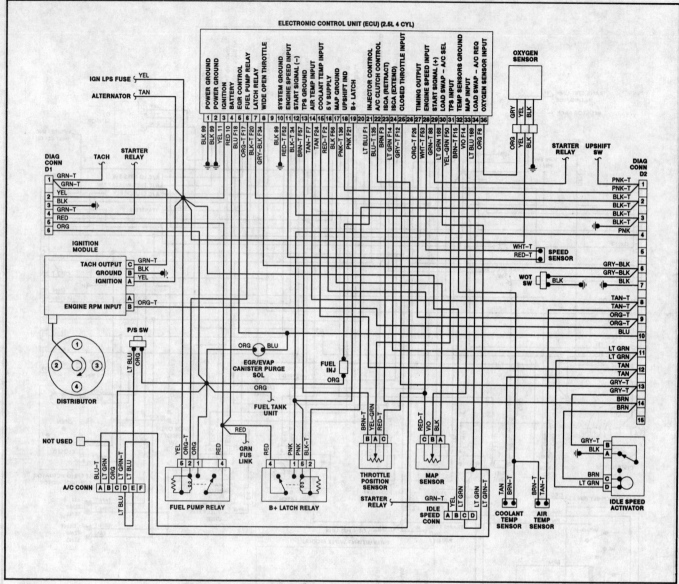

Fig. 19: 1989 Wrangler Throttle Body Injection Wiring Diagram

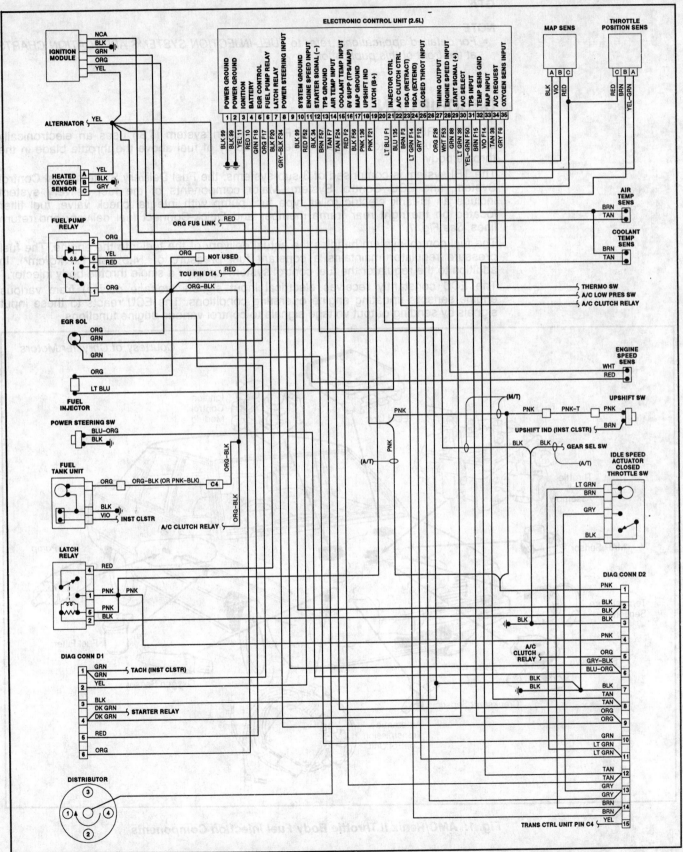

ELECTRONIC CONTROL UNIT (2.5L)

Fig. 20: 1989 Cherokee, Comanche & Wagoneer Throttle Body Injection Wiring Diagram

GTA

NOTE
- *For detailed applications, refer to FUEL INJECTION SYSTEMS APPLICATION CHARTS at the front of this publication.*

DESCRIPTION

The AMC/RENIX II Throttle Body Fuel Injection system (TBI) uses an electronically controlled fuel injector to inject a metered spray of fuel above the throttle blade in the throttle body.

The TBI system is comprised of 3 sub-systems: the Fuel Delivery System, Fuel Control System, and ECU Control System. Major components of the fuel delivery system include an in-tank electric roller type fuel pump with integral check valve, fuel filter located on the right rear frame member and Quick-Connect fuel delivery and return lines. *See Fig. 1.*

The fuel control system handles the actual delivery of the fuel into the engine. The fuel pressure regulator maintains a constant fuel pressure of 14.5 psi (1.0 kg/cm²). In addition to the regulator the fuel control system includes a single throttle body injector.

The ECU constantly receives electrical input signals (voltage pulses) from various engine sensors updating engine operating conditions. The ECU reacts to these input signals by sending output voltage signals to control various engine functions.

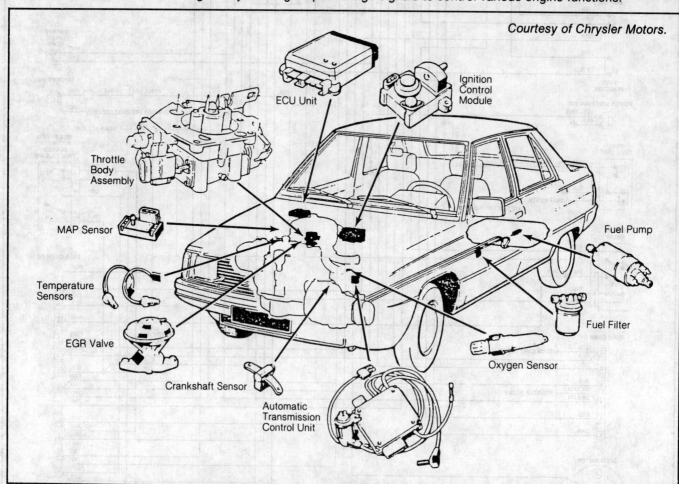

Courtesy of Chrysler Motors.

Fig. 1: AMC/Renix II Throttle Body Fuel Injection Components

OPERATION

FUEL DELIVERY

Power to the fuel pump is supplied by the fuel pump relay, located in engine compartment on inner right fender panel. The relay receives current when ignition switch is in the "START" or "RUN" position. A ground is then supplied by the ECU, closing the relay contact points and energizing the fuel pump.

The fuel pump is an electrical roller type pump which is located in the fuel tank. It contains an integral check valve which is designed to maintain fuel pressure in the system after the engine has stopped.

FUEL CONTROL

The fuel pump delivers fuel in excess of the maximum required by the engine. Excess fuel flows back to the fuel tank through the pressure regulator via the fuel return line. The fuel pressure regulator is mechanically operated and not controlled by the ECU.

The fuel pressure regulator is an integral part of the throttle body. The pressure regulator has a spring chamber that is vented to the same pressure as the tip of the injector. *See Fig. 2.*

Because the differential pressure between the injector nozzle and the spring chamber is the same, volume of fuel injected is dependent only on the length of time the injector is energized.

Courtesy of Chrysler Motors.

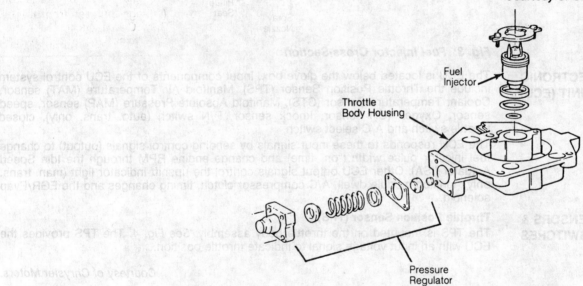

Fig. 2: Throttle Body Housing with Injector & Pressure Regulator

The fuel injector is an electrically operated solenoid mounted in the throttle body so that fuel is injected into the incoming airflow. *See Fig. 3.* When electric current is supplied to the injector a magnetic field draws the pintle assembly back a short distance away from the pintle seat, counteracting the action of the return spring and opening a small orifice at the end of the injector. Fuel supplied to the injector is forced around the pintle valve and through this opening, resulting in a fine spray of fuel. Since fuel pressure at injector is kept constant, volume of fuel injected is dependent only on the length of time that injector is energized by the ECU.

Courtesy of Chrysler Motors.

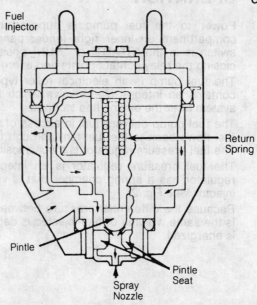

Fig. 3: Fuel Injector Cross-Section

ELECTRONIC CONTROL UNIT (ECU)	The ECU is located below the glove box. Input components of the ECU control system include the Throttle Position Sensor (TPS), Manifold Air Temperature (MAT) sensor, Coolant Temperature Sensor (CTS), Manifold Absolute Pressure (MAP) sensor, speed sensor, Oxygen (O_2) sensor, knock sensor, P/N switch (auto. trans. only), closed throttle switch and A/C select switch.

The ECU responds to these input signals by sending control signals (output) to change fuel injector pulse width ("on" time) and change engine RPM through the Idle Speed Actuator (ISA). Other ECU output signals control the upshift indicator light (man. trans. only), ignition module dwell, A/C compressor clutch, timing changes and the EGR/Evap solenoid.

DATA SENSORS & SWITCHES

Throttle Position Sensor (TPS)

The TPS is mounted on the throttle body assembly. *See Fig. 4.* The TPS provides the ECU with an input voltage signal to indicate throttle position.

Courtesy of Chrysler Motors.

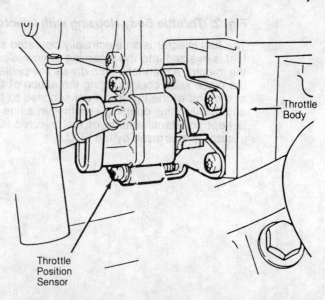

Fig. 4: Throttle Position Sensor Mounting Location

At minimum throttle opening (idle speed), a signal input of approximately one volt is transmitted to the ECU. As throttle opening increases, signal input voltage increases to a maximum of 5 volts at higher throttle openings.

Models with automatic transmissions have a dual TPS. The second TPS sends throttle position signals to the transmission ECU.

Manifold Air Temperature (MAT) Sensor

The MAT sensor is installed in the intake manifold. This sensor provides a varying voltage signal which is interpreted by the ECU as the relative temperature of the air/fuel mixture in the intake manifold. The ECU compensates for air density changes during high temperature operation.

Coolant Temperature (CTS) Sensor

The CTS is installed at the rear of the intake manifold in the water passage and provides a varying voltage signal to the ECU. The ECU determines cooling system temperature from this signal. When the engine is cold the ECU responds by enriching the air/fuel mixture.

This sensor also compensates for fuel condensation in the intake manifold, controls engine warm-up idle speed, increases the ignition advance and inhibits the EGR when the coolant is cold.

Manifold Absolute Pressure (MAP) Sensor

The MAP sensor is located on the plenum chamber rear panel. The MAP sensor detects vacuum in the intake manifold as well as ambient atmospheric pressure. This information is supplied to the ECU as an indication of engine load. A vacuum line from the throttle body supplies the sensor with manifold vacuum information. The ECU sends the sensor a constant 5 volts. Depending on manifold and barometric pressures, the sensor returns all or part of this voltage.

Speed Sensor

The speed sensor is attached to the transmission drive plate housing. This sensor detects flywheel drive plate teeth as they pass during engine operation and provides engine speed and crankshaft angle information to the ECU.

The flywheel drive plate has a large trigger tooth and notch located 90 degrees and 12 small teeth before each top dead center (TDC) position. *See Fig. 5.*

When a small tooth and notch pass the magnet core in the sensor, the concentration and then collapse of the magnetic field induces a small voltage spike into the sensor pick-up coil windings. These small voltage spikes enable the ECU to count the teeth as they pass the sensor. When a large trigger tooth and notch pass the magnet core in the sensor, the increased concentration and then collapse of the magnetic field induces a higher voltage spike into the sensor pick-up coil windings. The higher voltage spike

Courtesy of Chrysler Motors.

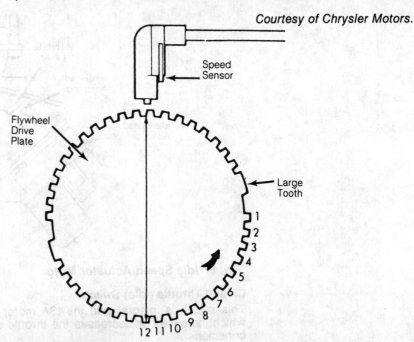

Fig. 5: Speed Sensor Operation

indicates to the ECU that a piston will be at the TDC position 12 teeth later. The ignition timing for the appropriate cylinder is either advanced or retarded as necessary by the ECU according to the sensor inputs.

Oxygen Sensor (O_2)

The O_2 sensor is located in the exhaust pipe. The voltage output from this sensor, which varies with the oxygen content in the exhaust gas, is supplied to the ECU.

If the oxygen content of exhaust gases is high (lean air/fuel mixture), the voltage signal to the ECU is low. As oxygen content decreases (mixture becomes richer), signal voltage increases.

The ECU utilizes this signal as a reference voltage to vary the air/fuel mixture to obtain a ratio of 14.7:1. This ratio is the most efficient for combustion and catalytic converter operation.

The O_2 sensor is equipped with a heating element that keeps the sensor at the proper operating temperature during all operating modes. Maintaining correct sensor temperature at all times allows the system to enter "closed loop" operation sooner, and to remain in "closed loop" operation during periods of extended idle.

Knock Sensor

The knock sensor is located in the cylinder head. This sensor provides an input to the ECU that indicates detonation (knock) during engine operation. When detonation (knock) occurs, the ECU retards the ignition advance to eliminate the detonation (knock) at the applicable cylinder.

Idle Speed Actuator (ISA) Motor

The ISA motor is an electrically driven actuator that changes the throttle stop angle by acting as a moveable idle stop. It controls engine idle speed and maintains a smooth idle during sudden engine deceleration.

The engine idle speed and engine deceleration throttle stop angle are controlled by the electric motor driven actuator. *See Fig. 6.*

The ECU controls the ISA motor by providing the appropriate voltage outputs to produce the idle speed or throttle stop angle required for the particular engine operating condition. Throttle stop angle is determined by input information from the A/C clutch request (on or off), Park/Neutral switch and TPS.

Under normal engine operating conditions, engine idle is maintained at a pre-programmed RPM which may vary slightly due to engine operating conditions. Under certain engine deceleration conditions, the throttle is held slightly open. For cold engine starting, the throttle is held open for a longer period to provide adequate engine warm-up prior to normal operation.

Courtesy of Chrysler Motors.

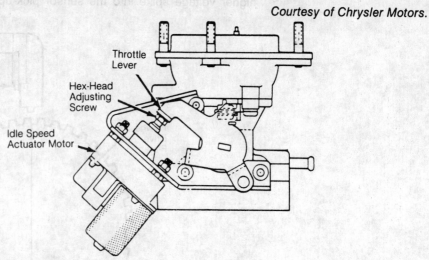

Fig. 6: Idle Speed Actuator Motor

Closed Throttle (Idle) Switch

This switch is integral with the ISA motor and provides a voltage signal to the ECU which increases or decreases the throttle stop angle in response to engine operating conditions.

Upshift Indicator Light (M/T Only)

On manual transmission models, the ECU controls the upshift indicator light. The indicator light is normally illuminated when the ignition switch is turned on without the engine running. The light is turned off when the engine is started.

The indicator light will be illuminated during engine operation in response to engine load and speed. If the gears are not shifted, the ECU will turn the light off after 3 to 5 seconds. A switch located on the transmission prevents the light from being illuminated when the transmission is shifted to the highest gear.

A/C Controls

The A/C inputs indicate to the ECU that the A/C switch is in the "ON" position and when the compressor clutch must engage to lower passenger compartment temperature. The ECU changes the engine idle speed using the ISA motor to compensate for A/C compressor operation load. A/C compressor clutch relay is located on the right inner fender panel. The ECU controls the A/C compressor clutch via this relay.

EGR/Canister Purge Solenoid

When EGR/Canister purge solenoid is energized by ECU, it prevents vacuum from activating the EGR valve and canister purge. Solenoid is energized when EGR/Canister purge action is undesirable, during engine warm up, closed throttle, wide open throttle and rapid acceleration/deceleration. If solenoid wire connector is disconnected, EGR and canister will operate at all times.

PRELIMINARY CHECKS

The following systems and components must be in good condition and operating properly before assuming a fuel injection system malfunction:
- Air filter.
- All support systems and wiring.
- Battery connections and specific gravity.
- Compression pressure.
- Electrical connections on components and sensors.
- Emission control devices.
- Ignition system.
- All vacuum line, fuel hose and pipe connections.

NOTE
- *The ECU is an extremely reliable part and must be the final component replaced if doubt exists.*

SYSTEM TESTING   **Fuel System Test**

NOTE
- *In order to pressure test this system at the throttle body a special Pressure Fitting (8983-501-572) needs to be installed.*

Courtesy of Chrysler Motors.

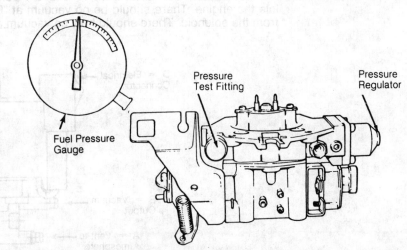

Fig. 7: Testing Fuel Pressure Regulator

Connect an accurate fuel pressure gauge to the fuel body pressure test fitting. *See Fig. 7.* Start engine. Raise engine speed to approximately 2000 RPM. The pressure should read 14.5 psi (1.0 kg/cm²). If the specification is not correct, check the fuel filter for blockage. If fuel filter is ok it will be necessary to adjust the pressure regulator.

With fuel pressure gauge installed, turn the Torx adjusting screw on the side of the pressure regulator to obtain desired pressure reading. Turning the screw inward increases pressure, outward decreases pressure.

TPS Test

See THROTTLE POSITION SENSOR TEST PROCEDURE CHART in this article.

MAT Sensor Test

The MAT sensor is located in the intake manifold. Disconnect the wire harness connector from the MAT sensor. Test the resistance of the sensor with a high impedance digital ohmmeter. The resistance should be less than 1000 ohms with the engine warm. Replace the sensor if the resistance is not within the specified range.

Test the resistance of the wire harness between the ECU wire harness connector terminal D-3 and the sensor connector terminal. Repeat the test with terminal C-8 at the ECU and the sensor connector terminal. Repair the wire harness if the resistance is greater than one ohm.

CTS Test

Disconnect the wire harness connector from the CTS switch. Test the resistance of the sensor. The resistance should be below 1000 ohms with the engine warm.

Test the resistance of the wire harness between ECU terminal D-3 and the sensor connector terminal. Repeat the test at terminal C-10 of the ECU and the sensor connector terminal. Repair the wire harness if an open circuit is indicated.

MAT SENSOR & CTS TEMPERATURE-TO-RESISTANCE VALUES

°F	°C (Approximate)	Ohms
212	100	185
160	70	450
100	38	1,600
70	20	3,400
40	4	7,500
20	-7	13,500
0	-18	25,000
-40	-40	100,700

EGR/Evap Solenoid Test

Verify that vacuum is present at vacuum fitting "C" of the EGR solenoid. *See Fig. 8.* Remove vacuum connector from "A" and "B". Connect a vacuum gauge to "B". Start and idle the engine. There should be no vacuum at "B". Disconnect electrical connector "D" from the solenoid. There should now be vacuum at "B".

Courtesy of Chrysler Motors.

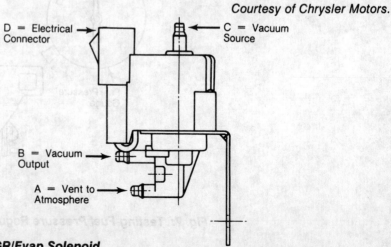

D = Electrical Connector

C = Vacuum Source

B = Vacuum Output

A = Vent to Atmosphere

Fig. 8: Testing EGR/Evap Solenoid

MAP Sensor Test

Inspect the MAP sensor hoses and connections. Repair as necessary. With the ignition on and the engine off, test the MAP sensor output voltage at the MAP sensor connector terminal "B" (marked on sensor body). *See Fig. 9.* The output voltage should be 4 to 5 volts. To verify the wiring harness condition, test ECU terminal C-6 for the same voltage described.

Test MAP sensor supply voltage at the sensor connector terminal "C" with the ignition on. The voltage should be 4.5-5.5 volts. The same voltage should also be at terminal C-14 of the ECU wire harness connector.

Test the MAP sensor ground circuit at the ECU connector between terminal D-3 and terminal B-12 with an ohmmeter. If the ohmmeter indicates an open circuit, inspect for a defective sensor ground connection, located on the flywheel/drive plate housing near the starter motor. If the ground connection is good the ECU may need to be replaced.

Courtesy of Chrysler Motors.

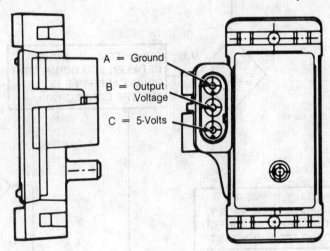

A = Ground

B = Output Voltage

C = 5-Volts

Fig. 9: Testing MAP Sensor

Speed Sensor Test

Disconnect the speed sensor connector from the ignition control module. Place an ohmmeter between terminals "A" and "B" (marked on connector). Reading should be 125-275 ohms with the engine hot. Replace sensor if readings are not within specifications.

O₂ Sensor Test

Disconnect the O₂ sensor connector. Connect an ohmmeter to terminals "A" and "B" of the O₂ sensor connector. *See Fig. 10.* Resistance should be 5-7 ohms. Replace the sensor if the reading is not as specified.

Courtesy of Chrysler Motors.

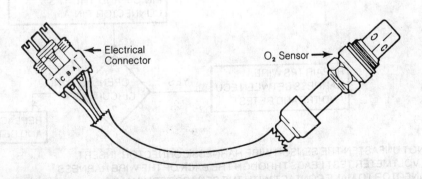

Electrical Connector

O₂ Sensor

Fig. 10: Testing O₂ Sensor

Courtesy of Chrysler Motors.

THROTTLE POSITION SENSOR TEST PROCEDURE CHART

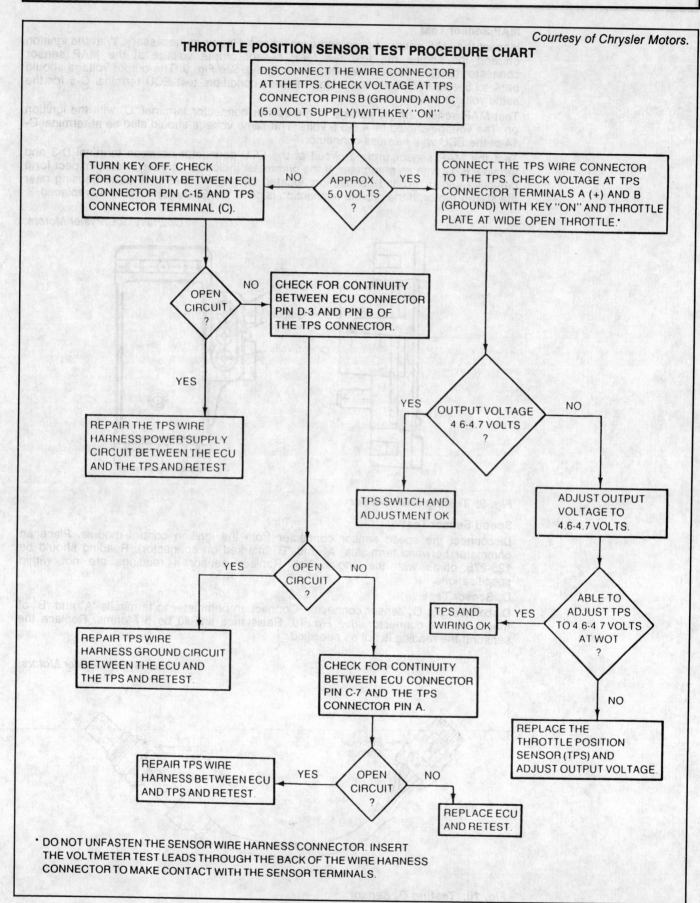

DISCONNECT THE WIRE CONNECTOR AT THE TPS. CHECK VOLTAGE AT TPS CONNECTOR PINS B (GROUND) AND C (5.0 VOLT SUPPLY) WITH KEY "ON".

APPROX. 5.0 VOLTS?

NO → TURN KEY OFF. CHECK FOR CONTINUITY BETWEEN ECU CONNECTOR PIN C-15 AND TPS CONNECTOR TERMINAL (C).

YES → CONNECT THE TPS WIRE CONNECTOR TO THE TPS. CHECK VOLTAGE AT TPS CONNECTOR TERMINALS A (+) AND B (GROUND) WITH KEY "ON" AND THROTTLE PLATE AT WIDE OPEN THROTTLE.*

OPEN CIRCUIT?

NO → CHECK FOR CONTINUITY BETWEEN ECU CONNECTOR PIN D-3 AND PIN B OF THE TPS CONNECTOR.

YES → REPAIR THE TPS WIRE HARNESS POWER SUPPLY CIRCUIT BETWEEN THE ECU AND THE TPS AND RETEST.

OUTPUT VOLTAGE 4.6-4.7 VOLTS?

YES → TPS SWITCH AND ADJUSTMENT OK

NO → ADJUST OUTPUT VOLTAGE TO 4.6-4.7 VOLTS.

OPEN CIRCUIT?

YES → REPAIR TPS WIRE HARNESS GROUND CIRCUIT BETWEEN THE ECU AND THE TPS AND RETEST.

NO → CHECK FOR CONTINUITY BETWEEN ECU CONNECTOR PIN C-7 AND THE TPS CONNECTOR PIN A.

ABLE TO ADJUST TPS TO 4.6-4.7 VOLTS AT WOT?

YES → TPS AND WIRING OK

NO → REPLACE THE THROTTLE POSITION SENSOR (TPS) AND ADJUST OUTPUT VOLTAGE.

OPEN CIRCUIT?

YES → REPAIR TPS WIRE HARNESS BETWEEN ECU AND TPS AND RETEST.

NO → REPLACE ECU AND RETEST.

* DO NOT UNFASTEN THE SENSOR WIRE HARNESS CONNECTOR. INSERT THE VOLTMETER TEST LEADS THROUGH THE BACK OF THE WIRE HARNESS CONNECTOR TO MAKE CONTACT WITH THE SENSOR TERMINALS.

Knock Sensor Test

With engine at normal operating temperature, check engine timing to verify proper adjustment. Set to specification if necessary. Connect a timing light to the No. 1 cylinder. Increase engine speed to approximately 1600 RPM and hold. Knock on the engine next to the knock sensor with a metallic object. The timing should retard.

ECU CONNECTOR PINS

ECU CONNECTOR PIN IDENTIFICATION
24 PIN CONNECTOR TBI

Terminal No.	Wire Function
A1	Not Used
A2	Not Used
A3	Not Used
A4	Not Used
A5	Fuel Pump Relay
A6	Not Used
A7	Not Used
A8	Upshift Lamp
A9	Latch Relay
A10	EGR/Evap Solenoid
A11	Not Used
A12	A/C Relay
B1	Injector
B2	Not Used
B3	ISA Motor Reverse
B4	ISA Motor Forward
B5	Not Used
B6	Not Used
B7	Battery Pos.
B8	Ignition
B9	Not Used
B10	Latched B Pos.
B11	Ground
B12	Ground

ECU CONNECTOR PIN IDENTIFICATION
32 PIN CONNECTOR TBI

Terminal No.	Wire Function
C1	Speed Sensor Pos.
C2	A/C Clutch Request
C3	Start
C4	Park/Neutral Switch
C5	Not Used
C6	MAP Sensor
C7	Throttle Position Sensor
C8	Air Temp. Sensor
C9	Not Used
C10	Coolant Temp Sensor
C11	Injection Power Supply
C12	Data Output to D2-1
C13	Not Used
C14	MAP Sensor Supply Voltage
C15	TPS Supply Voltage
C16	Not Used
D1	Speed Sensor Ground
D2	A/C Select
D3	Sensor Ground
D4	Not Used
D5	Not Used
D6	Not Used
D7	Not Used
D8	Knock Sensor Ground
D9	Oxygen Sensor Input
D10	Injector Power Supply
D11	Not Used
D12	Fuel Flow Meter
D13	Ignition Module Spark/Dwell
D14	Not Used
D15	Not Used
D16	Knock Sensor

ADJUSTMENTS

NOTE
- *On-car adjustment procedures should not be necessary during normal vehicle operation or maintenance. Adjustment of the listed components should only be required when a faulty component is replaced with a new one.*

IDLE SPEED ACTUATOR MOTOR

1) With air cleaner removed, air conditioner off (if equipped) and engine at normal operating temperature, connect a tachometer to terminals D-1 (pos) and D-3 (neg) of the small diagnostic connecter located on the heater plenum. *See Fig. 11.* Turn ignition off and observe ISA motor plunger. The plunger should move to fully extended position.

2) Disconnect the ISA motor wire connector and start the engine. Idle speed should be 3300-3700 RPM. If not, turn hex head adjusting nut on plunger until correct idle is obtained.

3) Hold the closed throttle switch plunger all the way in while opening the throttle. Release the throttle. The throttle lever should not make contact with the plunger. If contact is made, inspect throttle linkage and/or cable for binding or damage. Repair as necessary.

4) Reconnect the ISA motor wire connector and turn ignition off for 10 seconds. Motor should move to fully extended position. Start the engine. Engine should idle at approximately 3500 RPM for a short time and then fall to normal idle. Turn off engine and remove tachometer.

5) When final adjustments to the ISA motor have been made, apply a thread sealer to adjustment screw threads to prevent movement. Install air cleaner. Since step **3)** may set a trouble code, remove the negative battery cable for 10 seconds to clear ECU memory.

CLOSED THROTTLE SWITCH

NOTE
- *It is important that all testing be done with the idle speed control motor plunger in the fully extended position, as it would be after a normal engine shut down. If it is necessary to extend the motor plunger to test the switch, an ISA motor failure can be suspected. Refer to ISA MOTOR ADJUSTMENT.*

1) With ignition "ON", test switch voltage at small diagnostic connector D2 between pin 13 and pin 7 (ground). Voltage should be close to zero at closed throttle and greater than 2 volts off closed throttle position.

2) If the voltage is always zero, test for a short circuit to ground in the wiring harness or switch. Test for an open circuit between pin 20 of ECU connector J2 and throttle switch.

3) If voltage is always more than 2 volts, test for an open circuit in the wiring harness between the ECU and switch connector. Check for open circuit between the switch connector and ground. Repair or replace wiring harness as needed.

Courtesy of Chrysler Motors.

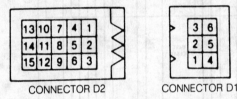

CONNECTOR D2   CONNECTOR D1

FENDER

Connector D2
1. ECU Data Output
2. System Power Relay
3. Park/Neutral Switch
4. System Power (Batt. Pos.)
5. A/C Clutch
6. WOT Switch
7. Ground
8. MAT Sensor
9. Ignition Power Module
10. EGR Valve/Canister Purge Solenoid
11. ISA Motor Forward
12. Coolant Temp. Switch
13. Closed Throttle Switch
14. ISA Motor Reverse
15. Auto. Trans. Diagnosis

Connector D1
1. Tach (RPM) Voltage (Input)
2. Ignition
3. Ground
4. Starter Motor Relay
5. Battery
6. Fuel Pump

Fig. 11: AMC/RENIX II Fuel Injection Diagnostic Connectors

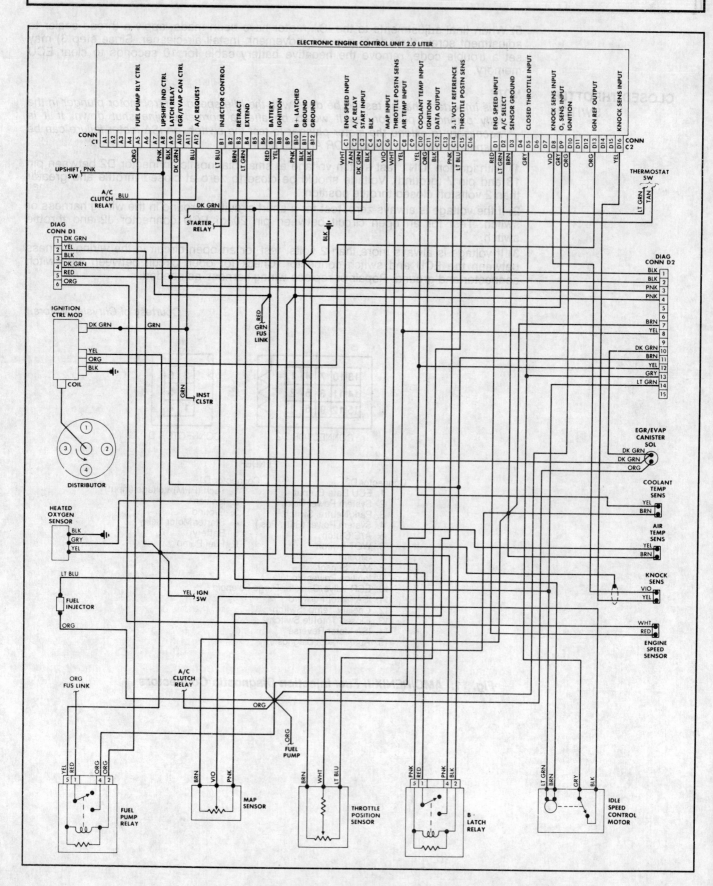

Fig. 12: AMC/RENIX II Fuel Injection System Wiring Diagram

Eagle: Medallion
Ford Motor Co.: Merkur XR4Ti, Scorpio,
 Tracer, Festiva
General Motors: Sprint Turbo

DESCRIPTION

The Bosch AFC (L-Jetronic) fuel injection system is used on all models. However, variations may exist between model applications. This article covers the Bosch AFC system in general, with manufacturers' differences noted under SPECIAL FEATURES.

The Bosch Airflow Controlled (AFC) fuel injection system is an electronically controlled system operated by incoming airflow. Some vehicles are equipped with a potentiometer to measure incoming air flow, while other vehicles use a hot wire type airflow sensor.

The AFC fuel injection system also contains a feedback system which measures oxygen content of exhaust gases and maintains the air/fuel ratio at about 14.7:1.

The fuel injection system consists of an electric fuel pump, fuel pressure regulator, fuel damper, fuel injectors, cold start injector, Electronic Control Unit (ECU), and airflow meter. In addition, an air temperature sensor, throttle position sensor, coolant temperature sensor, oxygen sensor, catalytic converter, auxiliary air valve, idle speed control valve, throttle body, and electrical relays are used.

NOTE
- *Not all models use all components.*

OPERATION

ELECTRIC FUEL PUMP(S)

Fuel under pressure from electric fuel pump flows through a fuel damper, fuel filter, injector fuel rail and fuel pressure regulator. Fuel pump(s) may be located on frame rail, in fuel tank or both. Electrical power for fuel pump operation during cranking mode is provided from starter relay via the fuel pump relay (if equipped) and ECU.

FUEL PRESSURE REGULATOR

The pressure regulator is a sealed unit which is divided by a diaphragm into 2 chambers (fuel and spring chambers). The fuel chamber receives fuel through the inlet side from the injector fuel rail. The spring chamber is connected to intake manifold vacuum.

At idle, intake manifold vacuum is high. The diaphragm is pulled down by intake manifold vacuum. Any excessive fuel is returned to the fuel tank. As the throttle is depressed, intake manifold vacuum decreases. The regulator spring overcomes manifold vacuum increasing fuel pressure.

FUEL INJECTORS

A fuel rail links the fuel pressure regulator with the fuel injectors. Each cylinder is provided with a solenoid-operated injector which sprays fuel toward the back of each intake valve.

ELECTRONIC CONTROL UNIT (ECU)

All components of the control system are electrically connected to the ECU. The ECU is a pre-programmed computer which receives and interprets data from various sensors to calculate the amount of fuel required by the engine to maintain efficiency with minimum exhaust emissions. The oxygen sensor informs the ECU of oxygen content of exhaust gases and the ECU constantly adjusts the air/fuel ratio by controlling the injector "on" time.

An automatic function of the ECU is to provide fuel enrichment whenever engine is cranked, regardless of engine temperature. This is activated by a direct electrical connection from the starter circuit to the ECU (most models). The ECU is a sealed unit, and no service is required.

AIRFLOW METER

Hot Wire Type – The airflow meter continually measures temperature, amount, density, and speed of air entering engine intake system. The meter consists of a platinum wire filament located within intake air stream.

The wire filament is kept at a constant temperature above that of air entering engine regardless of composition of air entering engine. The airflow meter sends a temperature related signal to be processed by the ECU. *See Fig. 1.*

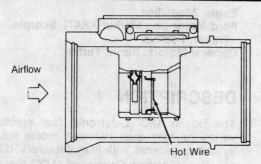

Fig. 1: Hot Wire Airflow Meter

Potentiometer Type – This airflow meter uses a movable vane connected to a potentiometer. As air entering the engine moves the vane, the potentiometer is moved informing the ECU on the amount of air entering the engine. Some potentiometer airflow meters use an air temperature sensor located inside the airflow meter air passage. *See Fig. 2.*

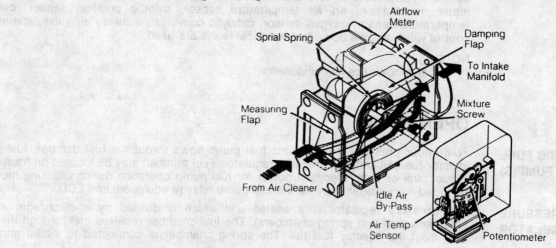

Fig. 2: Potentiometer Airflow Meter

AIR TEMPERATURE SENSOR

The air temperature sensor is an integral component of the airflow meter which converts temperature of incoming air into electrical signals. These signals are received by the ECU and processed to adjust the amount of fuel delivered by the injectors. The air temperature sensor is not serviceable.

THROTTLE POSITION SENSOR (TPS)

A contact-type TPS is installed on the throttle chamber. It converts throttle position into electrical signals to inform ECU of throttle position. Signals are sent to ECU when throttle is fully open or at idle. *See Fig. 3.* Some models send a specific signal to ECU, depending on throttle angle. The open contacts prevent loss of power during sudden acceleration/deceleration by signaling ECU of the required fuel enrichment.

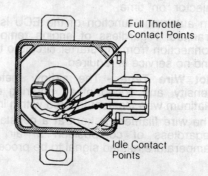

Fig. 3: Contact-Type Throttle Position Sensor

COOLANT TEMPERATURE SENSOR

This sensor provides ECU with engine temperature information relating to warm-up enrichment operation. Some models use a dual-sensor element which also signals the ignition computer (if equipped).

During warm-up period after a cold engine start, additional fuel is required to maintain engine performance. As engine temperature increases, the ECU decreases fuel enrichment until engine reaches normal operating temperature.

ELECTRICAL RELAYS

The various relays used with the electronic controls of the AFC injection system control power to injectors, fuel pump, ECU, and cold start system. The electrical relays may consist of one component for all relays or a combination of individual relays.

AUXILIARY AIR VALVE

Most models with Bosch AFC fuel injection use an Auxiliary Air Valve (AAV) to shorten engine warm-up time. The AAV supplies additional air into the intake system which increases engine RPM during a cold start.

The AAV consists of an electrically heated bi-metallic strip, movable disc, and air by-pass channel. The heater coil on the bi-metallic strip is energized by the fuel pump relay. Control of the valve is based upon engine temperature; the air by-pass channel is open when engine is cold and gradually closes as temperature rises. At predetermined temperatures, air by-pass channel is blocked and additional airflow stops. *See Fig. 4.*

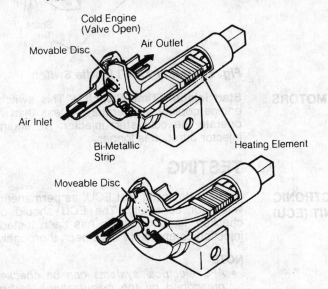

Fig. 4: Auxiliary Air Valve

COLD START INJECTOR

Most models use a cold start injector which delivers additional fuel, and a start injector time switch which controls operation of the cold start injector. The start injector time switch limits cold start injection to 1-12 seconds, depending upon engine coolant temperature. When engine coolant temperature rises above a specified point, bi-metallic contact breaks ground circuit of cold start injector and cold start enrichment is by-passed.

SPECIAL FEATURES

FORD MOTOR CO.

Inertia Switch – When impact forces exceed the magnetic force that retains the ball, it triggers the contact that opens the electric circuit. When this circuit is opened, the fuel pumps stop.

If the inertia switch is tripped, the engine stops running and cannot be restarted until it is reset. On Merkur XR4Ti, to reset switch, open liftgate and fold floor covering back for access to spare tire well. Press button on top of the inertia switch. *See Fig. 5.*

On Scorpio, open liftgate and press reset switch located near liftgate lock striker. On Tracer and Festiva 3 and 5 door models, open liftgate and reset switch located on left side of spare tire well. On Tracer station wagons, switch is located in vehicle jack storage compartment.

NOTE
- *DO NOT reset the inertia switch until the complete fuel system has been inspected for leaks.*

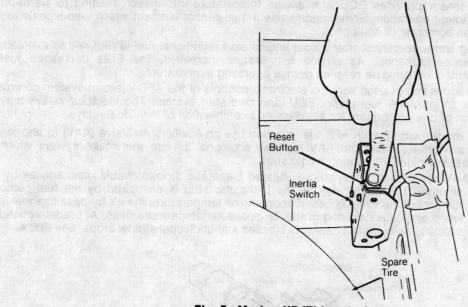

Courtesy of Ford Motor Co.

Reset Button

Inertia Switch

Spare Tire

Fig. 5: Merkur XR4Ti Inertia Switch

GENERAL MOTORS

Start Injector Time Switch – This switch has a bi-metallic coil and heat coil. When engine coolant temperature is less than 71°F (22°C) the start injector time switch will operate the cold start injector. As engine coolant temperature is colder, cold start injector duration is longer.

TESTING

ELECTRONIC CONTROL UNIT (ECU)

Do not attempt to test ECU, as permanent damage could result. It is possible to check wires for continuity. The ECU should only by judged faulty after compression is checked, ignition system has been tested and found problem-free, and all other fuel injection components have been thoroughly tested (including wiring).

NOTE
- *AFC electrical systems can be checked by using Electronic Fuel Injection testers prescribed by the manufacturer. Instructions for use of testers must be followed carefully to prevent damage to system.*

FUEL INJECTORS & RESISTORS

1) Connect tachometer to engine. Start engine and run at idle. Remove harness connector from injectors one at a time. Engine idle speed should drop 100-300 RPM as each injector is disconnected. If engine idle speed does not drop, check the wiring connector, injector resistance or injection signal from the computer.

2) Disconnect electrical connector from each injector. Measure injector resistance. See INJECTOR RESISTANCE SPECIFICATIONS table. If injector is not to specification, replace injector.

INJECTOR RESISTANCE SPECIFICATIONS

Application	Ohms
Eagle	14-18
Ford Motor Co.	
Merkur XR4Ti	2-2.7
Scorpio	16-18
Tracer	
Built Before 11-2-87	1-3
Built After 11-2-87	11-15
Festiva	11-15

3) On Tracer & Festiva models, check injector resistors. Disconnect resistor block. Measure resistance between terminal "B" and terminal No. 1, 2, 3 and 4. If reading is not 6 ohms, replace resistor block.

FUEL PRESSURE

CAUTION
- *Constant fuel pressure is maintained in fuel lines and component parts at all times. Relieve pressure before attempting to open system for testing. Do not allow fuel to flow onto engine or electrical parts or allow an open flame in area while testing fuel system components.*

Medallion

1) Relieve fuel system pressure. Connect a fuel pressure gauge between fuel pressure regulator and fuel rail. Disconnect and plug vacuum hose from fuel pressure regulator. Connect a vacuum pump to pressure regulator. Start engine.

2) With engine idling, fuel pressure should be 33-39 psi (2.3-2.7 kg/cm²). Apply 15 in. Hg to fuel pressure regulator. Fuel pressure should drop to 26-32 psi (1.8-2.2 kg/cm²).

Merkur XR4Ti

1) Relieve fuel system pressure. Connect Fuel Pressure Gauge (T85L-9974-A) and Adapter Hose (T85L-9974-A) to Schrader valve on fuel manifold. Start engine. If the engine will not start, remove fuel pump relay from fuse panel. Connect a jumper wire from relay terminals No. 87 to No. 30. *See Fig. 6.* With engine idling, fuel pressure should be 35-45 psi (2.5-3.2 kg/cm²).

2) If fuel pressure is not within specification, check for plugged fuel filter, restricted fuel line, leaking fuel pressure regulator or faulty fuel pump. Turn engine off. Fuel system should hold at least 30 psi (2.1 kg/cm²).

Courtesy of Ford Motor Co.

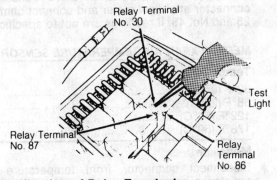

Fig. 6: Identification of Relay Terminals

Scorpio

1) Relieve fuel system pressure. Disconnect fuel pressure regulator fuel return line. Connect Fuel Pressure Gauge (T80L-9974-B) to Schrader vavle on fuel rail.

2) Connect a jumper wire to Self-Test connector Black/Red wire. Turn ignition switch to "RUN" position. Connect other end of jumper wire to ground for one minute.

3) Fuel pressure should be 43.5 psi (3.0 kg/cm²). Immediately after fuel pump shutdown, fuel pressure should be 30 psi (2.0 kg/cm²). If fuel pressure is not to specification, check fuel pressure regulator, restricted fuel lines, leaking fuel injectors or malfunctioning fuel pump.

Tracer & Festiva

1) Relieve fuel system pressure. Connect Fuel Pressure Gauge (014-00447) and Adapter (D87C9974-A) between fuel filter and fuel rail. Close adapter valve. Turn ignition on.

2) Connect a jumper wire between 2-pin fuel pump check connector located on cowl (near windshield wiper motor). Fuel pressure should be more than 60 psi (4.2 kg/cm²). If fuel pressure is less than specified, go to step **4)**. If pressure is okay, open valve on adapter and go to next step.

3) With jumper wire connected to fuel pump check connector, start engine. With engine running a various speeds, fuel pressure should be 28-36 psi (2.0-2.5 kg/cm²). Fuel system is functioning properly.

4) If fuel pressure is less than 60 psi (4.2 kg/cm²), remove fuel pressure gauge. Connect fuel pressure gauge to fuel pump outlet connection, located under rear seat.

5) With fuel pump check connector jumpered, turn ignition on. If fuel pressure is now to specification, check fuel lines for restrictions. If fuel pressure is still low, check fuel filter. Connect a voltmeter to fuel pump connector White/Yellow wire. Turn ignition on. Voltmeter should read 12 volts.

6) If voltage is not to specification, check fuel pump electrical wiring. If voltage is okay, check for continuity between 2 Black wires on fuel pump connector and ground. If continuity exists, replace fuel pump. Check fuel pump ground connection if Black wires do not show continuity.

Sprint Turbo

1) Relieve fuel system pressure. Connect fuel pressure gauge between fuel filter and fuel rail with "T" Fitting (J 34730-1-75). Start engine. With engine idling, fuel pressure should be 25-33 psi (1.7-2.3 kg/cm²).

2) Disconnect vacuum hose from fuel pressure regulator. With engine idling, fuel pressure should be 35-43 psi (2.5-3.0 kg/cm²). Turn engine off. With engine off, fuel pressure should be 22-32 psi (1.5-2.2 kg/cm²).

3) Connect a vacuum pump to fuel pressure regulator. Start engine. Apply 19 in. Hg to fuel pressure regulator. With engine idling, fuel pressure should be 34-42 psi (2.4-2.9 kg/cm²).

4) If fuel pressure is less than specified with engine idling, check for leakage at fuel feed line, restricted fuel filter or fuel line, defective fuel pump or fuel pressure regulator. If fuel pressure is more than specified with engine idling, check fuel pressure regulator, restricted pressure regulator vacuum hose or restricted fuel lines. If fuel pressure is less than specified with engine off, check for defective check valve in fuel pump, leakage in fuel feed hose, defective fuel pressure regulator or leaking injector.

AIR TEMPERATURE SENSOR

Merkur XR4Ti

On models with air temperature sensor, turn ignition off. Disconnect electrical connector at airflow meter and connect ohmmeter between break-out box terminals No. 25 and No. 46. If readings are not to specification, replace airflow meter.

MERKUR XR4Ti AIR TEMPERATURE SENSOR RESISTANCE

Temperature	Ohms
32°F (0°C)	5500-6100
68°F (20°C)	2000-3000
122°F (50°C)	760-970
178°F (50°C)	270-380

Medallion

Disconnect connector from temperature sensor. Using an ohmmeter, measure resistance across sensor terminals. See MEDALLION AIR TEMPERATURE SENSOR RESISTANCE table.

MEDALLION AIR TEMPERATURE SENSOR RESISTANCE

Temperature	Ohms
31-33°F (0-1°C)	254-266
67-69°F (19-20°C)	283-297
103-105°F (39-40°C)	315-329

Scorpio

Disconnect connector from temperature sensor. Using an ohmmeter, measure resistance across sensor terminals. See SCORPIO AIR TEMPERATURE SENSOR RESISTANCE table.

SCORPIO AIR TEMPERATURE SENSOR RESISTANCE

Temperature	Ohms
50°F (10°C)	59
65°F (18°C)	40
220°F (104°C)	1840

Sprint Turbo, Tracer & Festiva

See AIRFLOW METER in this article.

AIRFLOW METER

Sprint Turbo

Disconnect airflow meter connector. Using an ohmmeter, check resistance across meter terminals. See Fig. 7. See SPRINT TURBO AIRFLOW METER RESISTANCE table.

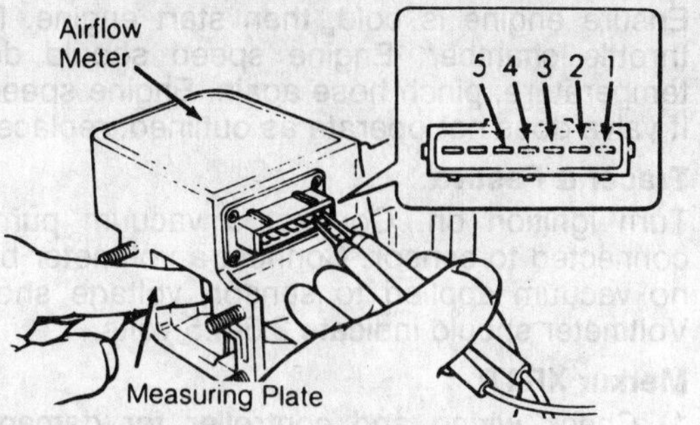

Courtesy of General Motors Corp.

Airflow
Meter

5 4 3 2 1

Measuring Plate

Fig. 7: Sprint Turbo Airflow Meter Terminals

SPRINT TURBO AIRFLOW METER RESISTANCE

Terminals	Ohms
5-3	100-300
5-4	200-400
5-1	
-4°F (-20°C)	10,000-20,000
32°F (0°C)	4000-7000
68°F (20°C)	2000-3000
104°F (40°C)	900-1300
5-2	
Measuring Plate Fully Closed	20-400
Measuring Plate Fully Opened	20-1000

Tracer & Festiva

Disconnect airflow meter connector. Using an ohmmeter, check resistance across meter terminals. *See Fig. 8.* See TRACER & FESTIVA AIRFLOW METER RESISTANCE table.

TRACER & FESTIVA AIRFLOW METER RESISTANCE

Terminals	Ohms
E_2-Vc	100-300
E_2-Vb	200-400
E_2-THA (Air Temp. Sensor)	
-4°F (-20°C)	10,000-20,000
32°F (0°C)	4000-7000
68°F (20°C)	2000-3000
104°F (40°C)	900-1300
140°F (60°C)	400-700
E_1-Fc	
Measuring Plate Fully Closed	Infinity
Measuring Plate Fully Open	0
E_2-Vs	
Measuring Plate Fully Closed	20-400
Measuring Plate Fully Open	20-1000

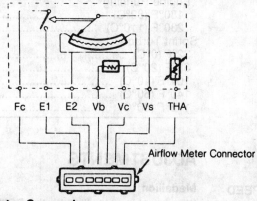

Fc E1 E2 Vb Vc Vs THA

Airflow Meter Connector

Fig. 8: Airflow Meter Connector

BOSCH AIRFLOW CONTROLLED SYSTEM

AUXILIARY AIR VALVE

Ensure engine is cold, then start engine. Pinch rubber hose between air valve and throttle chamber. Engine speed should decrease. After engine reaches operating temperature, pinch hose again. Engine speed should not decrease more than 50 RPM. If valve does not operate as outlined, replace auxiliary air valve.

ALTITUDE COMPENSATOR SENSOR

Tracer & Festiva

Turn ignition on. Connect a vacuum pump to pressure sensor. Leave connector connected to sensor. Connect a voltmeter between Light Green wire and ground. With no vacuum applied to sensor, voltage should be zero. Apply 30 in. Hg to sensor. Voltmeter should indicate 3.5-4.5 volts.

IDLE SPEED CONTROLLER (ISC) VALVE

Merkur XR4Ti

1) Check wiring and controller for damage or deterioration. Repair as necessary. Connect test light to idle speed actuator between Red wire and ground. Turn ignition on. If light is on, go to next step. If light is off, repair open in power feed circuit.

2) Connect jumper wire from actuator Gray wire to ground and turn ignition on. If speed controller operates, go to next step. If not, replace idle speed controller.

3) Using an ohmmeter, check for continuity between wire at ECA pin No. 21 and Gray wire at controller. If no continuity, repair wiring. If continuity is present, idle speed controller circuit is okay.

COLD START INJECTOR

All Models

Disconnect cold start injector connector. Using an ohmmeter, measure resistance between injector terminals. See COLD START INJECTOR RESISTANCE table.

COLD START INJECTOR RESISTANCE

Application	Ohms
Sprint Turbo	2-4

COOLANT TEMPERATURE SENSOR

All Models

Disconnect coolant temperature sensor connector. Using an ohmmeter, measure resistance between sensor terminals. See COOLANT TEMPERATURE SENSOR RESISTANCE table.

COOLANT TEMPERATURE SENSOR RESISTANCE

Temperature	Ohms
Eagle Medallion	
67-69°F (19-20°C)	283-297
175-179°F (79-81°C)	383-397
193-195°F (89-90°C)	403-417
Merkur XR4Ti, Tracer & Festiva	
50°F (10°C)	58,750
65°F (18°C)	40,500
180°F (82°C)	3600
220°F (104°C)	1840
Scorpio	
Engine Off	
140°F (60°C)	7700
240°F (115°C)	1300
Engine Running	
180°F (82°C)	4550
230°F (110°C)	1550
Sprint Turbo	
32°F (0°C)	6
68°F (20°C)	2.5
104°F (40°C)	1.5
140°F (60°C)	.5
176°F (80°C)	.2

ADJUSTMENTS

IDLE SPEED

Medallion

Idle speed is controlled by the ECU through the idle speed regulator.

Festiva & Tracer

1) Switch off all accessories. Set parking brake and block drive wheels. Connect tachometer to engine. Warm engine to normal operating temperature.

2) Unplug electric fan motor connector (if equipped). Adjust idle speed to specification by turning throttle adjusting screw.

FESTIVA & TRACER IDLE SPEED SPECIFICATIONS

Application	Man. Trans. RPM	Auto. Trans. RPM
Festiva	700-759	
Tracer	800-900	[1] 950-1050

[1] – Auto. Trans. in "N".

Merkur XR4Ti

Idle speed adjustment procedure is only to be performed if curb idle speed is not within specification. Curb idle speed is controlled by the EEC-IV processor assembly and idle speed control assembly. If idle RPM is not correct after performing this procedure, EEC-IV diagnosis is required.

1) Place transmission in Neutral. Turn A/C off. Start engine and warm up to operating temperature. Turn engine off and disconnect harness connector at Idle Speed Control (ISC).

2) Start engine and operate at 2000 RPM for 2 minutes. If electric radiator cooling fan comes on, disconnect fan wiring. Allow engine to idle. Check idle speed. If necessary, adjust idle speed by adjusting throttle plate screw. *See Fig. 9.*

Courtesy of Ford Motor Co.

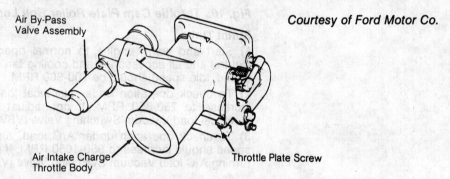

Air By-Pass Valve Assembly

Air Intake Charge Throttle Body

Throttle Plate Screw

Fig. 9: Merkur XR4Ti Throttle Plate Adjusting Screw Location

3) Turn engine off and reconnect ISC harness connector. Connect cooling fan wiring. Operate throttle plate and ensure it is not sticking in throttle bore. If idle speed is too high, turn ignition off and restart engine.

MERKUR XR4Ti IDLE SPEED SPECIFICATIONS

Application	Idle RPM
ISC Connected	825-975
ISC Disconnected	700-800

Scorpio

NOTE
 • *This procedure is necessary only if vehicle stalls at idle, has an idle surge with vehicle stopped or high idle speed.*

1) Ensure all electrical and vacuum connections are clean and tight. Ensure there are no vacuum leaks and ignition timing is set to specification.

2) Set Electronic Automatic Temperature Control (ECAT) system in "ECON" position. Place automatic transmission in "P" or manual transmission in Neutral.

3) Start engine and warm to operating temperature. Turn ignition off. Disconnect idle speed control valve electrical connector. Start engine and run at 2000 RPM for 30 seconds.

4) Return engine to idle. Check that idle speed is 675-725 RPM. If idle speed is not to specification, adjust by loosening throttle cam plate roller bolt before adjusting throttle stop screw to specified RPM. *See Fig. 10.*

5) Tighten throttle cam plate roller bolt checking that throttle stop still contacts throttle stop screw when throttle is opened. Check that throttle cam plate contacts roller without excessive throttle pedal play.

6) Raise engine speed to 2000 RPM for 30 seconds. Turn engine off. Connect idle speed control valve electrical connector. Open throttle valve to wide open throttle and ensure throttle plate is not binding in throttle bore.

Courtesy of Ford Motor Co.

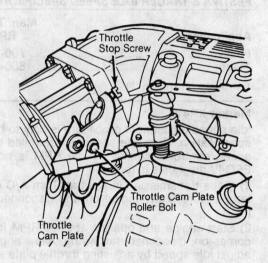

Fig. 10: Throttle Cam Plate Roller Bolt Location

Sprint Turbo

1) Start and warm engine to normal operating temperature. Ensure transaxle is in Neutral and all accessories and cooling fan are off. Connect tachometer and check idle speed. Idle speed should be 700-800 RPM.

2) To check operation under electrical loads, turn headlights on. Idle speed should increase to 750-850 RPM. If not, adjust idle-up speed to specification by turning electrical load Vacuum Switching Valve (VSV) adjustment screw. *See Fig. 11.*

3) To check operation under A/C load, turn blower motor and A/C system on. Idle speed should increase to 950-1050 RPM. If not, adjust idle-up speed to specification by turning A/C load Vacuum Switching Valve (VSV) adjustment screw.

Courtesy of General Motors Corp.

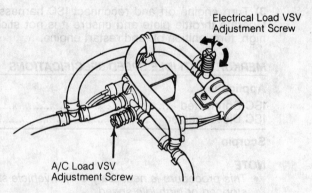

Fig. 11: Sprint Turbo Idle-Up RPM Adjustment

IDLE MIXTURE
THROTTLE POSITION
SENSOR (TPS)

Idle mixture is controlled by the ECU on all models and is not adjustable.

Medallion

1) To adjust TPS, loosen 2 attaching screws. Using an ohmmeter, measure resistance between terminals B and C with throttle closed. Ohmmeter should read zero ohms. Connect ohmmeter between terminals A and B with throttle closed. Ohmmeter should indicate infinity.

2) Open throttle partially. Connect an ohmmeter between terminals B and C. Ohmmeter should indicate infinity. Connect ohmmeter between terminals A and B. Ohmmeter should indicate infinity.

3) Open throttle valve to wide open throttle. Connect an ohmmeter between terminals B and C. Ohmmeter should indicate infinity. Connect ohmmeter between terminals A and B. Ohmmeter should indicate zero ohms. TPS is properly adjusted when switch clicks just as throttle is opened.

Merkur XR4Ti

1) Install EEC-IV Breakout Box (Rotunda T83L-50-EEC-IV) to vehicle. Using a digital voltmeter on 20-volt scale, connect positive lead to test Pin 47 and negative lead to test Pin 46. Turn ignition on. Loosen TPS attaching screws.

2) Adjust TPS so digital voltmeter reads one volt with throttle closed. Tighten TPS attaching screws. Observe digital voltmeter while opening throttle slowly to wide open throttle. Digital voltmeter should read from one volt (throttle closed) to 4 volts (wide open throttle) and back to one volt.

Scorpio

TPS is not adjustable.

Sprint Turbo

1) Insert .023" (.6 mm) feeler gauge between throttle stop screw and lever. Move TPS until continuity exists between terminals No. 1 and 2. *See Fig. 12.* Tighten TPS screws.

2) Insert .032" (.8 mm) feeler gauge between throttle stop screw and lever. No continuity should exist between terminals. Now insert a .016" (.4 mm) feeler gauge between throttle stop screw and lever. Continuity should exist between terminals. If readings are incorrect, repeat TPS adjustment procedure.

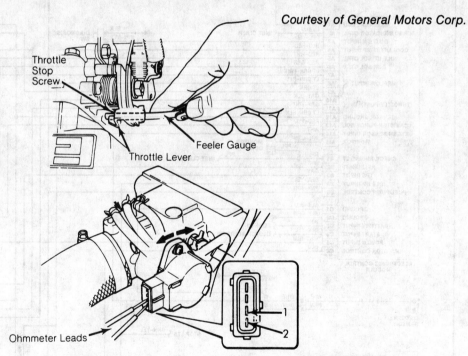

Courtesy of General Motors Corp.

Fig. 12: Sprint Turbo TPS Adjustment

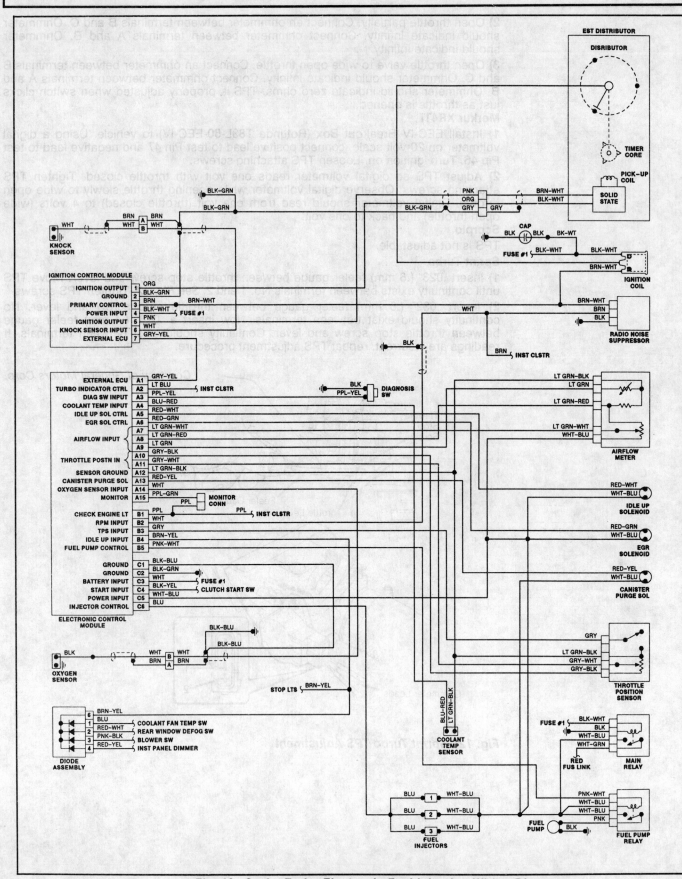

Fig. 13: Sprint Turbo Electronic Fuel Injection Wiring Diagram

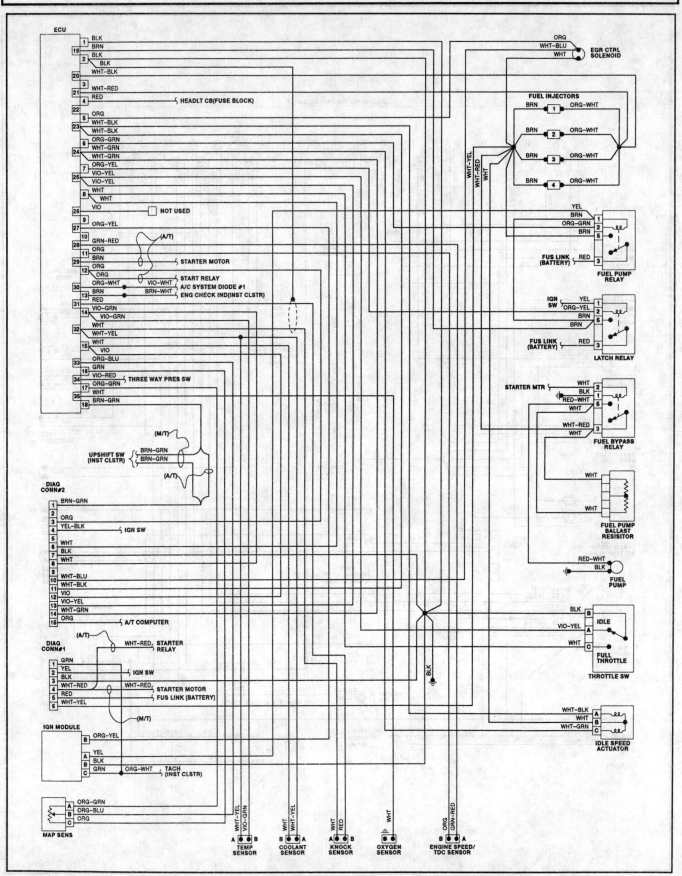

Fig. 14: Medallion Electronic Fuel Injection Wiring Diagram

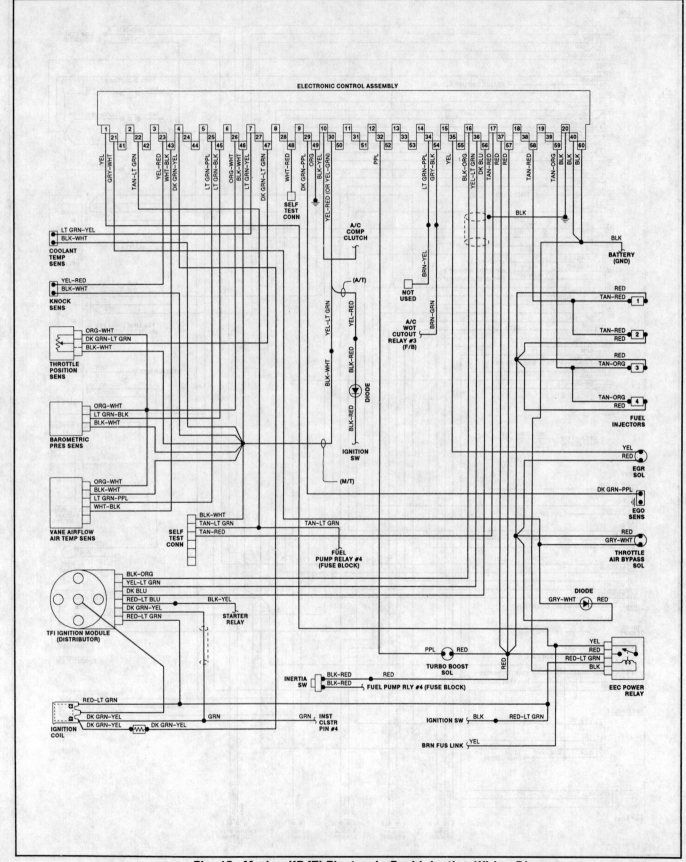

Fig. 15: Merkur XR4Ti Electronic Fuel Injection Wiring Diagram

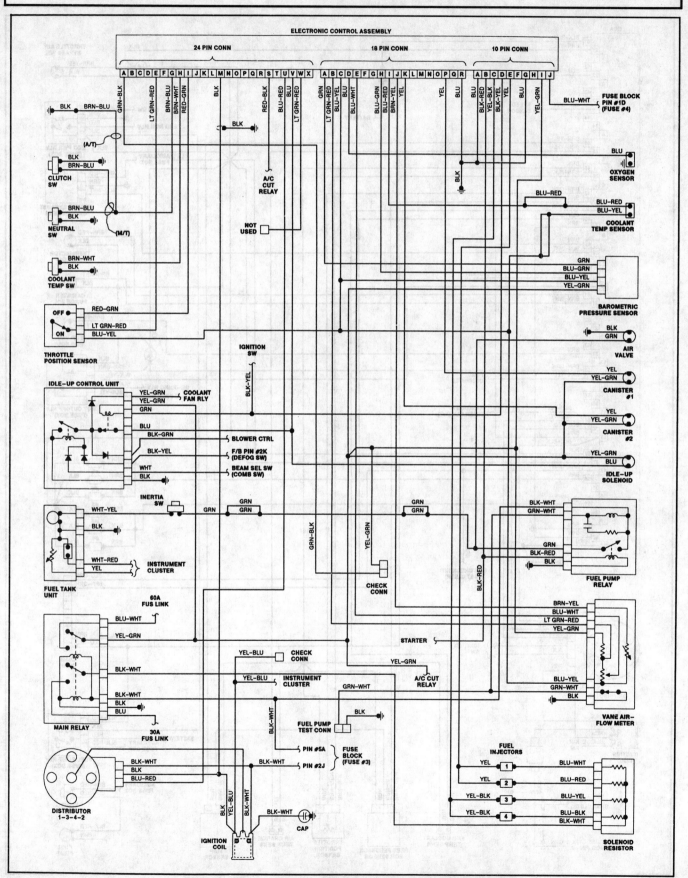

Fig. 16: Tracer Electronic Fuel Injection Wiring Diagram

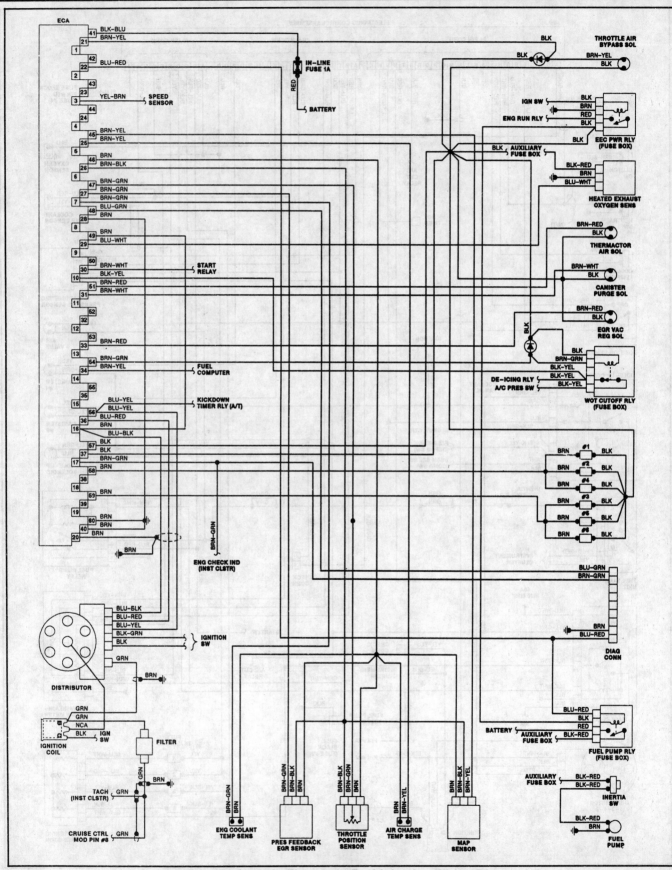

Fig. 17: Scorpio Electronic Fuel Injection Wiring Diagram

Chrysler Motors Imports: Colt, Conquest

NOTE
- *For detailed applications, refer to FUEL INJECTION SYSTEMS APPLICATION CHARTS at the front of this publication.*

DESCRIPTION

The Electronically Controlled Injection (ECI) system is a computerized emission, ignition and fuel control system. The ECI system controls engine operation and lowers exhaust emissions while maintaining good fuel economy and driveability. The Electronic Control Unit (ECU) is the "brain" of the ECI system. The ECU controls many engine related systems to constantly adjust engine operation.

The ECI system consists of the following sub-systems: Fuel control, data sensors, Electronic Control Unit (ECU), Idle Speed Control (ISC), emission control, fuel cut-off and catalytic converters.

NOTE
- *Primary sub-systems which affect fuel system operation will be covered in this article: Fuel supply system, fuel injection mixer, fuel pressure regulator, ECU, data sensors and fuel cut-off system.*

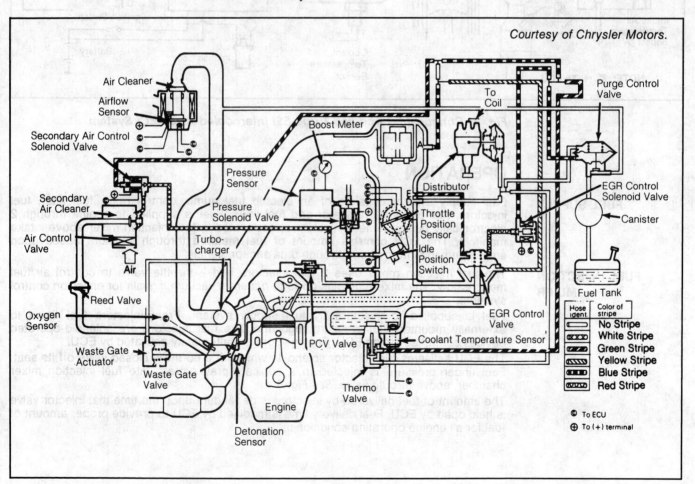

Courtesy of Chrysler Motors.

Hose ident	Color of stripe
	No Stripe
	White Stripe
	Green Stripe
	Yellow Stripe
	Blue Stripe
	Red Stripe

ⓒ To ECU
⊕ To (+) terminal

Fig. 1: Colt & Conquest ECI System

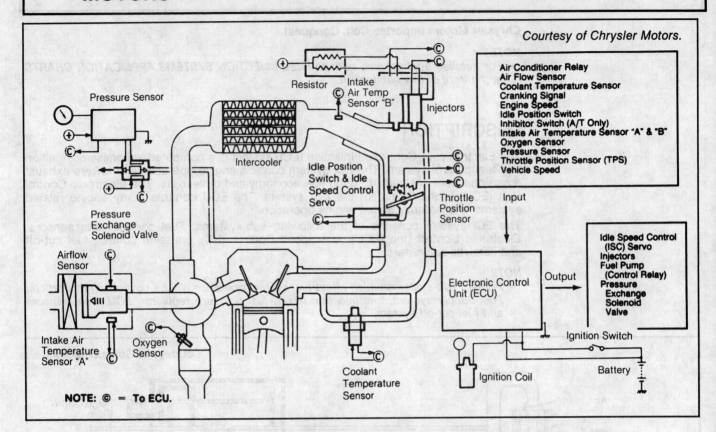

Courtesy of Chrysler Motors.

Fig. 2: Conquest Intercooled Turbo ESI Intercooled Turbo ECI System

OPERATION

FUEL SUPPLY

Fuel supply system consists of an electric fuel pump, control relay, fuel filter, fuel injectors, fuel pressure regulator and fuel lines. Fuel is supplied to engine through 2 electronically pulsed (timed) injector valves located in fuel injection mixer above intake manifold. The ECU controls amount of fuel metered through injectors based upon engine demand information, through data sensor signals.

FUEL INJECTION MIXER

The fuel injection mixer uses 2 fuel injectors and a throttle valve to control air/fuel mixture flow. The mixer contains ports to generate vacuum signals for emission control systems.

Fuel is supplied to injectors by fuel supply system. From injectors fuel flows to externally mounted fuel pressure regulator. The fuel injectors are solenoid-operated devices controlled by the ECU. The injectors are alternately activated by ECU.

The ECU activates fuel injector solenoids which lift a normally closed valve off its seat. Fuel under pressure is injected in a conical spray pattern into fuel injection mixer chamber, above throttle valve. *See Fig. 3.*

The amount of fuel delivered by injectors is dependent upon the time that injector valve is held open by ECU. Fuel delivery time is modified by ECU to provide proper amount of fuel for all engine operating conditions.

Courtesy of Chrysler Motors.

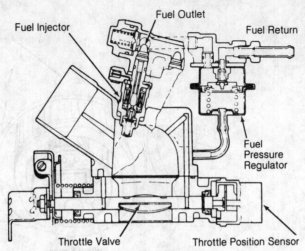

Fig. 3: Sectional View of Fuel Injection Mixer

FUEL PRESSURE REGULATOR

The fuel pressure regulator is externally mounted on fuel injection mixer. Fuel flows from top of fuel injectors to pressure regulator. The regulator is a diaphragm-operated relief valve with injector pressure on one side and pressure on the other side. The pressure regulator maintains a constant pressure drop across injectors throughout all engine operating conditions.

ELECTRONIC CONTROL UNIT (ECU)

The ECU receives various signals from data sensors and switches. These signals are processed by the ECU for controlling fuel delivery. The frequency and duration of injection (fuel delivery time) is controlled by ECU. Fuel delivery time is modified for such operating conditions as engine cranking, cold starting, altitude, acceleration and deceleration.

When ignition switch is turned to "START" position, ECU calculates fuel delivery based primarily upon coolant temperature and throttle position. ECU sends an electrical signal to injectors to provide fuel for prescribed period of time. After ignition is released from "START" position and engine speed is above a specified RPM, ECU changes enrichment signal.

Immediately after engine starts, the ECU issues electrical signals to injectors to provide stable combustion. During engine warm-up, the ECU monitors all data sensor information and provides a richer mixture until coolant temperature reaches a preset value.

When coolant temperature exceeds preset value, ECU processes other data sensor information and issues appropriate electrical signals to injectors. This period of time is referred to as open loop mode of operation. The ECU controls fuel delivery based upon open loop programmed information until the oxygen sensor is warm enough to send modifying signals to ECU.

When the oxygen sensor is warm enough, the ECU accepts oxygen sensor information and uses it for controlling fuel delivery. When the ECU is accepting oxygen sensor information, this is referred to as closed loop mode of operation.

During closed loop operation, the ECU stores the mean values of feedback signals used to maintain 14.7:1 air/fuel ratio. During open loop operation, the ECU uses these mean values to modify pre-programmed information. By doing this, the ECU can more closely control exhaust emissions even when engine is in open loop mode of operation.

DATA SENSORS

Each sensor furnishes electrical impulses to ECU. The ECU computes fuel delivery and spark timing necessary to maintain desired air/fuel mixture, thus controlling amount of fuel delivered to engine. Data sensors are interrelated to each other. Operation of each sensor is as follows:

Airflow Sensor

This sensor, mounted in air cleaner, measures airflow rate through the air cleaner and sends a proportionate electrical signal to ECU. The ECU uses airflow sensor information for controlling fuel delivery and secondary air management.

On Conquest intercooled models, an atmospheric pressure pick-up nipple (leading to atmospheric pressure switching solenoid valve) has been added to air cleaner and locations of oil separator and breather nipple have been changed. *See Fig. 4.*

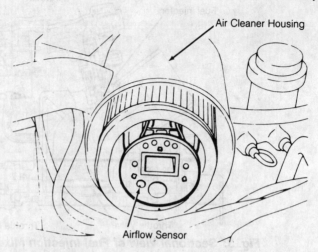

Air Cleaner Housing

Airflow Sensor

Fig. 4: Air Cleaner & Airflow Sensor Assembly

Intake Air Temperature Sensor

This sensor, located on airflow sensor, is a resistor-based sensor that detects temperature of incoming air and sends an electrical signal to ECU. The ECU uses air temperature sensor information for controlling fuel delivery.

On Conquest intercooled models, an intake air temperature sensor "B" has been installed to the air intake pipe to obtain optimum air/fuel mixture according to intake air temperatures when engine is operating under high load.

Pressure Sensor or Baromeric Pressure Sensor

On 1987-89 models, this sensor is located on the airflow sensor. It senses barometric pressure. Barometric pressure changes due to weather and/or altitude. This information is sent to ECU for controlling air/fuel ratio and ignition timing. On 1984-86 models, this firewall mounted sensor is connected to an electrically controlled solenoid valve. The solenoid valve has 2 hoses connected to it; 1 hose connects above throttle valve and other hose connects below throttle valve.

The solenoid valve is activated by ECU whenever ignition switch is turned to "ON" or "START" positions, for a specific period of time. When activated, the solenoid measures ambient barometric pressure from above throttle valve.

After a predetermined period of time, solenoid valve is deactivated by ECU. When deactivated, the solenoid measures intake manifold pressure below throttle valve. The ECU compares barometric pressure and intake manifold pressure and an absolute value is used to determine fuel delivery and ignition timing.

Coolant Temperature Sensor (CTS)

The coolant temperature sensor is installed in intake manifold. This sensor is a thermistor which converts temperature of engine coolant to electrical signal for use by ECU. The ECU uses coolant temperature information for controlling fuel delivery time, EGR and air injection system.

Engine Speed

Engine speed signal is received from negative side of ignition coil. Electrical signals from ignition coil are sent to ECU where time between signals is used to calculate engine speed. This information is used by ECU for controlling fuel delivery time, EGR and air injection system.

Oxygen Sensor (O₂)

This sensor, mounted in exhaust system sends an output voltage to the ECM which varies according to the oxygen content in exhaust gas stream. This signal is used by the ECU to control fuel delivery time. This is called closed loop operation. During engine warm-up, when oxygen sensor signals are not being interpreted by ECU, is open loop operation.

Throttle Position Sensor (TPS)

This sensor is mounted on fuel injection mixer. The sensor, a rotary potentiometer, signals ECU of changes in throttle valve position. This information is used for controlling fuel delivery time.

Idle Position Switch

This switch, part of the Idle Speed Control (ISC) servo, is activated when throttle valve is closed (idle position). When throttle valve is at any other position, the switch is deactivated. This information is used by the ECU for controlling fuel delivery time (during deceleration) and air injection system. This switch is also used as an idle speed adjusting device.

Detonation Sensor

The detonation sensor (knock sensor), located in cylinder block, converts engine vibration (knock) into an electrical signal. This signal is processed by the Electronic Spark Control (ESC) ignitor and relayed to ECU for determining amount of ignition timing retard. The ECU sends a signal to ESC ignitor to modify ignition timing. Ignition timing is retarded only during period of knock.

NOTE

- *Detonation sensor is strong enough to withstand engine vibration, but excessive impact with hammer, wrench, etc., can damage sensor.*

Vehicle Speed Sensor (VSS)

This sensor, used on 1987-89 models, is built in the speedometer. It converts transmission speedometer gear revolutions into pulse signals, which are sent to the ECU.

Motor Position Sensor (MPS)

The MPS is located inside the ISC servo. It senses ISC servo plunger position and signals the ECU. The ECU uses this information to control idle speed.

Inhibitor Switch

This switch is used on automatic transmission models only. The switch senses if transmission is in Neutral or Park. Based on this signal, the ECU measures automatic transmission load and drives the ISC servo to maintain optimum idle speed.

FUEL CUT-OFF

Two different fuel cut-off systems are used to change fuel delivery rate to engine:

Deceleration Fuel Cut-Off

During vehicle operation, idle position switch not at idle position, fuel delivery is determined by ECU responding to throttle valve closing speeds.

To decrease HC emissions during vehicle deceleration, fuel delivery time is decreased by ECU changing injection interval. When engine is operated under predetermined conditions, injection interval is changed from once every 3 pulses of airflow sensor to once every 6 pulses of airflow sensor.

Over Boost Fuel Cut-Off

This fuel cut-off system protects the engine during turbocharger operation. When the pressure sensor detects higher manifold pressure than the predetermined value stored in ECU memory, the ECU changes fuel delivery rate. When pressure value is exceeded, the fuel injectors are energized according to ignition spark timing.

COLD MIXTURE HEATER

Cold mixture heater is a Positive Temperature Coefficient (PTC) heater loacted on intake manifold. When engine coolant temperature is below 158°F (70°C), ECU energizes cold mixture heater to operate PTC heater. Fuel is then heated and atomized before entering combusion chamber.

TESTING & DIAGNOSIS

SELF-DIAGNOSTIC SYSTEM

Pretest Inspection

The diagnostic system, which monitors all input signals from each sensor, is an integral part of the ECU. If an abnormal input signal occurs, that item is memorized by the ECU. There are 6 diagnostic codes which can be confirmed using a voltmeter (Code No. 4 not used in this application). See DIAGNOSTIC CODES table.

If 2 or more systems are non-functional, they are indicated in order of increasing code number. Indication is made by deflection of voltmeter pointer. A constant 12 volts indicates system is normal. If system is abnormal, voltmeter will alternate between 0-12 volts every .4 second to display codes. After indication of zero volts for 2 seconds, the next code (if any) will be indicated. System malfunctions encountered are identified as either hard failures or intermittent failures as determined by the ECU.

"Hard Failures"

Hard failures cause "malfunction" light to illuminate and remain on until the malfunction is repaired. If light comes on and remains on (light may flash) during vehicle operation, cause of malfunction must be determined using diagnostic (code) charts. If a sensor fails, control unit will use a substitute value in its calculations to continue engine operation. In this condition, vehicle is functional, but loss of good driveability will most likely be encountered.

"Intermittent Failures"

Intermittent failures may cause "malfunction" light to flicker or illuminate and go out after the intermittent fault goes away. The corresponding trouble code, however, will be retained in control unit memory. If related fault does not reoccur within a certain time frame, related trouble code will be erased from control unit memory. Intermittent failures may be caused by sensor, connector or wiring related problems.

NOTE

- *ECU abnormal diagnostic memory is kept by direct power from battery. Memory is not erased by turning ignition off, but will be erased if battery or ECU is disconnected. Oxygen sensor memory is erased when ignition is turned off. To diagnose oxygen sensor, drive vehicle a good distance and keep engine running.*

ENTERING SELF-DIAGNOSTICS

Self-Diagnostic Test

Turn ignition switch to "OFF" position. Connect voltmeter to self-diagnostic output harness connector located in glove box. Proceed to RETRIEVING CODES.

RETRIEVING CODES

Turn ignition switch to "ON" position and ECU will display code(s) as pulses on voltmeter. Record any diagnostic codes and perform necessary component repairs. Proceed to CLEARING CODES.

CLEARING CODES

After checking and repair, turn ignition off and disconnect negative battery cable for 15 seconds or more to erase ECU code memory.

DIAGNOSTIC CODES

Code	Diagnostic Item
1	Oxygen Sensor & Computer
2	Engine Speed Sensor
3	Airflow Sensor
4	Not Used
5	Throttle Position Sensor
6	ISC Motor Position Sensor
7	Coolant Temperature Sensor

NOTE

- *The ECI system requires a special tester (Chrysler Motors ECI Checker - MD998451) to be fully diagnosed. However, some checks of individual components may be made using regular shop test equipment. Components covered in this article pertain only to fuel portion of system.*

FUEL INJECTORS

1) Disconnect secondary wire from ignition coil. Remove air intake pipe. To check injector operation, carefully look through air inlet of injection mixer with ignition in "ST" position. Unless injection is very poor, injector state should be considered normal. Turn ignition off and check injector for leakage.

2) To check operation of other injector, unplug connector from known good injector and install on other injector. Ensure wiring harness has adequate slack so engine torque will not damage harness. Check injector using step **1)**.

3) Using an ohmmeter, measure resistance between injector terminals. If resistance is not 2-3 ohms at 68°F (20°C), replace injector.

FUEL PUMP

1) Turn ignition off. Apply battery voltage to fuel pump "CHECK" connector. Listen for sound of fuel pump. Remove fuel cap is pump cannot be heard. Pinch fuel line to check if fuel pressure exist inside fuel line.

IDLE SWITCH (ISC SERVO)

Ensure ignition is off. Disconnect ISC motor connector. Using an ohmmeter, check continuity between terminal No. 2 and ground on ISC servo assembly. *See Fig. 5.* With throttle pedal depressed, there should be no continuity. Release throttle pedal and ensure there is continuity. If switch fails either test, replace ISC servo.

Courtesy of Chrysler Motors.

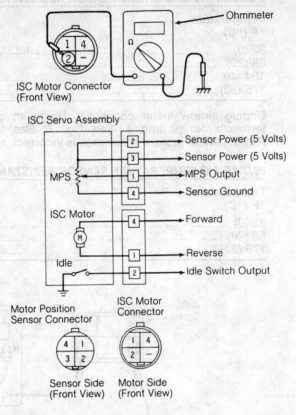

Fig. 5: Idle Switch/ISC Servo Assembly Connector Identification

ISC SERVO MOTOR

Using an ohmmeter, check resistance between terminals No. 1 and 4 of ISC servo assembly. *See Fig. 5.* Resistance should be 5-11 ohms at 68°F (20°C). Check for continuity between terminals No. 1 and 4, and ground. Tape 4 flashlight batteries (6 volts) together. Apply positive and negative voltage to terminals No. 1 and 4. ISC servo should extend and retract. If ISC servo fails any of the test, replace ISC servo.

THROTTLE POSITION SENSOR (TPS)

Ensure ignition is off. Unplug TPS connector. Using an ohmmeter, measure resistance between terminals No. 1 and 3. *See Fig. 6.* Resistance should be 3.5-6.5 ohms. Measure resistance between terminals No. 1 and 2. As throttle valve is rotated from closed to wide open position, resistance should make a smooth transition from 500 ohms to 3500-6500 ohms. If TPS fails any of these test, replace TPS.

Courtesy of Chrysler Motors.

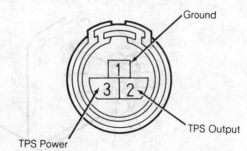

Fig. 6: Throttle Position Sensor (TPS) Connector Identification

COOLANT TEMPERATURE SENSOR (CTS)

Remove CTS from intake manifold. Immerse sensor probe portion in water. Hold sensor housing .12" (3 mm) away from water surface. Gradually warm water and measure resistance between terminal and ground. See COOLANT TEMPERATURE SENSOR RESISTANCE table. If resistance is not as specified, replace CTS.

COOLANT TEMPERATURE SENSOR RESISTANCE

Temperature °F (°C)	Ohms
32 (0)	5900
68 (20)	2450
104 (40)	1100
176 (80)	300

INTAKE AIR TEMPERATURE SENSOR

Unplug airflow meter connector. Using an ohmmeter, measure resistance between terminals No. 2 and 4. *See Fig. 7.* See INTAKE AIR TEMPERATURE SENSOR RESISTANCE table. If resistance is incorrect, replace sensor.

INTAKE AIR TEMPERATURE SENSOR RESISTANCE

Temperature °F (°C)	Resistance (Ohms)
32 (0)	6000
68 (20)	2650-2700
176 (80)	400

Courtesy of Chrysler Motors.

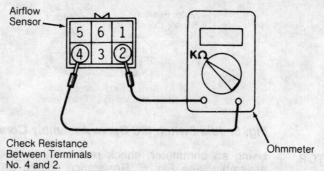

Fig. 7: Measuring Intake Air Temperature Sensor Resistance

EFI CONTROL RELAY

Unplug EFI control relay connector. Using an ohmmeter, measure for no continuity between terminals No. 1 and 7, then terminals 3 and 7. *See Fig. 8.* Apply 12 volts to terminal No. 8 (positive) and No. 4 (negative). Measure for continuity between terminals No. 3 and 7. Apply positive voltage (12 volts) to terminal No. 6 and negative to terminal No. 4. Check for continuity between terminals No. 1 and 7. If relay fails any of the test, replace relay.

Courtesy of Chrysler Motors.

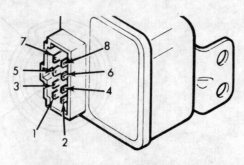

1. Fuel Pump
2. ECU
3. Resistor
4. Ground
5. ECU
6. Ignition Switch "ST"
7. Battery
8. Ignition Switch "IG"

Fig. 8: EFI Control Relay Terminal Positions

ADJUSTMENTS

IDLE SPEED CONTROL SYSTEM & TPS

1984-86 Models

1) Run engine at fast idle until coolant temperature is 185-205°F (85-90°C). Turn off engine. Disconnect accelerator cable from throttle lever of injection mixer. Loosen the 2 screws installing the throttle position sensor.

2) Turn throttle position sensor fully clockwise. Tighten screws. Turn ignition switch to the "ON" position for more than 15 seconds, then turn ignition switch to "OFF" position. This will set ISC servo to specified position. Disconnect ISC servo harness connector. Start engine. Check engine speed and adjust to specification. *See Fig. 9.*

ENGINE IDLE SPEEDS

Application	RPM
Colt	600
Conquest	750

Courtesy of Chrysler Motors.

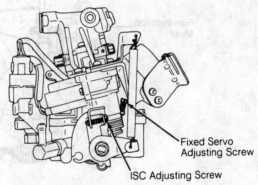

Fig. 9: ISC Servo Idle Adjusting Screw

3) Stop engine. To read output voltage of TPS, insert test probe from rubber cap side of TPS connector. Measure voltage from leads Green/White (TPS output) and Green/Black wire (ground) of body side harness.

4) Turn ignition switch to "ON" position without starting engine. Read TPS output voltage. If measurement of output voltage is not within .45-.51 volts, adjustment is needed. Loosen TPS mounting screws and turn sensor to obtain an output voltage of .45-.51 volts.

5) Immediately open and close throttle 1 or 2 times and recheck adjustment. If correct remove adapter, voltmeter and test probes and reconnect ISC servo harness connector.

1987-89

1) Ensure engine coolant temperature is 185-205°F (85-95°C). Turn all accessories off. Place transmission in Neutral or Park. Place steering wheel in straight ahead position and loosen accelerator cable. Turn ignition off. Unplug ECU connectors. Install ECI checker and Connector (MD998452) to system.

2) Connect a voltmeter to ECI checker marked "EXTENSION". Set checker to "EXTENSION" position. Set check switch to "6" on Colt and Colt Turbo or "7" on Conquest. Place select switch to "A" position on all models. Turn ignition on for 15 seconds, but do not start engine. Turn ignition off. Unplug ISC connector.

3) Place ISC motor in idle position. Open throttle valve, by hand 2-3 times and allow throttle plate to snap back after each opening. Loosen fixed servo adjusting screw. *See Fig. 9.* Start engine and allow engine to idle. Check idle speed. Idle speed should be 700 RPM on Colt and 850 RPM on Conquest.

4) Output voltage should be .9 volts on all models. If idle speed and output voltage is not as specified, using ISC adjusting screw, set to specification. *See Fig. 9.* After specified idle speed and output voltages have been obtained, turn engine off.

5) On Colt models, turn ignition on, but do not start engine. Check TPS output voltage. If voltage is not .48-.52 volt, adjust TPS to specification. Rotate TPS clockwise to increase voltage.

6) On Conquest, place select switch on checker to "6" position. Turn ignition on, but do not start engine. Check TPS output voltage. If voltage is not .48-.52 volt, adjust TPS to specification. Rotate TPS clockwise to increase voltage.

7) On all models, turn ignition off. Remove all test equipment and reconnect connectors. Start engine and ensure idle speed is correct. If idle speed is not correct, repeat procedure. If idle speed is correct, turn ignition off. Disconnect battery cable for 5-6 seconds to erase diagnosis memory.

ACCELERATOR CABLE FREE PLAY

1) Keep ignition key at "ON" position for at least 15 seconds before adjustment. Confirm that accelerator inner cable has no free play. Adjust to specification.

2) If cable shows excessive free play adjust as follows: Check routing of accelerator cable to confirm that there is no sharp twists or bends. Use adjusting nut to make adjustment. Adjust free play to 0-.04" (0-1 mm) at pedal. *See Fig. 10.* On turbocharged vehicles, check that idle control switch touches the stopper after idle speed control adjustment.

Courtesy of Chrysler Motors.

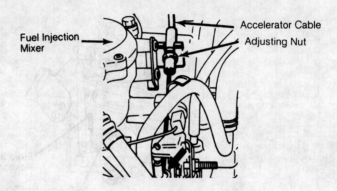

Fig. 10: Accelerator Cable Adjustment

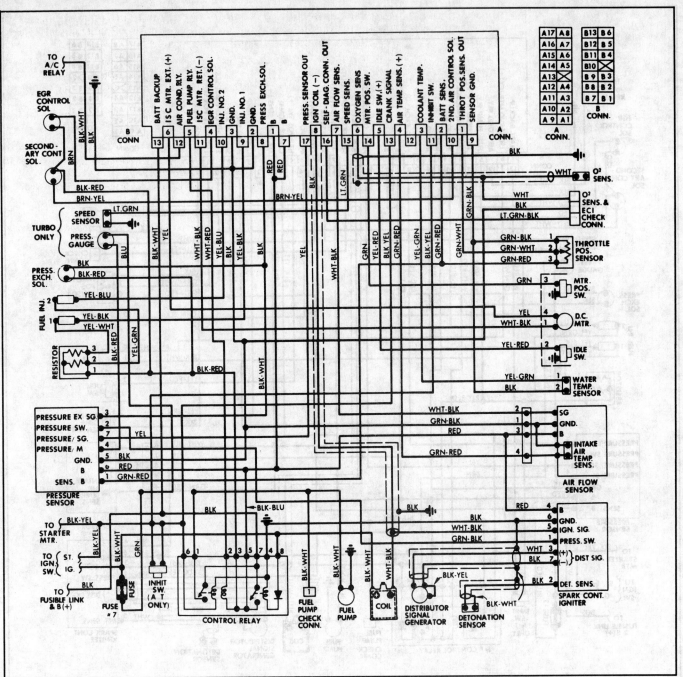

Fig. 11: 1983-87 Colt ECI Fuel Injection Wiring Diagram

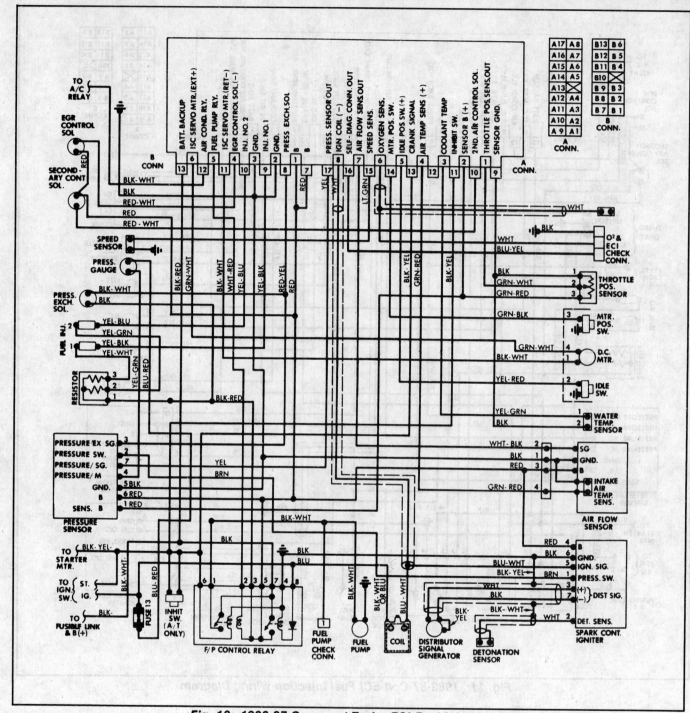

Fig. 12: 1983-87 Conquest Turbo ECI Fuel Injection Wiring Diagram

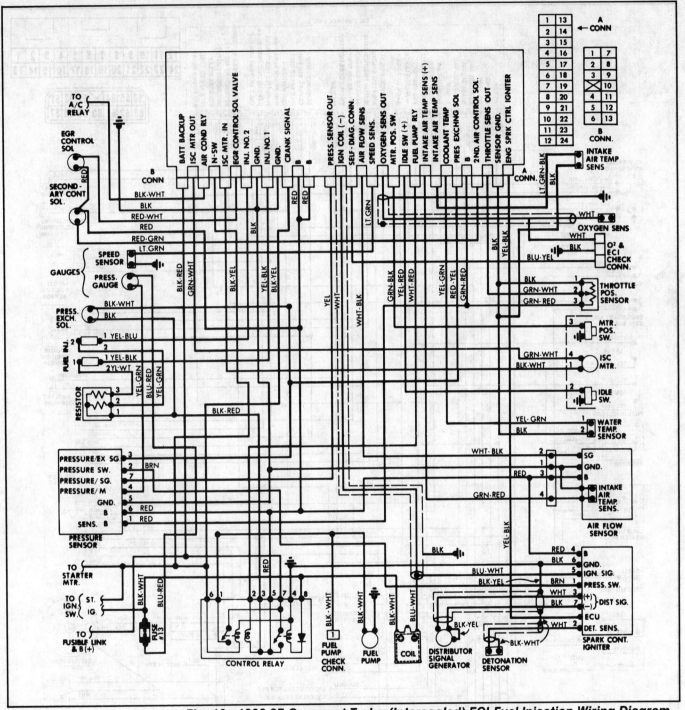

Fig. 13: 1983-87 Conquest Turbo (Intercooled) ECI Fuel Injection Wiring Diagram

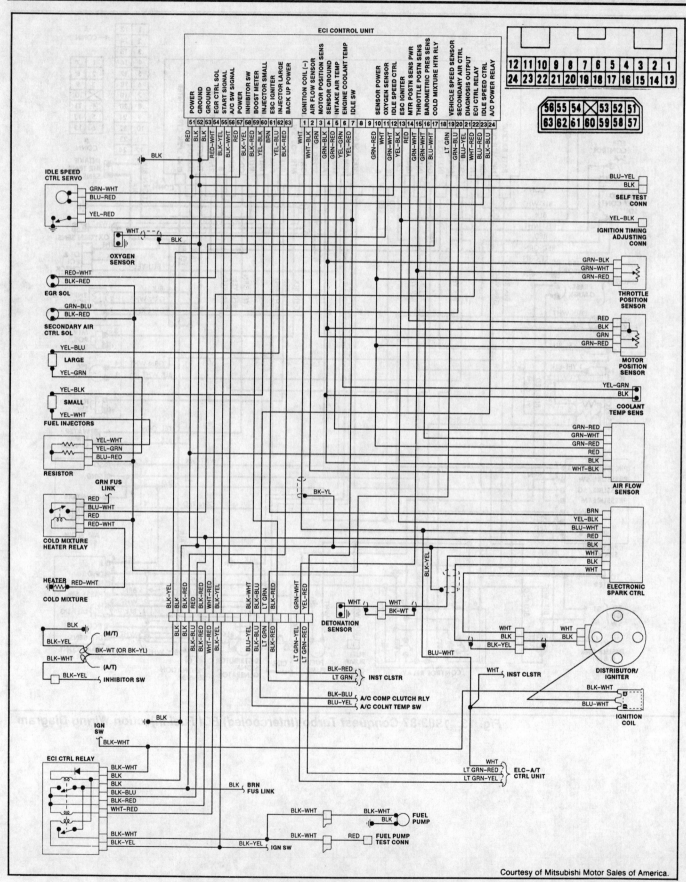

Courtesy of Mitsubishi Motor Sales of America.

Fig. 14: 1988 Colt ECI Fuel Injection Wiring Diagram

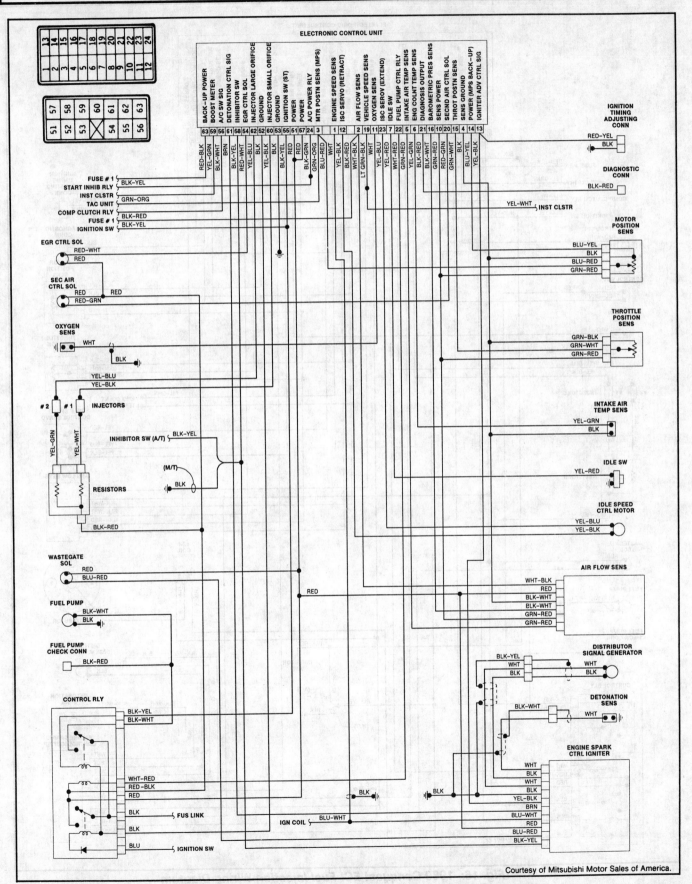

Fig. 15: 1988 Conquest ECI Fuel Injection Wiring Diagram

Courtesy of Mitsubishi Motor Sales of America.

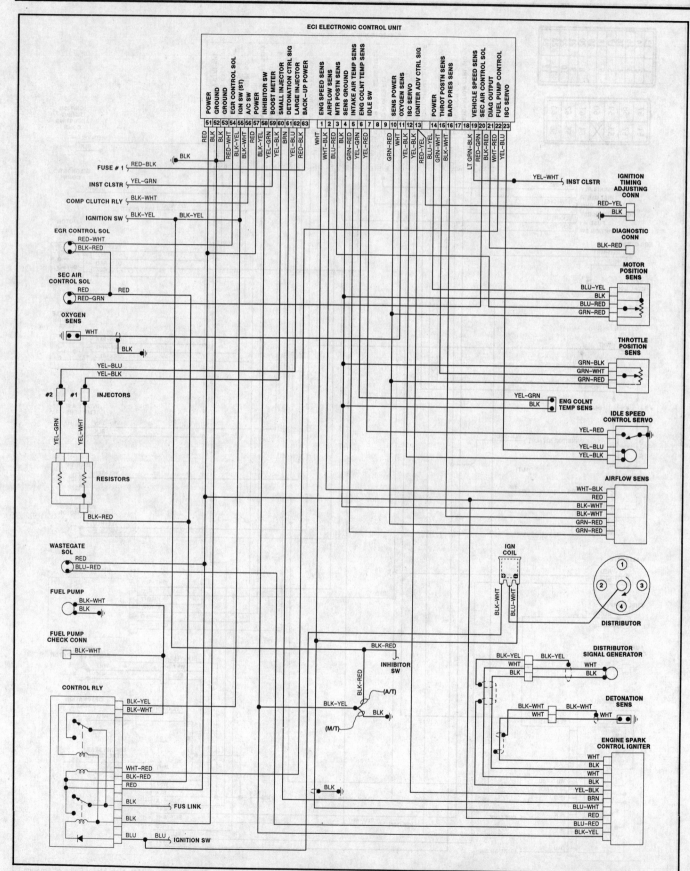

Fig. 16: 1989 Conquest ECI Fuel Injection Wiring Diagram

MITSUBISHI MULTI-POINT FUEL INJECTION

Colt, Colt Vista, Colt Turbo, Raider

DESCRIPTION

Multi-Point Fuel Injection (MPFI) is a system that controls the air/fuel mixture to the optimum ratio. Engine operating conditions are determined at the Electronic Control Unit (ECU) according to input signals from the various sensors and the injectors idle speed and ignition timing.

NOTE
- *Primary sub-systems which affect operation of the fuel system will be covered in this article.*

OPERATION

AIR/FUEL RATIO CONTROL SYSTEM

Air/fuel ratio system consists of an electric fuel pump, control relay, fuel filter, injectors, fuel pressure regulator and fuel lines. Correct air/fuel ratio is achieved by controlling on time of each injector based on data from various sensors.

FUEL PRESSURE REGULATOR

The fuel pressure regulator is located on the fuel delivery pipe. Fuel flows from fuel injectors to pressure regulator. The regulator maintains constant injector fuel pressure at 36.3 psi (2.56 kg/cm²), higher than manifold inside pressure. Excess fuel is returned through the return pipe to the fuel tank.

ELECTRONIC CONTROL UNIT (ECU)

The ECU calculates injection timing and rate according to the signals received from various engine sensors. The sensors convert such conditions as intake air, amount of oxygen in the exhaust gas, coolant temperature, intake air temperature, engine RPM, and vehicle driving speed into electrical signals, which are sent to the ECU.
Analyzing these signals, the ECU determines amount of fuel to inject according to driving conditions. Fuel injection is sequential injection type. During idle, the Idle Speed Control (ISC) servo is driven according to engine load to assure stable idling.

DATA SENSORS

Each sensor furnishes electrical impulses to ECU. The ECU computes fuel delivery and spark timing necessary to maintain desired air/fuel mixture. Data sensors are interrelated to each other. Operation of each sensor is as follows:

Airflow Sensor

Airflow sensor, mounted inside the air cleaner assembly, is designed to measure airflow rate through air cleaner and sends a proportionate electrical signal to ECU. The ECU uses the signal to determine basic fuel injection duration. *See Fig. 1.*

Courtesy of Chrysler Motors.

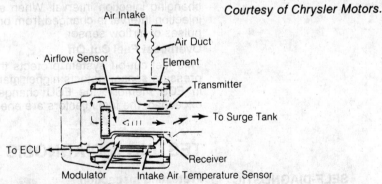

Fig. 1: Multi-Point Injection Airflow Sensor

Atmospheric Pressure Sensor

Pressure sensor is installed on airflow sensor. It senses atmospheric pressure and converts it to voltage which is sent to ECU. The ECU computes altitude and corrects air/fuel ratio and ignition timing.

Coolant Temperature Sensor (CTS)

The coolant temperature sensor is a thermistor which converts engine coolant temperature to an electrical signal for use by ECU. The ECU uses coolant temperature information for controlling fuel delivery time, EGR and air injection system.

Engine Speed Sensor

Engine speed signal is received from ignition coil. Electrical signals from ignition coil are sent to ECU where time between signals is used to calculate engine speed. This information is used by ECU for controlling fuel delivery time, EGR and idle speed.

Idle Position Switch

The Idle Position Switch (IPS) is mounted on the throttle body. When throttle valve is closed, the switch is activated. When throttle valve is at any other position, the switch is deactivated. This information is used by the ECU for controlling fuel delivery time (during deceleration) and idle speed. This switch is also used as an idle speed adjusting device.

Intake Air Temperature Sensor

The intake air temperature sensor is mounted inside of air cleaner. This sensor measures temperature of incoming air and supplies air density information to ECU. The ECU uses air temperature sensor information for controlling fuel delivery.

Inhibitor Switch

The inhibitor switch, used on automatic transaxle models only, informs the ECU of the transaxle position (Park, Neutral or Drive). Base on this signal, the ECU measures the load and drives the ISC servo to maintain optimum idle speed.

Motor Position Sensor (MPS)

Motor position sensor is incorporated in the ISC servo. It is a variable resistor type sensor. The MPS senses ISC servo plunger position and sends a signal to ECU. The ECU then controls throttle valve opening and idle speed.

Oxygen Sensor (O_2)

The oxygen sensor is located in the exhaust system. Output voltage of oxygen sensor varies with oxygen content in exhaust gas stream. Oxygen sensor signal is used to control fuel delivery time.

Throttle Position Sensor

The Throttle Position Sensor (TPS) is a potentiometer attached to the throttle body. The sensor signals ECU of changes in throttle valve position. This information is used to control idle speed and fuel delivery time.

Vehicle Speed Sensor

The vehicle speed sensor uses a reed switch. Speed sensor in speedometer converts speedometer gear revolution (vehicle speed) into pulse signals which are sent to ECU.

FUEL CUT-OFF There are different fuel cut-off systems used to change fuel delivery rate to engine.

Deceleration Fuel Cut-Off

During vehicle operation, when idle position switch is off idle position, fuel delivery is determined by ECU responding to throttle valve closing speeds.

To decrease HC emissions during vehicle deceleration, fuel deivery time is decrease by changing injection interval. When engine is operated under predetermined conditions, injection interval is changed from once every 3 pulses of airflow sensor to once every 6 pulses of airflow sensor.

Overboost Fuel Cut-Off

This fuel cut-off system protects the engine during turbocharger operation. When the pressure sensor detects higher manifold pressure than the predetermined value stored in ECU memory, the ECU changes fuel delivery rate. When the pressure value is exceeded, the fuel injectors are energized according to ignition spark timing.

TESTING & DIAGNOSIS

SELF-DIAGNOSTIC SYSTEM

Pretesting Inspection

If MPFI system components fail, interruption of fuel supply will result. Engine may be hard to start or not start at all. Unstable idle and/or poor driveability will be noticed. The Self-Diagnostic system monitors all input signals from each sensor. If an abnormal input signal occurs, that item is memorized by the ECU and given a code number. See DIAGNOSTIC FAULT CHARTS. Fault codes can be confirmed by using a voltmeter.

If 2 or more systems are non-functional, they are indicated by order of increasing code number. Indication is made by 12 volt pulses of voltmeter pointer. A constant repetition of short 12 volt pulses indicates system is normal. If system is abnormal, voltmeter will pulse between zero and 12 volts.

NOTE
- *ECU diagnostic memory is kept by direct power supply from the battery. Memory is not erased by turning off ignition but is erased if battery or ECU is disconnected.*

ENTERING SELF-DIAGNOSTICS

Self-Diagnostic Test (Using Voltmeter)

Turn ignition switch to "OFF" position. Connect voltmeter between PFI self-diagnostic output and ground of diagnostic connector, located in glove box or junction box by left side kick panel (except Colt Wagon). Colt Wagon diagnostic connector is located under dash at right side of center console (no illustration available). Turn ignition switch to "ON" position and disclosure of ECU memory will begin.

RETRIEVING CODES

NOTE
- *Oxygen sensor memory is erased when ignition switch is turned off. To diagnose oxygen sensor, drive vehicle a good distance and keep engine running.*

1) Signals will appear on voltmeter as long and short 12-volt pulses. Long pulses represent tens; short pulses represent ones. For example 4 long pulses and 3 short pulses indicates code 43. See DIAGNOSIS FAULT CHART. After recording abnormal code(s), perform necessary repair.

2) After repair, turn ignition off and disconnect negative battery cable for 15 seconds to erase ECU memory. Reconnect power supply and repeat self-diagnostics to confirm repair.

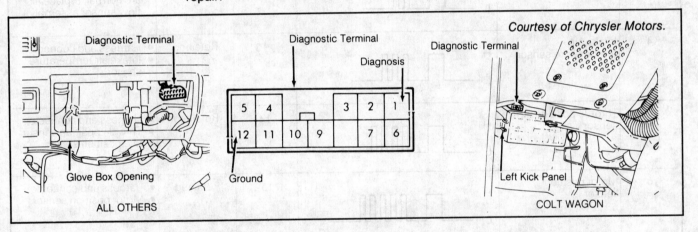

Fig. 2: Self-Diagnostic Connector Locations & Terminal Identification

CLEARING CODES

Turn ignition off and disconnect negative battery cable for 15 seconds to erase ECU memory. Reconnect power supply and repeat self-diagnostics to confirm codes are erased.

CODE CHARTS *Courtesy of Chrysler Motors.*

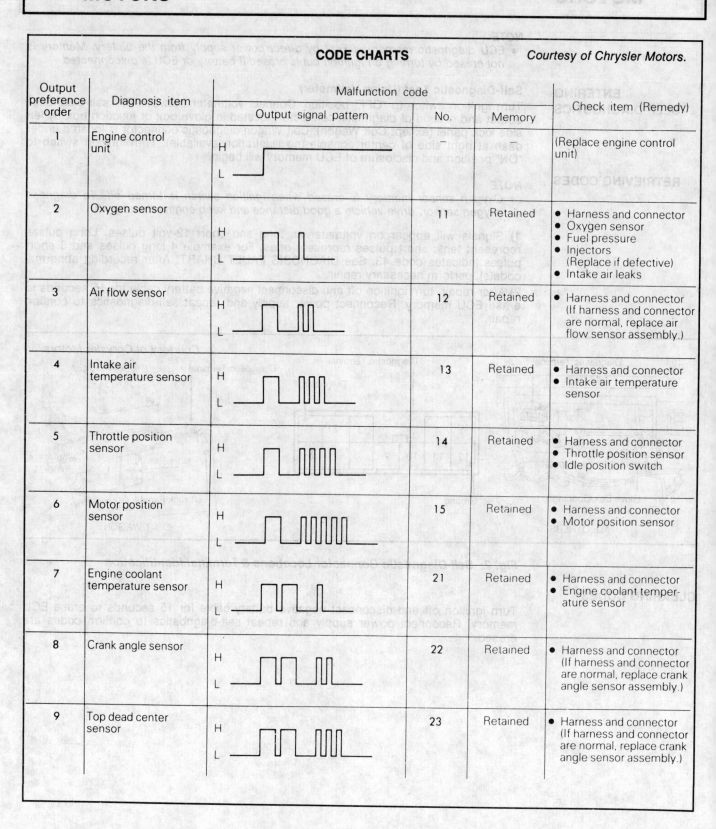

Output preference order	Diagnosis item	Malfunction code			Check item (Remedy)
		Output signal pattern	No.	Memory	
1	Engine control unit	H L	—		(Replace engine control unit)
2	Oxygen sensor	H L	11	Retained	• Harness and connector • Oxygen sensor • Fuel pressure • Injectors (Replace if defective) • Intake air leaks
3	Air flow sensor	H L	12	Retained	• Harness and connector (If harness and connector are normal, replace air flow sensor assembly.)
4	Intake air temperature sensor	H L	13	Retained	• Harness and connector • Intake air temperature sensor
5	Throttle position sensor	H L	14	Retained	• Harness and connector • Throttle position sensor • Idle position switch
6	Motor position sensor	H L	15	Retained	• Harness and connector • Motor position sensor
7	Engine coolant temperature sensor	H L	21	Retained	• Harness and connector • Engine coolant temperature sensor
8	Crank angle sensor	H L	22	Retained	• Harness and connector (If harness and connector are normal, replace crank angle sensor assembly.)
9	Top dead center sensor	H L	23	Retained	• Harness and connector (If harness and connector are normal, replace crank angle sensor assembly.)

CODE CHARTS

Courtesy of Chrysler Motors.

Output preference order	Diagnosis item	Malfunction code			Check item (Remedy)
		Output signal pattern	No.	Memory	
10	Vehicle speed sensor (reed switch)	H L	24	Retained	• Harness and connector • Vehicle speed sensor (reed switch)
11	Barometric pressure sensor	H L	25	Retained	• Harness and connector (If harness and connector are normal, replace barometric pressure sensor assembly.)
12	Detonation sensor	H L	31	Retained	• Harness and connector (If harness and connector are normal, replace detonation sensor.)
13	Injector	H L	41	Retained	• Harness and connector • Injector coil resistance
14	Fuel pump	H L	42	Retained	• Harness and connector • Control relay
15	EGR	H L	43	Retained	• Harness and connector • EGR thermo sensor • EGR valve • EGR valve control solenoid valve • EGR valve control vacuum
16	Ignition coil	H L	44	Retained	• Harness and connector • Ignition coil • Power transistor
17	Normal state	H L			—

TURBOCHARGER

Wastegate Actuator Operation Test

Apply a pressure of 9.2 psi to wastegate actuator to ensure actuator rod moves. Do not apply more than 10.3 psi to wastegate actuator or attempt to adjust wastegate valve.

ENGINE SENSORS & SWITCHES

Coolant Temperature Sensor

1) Remove coolant temperature sensor from intake manifold. Place end of sensor in water. Do not allow sensor to touch container. Terminal connector portion of sensor should be .12" (3.0 mm) above water.

2) Gradually heat water and read resistance values across terminal connectors. See COOLANT TEMPERATURE SENSOR RESISTANCE table. If not within specifications, replace sensor.

COOLANT TEMPERATURE SENSOR RESISTANCE

Temperature °F (°C)	Resistance Ohms
32 (0)	5900
68 (20)	2500
104 (40)	2700
176 (80)	300

Idle Switch

Disconnect the ISC servo connector. On Colt 1.5L and Colt Wagon, check for continuity between terminal No. 3 and ground. *See Fig. 3.* On Colt 1.6L, check at terminal No. 1 or terminal No. 2 on Colt Vista and Raider. With accelerator pedal depressed, no continuity should be present. With accelerator released, continuity should be present. If out of specification, replace ISC servo assembly.

Courtesy of Chrysler Motors.

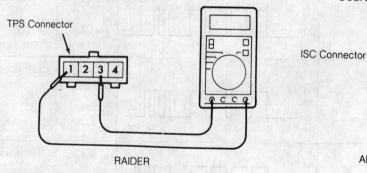

Fig. 3: Checking Idle Switch Continuity

Intake Air Temperature Sensor

Disconnect airflow sensor connector. Check intake air temperature sensor by measuring resistance between terminals No. 4 and 6. *See Fig. 4.* For specifications, see INTAKE AIR TEMPERATURE SENSOR RESISTANCE table. Replace airflow sensor assembly if not to specification.

Courtesy of Chrysler Motors.

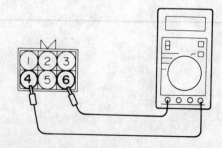

Fig. 4: Checking Intake Air Temperature Sensor

INTAKE AIR TEMPERATURE SENSOR RESISTANCE

Temperature °F (°C)	Resistance Ohms
32 (0)	6000
68 (20)	2700
176 (80)	400

Motor Position Sensor

1) Disconnect motor position sensor connector. Measure resistance between terminals No. 3 and 1. Resistance should be 4000-6000 ohms. Disconnect ISC motor connector. *See Fig. 5.*

2) Attach ohmmeter between terminals No. 1 and 2 of motor position sensor. Connect a 6-volt DC power supply between terminals No. 1 and 2 of ISC motor connector to operate ISC motor.

3) Ensure motor position sensor resistance changes smoothly when motor is extended or retracted. Resistance should change from 500 ohms to approximately 4000-6000 ohms. If not to specification, replace ISC servo assembly.

Courtesy of Chrysler Motors.

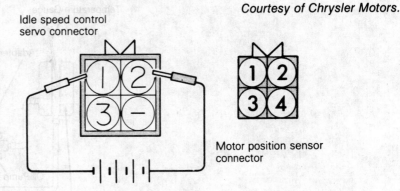

Idle speed control servo connector

Motor position sensor connector

Fig. 5: Checking Motor Position Sensor

Oxygen Sensor

Warm engine until coolant temperature is 185-205°F (85-95°C). Remove oxygen sensor connector. Connect voltmeter. *See Fig. 6.* With engine running at greater than 1300 RPM, measure output voltage. Voltage should be approximately one volt. If not, replace oxygen sensor.

Courtesy of Chrysler Motors.

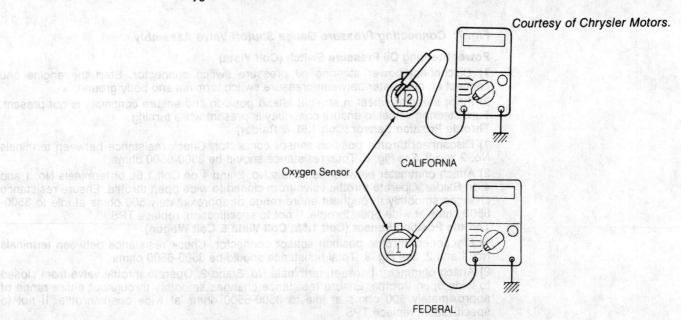

CALIFORNIA

Oxygen Sensor

FEDERAL

Fig. 6: Testing Oxygen Sensor

Power Steering Oil Pressure Switch (Colt & Raider)

1) Disconnect pressure hose from oil pump and connect in series a pressure gauge shutoff valve assembly capable of handling 250 psi (18 kg/cm²), 2 tee-in pressure hoses and appropriate adapters. *See Fig. 7.*

2) Bleed air from system by grounding coil wire and cranking engine while turning steering wheel all the way from left to right 6 times. Start engine and turn steering wheel back and forth to raise fluid temperature to approximately 122-140°F (50-60°C).

3) With engine idling, disconnect connector from oil pressure switch and connect an ohmmeter to pressure switch terminal and engine ground.

4) Gradually close shutoff valve of pressure gauge to increase hydraulic pressure. Ensure pressure switch is closed (shows continuity) at a pressure of 213-248 psi (15-17 kg/cm²).

5) Gradually open shutoff valve and ensure switch is open (no continuity) at 100-171 psi (7-12 kg/cm²). Remove testing equipment. Bleed air from system by grounding coil wire and cranking engine while turning steering wheel all the way from left to right 6 times.

Courtesy of Chrysler Motors.

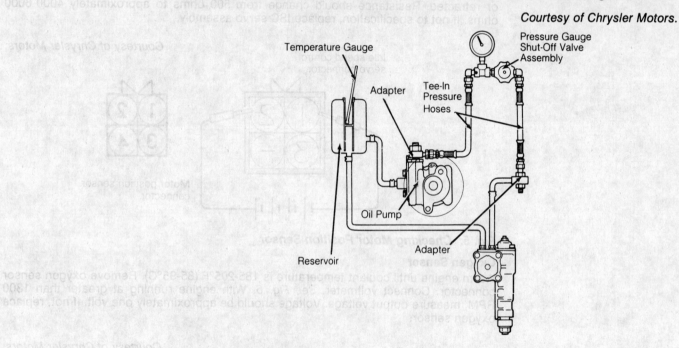

Fig. 7: Connecting Pressure Gauge Shutoff Valve Assembly

Power Steering Oil Pressure Switch (Colt Vista)

1) Disconnect power steering oil pressure switch connector. Start the engine and connect an ohmmeter between pressure switch terminal and body ground.

2) Place steering wheel in straight ahead position and ensure continuity is not present. Turn steering wheel to ensure continuity is present while turning.

Throttle Position Sensor (Colt 1.6L & Raider)

1) Disconnect throttle position sensor connector. Check resistance between terminals No. 2 and 3. *See Fig. 8.* Total resistance should be 3500-6500 ohms.

2) Attach ohmmeter between terminals No. 2 and 4 on Colt 1.6L or terminals No. 1 and 4 on Raider. Operate throttle valve from closed to wide open throttle. Ensure resistance changes smoothly throughout entire range of approximately 500 ohms at idle to 3500-6500 ohms at wide open throttle. If not to specification, replace TPS.

Throttle Position Sensor (Colt 1.5L, Colt Vista & Colt Wagon)

1) Disconnect throttle position sensor connector. Check resistance between terminals No. 1 and 2. *See Fig. 8.* Total resistance should be 3500-6500 ohms.

2) Attach ohmmeter between terminals No. 3 and 2. Operate throttle valve from closed to wide open throttle. Ensure resistance changes smoothly throughout entire range of approximately 500 ohms at idle to 3500-6500 ohms at wide open throttle. If not to specification, replace TPS.

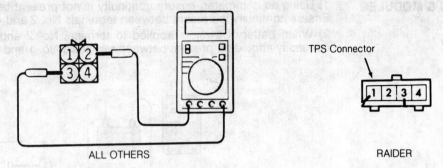

Fig. 8: Checking Throttle Position Sensor

Transmission Inhibitor Switch

Disconnect inhibitor switch connector at transmission. Operate inhibitor switch control lever to check continuity between terminals of inhibitor switch connector. *See Fig. 9.*

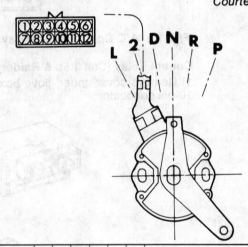

Terminal No.	P	R	N	D	2	L	Connected circuits
1					○		
2			○				
3	○						
4	○	○	○	○	○	○	Ignition switch "ON" terminal
5						○	
6				○			
7		○					
8	○		○				Ignition switch "ST" terminal
9	○		○				Starter motor "S" terminal
10		○					Ignition switch "ON" terminal
11		○					Backup lamp

When shift lever is in position marked in table, boxes with circles should show continuity.

Fig. 9: Checking A/T Transmission Inhibitor Switch

RELAYS, SOLENOIDS, MOTORS & MODULES

A/C Compressor Power Relay

1) Using an ohmmeter, ensure continuity is not present between terminals No. 1 and 3. Ensure continuity is present between terminals No. 2 and 4.

2) When battery power is applied to terminal No. 2 and terminal No. 4 is grounded, continuity should be present between terminals No. 1 and 3. *See Fig 10.*

Courtesy of Chrysler Motors.

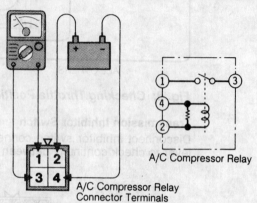

Fig. 10: A/C Compressor Relay Terminal Identification

Control Relay (Colt 1.5L & Raider)

1) Remove cover under glove box and remove glove box. Disconnect control relay and 10-pin connector.

Courtesy of Chrysler Motors.

COLT 1.5L & RAIDER

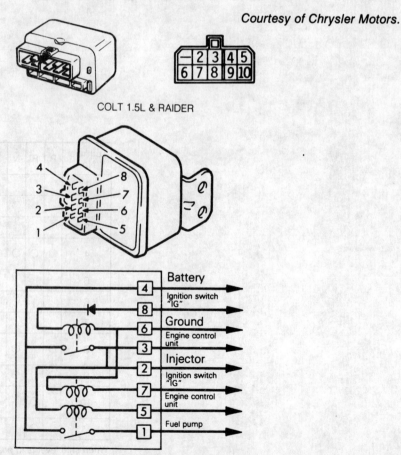

ALL OTHERS

Fig. 11: Control Relay Terminal Identification

2) While applying battery voltage to terminal No. 10, ensure battery voltage is not present at terminal No. 4 and terminal No. 5. *See Fig. 11.* While applying battery voltage to terminal No. 10 and grounding terminal No. 8, ensure battery voltage is present at terminals No. 4 and 5.

3) While grounding terminal No. 6, ensure continuity is not present at terminals No. 2 and 3. While applying battery voltage to terminal No. 9 and grounding terminal No. 6, ensure continuity is present at terminals No. 3 and 2.

4) To check for short to ground, apply battery voltage to terminal No. 3, ensure there is no voltage at terminal No 2. While applying battery voltage to terminal No. 3 and grounding terminal No. 7, battery voltage should be present at terminal No. 2. Replace control relay if it fails test procedure.

Control Relay (All Others)

1) Disconnect electrical connector. Using an ohmmeter, ensure approximately 95 ohms of resistance is present between relay terminals No. 3 and 5, then 2 and 5. *See Fig. 11.*

2) Using an ohmmeter, ensure approximately 35 ohms of resistance is present between relay terminals No. 6 and 7. Ensure there is infinite resistance between relay terminals No. 1 and 4. Ensure there is infinite resistance between terminals No. 2 and 4.

3) Place positive lead of ohmmeter on terminal No. 6 and negative lead on terminal No. 8, ensure there is infinite resistance. Using an ohmmeter, place positive lead of ohmmeter on terminal No. 8 and negative lead on terminal No. 6, ensure there is zero resistance.

4) While grounding terminal No. 6 and applying battery voltage to terminal No. 8, ensure there is zero resistance between terminals No. 2 and 4.

5) While grounding terminal No. 6 and applying battery voltage to terminal No. 7, ensure there is zero resistance between terminals No. 1 and 4.

ISC Motor Resistance Test (Colt 1.5L, Colt Vista & Colt Wagon)

Disconnect ISC motor connector. Using an ohmmeter, ensure resistance between terminals No. 1 and 2 is 5-35 ohms at a temperature of 68°F (20°C). *See Fig. 12.*

ISC Motor Resistance Test (Colt 1.6L & Raider)

1) At ISC motor connector, measure resistance between terminals No. 2 and 1 or terminals No. 2 and 3 on Colt or terminals No. 1 and 3 on Raider. *See Fig. 12.* Resistance should be 28-32 ohms at a temperature of 68°F (20°C).

2) Measure resistance between terminals No. 5 and 6, or between terminals No. 4 and 5 of ISC motor connector. Resistance should be 28-32 ohms at a temperature of 68°F (20°C).

ISC Motor Operation Test (Colt 1.5L, Colt Vista & Colt Wagon)

Disconnect ISC motor connector. Connect four 1.5-volt DC batteries (6 volts total) between terminals No. 1 and 2 of ISC motor connector to operate ISC motor. *See Fig. 13.* If motor does not operate, replace ISC servo assembly.

ISC Motor Operation Test (Colt 1.6L & Raider)

Remove ISC motor. With 6 volts applied to terminals No. 2 and 5 of the connector, apply and remove a ground in sequence to terminals No. 3 and 6, No. 1 and 6, No. 1 and 4, No. 3 and 4, and No. 3 and 6. Reverse grounding sequence. If motor does not operate when grounding any of the mentioned terminals, replace ISC motor. *See Fig. 12.*

Courtesy of Chrysler Motors.

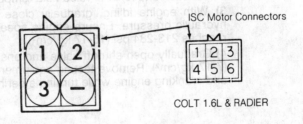

ISC Motor Connectors

COLT 1.5L, COLT VISTA, COLT WAGON

COLT 1.6L & RADIER

Fig. 12: ISC Motor Connectors

FUEL DELIVERY SYSTEM

Fuel Pressure Control Solenoid Valve (Colt 1.6L Turbo)

1) Mark and remove vacuum hoses from solenoid valve. Disconnect wiring harness. Connect a vacuum pump to nipple where Black vacuum hose was connected and leave connected throughout testing procedure. *See Fig. 13.*

2) Apply vacuum to ensure vacuum leaks. Plug nipple where Blue-striped hose was connected and apply vacuum to ensure vacuum holds. Remove plug from Blue-striped hose nipple. Apply battery voltage to one terminal of valve and ground the other. Apply vacuum to ensure vacuum holds.

3) Using an ohmmeter, check resistance across terminals of solenoid. Resistance should be 36-46 ohms at a temperature of 20°C (68°F).

Courtesy of Chrysler Motors.

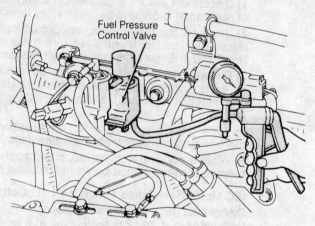

Fig. 13: Fuel Pressure Control Valve (Colt 1.6L Turbo)

FUEL CONTROL

Fuel Injectors Operating Check

Using a stethoscope, check the operating sound of injectors during engine cranking or idling. Ensure operational sound increases when engine RPM is increased.

Fuel Injector Resistance Check (Colt Turbo)

With injector electrical connector disconnected, measure resistance between injector terminals. Resistance should be 2-3 ohms at a temperature of 68°F (20°C).

Fuel Injector Resistance Check (All Others)

With injector electrical connector disconnected, measure resistance between injector terminals. Resistance should be 13-16 ohms at a temperature of 68°F (20°C).

IDLE CONTROL SYSTEM

Power Steering Idle-Up System Check (Colt Vista & Colt Wagon)

1) Disconnect pressure hose from oil pump and connect in series a pressure gauge shutoff valve assembly capable of handling 300 psi (21 kg/cm²), 2 tee-in pressure hoses, and appropriate adapters.

2) Bleed air from system by grounding coil wire and cranking engine while turning steering wheel all the way from left to right 6 times. Start engine and turn steering wheel back and forth to raise fluid temperature to approximately 122-140°F (50-60°C).

3) With engine idling, gradually close shutoff valve of pressure gauge to increase hydraulic pressure. Ensure engine speed increased 200-250 RPM when fluid pressure reaches 213-284 psi (15-17 kg/cm²).

4) Gradually open shutoff valve and ensure engine speed returns to idle at 100-284 psi (7-17 kg/cm²). Remove testing equipment. Bleed air from system by grounding coil wire and cranking engine while turning steering wheel all the way from left to right 6 times.

ADJUSTMENTS

IDLE SPEED CHECK

1) Ensure engine is at operating temperature and all accessories are off. Place transmission in Neutral and front wheels in a straight-ahead position.

2) Ensure ignition timing is adjusted to specification. Raise engine speed to 2000-3000 RPM for more than 5 seconds. Run engine at idle for 2 minutes. Check engine idle speed.

3) Refer to IDLE SPEED SPECIFICATIONS table. If idle speed is not within specification, diagnosis of Idle Speed Control (ISC) motor is necessary.

BASIC IDLE SPEED

1) Ensure engine is at operating temperature and all accessories are off. Place transmission in Neutral and front wheels in a straight-ahead position. Turn engine off. Ground ignition timing adjustment connector. Using a jumper wire, ground terminal No. 10 of self-diagnostic connector. Start engine.

2) Check basic idle speed. Refer to IDLE SPEED SPECIFICATIONS table. If basic idle speed is not to specification, check throttle valve for dirt accumulation before adjusting basic idle speed.

3) If throttle valve is clean, check idle switch for proper adjustment. Refer to IDLE SWITCH adjustment procedure in this article. If idle switch is okay, check fast idle air valve for malfunction. If idle switch and fast idle air valve are okay, adjust basic idle by turning air by-pass screw located on throttle valve assembly.

IDLE SPEED SPECIFICATIONS

Application	RPM
Colt & Colt Turbo	
Curb Idle	650-850
Base Idle	700-800
Colt Vista	
Curb Idle	600-800
Colt Wagon	
Curb Idle	600-800
Raider	
Curb Idle	600-800
Base Idle	650-750

IDLE SWITCH

1) Loosen tension in throttle cable. Disconnect idle switch connector. Loosen lock nut on idle switch. Turn idle switch counterclockwise until throttle valve fully closes. Connect a DVOM between Black/Yellow and Green/Black wire on, at idle switch, on Raider or idle switch terminal and ground on all other models. On Raider, place a .025" (.64 mm) feeler gauge between throttle valve and throttle lever. Turn TPS on Raider or idle switch on all other models clockwise until DVOM reads continuity.

2) On Raider, turn TPS counter clockwise until there is no continuity. Readjust TPS. On all others, turn idle switch another 15/16 of a turn more from when ohmmeter reads continuity. Hold idle switch while tightening idle switch lock nut. Adjust throttle cable, basic idle speed and throttle position sensor.

THROTTLE POSITION SENSOR (TPS)

NOTE
- *All models except Colt Vista and Colt Wagon can be adjusted using harness adapters and/or ECI checkers. If these tools are used, follow manufacturer's instructions.*

1) Loosen throttle cable. With ignition off, connect DVOM between wires specified in FUEL INJECTED ENGINE TPS WIRE IDENTIFICATION table. Turn ignition on. Do not start engine.

2) Check TPS voltage. If reading is not .48-.52 volt, adjust TPS until correct voltage is obtained. Turn ignition off and disconnect DVOM. Adjust throttle cable. Disconnect vehicle battery for 10 seconds or more. Start engine and check idle speed.

FUEL INJECTED ENGINE TPS WIRE IDENTIFICATION

Application	Wire Colors
Colt	Green/Blue & Green/White
Colt Vista & Colt Wagon	Green/White & Black
Radier	Green/Black & Green/White

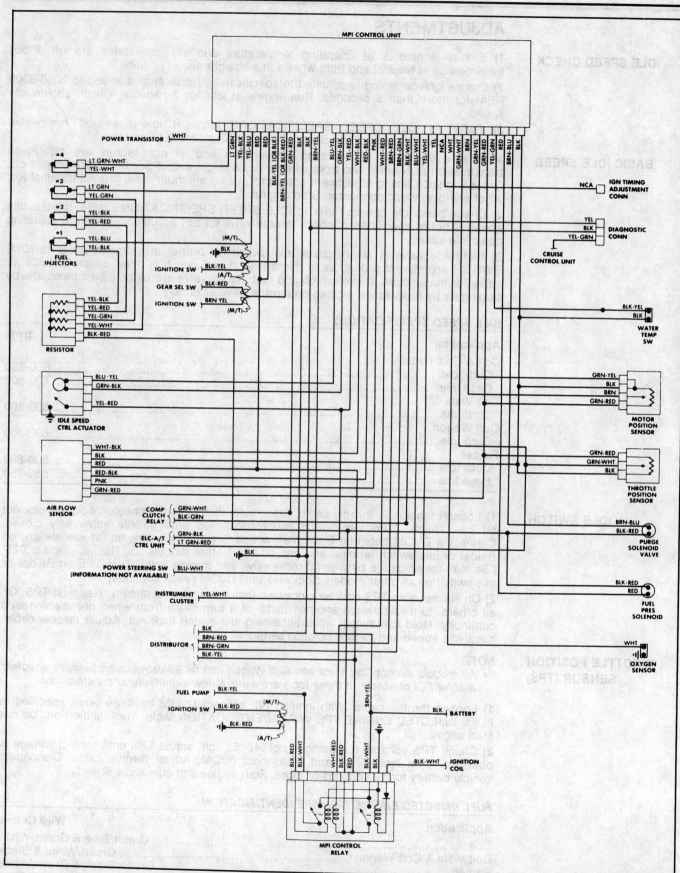

Fig. 14: 1987-88 Colt Vista MPFI Wiring Diagram

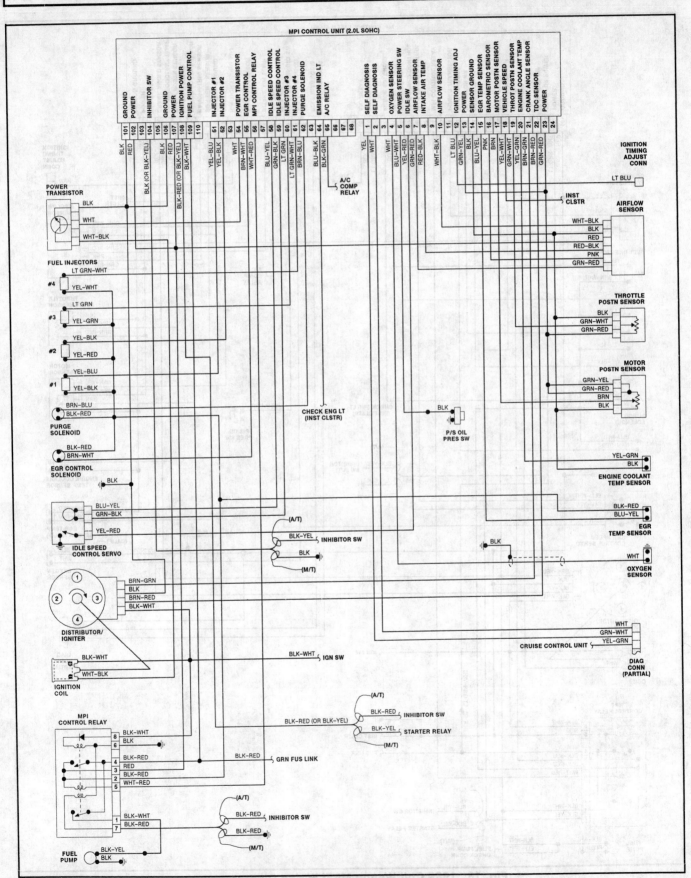

Fig. 15: 1989 Colt Vista MPFI Wiring Diagram

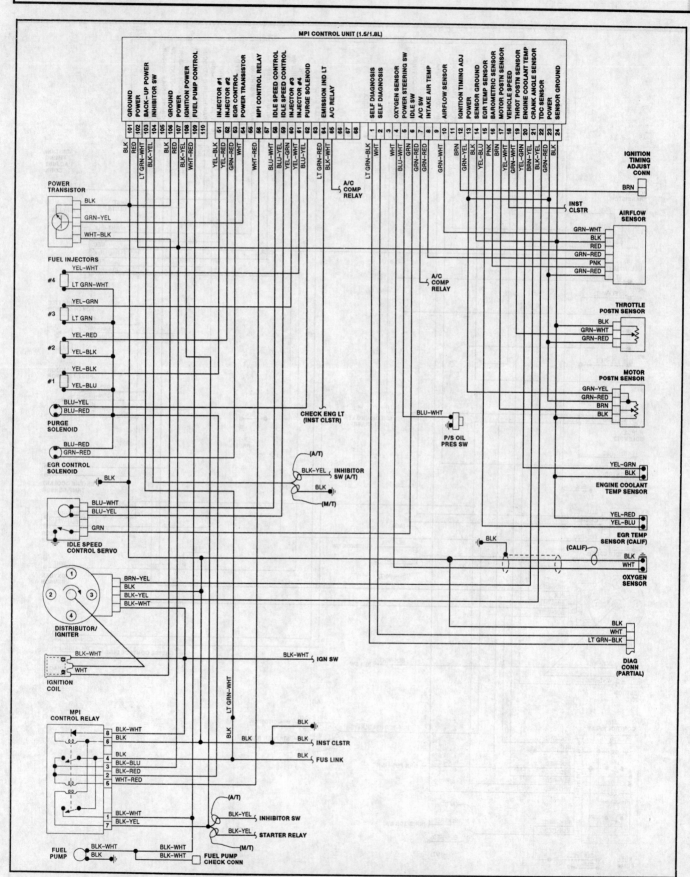

Fig. 16: 1989 Colt Wagon MPFI Wiring Diagram

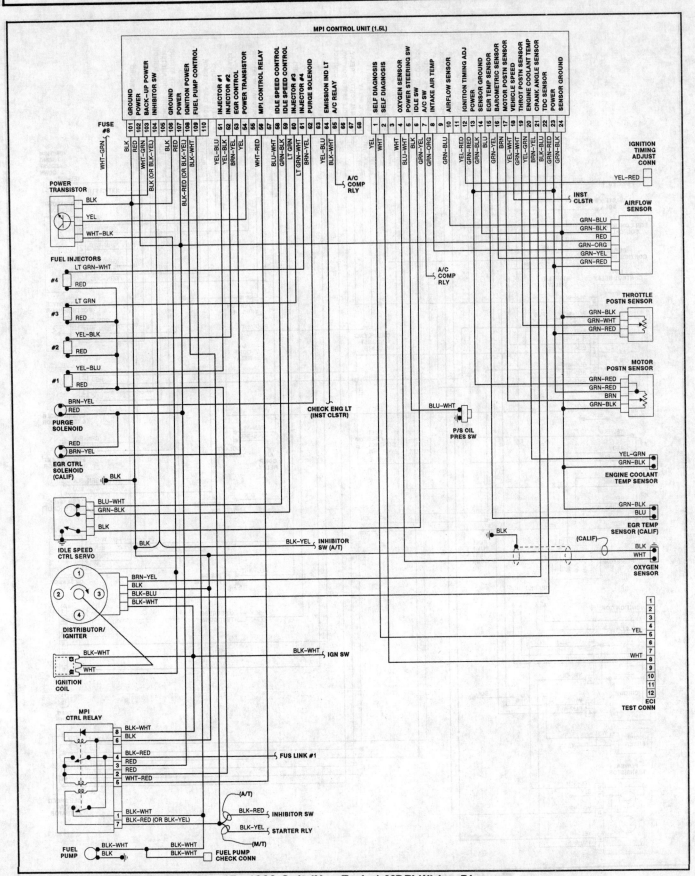

Fig. 17: 1989 Colt (Non-Turbo) MPFI Wiring Diagram

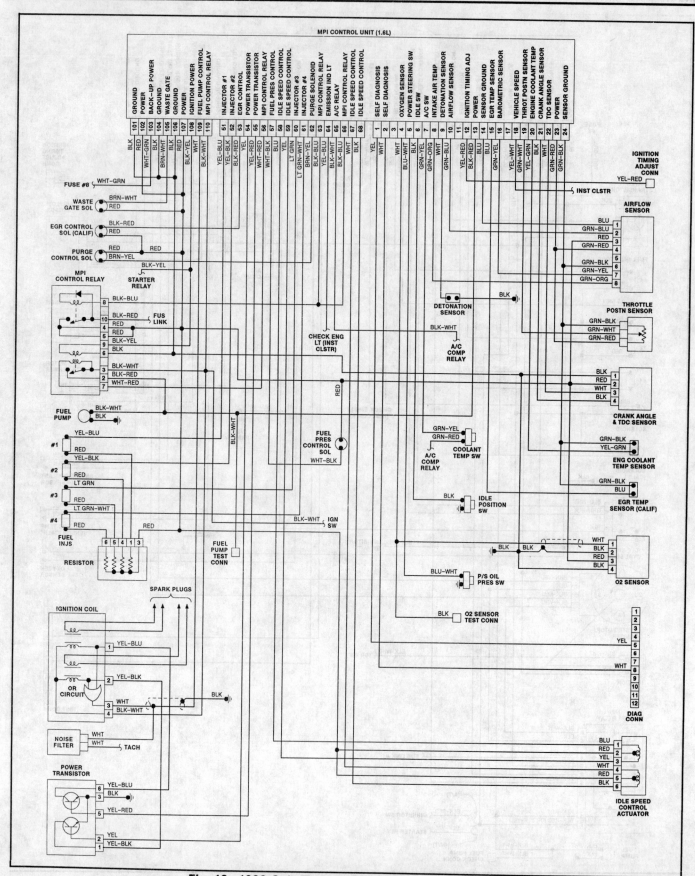

Fig. 18: 1989 Colt (Turbo) MPFI Wiring Diagram

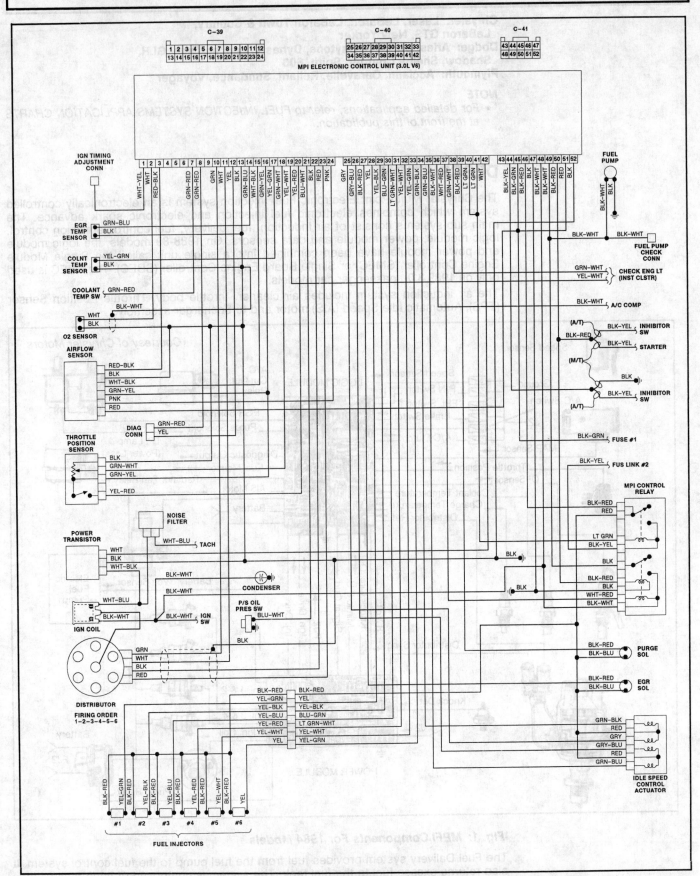

Fig. 19: 1989 Raider MPFI Wiring Diagram

MULTI-POINT FUEL INJECTION

Chrysler: Laser, LeBaron, LeBaron Town & Country,
LeBaron GTS, New Yorker
Dodge: Aries, Caravan, Daytona, Dynasty, Lancer, Omni GLH,
Shadow, Shelby Charger, Spirit, 600
Plymouth: Acclaim, Caravelle, Reliant, Sundance, Voyager

NOTE
* *For detailed applications, refer to FUEL INJECTION SYSTEMS APPLICATION CHARTS at the front of this publication.*

DESCRIPTION

The Chrysler Multi-Point Electronic Fuel Injection system is an electronically controlled system which combines electronic fuel injection and electronic spark advance. The main sub-systems consist of air induction, fuel delivery, fuel control, emission control, logic module, power module and data sensors. On 1988-89 models, the logic module and power module have been combined into a single unit called the Single Module Engine Controller (SMEC) or Single Board Engine Controller (SBEC). The SBEC is used only on 1989 3.0L passenger car models.

The air induction system includes air cleaner, throttle body, Throttle Position Sensor (TPS), Automatic Idle Speed (AIS) motor and turbocharger assembly.

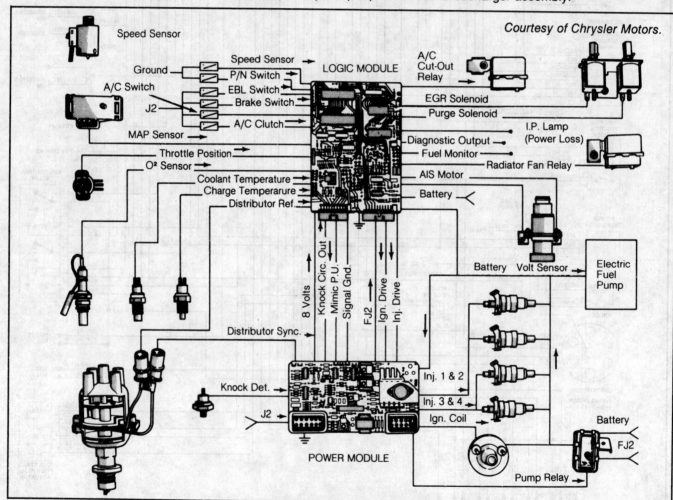

Fig. 1: MPFI Components For 1984 Models

The Fuel Delivery system provides fuel from the fuel pump to the fuel control system. It also returns excess fuel to the fuel tank. The system is composed of an in-tank electric fuel pump, fuel filter, 2 check valves and return line. Power is provided to operate the fuel pump through an Automatic Shutdown (ASD) relay inside the power module.

The ASD relay also controls the ignition coil, fuel injectors and portions of the power module.

The Fuel Control system handles the actual delivery of fuel into the engine. The fuel pressure regulator maintains constant fuel pressure. In addition to the regulator, the system consists of the fuel rail and 4 or 6 fuel injectors.

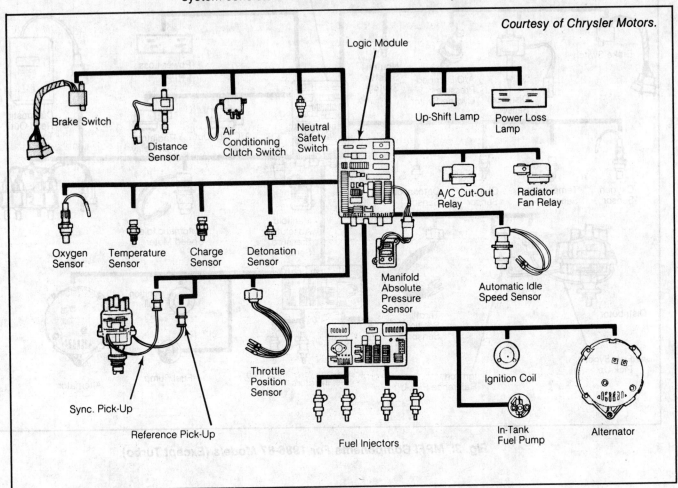

Courtesy of Chrysler Motors.

Fig. 2: MPFI Components For 1985 Models

The emission system, although directly operated by the SMEC/SBEC, is not unique to the multi-point fuel injection engine. Emission system controls include EGR, aspirated air system, evaporative emission control and crankcase ventilation.

The logic module is a digital microprocessor computer. This module receives input signals from various switches and sensors. It then computes the fuel injector pulse width, spark advance, ignition coil dwell, idle speed, canister purge cycles, alternator field control, feedback control and boost level from this information.

The power module is a small computer that contains necessary circuitry to power the ignition coil and fuel injectors. This module supplies most of the operating current for the entire system.

Data sensors provide the logic module with engine operating information. The computer analyzes this information and corrects air/fuel ratio, ignition timing, and emission control as needed to maintain efficient engine operation.

The SMEC (1988-89) and SBEC (1989 3.0L) contains the circuits necessary to operate the ignition coil, fuel injector, and alternator field. It computes the fuel injector pulse width, spark advance, ignition coil dwell, idle speed, purge, cooling fan operation and alternator charge rate. An externally-mounted ASD relay is turned on and off by the SMEC/SBEC. Distributor pick-up signal is sent to the SMEC/SBEC. The SMEC/SBEC contains a voltage converter which converts battery voltage to 8 volts output (on 1988 models) and 9 volts output (on 1989 models).

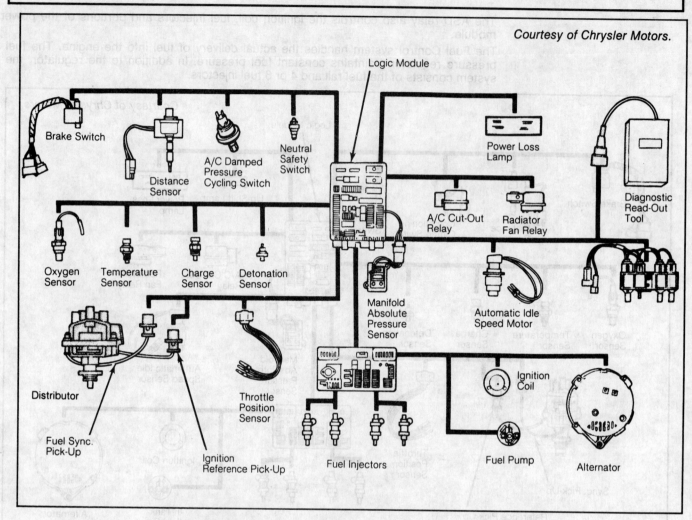

Courtesy of Chrysler Motors.

Fig. 3: MPFI Components For 1986-87 Models (Except Turbo)

OPERATION

AIR INDUCTION

On turbo models, air is drawn through the air cleaner and forced through the throttle body by the turbocharger. Maximum turbocharger boost of 7.25 psi (.5 kg/cm²) is controlled by a calibrated wastegate on 1984 models. On 1985 and later models, wastegate is controlled by the SMEC/SBEC and can increase turbo boost to higher levels for brief periods of time. The amount of air entering the engine is controlled by a cable operated throttle valve in the throttle body.

On all models, the throttle body houses the TPS and AIS motor. The TPS is an electrical resistor which is connected to the throttle valve. The TPS transmits a signal to the logic module in relation to throttle valve angle. From this signal, the SMEC/SBEC calculates fuel injector "on" time to provide adequate air/fuel mixture.

The AIS motor controls the flow of air through the throttle body during engine idle. The motor opens and closes an air by-pass on the back of the throttle body to increase or decrease idle speed as engine load varies. The SMEC/SBEC monitors the AIS motor and issues a change command to the injectors to increase or decrease the amount of fuel injected.

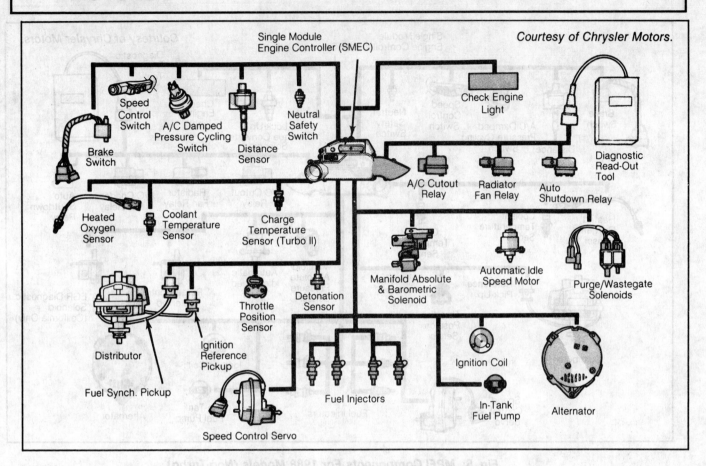

Courtesy of Chrysler Motors.

Fig. 4: MPFI Components For 1988 Models (Turbo)

FUEL DELIVERY

Fuel is drawn through a filter sock on one end of the fuel pump and is forced out the opposite end. Two check valves are used. One relieves internal fuel pump pressure and regulates maximum fuel pressure output.

The other check valve, located near pump outlet, restricts fuel flow in either direction when pump is not running. Fuel not used by the engine is returned to the tank by a fuel return line.

Power to the fuel pump is supplied through the ASD relay of the SMEC/SBEC. When the SMEC/SBEC receives a signal from the distributor during engine cranking, the module grounds the ASD motor, closing the contacts of the ASD. When the contacts are closed, power is applied to the fuel pump.

Whenever the distributor signal is lost, the SMEC/SBEC opens the ASD motor, which cuts power supply to fuel pump. This is a safety feature designed to prevent fuel flow should the vehicle be involved in an accident or if the engine stalls.

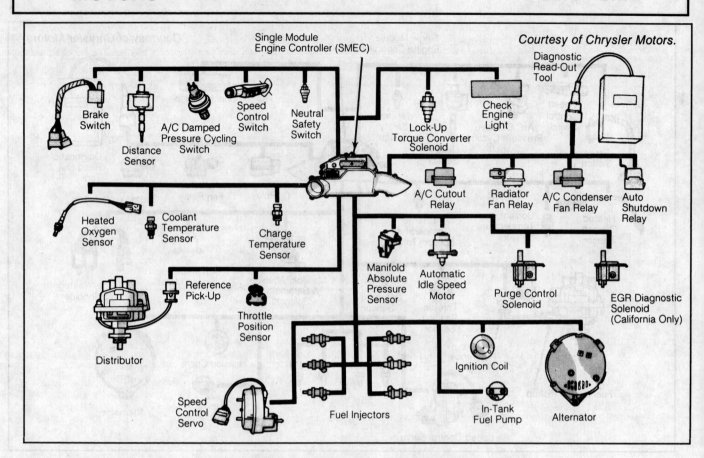

Courtesy of Chrysler Motors.

Fig. 5: MPFI Components For 1988 Models (Non-Turbo)

FUEL CONTROL   The fuel control system handles the actual delivery of fuel into the engine. Fuel from the pump enters the fuel rail, passes injectors, and enters fuel pressure regulator.

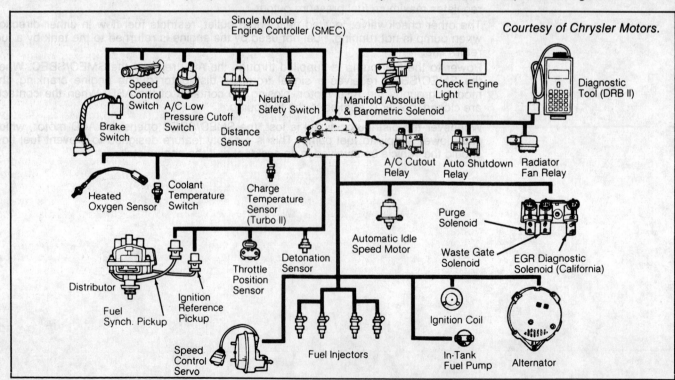

Courtesy of Chrysler Motors.

Fig. 6: MPFI Components For 1989 2.2L & 2.5L Turbo Models

Fuel injectors are electrically operated solenoid valves which are powered by the power module and controlled by the SMEC/SBEC. The SMEC/SBEC determines injector pulse width, or "on" time. The SMEC/SBEC sends this information to the power module which energizes the injectors for the required period of time.

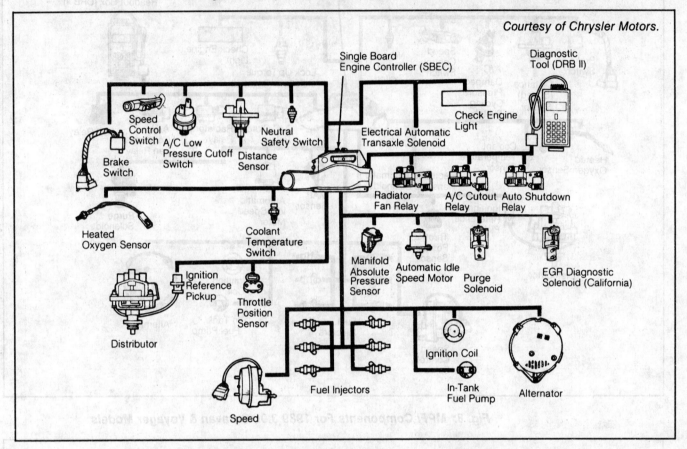

Courtesy of Chrysler Motors.

Fig. 7: MPFI Components For 1989 3.0L Passenger Car Models

EMISSION CONTROL

The SMEC/SBEC controls the purge and EGR solenoids. When engine temperature reaches 145°F (61°C), the SMEC/SBEC de-energizes the purge solenoid. When de-energized, ported vacuum flows to canister purge valve and fuel vapors vent to throttle body.

LOGIC MODULE

The logic module is a digital microprocessor computer. The module receives input signals from various sensors, switches and components. This information is used by the computer to calculate fuel injector pulse width ("ON" time), spark advance, ignition dwell, idle speed, canister purge and EGR flow.

The EGR solenoid is operated in the same manner when engine temperature reaches 70°F (21°C). No EGR occurs during idle or wide open throttle. The module also monitors and tests input signals for accuracy. This monitoring is called "On Board Diagnostics". If a malfunction is found in a major sub-system, the information is stored in the module under a specific fault code for later diagnosis. When a fault is detected, the module energizes the power loss light on the instrument panel.

When this light is illuminated, the logic module goes into the "Limp-In Mode" of operation. In this mode, the logic module is substituting information to allow the vehicle to be driven in for repairs. A significant loss of driveability usually occurs in this mode.

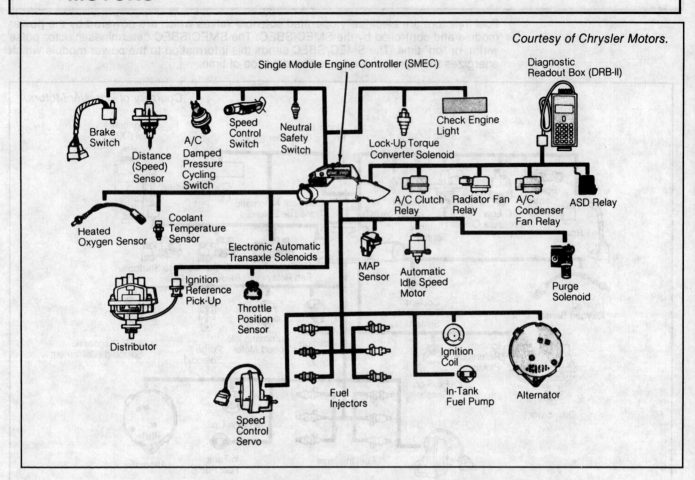

Courtesy of Chrysler Motors.

Single Module Engine Controller (SMEC)

Diagnostic
Readout Box (DRB-II)

Brake
Switch

Distance
(Speed)
Sensor

A/C
Damped
Pressure
Cycling
Switch

Speed
Control
Switch

Neutral
Safety
Switch

Check Engine
Light

Lock-Up Torque
Converter Solenoid

A/C Clutch
Relay

Radiator Fan
Relay

A/C
Condenser
Fan Relay

ASD Relay

Heated
Oxygen Sensor

Coolant
Temperature
Sensor

Electronic Automatic
Transaxle Solenoids

MAP
Sensor

Automatic
Idle Speed
Motor

Purge
Solenoid

Distributor

Ignition
Reference
Pick-Up

Throttle
Position
Sensor

Fuel
Injectors

Ignition
Coil

In-Tank
Fuel Pump

Alternator

Speed
Control
Servo

Fig. 8: MPFI Components For 1989 3.0L Caravan & Voyager Models

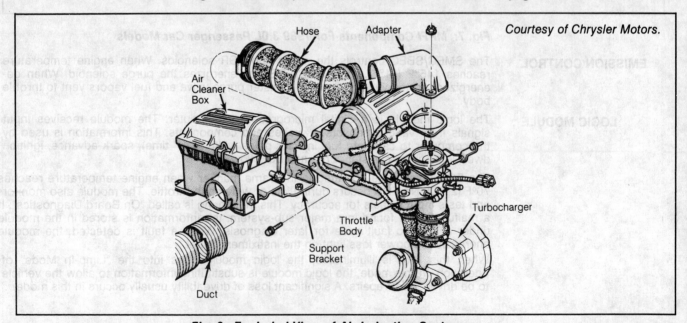

Courtesy of Chrysler Motors.

Hose

Adapter

Air
Cleaner
Box

Turbocharger

Throttle
Body

Support
Bracket

Duct

Fig. 9: Exploded View of Air Induction System

POWER MODULE

The power module contains necessary circuitry to power the ignition coil and fuel injectors. The ASD relay, inside the power module, is energized by the power module to activate the fuel pump, ignition coil and power module itself.

This computer also receives distributor signals. If no distributor signal is received by the computer, the ASD relay is not activated and power to the fuel pump and ignition coil is cut. The power module contains a voltage converter which reduces battery voltage to a constant 8-volt output (on 1988 models) and 9-volt output (on 1989 models) to power the logic module and distributor.

The Single Module Engine Controller (SMEC), introduced on 1988 models, performs the functions formerly performed by the logic module and power module. The ASD relay is external on the SMEC.

SINGLE MODULE ENGINE CONTROLLER

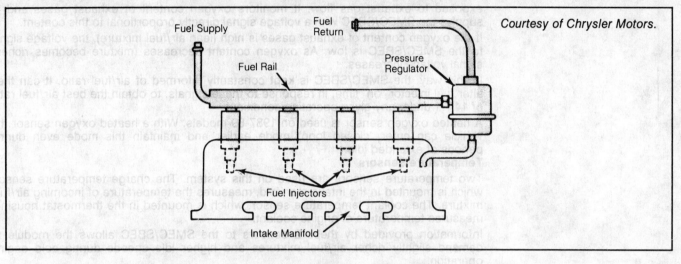

Courtesy of Chrysler Motors.

Fig. 10: Fuel Control Components

SINGLE BOARD ENGINE CONTROLLER

The Single Board Engine Controller (SBEC), introduced on 1989 3.0L passenger car models, performs the functions formerly performed by the logic module and power module. The ASD relay is external on the SBEC.

DATA SENSORS & SWITCHES

Manifold Absolute Pressure (MAP) Sensor

The MAP sensor is located in the passenger compartment, just above the logic module on 1984 models. On 1985-86 models, it is located in the logic module. *See Fig. 11.* On 1987-89 models, it is located in the engine compartment on the right shock tower. It monitors manifold vacuum via a vacuum line from the throttle body to the sensor.

The sensor supplies the SMEC/SBEC with an electrical signal which keeps the module informed of manifold vacuum conditions and barometric pressure. This information is combined with data supplied by other sensors to determine correct air/fuel ratio.

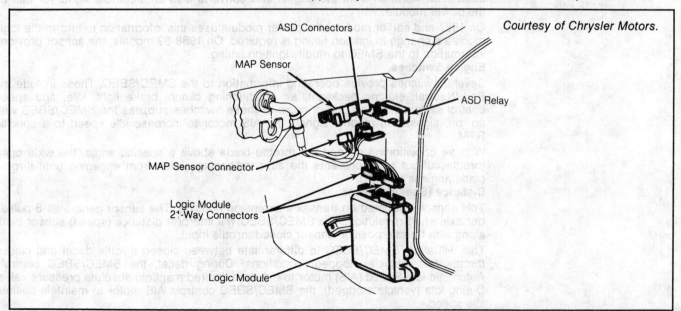

Courtesy of Chrysler Motors.

Fig. 11: Location of Logic Module, MAP Sensor & ASD Relay on 1984 Models

Barometric Pressure Solenoid (2.5L)

The barometric pressure read solenoid is controlled by the SMEC. The solenoid is in the MAP sensor vacuum line, next to the MAP sensor. The solenoid controls whether manifold pressure or atmospheric pressure is supplied to the MAP sensor.

Oxygen Sensor

The oxygen (O_2) sensor is mounted in the exhaust manifold or turbo outlet where it is exposed to exhaust gas flow. It monitors oxygen content of exhaust gases and to supplys the SMEC/SBEC with a voltage signal directly proportional to this content.

If the oxygen content of exhaust gases is high (lean air/fuel mixture), the voltage signal to the SMEC/SBEC is low. As oxygen content decreases (mixture becomes richer), signal voltage increases.

In this way, the SMEC/SBEC is kept constantly informed of air/fuel ratio. It can then alter fuel injector "on" time, in response to these signals, to obtain the best air/fuel ratio of 14.7:1 under any given operating conditions.

A heated oxygen sensor is used on 1987-89 models. With a heated oxygen sensor the vehicle can enter "closed loop" mode earlier and maintain this mode even during periods of extended idle.

Temperature Sensors

Two temperature sensors are used on this system. The charge temperature sensor, which is mounted in the intake manifold, measures the temperature of incoming air/fuel mixture. The coolant temperature sensor, which is mounted in the thermostat housing, measures temperature of engine coolant.

Information provided by these 2 sensors to the SMEC/SBEC allows the module to demand slightly richer air/fuel mixtures and higher idle speeds during cold engine operation.

If coolant temperature sensor should fail, the information supplied by the charge temperature sensor is sufficient to determine engine operating temperature and engine warm-up cycles until the coolant temperature sensor can be replaced.

Throttle Position Sensor (TPS)

The TPS is an electric resistor which is activated by throttle shaft movement. It is mounted on the throttle body and senses the angle of throttle blade opening.

A voltage signal is produced by the sensor which varies with this angle. This signal is transmitted to the SMEC/SBEC where it is used to adjust air/fuel ratio during acceleration, deceleration, idle, and wide open throttle.

Knock Sensor

The knock sensor is mounted on the rear of the valve cover on 1985 and earlier models, and on the intake manifold on later models. The knock sensor picks up detonation vibration from the engine and converts it to an electrical signal for use by the power module.

On 1987 and earlier models, the power module uses this information to inform the logic module a change in ignition timing is required. On 1988-89 models, the sensor provides information to the SMEC to modify ignition timing.

Engine Switches

Several switches provide operating information to the SMEC/SBEC. These include the Park/Neutral, electric backlight, air conditioning clutch, brake light, idle, and speed control switches. When one or more of these switches supplies the SMEC/SBEC with an "on" signal, the module signals the AIS motor to increase idle speed to a specific RPM.

With air conditioning on and the throttle blade above a specific angle, the wide open throttle cut-out relay prevents the air conditioning clutch from engaging until throttle blade angle is reduced.

Distance (Speed) Sensor

This sensor is located on transaxle extension housing. The sensor generates 8 pulses per axle shaft revolution. The SMEC/SBEC will interpret distance (speed) sensor input along with throttle position sensor closed throttle input.

This will allow SMEC/SBEC to differentiate between closed throttle decel and closed throttle idle (vehicle stopped) conditions. During decel, the SMEC/SBEC controls Automatic Idle Speed (AIS) motor to maintain desired manifold absolute pressure valve. During idle (vehicle stopped), the SMEC/SBEC controls AIS motor to maintain desired idle speed.

AUTOMATIC IDLE SPEED MOTOR (AIS)

The automatic idle speed motor is operated by the SMEC/SBEC. Data from throttle position sensor, speed sensor, coolant temperature sensor, and various switch operations, are used by the SMEC/SBEC to adjust engine idle. The AIS adjusts the air portion of the air fuel mixture through an air by-pass as an increase or decrease is needed.

TROUBLE SHOOTING

PRELIMINARY CHECKS

Most Chrysler Motors MPFI driveability problems are due to faulty or poor wiring, or loose and/or leaking hose connections. To avoid unnecessary component testing, a visual check should be performed before beginning trouble shooting:

- Air ducts to air cleaner, from the air cleaner to turbocharger and from turbocharger to throttle body.
- All component electrical connections must be clean, tight and unbroken.

Check vacuum lines for secure, leak-free connections in these areas:

- Throttle body.
- EGR and purge solenoids (located on a common bracket at right rear corner of engine compartment).
- Vapor canister.
- PCV valve to turbocharger vacuum port.
- Back pressure transducer.
- MAP sensor.

Ensure the following electrical connectors are securely attached:

- 21-pin and 25-pin connectors at logic module (1987 & earlier models).
- 12-pin and 10-pin connectors at power module (1987 & earlier models).
- 14-pin and 60-pin connectors at SMEC (1988 models) and SBEC (1989 3.0L).
- MAP sensor connector.
- ASD relay connectors (where applicable).
- EGR and purge solenoid connectors.
- Speed sensor connector (located in-line with speedometer cable).
- Charge temperature sensor connector.
- Throttle body connector.
- Fuel injector connector.
- O₂ sensor connector.
- Coolant temperature sensor connector.
- Distributor connector.
- Knock sensor conector (located at rear of valve cover).
- Radiator fan relay connectors (1989 models).
- Harness ground eyelet mounting to intake manifold.

FAULT CODES

The Chrysler Motors multi-point fuel injection system is equipped with a self-diagnostic capability which stores certain "fault codes" in the SMEC/SBEC when system malfunctions occur. These codes may be recalled to aid in system diagnosis. See ENTERING ON-BOARD DIAGNOSIS in TESTING & DIAGNOSIS.

If the problem is repaired or no longer exists, the SMEC/SBEC cancels out the code. On 1987 and earlier models, codes are cancelled after 30 ignition on/off cycles. On 1988-89 models, codes are cancelled after 50-100 on/off cycles. The following list presents these codes and the system malfunctions they represent.

Code 11

No distributor signal to SMEC/SBEC since restoration of battery voltage.

Code 12

Problem in standby memory. Appears if direct battery feed is interrupted to SMEC/SBEC.

Code 13

Problem with MAP sensor pneumatic system. Appears if sensor vacuum level does not change between start and start/run transfer speed (500-600 RPM).

Code 14

Problem with MAP sensor electrical system. MAP sensor signal outside range of .02-4.9 volts.

Code 15

Problem with vehicle distance or speed sensor circuit.

Code 16

Loss of battery voltage.

Code 17

Engine running too cool, check cooling system.

Code 21

Problem with O₂ Sensor feedback circuit. This occurs if engine temperature is above 170°F (77°C), engine speed is above 1500 RPM, but no O₂ sensor signal for more than 5 seconds (on 1988 models) and no O₂ sensor signal for a 2 minute period (on 1989 models).

Code 22

Problem with coolant temperature sensor circuit. Appears if temperature sensor indicates an incorrect temperature or a temperature that changes too fast to be real.

Code 23

Problem with charge temperature sensor circuit. Appears if charge temperature sensor indicates an incorrect temperature or a temperature that changes too fast to be real.

Code 24

Problem with TPS circuit. Appears if sensor signal is below .16 volts or above 4.7 volts.

Code 25

Problem with AIS control circuit. Appears if proper voltage from AIS system is not present. An open harness or motor will not activate this code.

Code 26

Indicates a problem with injector No. 1 circuit.
Or indicates a problem with injector No. 2 or No. 3 circuits (1989 3.0L).

Code 27

Indicates a problem with injector No. 1 circuit.
Or indicates a problem with injector No. 2 or No. 3 circuits (1989 3.0L).

Code 31

Problem with canister purge solenoid circuit. Appears when proper voltage at purge solenoid is not present (open or shorted system).

Code 32 (1987 & Earlier Models)

Problem with power loss light circuit. Appears when proper voltage to circuit is not present (open or shorted system).

Code 32 (1988-89 Models)

Problem in EGR circuit.

Code 33

Open or shorted circuit at air conditioning WOT cut-out relay circuit.

Code 34 (1987 & Earlier Models)

Open or shorted circuit at EGR solenoid.

Code 34 (1988-89 Models)

Problem in speed control solenoid driver circuit.

Code 35

Problem in cooling fan relay circuit. Code appears if fan is either not working or operates at the wrong time.

Code 36

Indicates a problem in wastegate solenoid circuit.

Code 37

Indicates a problem in barometric read solenoid circuit (FWD car only), or part throttle unlock solenoid driver circuit (FWD Van/Wagon only).

Code 41

Problem with charging system. Appears if battery voltage from ASD relay is below 11.75 volts.

Code 42

Problem in ASD relay circuit. Appears if, during cranking, battery voltage from ASD relay is not present for at least 1/3 of a second after first distributor pulse, or if battery voltage is present for more than 3 seconds after engine stalls (last distributor pulse).

Code 43

Problem in spark interface circuit. Appears if anti-dwell or injector control signal is not present between SMEC/SBEC and power module.

Code 44

On 1985 and earlier models code indicates problem in logic module. On 1986-87 models code indicates battery temperature is out of range. On 1988-89 models code indicates problem in logic module.

Code 45 (1987 & Earlier Models)

Problem in overboost shutoff circuit.

Code 45 (1989 Models)

MAP reading above overboost limit detected during engine operation.

Code 46

Indicates battery voltage too high.

Code 47
Indicates battery voltage is too low.

Code 51
On 1985 and earlier models code indicates problem in closed loop fuel system. Appears during closed loop conditions (air/fuel mixture being influenced by O_2 sensor signals) if O_2 signal is low or high for more than 2 minutes. On 1986-89 models code indicates O_2 feedback system is lean.

Code 52 (1986-89 Models)
O_2 feedback system is rich.

Codes 52 & 53 (1985 & Earlier Models)
Problem in logic module. Appears if an internal failure exists in the logic module.

Code 53 (1988-89 Models)
Internal module problem.

Code 54 (1986-89 Models)
Problem in distributor synchronization (FWD car), or high data rate (FWD Van/Wagon) circuit.

Code 54 (1985 & Earlier Models)
Problem in distributor synchronization pick-up circuit. Code appears if, at start/run transfer speed, reference pick-up signal is present but synchronization signal is not present at logic module.

Code 55
This is the "end of message" code. This code will always appear as the final code after all other fault codes have been displayed.

Code 88
This will be the first code displayed. It implies the start of the message and appears on the Diagnostic Read-Out or DRB II only.

DIAGNOSIS & TESTING

SYSTEM DIAGNOSIS

The self-diagnostic capabilities of this system can greatly simplify testing. If the logic SMEC/SBEC receives an incorrect signal or no signal from either the coolant temperature sensor, MAP sensor or TPS, the power loss light on the instrument panel is illuminated.

This light acts as a warning device to inform the operator that a malfunction in the system has occurred and immediate service is required. When certain malfunctions occur, the SMEC/SBEC enters the "Limp-In Mode".

In this mode, the SMEC/SBEC attempts to compensate for the failure of the particular component by substituting information from other sources. Ideally, this will allow the vehicle to be operated until proper repairs can be made.

If the power loss light comes on, or if certain driveability or engine performance difficulties exist, the probable source of these difficulties may be determined by entering "On-Board Diagnosis" and recording the fault codes as they are displayed.

Once these codes are known, refer to the TROUBLE SHOOTING section to determine the questionable circuit. Refer to the COMPONENT CONNECTOR IDENTIFICATION CHARTS to locate testing points for each circuit. Test circuits and repair or replace as needed.

ENTERING ON-BOARD DIAGNOSIS

1) Attach the Chrysler Diagnostic Read-Out Box (C-4805) or DRB II to diagnostic connector. The connector is located in the engine compartment near the right side strut tower.

2) If test box is not available, codes may be read by counting the flashes of the power loss light on the instrument panel.

3) Start engine (if possible). Move transmission shift lever through all positions, ending in Park. Turn A/C switch on, then off (if equipped).

4) Stop engine. Without starting engine again, turn the key on, off, on, off and on. Record fault codes as displayed on diagnostic read-out, or by counting flashes of the power loss light.

5) Codes displayed by the light are indicated by a series of flashes. For example, Code 23 is displayed as flash, flash, pause, flash, flash, flash. After a slightly longer pause, any other codes stored are displayed in numerical order.

6) The setting of a specific fault code is the result of a particular system failure, NOT a specific component. The existence of a particular code indicates the probable area of the malfunction, not necessarily the failed component itself.

SWITCH TEST

After all codes have been displayed and Code 55 has indicated end of message, actuate the following component switches. When these switches are activated and released, the digital display on the diagnostic read-out box must change. If not, check specific switch.

- Brake pedal
- Shift gear selector from Park to Reverse and back to Park
- A/C switch (if equipped)
- Rear defogger switch (if equipped)

CIRCUIT ACTUATION TEST MODE (ATM)

NOTE
- *Test can only be performed with Diagnostic Read-Out Box (C-4805).*

To put system into Circuit Actuation Test Mode, first attach diagnostic read-out box and call up fault codes. Wait for code 55 to appear on the display screen. Press ATM button on tool to activate display. If a specific ATM test is desired, hold the ATM button down until desired code appears.

The computer will continue to turn the selected circuit "on" and "off" for as long as five minutes or until the ATM button is pressed again or if the ignition switch is turned off.

If the ATM button is not pressed again, the computer will continue cycling the selected circuit for five minutes and then shut the system off. Turning the ignition switch off will also shut the test mode off.

CIRCUIT ACTUATION TEST DISPLAY CODES

Code 1
Spark Actuation-Once every 2 seconds.

Code 2
Injector Actuation-Once every 2 seconds.

Code 3
AIS Actuation-One pulse open; one pulse closed every 4 seconds.

Code 4
Radiator Fan Actuation-One pulse every 2 seconds.

Code 5
A/C Cut-Out Relay Actuation-One pulse every 2 seconds.

Code 6
ASD Relay Actuation-One click every 2 seconds.

Code 7
Purge Solenoid Actuation-One click every 2 seconds.

Code 8 (1987 & Earlier Models)
EGR Solenoid Actuation-One click every 2 seconds.

Code 8 (1988-89 Models)
Speed Control Vacuum and Vent Solenoid Actuation-Once every 2 seconds.

Code 9
Wastegate Solenoid Actuation-One click every 2 seconds.

Code 10
Barometric Read Solenoid Actuation-One click every 2 seconds.

Code 11
Alternator Full Field Actuation-One click every 2 seconds.

SENSOR READ TEST MODE

Using the diagnostic read-out tool put the system into the On-Board Diagnostic Test Mode and wait for Code 55 to appear on the display screen. Press the ATM button on the tool to activate the display. If a specific sensor read test is desired, hold the ATM button down until desired test code appears. Slide the read/hold switch to the "hold" position to display the corresponding sensor output level. See SENSOR READ TEST DISPLAY CODES.

SENSOR READ TEST DISPLAY CODES

Code 1
Battery Temperature Sensor-Units are volts x 10.

Code 2
Oxygen Sensor Voltage-Units are volts x 10.

Code 3
Charge Temperature Sensor-Units are volts x 10.

Code 4

Engine Coolant Temp. Sensor-Units are °F/10.

Code 5

Throttle Position Sensor-Units are volts x 10.

Code 6

Peak Knock Sensor Voltage-Units are volts.

Code 7

Battery Voltage-Units are volts.

Code 8

MAP Sensor Voltage-Units are volts x 10.

Code 9

Speed Control Switch

Cruise "OFF"-Display is blank.

Cruise "ON"-Display shows "00".

Cruise "SET"-Display shows "10".

Cruise "RESUME"-Display shows "01".

Code 10

Fault codes erase routine. Display will flash "0" for 4 seconds.

FUEL SYSTEM PRESSURE TEST

1) Prior to working on fuel system, fuel pressure must be relieved. Loosen gas cap to release any in-tank pressure. Remove wiring harness connector from any injector. Ground one terminal of that injector with a jumper wire.

2) Connect a jumper wire to other terminal of injector and touch positive terminal of battery for NO MORE than 5 seconds. Remove jumper wires and service fuel system.

3) To test fuel system pressure, remove fuel supply hose from throttle body. Connect Pressure Gauges (C-3292 and C-4799) between fuel filter hose and throttle body. Start engine.

4) Gauge should read 48 psi (3.3 kg/cm²) for 3.0L system and 55 psi (3.9 kg/cm²), for Turbo I and II system. If pressure is correct no further testing is required. Reinstall fuel hose using a new clamp.

5) If fuel pressure is too low, install pressure gauge between fuel filter hose and fuel line. Start engine. If pressure is now correct, replace fuel filter.

6) If no change is observed, gently squeeze return hose. If pressure increases, replace pressure regulator. If no change is observed, problem is either a plugged pump filter sock or defective fuel pump.

7) If fuel pressure is too high, remove fuel return hose from pressure regulator end. Connect another piece of hose and place other end of hose in clean container.

8) Start engine. If pressure is now correct, check for restricted fuel return line. If no change is observed, replace fuel regulator.

ADJUSTMENTS

AIS MOTOR

Monitoring and control of idle speed is performed by SMEC/SBEC and is not adjustable.

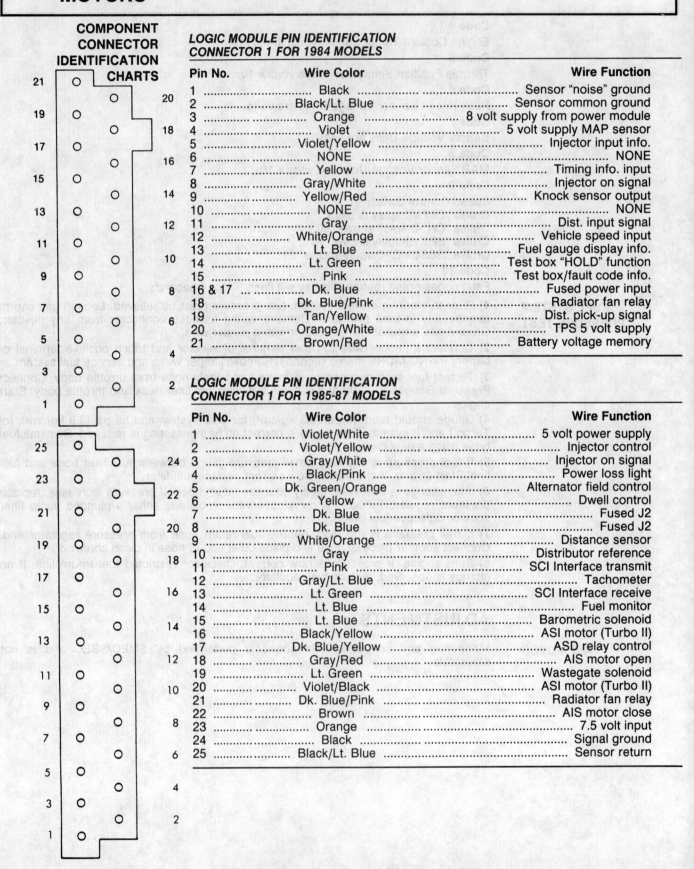

COMPONENT CONNECTOR IDENTIFICATION CHARTS

LOGIC MODULE PIN IDENTIFICATION
CONNECTOR 1 FOR 1984 MODELS

Pin No.	Wire Color	Wire Function
1	Black	Sensor "noise" ground
2	Black/Lt. Blue	Sensor common ground
3	Orange	8 volt supply from power module
4	Violet	5 volt supply MAP sensor
5	Violet/Yellow	Injector input info.
6	NONE	NONE
7	Yellow	Timing info. input
8	Gray/White	Injector on signal
9	Yellow/Red	Knock sensor output
10	NONE	NONE
11	Gray	Dist. input signal
12	White/Orange	Vehicle speed input
13	Lt. Blue	Fuel gauge display info.
14	Lt. Green	Test box "HOLD" function
15	Pink	Test box/fault code info.
16 & 17	Dk. Blue	Fused power input
18	Dk. Blue/Pink	Radiator fan relay
19	Tan/Yellow	Dist. pick-up signal
20	Orange/White	TPS 5 volt supply
21	Brown/Red	Battery voltage memory

LOGIC MODULE PIN IDENTIFICATION
CONNECTOR 1 FOR 1985-87 MODELS

Pin No.	Wire Color	Wire Function
1	Violet/White	5 volt power supply
2	Violet/Yellow	Injector control
3	Gray/White	Injector on signal
4	Black/Pink	Power loss light
5	Dk. Green/Orange	Alternator field control
6	Yellow	Dwell control
7	Dk. Blue	Fused J2
8	Dk. Blue	Fused J2
9	White/Orange	Distance sensor
10	Gray	Distributor reference
11	Pink	SCI Interface transmit
12	Gray/Lt. Blue	Tachometer
13	Lt. Green	SCI Interface receive
14	Lt. Blue	Fuel monitor
15	Lt. Blue	Barometric solenoid
16	Black/Yellow	ASI motor (Turbo II)
17	Dk. Blue/Yellow	ASD relay control
18	Gray/Red	AIS motor open
19	Lt. Green	Wastegate solenoid
20	Violet/Black	ASI motor (Turbo II)
21	Dk. Blue/Pink	Radiator fan relay
22	Brown	AIS motor close
23	Orange	7.5 volt input
24	Black	Signal ground
25	Black/Lt. Blue	Sensor return

SMEC 60-PIN IDENTIFICATION CONNECTOR FOR 1988 MODELS

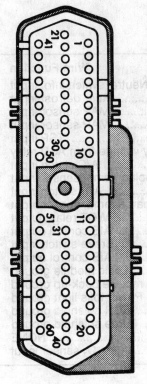

Pin No.	Wire Color	Wire Function
1	Dark Green/Red	Map Sensor
2	Black/Lt. Green	Detonation Sensor (Except 3.0L)
3	Tan/White	Coolant Sensor
4	Black/Lt. Blue	Sensor Return
5	Black/White	Signal Ground
6	NONE	NONE
7	White/Lt. Green	Speed Control Resume
8	Yellow/Red	Speed Control ON/OFF
9	Brown/Red	Speed Control Set
10	Dark Green/Black	Z1 Input
11	NONE	NONE
12	Dark Blue/White	FJ2
13	Violet/White	5-Volt Supply
14	Dark Green/Orange	Alternator Field/Voltage Regulator
15	Lt. Blue/Red	Ground
16	Lt. Blue/Red	Ground
17	Brown/White	AIS-1
18	Yellow/Black	AIS-2
19	Gray/Red	AIS-3
20	Violet/Black	AIS-4
21	Black/Red	Charge Temperature Sensor
22	Orange/Lt. Blue	Throttle Position Sensor
23	Black/Dark Green	Oxygen Sensor
24	NONE	NONE
25	NONE	NONE
26	Tan/Yellow	Fuel Synchronizer Pickup (Except 3.0L)
27	NONE	NONE
28	NONE	NONE
29	White/Pink	Brake Switch
30	Brown/Yellow	Park/Neutral Switch
31	Lt. Green	SCI Receive
32	NONE	NONE
33	Violet/Yellow	Injector Control No. 1
34	Yellow	Dwell Control
35	NONE	NONE
36	Lt. Blue/Black	Fuel Monitor
37	NONE	NONE
38	NONE	NONE
39	Lt. Green/Black	Wastegate Solenoid (Except 3.0L)
40	Gray/Yellow	EGR Solenoid (3.0L California)
41	Red	Direct Battery
42	NONE	NONE
43	NONE	NONE
44	NONE	NONE
45	Brown	A/C Clutch Input
46	NONE	NONE
47	Gray/Black	Reference Pickup
48	White/Orange	Vehicle Distance Pickup
49	Tan/Yellow	High Data Rate Pickup
50	Gray/Lt. Blue	Tachometer Signal
51	Pink	SCI Transmit
52	Orange	8-Volt Input (9-Volt Input, 3.0L)
53	Tan/Red	Speed Control Vacuum Solenoid
54	Pink/Black	Purge Solenoid
55	Orange/Black	Lock-Up Torque Converter (3.0L Only)
56	Dark Blue/Orange	A/C Wide Open Throttle Cut-Out
57	Dark Blue/Pink	Radiator Fan Relay
58	Dark Blue/Yellow	Auto Shut Down Relay
59	Black/Pink	Check Engine Light
60	Lt. Green/Red	Speed Control Vent Solenoid

SMEC 14-PIN IDENTIFICATION CONNECTOR FOR 1988 MODELS

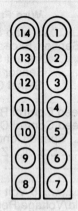

Pin No.	Wire Color	Wire Function
1	Orange	8-Volt Output (9-Volt Output, 3.0L)
2	Black/White	Ground
3	Dark Blue/Lt. Blue	FJ2 Output
4	Dark Blue	12 Volts
5	Yellow/White	Injector Control
6	Black	Ground
7	Black	Ground
8	Violet/Yellow	Injector Control No. 1
9	White	Injector Driver No. 1
10	Tan	Injector Driver No. 2
11	Dark Green/Orange	Voltage Regulator Signal
12	Black/Yellow	Ignition Coil Driver
13	Yellow	Anti-Dwell Signal
14	Dark Green	Voltage Regulator/Alternator Field Control

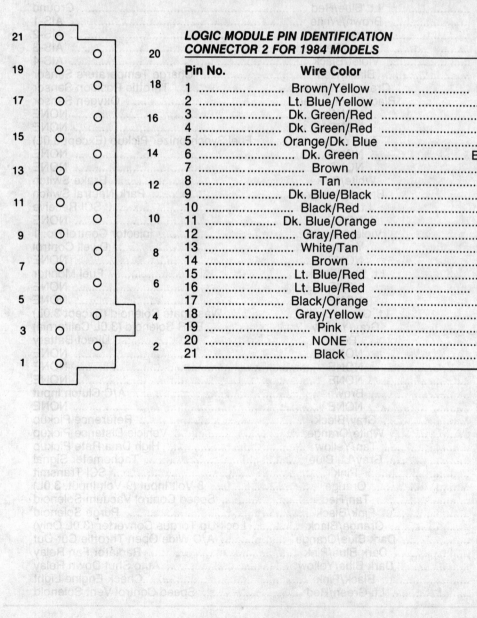

LOGIC MODULE PIN IDENTIFICATION CONNECTOR 2 FOR 1984 MODELS

Pin No.	Wire Color	Wire Function
1	Brown/Yellow	Park/Neutral switch info. input
2	Lt. Blue/Yellow	Rear defrost input
3	Dk. Green/Red	MAP sensor input
4	Dk. Green/Red	MAP sensor input
5	Orange/Dk. Blue	TSP input
6	Dk. Green	Battery voltage from ASD relay
7	Brown	A/C clutch input
8	Tan	Coolant temp. sensor input
9	Dk. Blue/Black	A/C switch input
10	Black/Red	Charge temp. sensor input
11	Dk. Blue/Orange	WOT relay output
12	Gray/Red	AIS control output
13	White/Tan	Brake switch input
14	Brown	AIS control output
15	Lt. Blue/Red	Logic module ground
16	Lt. Blue/Red	Logic module back-up ground
17	Black/Orange	Power loss light ground
18	Gray/Yellow	EGR solenoid ground
19	Pink	Purge solenoid ground
20	NONE	NONE
21	Black	O_2 sensor input

LOGIC MODULE PIN IDENTIFICATION
CONNECTOR 2 FOR 1985-87 MODELS

Pin No.	Wire Color	Wire Function
1	Orange/White	TPS 5 volts
2	Red/White	Battery standby
3	Dk. Blue/Orange	A/C cut-out relay
4	Black/Pink	Power loss light
5	Pink	Purge solenoid
6	Gray/Yellow	EGR Solenoid
7	Lt. Blue/Red	Power ground
8	Lt. Blue/Red	Power ground
9	Brown/Red	Speed control set
10	White	Speed control resume
11	Brown	A/C clutch
12	Brown/Yellow	Park/Neutral switch
13	White/Tan	Brake switch
14	White/Orange	Distance sensor
15	Yellow/Red	Speed control on/off
16	NONE	NONE
17	Tan/Yellow	Distributor switch
18	Black	Oxygen sensor
19	Dk. Green/Red	MAP signal
20	Red/Black	Battery temp. signal
21	Orange/Dk. Blue	TPS
22	Red/White	Battery sense
23	Tan	Coolant sensor
24	Black/Lt. Green	Detonation sensor
25	Black/Red	Charge sensor

POWER MODULE PIN IDENTIFICATION
12-PIN CONNECTOR FOR 1984 MODELS

Pin No.	Wire Color	Wire Function
1	Violet/Yellow	Injector control input
2	Black/Lt. Blue	Knock sensor ground
3	NONE	NONE
4	Black/Lt. Green	Knock sensor input
5	NONE	NONE
6	NONE	NONE
7	Gray/White	Injector on control
8	Tan/Yellow	Dist. synch. input
9	Yellow/Red	Knock sensor input
10	Yellow	Spark adv. info. from logic module
11	NONE	NONE
12	Orange	8 volt output. Supply voltage for Hall pick-up and logic module

POWER MODULE PIN IDENTIFICATION
12-PIN CONNECTOR FOR 1985-87 MODELS

Pin No.	Wire Color	Wire Function
1	Violet/Yellow	Injector control input
2	Black/Lt. Blue	Signal ground
3	Red/Black	Battery temp. sensor
4	NONE	NONE
5	Dk. Blue/Yellow	ASD relay control
6	Red/White	Battey sense and standby
7	NONE	NONE
8	Gray/White	Injector control
9	NONE	NONE
10	Yellow	Dwell control
11	Dk. Green/Orange	Alternator field control
12	Orange	8 volt output

POWER MODULE PIN IDENTIFICATION
10-PIN CONNECTOR FOR 1984 MODELS

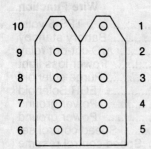

Pin No.	Wire Color	Wire Function
1	Black/Yellow	Ignition coil trigger
2	Dk. Blue	J2 feed from ignition
3	Dk. Blue	Fused J2
4	Tan	Injector 3 and 4 output
5	White	Injector 1 and 2 output
6	NONE	NONE
7	Dk. Blue/Yellow	ASD relay ground
8	Dk. Green	Power supply from power module
9	Black	Power module ground
10	Black	Power module back-up ground

POWER MODULE PIN IDENTIFICATION
10-PIN CONNECTOR FOR 1985-87 MODELS

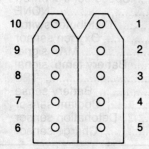

Pin No.	Wire Color	Wire Function
1	Black/Yellow	Coil neg. terminal
2	Dk. Blue	J2 ignition input
3	Dk. Blue	Fused J2 output
4	Pink	Direct battery
5	White	Injector
6	Dk. Green/Black	Switched battery
7	Tan	Injector
8	Dk. Green	Alternator field
9	Black	Power ground
10	Black	Power ground

NOTE

- *For 1989 connector identification, refer to appropriate wiring diagram.*

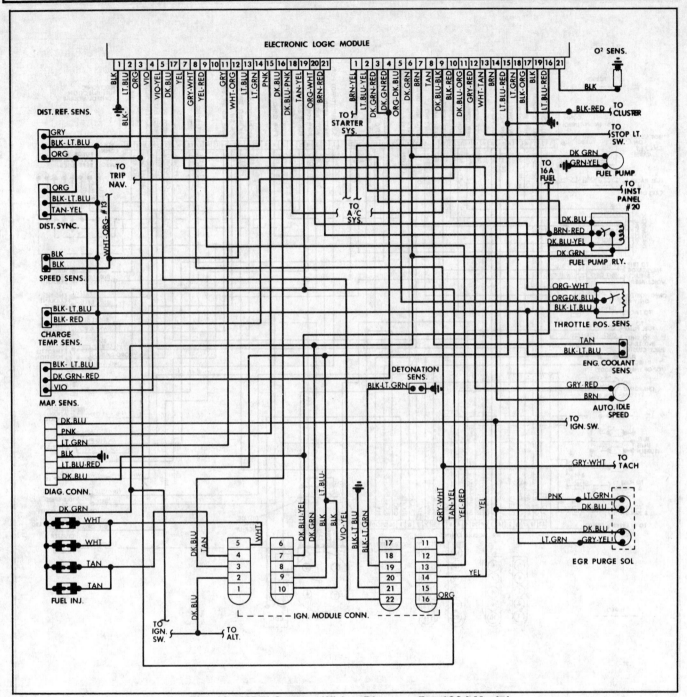

Fig. 12: MPFI System Wiring Diagram For 1984 Models

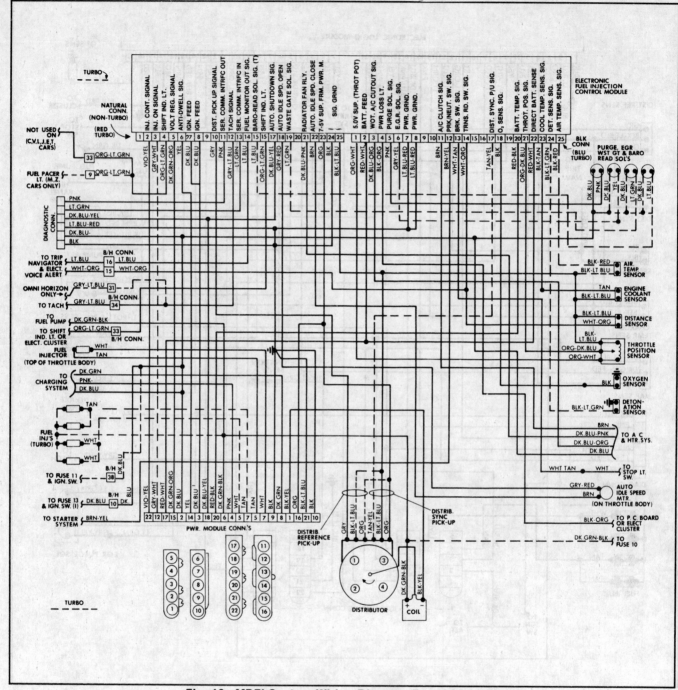

Fig. 13: MPFI System Wiring Diagram For 1985-86 Models

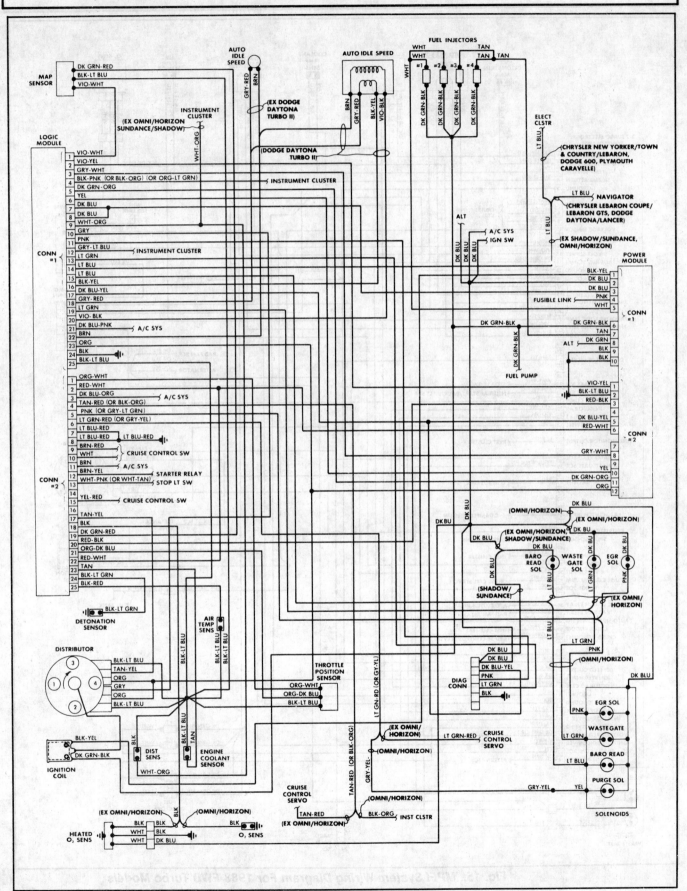

Fig. 14: MPFI System Wiring Diagram For 1987 Turbo Models

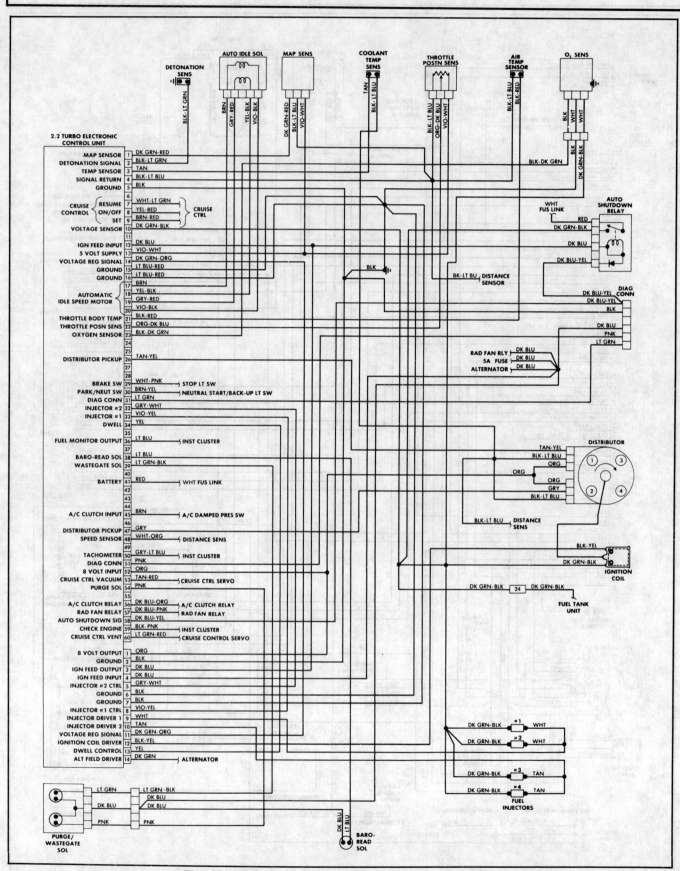

Fig. 15: MPFI System Wiring Diagram For 1988 FWD Turbo Models

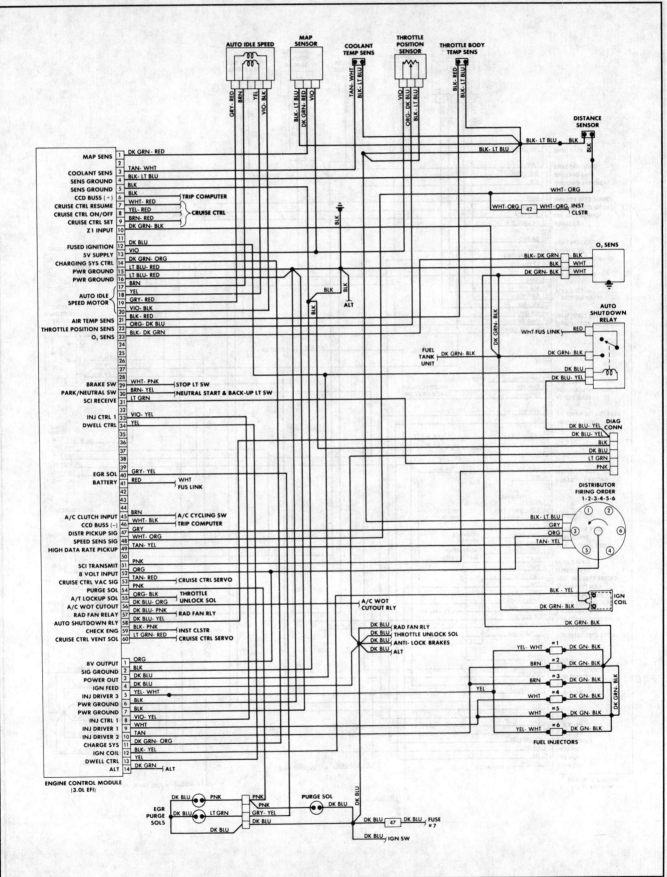

Fig. 16: MPFI System Wiring Diagram For 1988 New Yorker & Dynasty Models

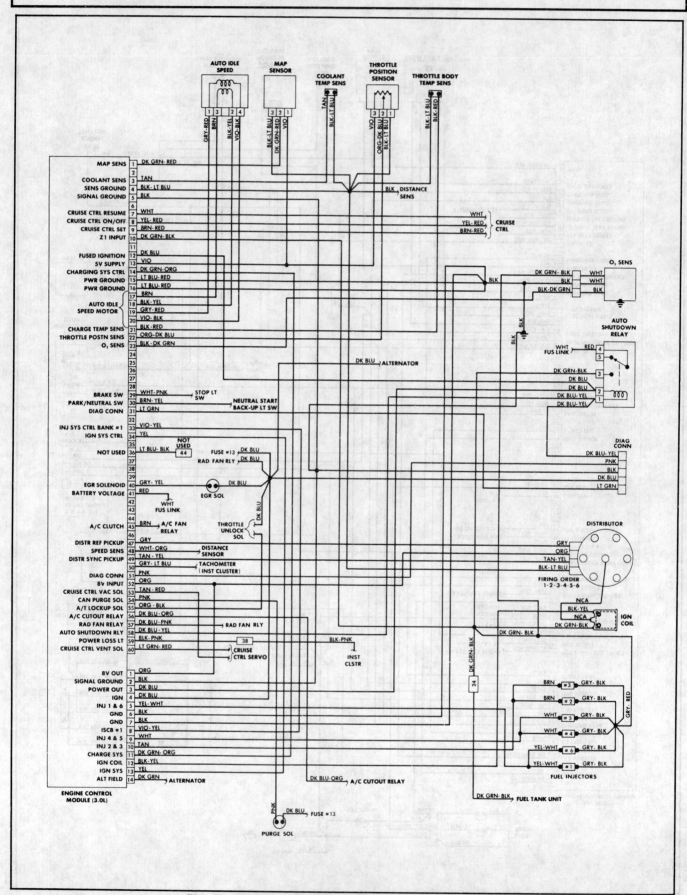

Fig. 17: MPFI System Wiring Diagram For 1988 Voyager 3.0L Models

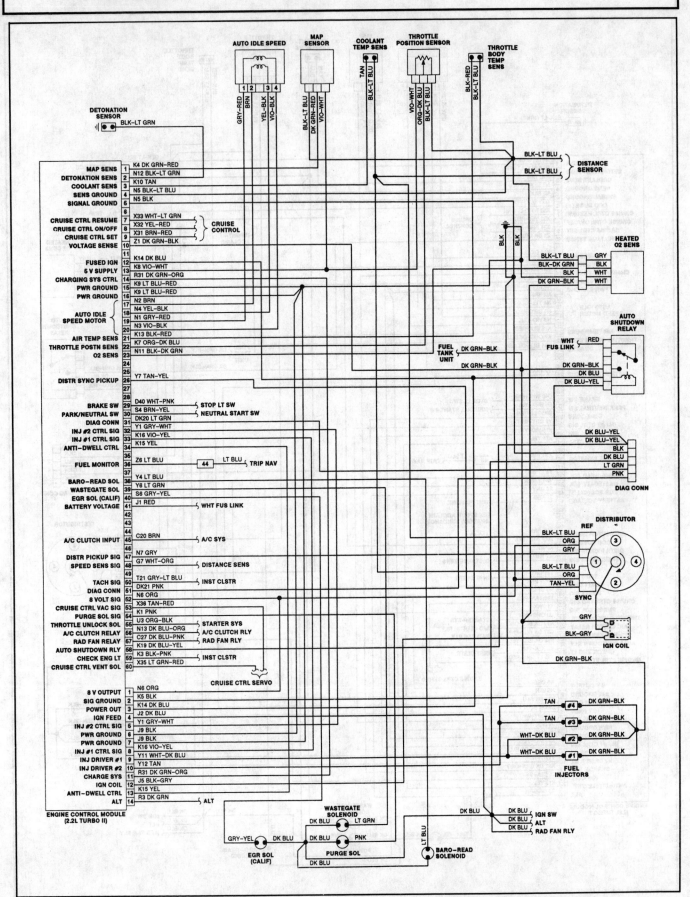

Fig. 18: MPFI System Wiring Diagram For 1989 2.2L Turbo II Models

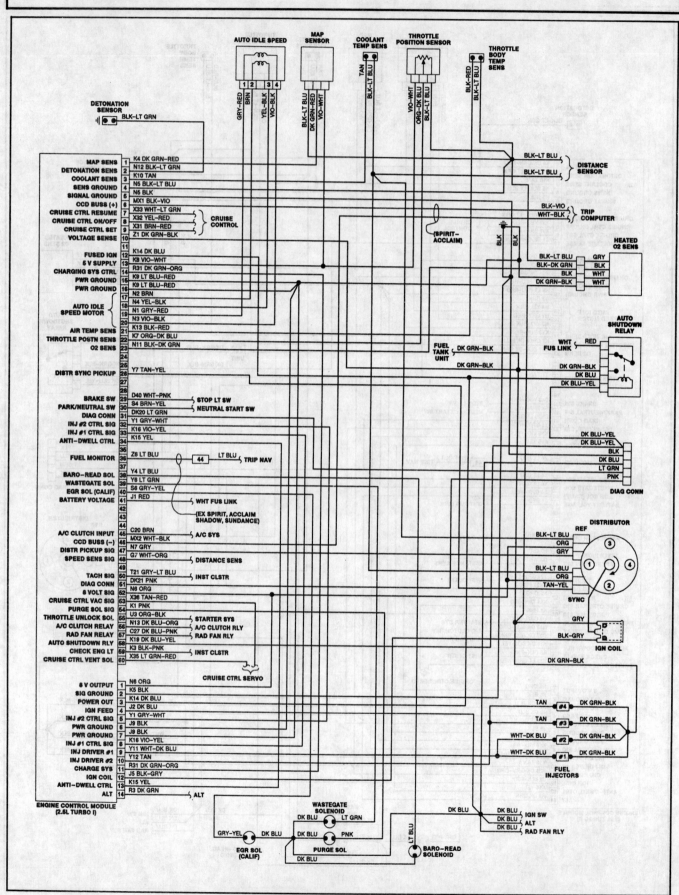

Fig. 19: MPFI System Wiring Diagram For 1989 2.5L Turbo I Models

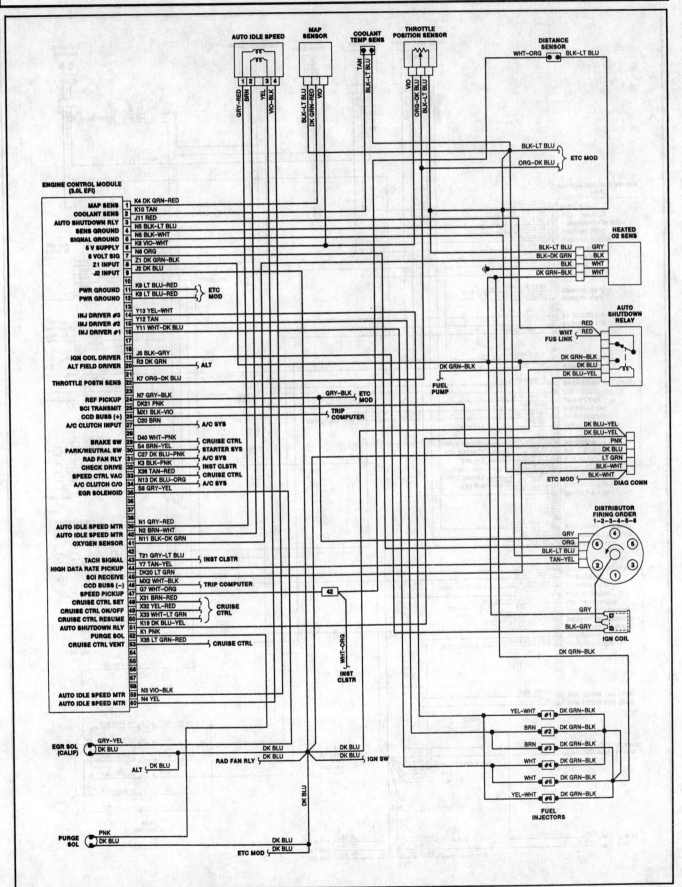

Fig. 20: MPFI System Wiring Diagram For 1989 Acclaim & Spirit Models

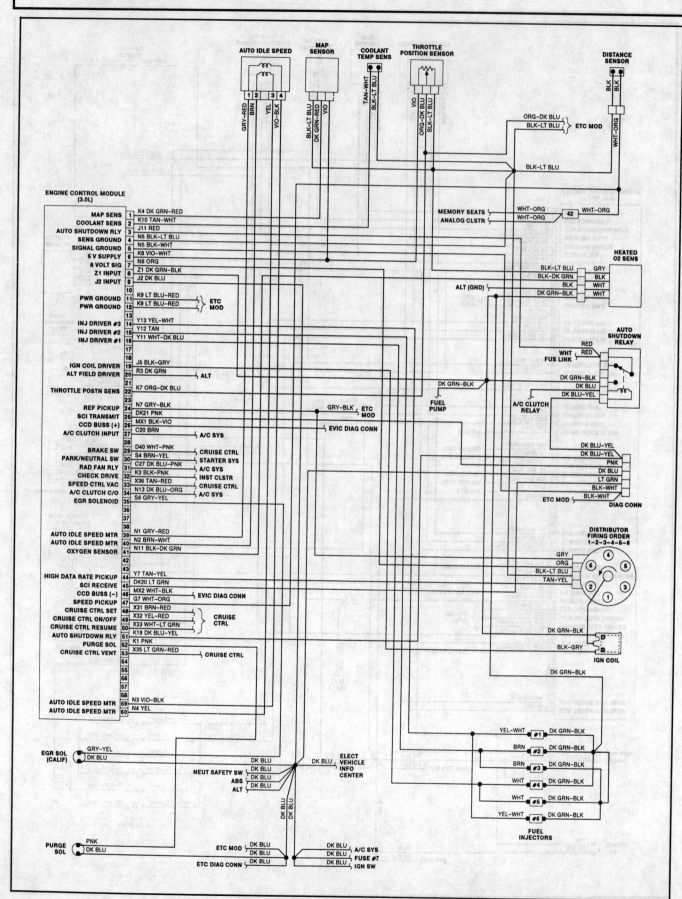

Fig. 21: MPFI System Wiring Diagram For 1989 Dynasty & New Yorker Models

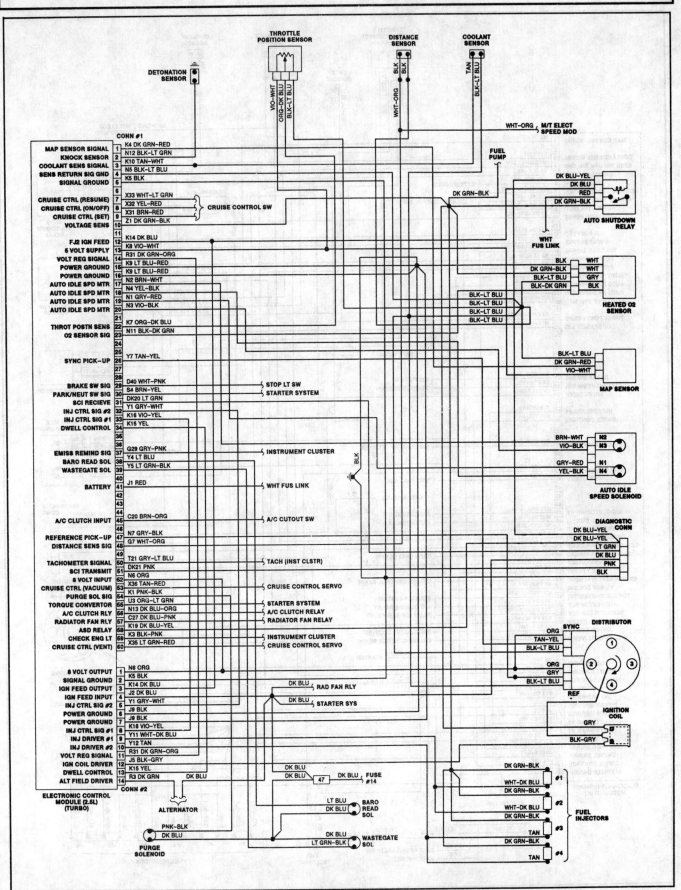

Fig. 22: MPFI System Wiring Diagram For 1989 Caravan & Voyager 2.5L Turbo I Models

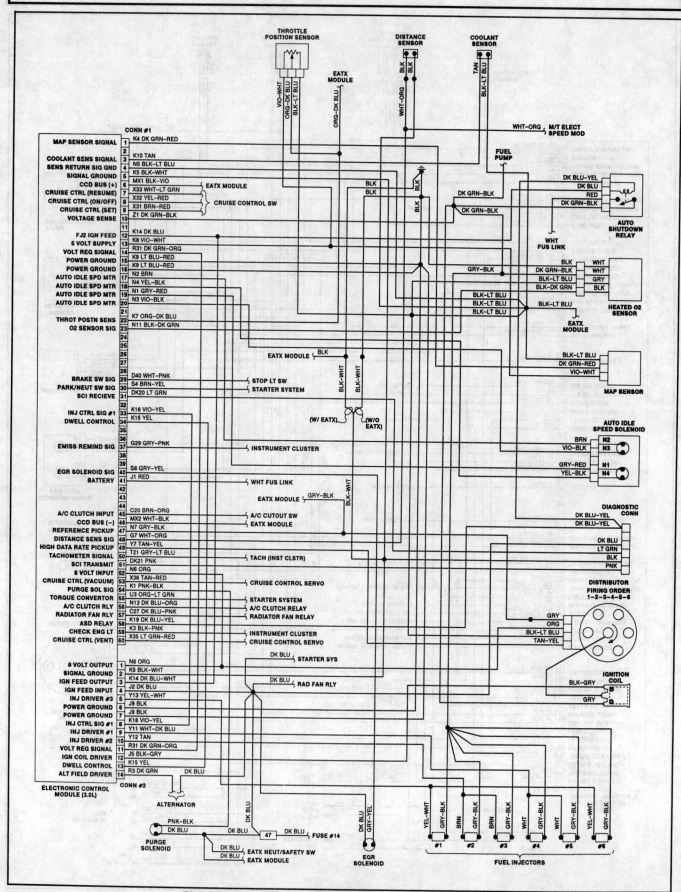

Fig. 23: MPFI System Wiring Diagram For 1989 Caravan & Voyager 3.0L Models

Chrysler: Laser, LeBaron, LeBaron GTS,
 New Yorker, Town & Country
Dodge: Aries, Caravan, Dakota, Daytona, Dynasty, Lancer,
 Omni America, Pickup, Ramcharger, Shadow, Spirit
 Van, 600
Plymouth: Acclaim, Caravelle, Horizon, Reliant, Sundance, Voyager

NOTE
- *For detailed applications, refer to FUEL INJECTION SYSTEMS APPLICATION CHARTS at the front of this publication.*

DESCRIPTION

The Chrysler Throttle Body Fuel Injection system is computer controlled and utilizes a throttle-body assembly with a single fuel injector. The use of 2 fuel injectors housed in a single throttle body are used on 3.9L, 5.2L and 5.9L (truck & van) models.

Fuel is supplied to the engine through an electronically controlled injector valve located in the throttle body assembly on top of the intake manifold. On 1987 and earlier models, power to the injector is supplied by the power module while the length of time the injector is left open is determined by the logic module. On 1988-89 models, the Single Module Engine Controller (SMEC) replaces both the power module and logic module.

The logic module is a digital, pre-programmed computer which controls ignition timing, emission control devices and idle speed in addition to air/fuel ratio. The amount of fuel to be metered through the injector is based on engine operating condition information as supplied by various engine sensors and switches. These include the Manifold Absolute Pressure (MAP) sensor, Throttle Position Sensor (TPS), oxygen sensor, coolant temperature sensor, throttle body temperature sensor and vehicle distance/speed sensor.

The logic module converts input signals from these sensors/switches into control signals. Control signals are sent to the power module which responds by issuing electrical signals to specific components to alter air/fuel ratio and/or ignition timing to meet the indicated conditions.

The Single Module Engine Controller (SMEC) performs the functions formally performed by the logic module and power module. It contains a 14-pin connector and a 60-pin connector. *See Fig. 6.*

OPERATION

FUEL SUPPLY

An electric fuel pump is located in the fuel tank as an integral part of the fuel gauge sending unit. This pump supplies fuel at 36.0 psi (2.5 kg/cm²) for all models through 1985, or 14.5 psi (1.0 kg/cm²) on 1986 and newer models, to the throttle body assembly. Power to the fuel pump is supplied by the power module via the Automatic Shut-Down (ASD) relay. The power module is supplied with an operating signal from the distributor. If this signal is not received, the ASD relay is not activated and power to the fuel pump is cut off.

THROTTLE BODY ASSEMBLY

Throttle Body

The throttle body is mounted on the intake manifold, in the same position as a conventional carburetor. It houses the fuel injector, pressure regulator, throttle position sensor, idle speed control actuator, throttle body temperature sensor and the automatic idle speed motor. Air flow control is via a throttle blade in the base of the throttle body, which is operated by conventional throttle linkage from the accelerator pedal. The throttle body chamber provides for the metering, atomization and distribution of fuel into the incoming air stream.

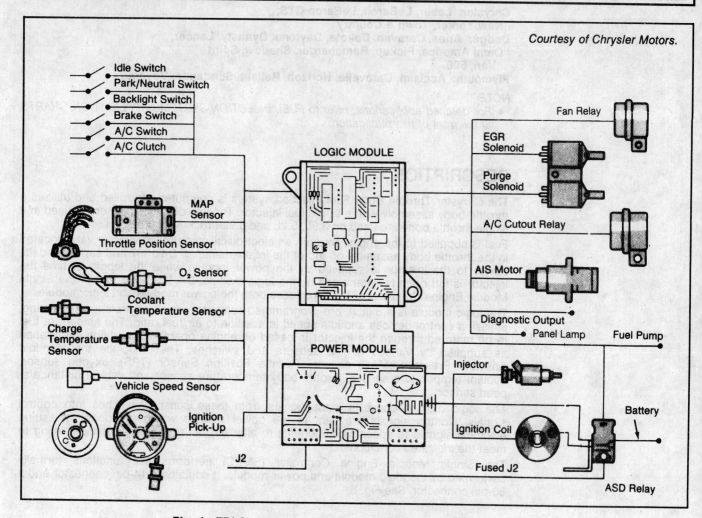

Courtesy of Chrysler Motors.

Fig. 1: TBI Components (1984 & Earlier Models)

Fuel Injector (1985 & Earlier Models)

The fuel injector is mounted in the throttle body so that fuel from the injector is directed into the incoming air stream. While power to the injector is supplied by the power module, it is controlled by indirect signal from the logic module/SMEC. When electric current is supplied to the injector, an integral armature and pintle valve move a short distance against a spring, opening a small orifice.

Fuel supplied to the injector is forced around the pintle valve and through this opening, resulting in a fine spray of fuel in the shape of a hollow cone. Since a constant pressure drop is maintained across the injector (by the pressure regulator), the length of time that this opening is maintained (injector "on" time) determines the amount of fuel entering the engine and, therefore, the air/fuel ratio.

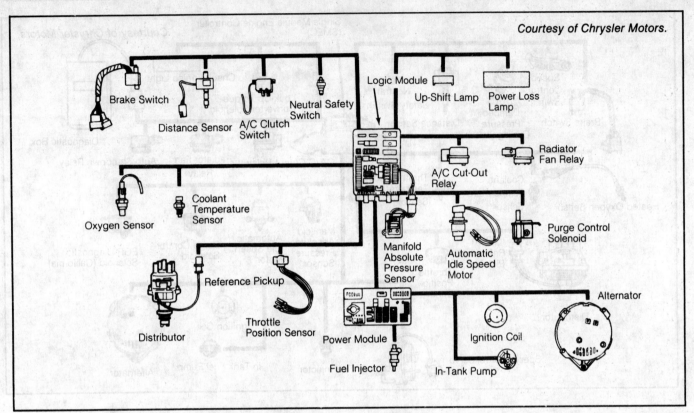

Courtesy of Chrysler Motors.

Fig. 2: TBI Components (1985 Models)

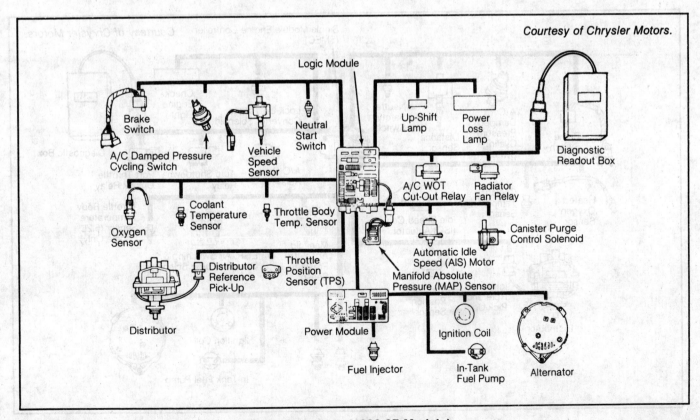

Courtesy of Chrysler Motors.

Fig. 3: TBI Components (1986-87 Models)

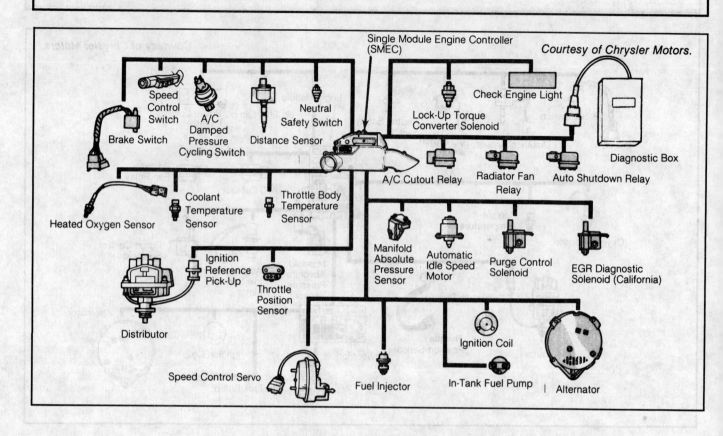

Fig. 4: TBI Components (1988 4-Cyl. Models)

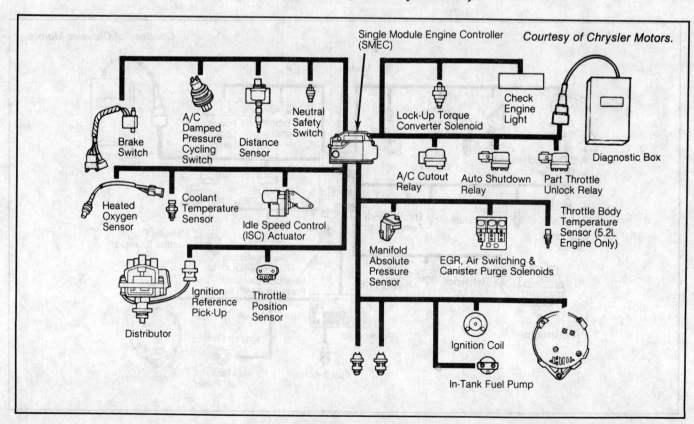

Fig. 5: TBI Components (1988 3.9L & 5.2L Models)

Courtesy of Chrysler Motors.

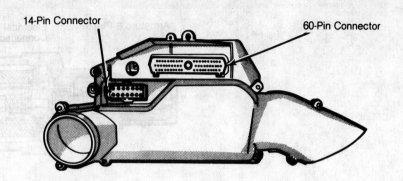

Fig. 6: Single Module Engine Controller (SMEC) Connector Locations

Fuel Injector (1986-89 Models)

The fuel injector used in 1986-89 models is an electronic solenoid driven by the Power Module/SMEC and controlled by the Logic Module/SMEC. The logic module/SMEC, based on sensor inputs, determines when and how long the Power Module should operate the injector. When electrical current is supplied to the injector a spring loaded ball is lifted from it's seat. This allows fuel to flow through six spray orifices and deflects off the sharp edge of the injector nozzle. This action causes the the fuel to form a 45 degree cone shaped spray pattern before entering the air stream in the throttle body. Fuel is supplied to injectors at a constant 14.5 psi (1.0 kg cm²).

Courtesy of Chrysler Motors.

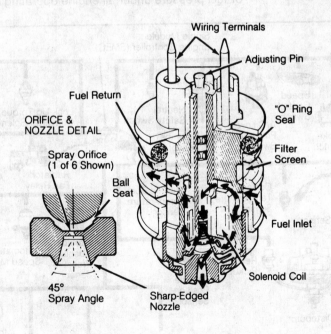

Fig. 7: Cross-Sectional View of Fuel Injector (All 1986-89 Models)

Courtesy of Chrysler Motors.

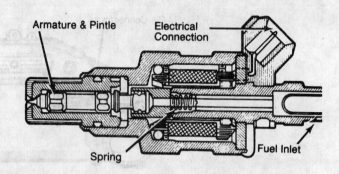

Fig. 8: Cross-Sectional View of Fuel Injector (1985 & Earlier Models)

Pressure Regulator

The pressure regulator is located downstream from the fuel injector, on the throttle body. The regulator is equipped with a spring loaded rubber diaphragm which covers a fuel return port. Fuel flow past the injector enters the regulator and is restricted from flowing back to the tank by the blocked return port. When fuel pressure reaches 36 psi (2.5 kg/cm²) for all models thru 1985, or 14.5 psi (1.0 kg/cm²) on 1986 and newer models, the spring is compressed enough to allow fuel into the return line. The diaphragm and spring assembly are constantly opening and closing to maintain fuel pressure at this constant value.

On 1985 and earlier models, further control of fuel pressure is provided by a vacuum line from the regulator to a vacuum port located in the throttle body, above the throttle plate. As venturi vacuum increases, less pressure is required to introduce the same amount of fuel into the air stream. This vacuum assist allows for additional fine tuning of fuel pressure under all engine operating conditions.

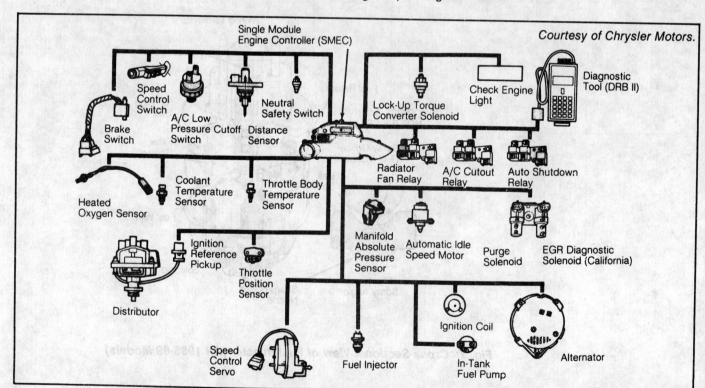

Fig. 9: TBI Components (1989 Passenger Car Models)

Fuel Pump

The Throttle Body Injection (TBI) fuel pump is an immersible in-tank pump with a permanent magnet electric motor. Voltage to operate fuel pump is supplied through the Auto Shutdown Relay (ASD).

A/C Cut-Out Relay

The A/C cut-out relay is connected in series with the A/C damped pressure switch, A/C switch and A/C fan relay. The relay remains in the energized ("ON") position during engine operation. When SMEC senses low idle speeds or wide open throttle, it de-energizes the relay and causes the compressor clutch to disengage.

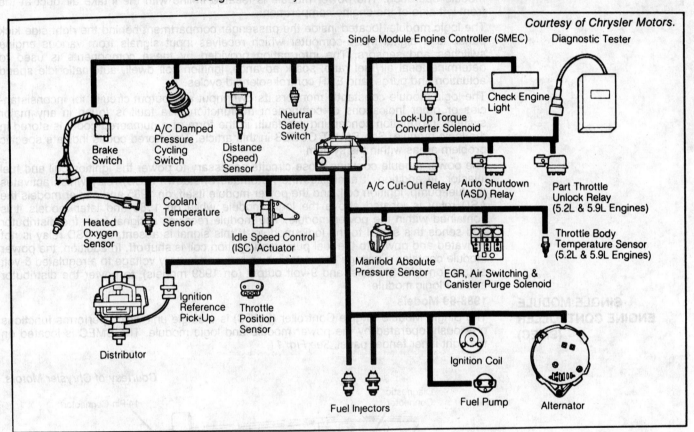

Courtesy of Chrysler Motors.

Fig. 10: TBI Components (1989 Light Truck Models)

Automatic Idle Speed Motor (AIS)

The AIS is mounted on the throttle body assembly and is controlled by an electrical signal from the logic module/SMEC. The logic module/SMEC uses sensor input to determine optimum engine idle speed for any idle condition. The AIS is then adjusted to allow a specific amount of air through an air by-pass in the throttle body. This by-pass is enlarged or restricted as an increase or decrease in engine idle speed is required to meet varying engine operating conditions. This results in a change in air/fuel ratio, which alters the oxygen content of the exhaust gases as detected by the O_2 sensor. The logic module/SMEC then changes the amount of fuel introduced into the intake charge to maintain an ideal air/fuel ratio.

Idle Speed Control (ISC) Actuator

The ISC actuator is controlled by the SMEC. The ISC actuator adjusts idle speed by physically moving the throttle lever. Through input from various sensors, the SMEC uses this data to determine engine load and barometric pressure. This information is combined with other data to determine correct air/fuel mixture and ignition timing.

Auto Shutdown (ASD) Relay

The ASD interrupts power to the fuel pump, fuel injectors and ignition coil, when the ignition switch is in "RUN" position and no distributor signal is present.

Purge Solenoid

The SMEC controls purge solenoid based upon engine temperature. When engine operating temperatures reaches 80ºF (27ºC) for 3.9L and 145ºF (61ºC) for all other models, fuel vapors will be purged through throttle body.

LOGIC & POWER MODULES

1987 & Earlier Models

Power to the ignition coil and fuel injector is supplied by the power module. The high current required to operate these devices requires that their power source (the power module) be isolated from the logic module to avoid electrical interference with logic module operation. The power module is located in-line with the intake air duct at the front left side of the engine compartment.

The logic module, located inside the passenger compartment behind the right side kick panel, is a digital micro-computer which receives input signals from various engine switches and sensors. The information provided by these components is used to determine ideal air/fuel ratio, spark advance, ignition coil dwell, automatic idle speed actuation and purge, and EGR control solenoid cycles.

The logic module constantly monitors its own input and output circuits for inconsistancies or other indications of component malfunction. If a fault is found in any major system, information concerning this fault, in the form of a numbered code, is stored in the logic module. See FAULT CODES in this article. The stored codes indicate specific problem areas within the system.

The power module contains those circuits necessary to power the ignition coil and fuel injector. In addition, it energizes the Automatic Shutdown (ASD) relay which activates the fuel pump, ignition coil, and the power module itself, on 1985 and earlier models the ASD relay is located above the logic module while on 1986 and later models it is contained within the power module. The module receives a signal from the distributor and sends this signal to the logic module. If this signal is absent, the ASD relay is not activated and power to the fuel pump and ignition coil is shut off. In addition, the power module contains a voltage converter which reduces battery voltage to a regulated 8-volt output (on 1988 models) and 9-volt output (on 1989 models), to power the distributor and the logic module.

SINGLE MODULE ENGINE CONTROLLER (SMEC)

1988-89 Models

The Single Module Engine Controller (SMEC) is a single unit which performs functions previously operated by the power module and logic module. The SMEC is located on the right inner fender panel. *See Fig. 11.*

Courtesy of Chrysler Motors.

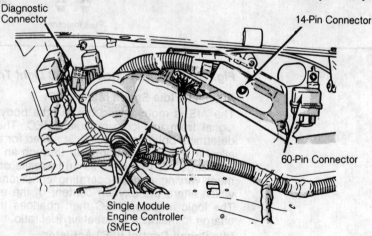

Fig. 11: Location of Single Module Engine Control

ENGINE SENSORS & SWITCHES

Manifold Absolute Pressure Sensor (MAP)

The MAP sensor is located in the passenger compartment on 1984 and earlier models, just above the logic module. *See Fig. 12.* On 1985-86 models the MAP sensor is mounted on the logic module. The MAP sensor on 1987 and 1989 models is located under the hood on the right shock tower. On 1988 RWD vans, the MAP sensor is located in left front of engine compartment. On 1989 3.9L, 5.2L and 5.9L (truck & van) models, the MAP sensor is connected to the throttle body. On all other models, the

MAP sensor is located next to the right front firewall. The MAP sensor monitors manifold vacuum by a vacuum line from the throttle body to the sensor. The sensor supplies the logic module/SMEC with an electrical signal which keeps the module informed of manifold vacuum conditions and barometric pressure. This information is combined with data supplied by other sensors to determine correct air/fuel ratio.

Courtesy of Chrysler Motors.

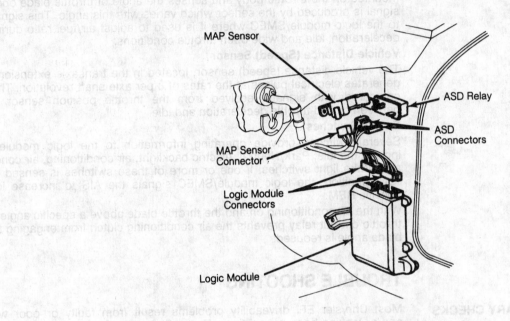

Fig. 12: Location of Logic Module, MAP Sensor & ASD Relay (1984 & Earlier Models)

Oxygen Sensor

The oxygen (O$_2$) sensor is installed into the exhaust manifold, just ahead of the tail-pipe connection, so that it is exposed to exhaust gas flow. Its function is to monitor the oxygen content of the exhaust and to supply the logic module/SMEC with a voltage signal which is directly proportional to this content. If exhaust oxygen content is high (lean air/fuel mixture) the voltage signal to the logic module/SMEC is low. As oxygen content decreases (mixture becomes richer) signal voltage increases. In this manner, the logic module/SMEC is kept constantly informed of air/fuel ratio. It can then alter fuel injector "on" time, in response to these signals, to obtain the best air/fuel ratio under any given condition.

A heated oxygen sensor is used on 1987-89 models. The utilization of a heated oxygen sensor makes it possible for the vehicle to enter the "closed loop" mode sooner and to remain in this operating mode even during periods of extended idle.

Temperature Sensors

Two temperature sensors are used on 1983 models. The coolant temperature sensor measures temperature of engine coolant. The charge temperature sensor, measures temperature of incoming air/fuel mixture. The charge temperature sensor is not used on 1984 and later models.

The coolant temperature sensor is mounted in the thermostat housing to monitor engine coolant (operating) temperature. It supplies the logic module/SMEC with a voltage signal which varies with coolant temperature. The charge temperature sensor, mounted in the intake manifold, supplies the logic module/SMEC with the same type of information as the coolant temperature sensor, only concerning the temperature of the incoming air/fuel mixture.

Information provided by the charge temperature sensor (if equipped), and the coolant temperature sensor allows the logic module/SMEC to demand slightly richer air/fuel mixtures and higher idle speeds during cold engine operation. If coolant temperature switch should malfunction, information supplied by the charge temperature sensor is sufficient to determine engine operating temperature and engine warm-up cycles until the coolant temperature sensor can be repaired or replaced. Coolant temperature sensor also functions as control for electric cooling fan.

Throttle Body Temperature Sensor

This sensor is mounted in the throttle body to measure fuel temperature. Fuel temperature input signal controls mixture used in hot start situations.

Throttle Position Sensor (TPS)

The TPS is an electric resistor which is activated by movement of the throttle shaft. It is mounted on the throttle body and senses the angle of throttle blade opening. A voltage signal is produced by the sensor which varies with this angle. This signal is transmitted to the logic module/SMEC where it is used to adjust air/fuel ratio during acceleration, deceleration, idle and wide open throttle conditions.

Vehicle Distance (Speed) Sensor

The vehicle distance (speed) sensor, located in the transaxle extension shaft housing, generates electrical pulses at the rates of 8 per axle shaft revolution. These signals are compared with signals received from the throttle position sensor to differentiate between closed throttle deceleration and idle.

Engine Switches

Several switches provide operating information to the logic module/SMEC. These include the idle, Park/Neutral, electric backlight, air conditioning, air conditioning clutch, and brake light switches. If one or more of these switches is sensed as being in the "ON" position, the logic module/SMEC signals the AIS to increase idle speed to a specific RPM.

With the air conditioning on and the throttle blade above a specific angle, the wide open throttle cut-out relay prevents the air conditioning clutch from engaging until the throttle blade angle is reduced.

TROUBLE SHOOTING

PRELIMINARY CHECKS

Most Chrysler EFI driveability problems result from faulty or poor wiring, or loose and/or leaking hose connections. To avoid unnecessary testing, a visual check should be performed before trouble shooting to help spot these common faults. A preliminary visual check should include:

- Air ducts to air cleaner and from air cleaner to throttle body
- Electrical connections at all components. (clean, tight and unbroken)

Check vacuum lines for secure connections:

- Throttle body (2 front, 2 rear)
- EGR (on 1985 and earlier) and/or purge solenoids
  (located on a common bracket at right rear corner of engine compartment)
- Vapor canister
- PCV valve to intake manifold vacuum port
- Back pressure transducer
- MAP sensor
- Heated air door
- EGR solenoid vacuum connector
- Air switching solenoid vacuum connector

Ensure the following electrical connectors are securely attached:

- Logic module/SMEC connectors
- MAP sensor connectors
- ASD relay connectors
- Power module connectors
- EGR and purge solenoid connectors
- Speed sensor connector
- Charge temperature sensor connector (1983 only)
- Radiator fan relay connector (1984 & 1989)
- AIS motor and TPS connectors
- Fuel injector connector
- O₂ sensor connector
- Coolant temperature sensor connector
- Distributor connector
- Throttle body temperature sensor connector
- A/C cut-out relay
- Ground eyelet on left cylinder head

- Alternator wiring harness
- Ground strap at engine block and on firewall near coil
- Ignition coil wiring harness connections
- Transaxle wiring harness connections
- Wiring connections at all relays
- Wiring connections at fuel pump
- 4-pin AIS motor (1989 2.5L truck & van models)
- 2-pin air switching solenoid connector (1989 3.9L, 5.2L & 5.9L)
- 2-pin canister purge solenoid connector (all 1989 truck & van models)
- 2-pin coolant temperature sensor connector (1989 2.5L truck & van models)
- 3-pin distributor reference connector (all 1989 truck & van models)
- 2-pin EGR diagnostic solenoid connector (1989 3.9L, 5.2L & 5.9L)
- 4-pin fuel injector connector (1989 3.9L, 5.2L & 5.9L)
- 3-pin heated O_2 sensor connector (1989 3.9L)
- 4-pin heated O_2 sensor connector (1989 5.2L & 5.9L)
- 4-pin ISC actuator connector (1989 3.9L, 5.2L & 5.9L)
- 2-pin fuel injector connector (1989 2.5L truck & van models)
- 3-pin MAP sensor connector (all 1989 truck & van models)
- 2-pin speed sensor connector at transmission (1989 3.9L, 5.2L & 5.9L)
- 2-pin throttle body temperature sensor connector
  (all 1989 truck & van models)
- 3-pin TPS connector (all 1989 truck & van models)
- 60-pin and 14-pin SMEC connectors

FAULT CODES

The Chrysler TBI system is equipped with a self-diagnostic capability which stores fault codes in the logic module/SMEC. These codes can be recalled to aid in system diagnosis. See ENTERING ON-BOARD DIAGNOSIS in DIAGNOSIS & TESTING section. The following list includes codes and system malfunctions.

NOTE
- *All fault codes are not used on all models and/or years.*

Code 11
No distributor signal to logic module/SMEC since restoration of battery voltage.

Code 12
Indicates a problem in stand-by memory circuit. Code appears if direct battery feed to logic module/SMEC is interrupted.

Code 13
Problem with MAP sensor pneumatic system. Appears if sensor vacuum level does not change between start and start/run transfer speed (500-600 RPM).

Code 14
Problem with MAP sensor electrical system. MAP sensor signal outside of .02-4.9 volt range.

Code 15
On 1985-89 models, code indicates a problem with distance sensor circuit. On earlier models, it indicates a problem with speed sensor circuit. Engine speed above 1470 RPM, sensor indicates less than 2 MPH. Code valid only if sensed while vehicle is moving.

Code 16
Loss of battery voltage sense.

Code 17
Problem with cooling system, engine running too cool.

Code 21
Problem with O_2 sensor feedback circuit. Occurs if engine temperature is above 170°F (77°C), engine speed is above 1500 RPM, but O_2 sensor stays rich or lean for more than 60 seconds.

Code 22
Problem with coolant temperature sensor circuit. Appears if sensor indicates an incorrect temperature or a temperature that changes too fast to be real.

Code 23
Problem with throttle body temperature sensor circuit. Appears if sensor indicates an incorrect temperature or a temperature that changes too fast to be real.

Code 24

Problem with TPS circuit. Appears if sensor signal is either below .16 volts or above 4.7 volts.

Code 25

Problem with AIS control circuit. Appears if proper voltage from AIS system is not present. An open harness or motor will not activate this code.

Code 26

Peak injector current has not been reached.

Code 27

Internal problem in the fuel circuit in the logic module/SMEC.

Code 31

Problem with canister purge solenoid circuit. Appears when proper voltage at the purge solenoid is not present (open or shorted system).

Code 32 (1987 and Earlier Models)

Problem with Power Loss Lamp circuit. Appears when proper voltage to the circuit is not present (open or shorted system).

Code 32 (1988-89 Models, California Only)

EGR diagnosis is necessary.

Code 33

Open or shorted circuit at air conditioning WOT cut-out relay circuit.

Code 34 (1987 and Earlier Models)

Open or shorted circuit at EGR solenoid.

Code 34 (1988-89 Models)

Problem in speed control solenoid driver circuit.

Code 35

Indicates a problem in the fan control relay circuit (FWD), or idle switch circuit (RWD).

Code 36

Problem in air switching solenoid circuit.

Code 37

Problem in part throttle unlock solenoid driver circuit

Code 41

Problem with charging system. Appears if battery voltage from the ASD relay is below 11.75 volts.

Code 42

Problem in the ASD relay circuit. Appears if, during cranking, battery voltage from ASD relay is not present for at least 1/3 of a second after first distributor pulse, or if battery voltage is present for more than 3 seconds after engine stalls (last distributor pulse).

Code 43

Problem in spark interface/injection coil control circuit.

Code 44

Problem in logic module/SMEC. Appears if an internal failure is present in module.

Code 46

Battery voltage too high.

Code 47

Battery voltage too low.

Code 51

On 1985-89 models, oxygen feedback system is lean. On earlier models, indicates a problem in standby memory. Appears if direct battery feed to logic module is interrupted. Code will disappear after about 30 ignition key on/off cycles once logic module receives a distributor signal.

Code 52 (1985-89 Models)

Oxygen feedback system is rich.

Code 53 (1985-89 Models)

Internal logic module/SMEC problem.

Codes 52, 53, 54 (1984 and Earlier Models)

Problem in logic module. Appears if an internal failure exists in logic module.

Code 55
This is the "end of message" code. This code will always appear as the final code after all other fault codes have been displayed.

Code 62
Attempt is unsuccessful to update EMR mileage in SMEC EEPROM (trucks only).

Code 63
Attempt is unsuccessful to write to an EEPROM location by SMEC.

Code 88
This will be the first code displayed. It implies the start of the message and appears on the diagnostic read-out box only.

DIAGNOSIS & TESTING

FUEL SYSTEM PRESSURE TEST

1) Fuel injection system holds constant pressure of 14.5 psi (1.0 kg/cm²). This pressure must be released before disconnecting any fuel carrying components. To release pressure in tank, open fuel tank cap slowly.

2) To release remaining pressure in system, disconnect 2-pin connector from fuel injector. Using jumper wire, connect injector harness ground terminal No. 1 to ground. Connect injector harness terminal No. 2 to positive battery terminal with second jumper wire.

3) DO NOT keep injector connected to positive battery terminal for longer than 5 seconds. Remove jumper wires. Fuel injection system can now be opened and fuel carrying components removed as pressure is fully released.

FUEL SYSTEM PRESSURE RELEASE

1) Release fuel system pressure. Disconnect 5/16" hose from engine fuel line assembly. Connect Fuel System Pressure Testers (C-3292 and C-4749) between fuel filter outlet hose and throttle body. Connect Diagnostic Readout Box II (DRB II). Turn ignition on. Use Actuate Outputs Test "Auto Shutdown Relay" on DRB II to activate fuel pump.

2) Check reading on testers after system is pressurized. The correct pressure reading should be 13.5-15.5 psi (.95-1.09 kg/cm²). If system pressure is correct, remove testers. Reconnect fuel hose to throttle body using a new hose clamp of same type as original equipment.

3) If pressure reading is low, move pressure tester so reading is taken between fuel supply line and fuel filter inlet hose. Repeat pressure test. If reading is correct, replace fuel filter.

4) If reading is still low, lightly squeeze fuel return hose. Replace fuel pressure regulator if pressure now increases. If pressure continues to read low, fuel pump is defective or fuel pump intake filter is plugged.

5) If pressure reading is high, disconnect fuel return hose at throttle body. Connect separate hose to throttle body/fuel pressure regulator with open end in clean container. Repeat pressure test. If reading is still high, check for plugged fuel injector. If injector is okay, replace fuel pressure regulator. If reading is correct, check all return lines for restrictions.

SYSTEM DIAGNOSIS

If properly utilized, the self-diagnostic capabilities of this system can greatly simplify testing.

If the logic module/SMEC receives an incorrect signal or no signal from either the coolant temperature sensor, MAP sensor or TPS, a power loss light on the instrument panel is illuminated. This light acts as a warning device to inform the operator that a malfunction has occurred and immediate service is required.

When certain malfunctions occur, the logic module/SMEC enters the "Limp In Mode". In this mode, the logic module/SMEC attempts to compensate for the failure of the particular component by substituting information from other sources. Ideally, this will allow the vehicle to be operated until proper repairs can be made.

If the power loss light comes on, or if certain driveability or engine performance difficulties exist, the probable source of these difficulties may be determined by entering "On Board Diagnosis" and recording fault codes as they are displayed.

Once these codes are known, refer to the TROUBLE SHOOTING section to determine the questionable circuit. Then use the CHRYSLER THROTTLE BODY FUEL INJECTION SYSTEM WIRING DIAGRAM and COMPONENT CONNECTOR IDENTIFICATION charts to locate testing points for each circuit. Test circuits and repair or replace as needed.

ENTERING ON-BOARD DIAGNOSIS

1) Attach Diagnostic Read-Out Box (C-4805) to diagnostic test connector. Connector is located in engine compartment near right strut tower on FWD models. On Dakota, it is located on right side of engine compartment next to right fender and firewall. On all other pickups, it is located near master cylinder. On RWD vans, diagnostic connector is located next to SMEC.

2) If test box is not available, codes may be determined by flashing Light Emitting Diode (LED) on logic module (1984 and earlier models) or flashing of power loss light (1985-89 models).

3) Start engine (if possible). Move transmission shift lever through all positions, ending in park. Turn A/C switch on, then off (if present).

4) Stop engine. Without starting engine again, turn key on, off, on, off and on. Record fault codes as displayed on diagnostic read-out box, or by counting flashes of LED/power loss light.

5) Codes displayed by the LED are indicated by a series of flashes. For example, Code 23 is displayed as flash, flash, pause, flash, flash, flash. After a slightly longer pause, any other codes stored are displayed in numerical order.

6) The setting of a specific fault code is the result of a particular system failure, not a specific component. Therefore, the existence of a particular code denotes the probable area of the malfunction, not necessarily the failed component itself.

Courtesy of Chrysler Motors.

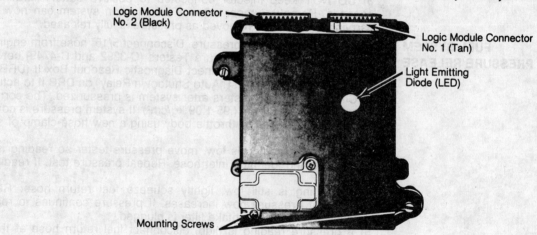

Logic Module Connector No. 2 (Black)

Logic Module Connector No. 1 (Tan)

Light Emitting Diode (LED)

Mounting Screws

Fig. 13: View of Logic Module Showing Location of Light Emitting Diode (LED)

COMPONENT CONNECTOR IDENTIFICATION CHARTS

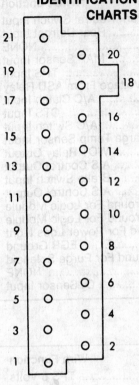

LOGIC MODULE PIN IDENTIFICATION
CONNECTOR 1 FOR 1984 AND EARLIER MODELS

Pin No.	Wire Color	Wire Function
1	Black	Sensor Input Ground
2	Black/Lt. Blue	Engine Sensor Common Ground
3	Orange	8 Volts From Power Module
4	Violet	5 Volt Supply MAP Sensor
5	Violet/Yellow	Injector Info. Input
6	Violet	NONE
7	Yellow	Power Module Timing Info.
8	Dk. Gray	NONE
9	Yellow/Red	NONE
10	NONE	NONE
11	Gray	Dist. RPM Info.
12	White/Orange	Vehicle Speed Input
13	Blue/Black	Fuel Gauge Display Info.
14	Lt. Green	"Hold" Function On Test Box
15	Pink	Test Box Fault Code Info.
16 & 17	Dk. Blue	Fused Power From Power Module
18	NONE	NONE
19	NONE	NONE
20	Orange/White	5 Volt TPS Power Supply
21	Brown/Red	Battery Voltage For Memory

LOGIC MODULE PIN IDENTIFICATION
CONNECTOR 1 FOR 1985-87 MODELS

Pin No.	Wire Color	Wire Function
1		5-Volt Function
2	Violet/Yellow	Injector Control
3	NONE	NONE
4	NONE	NONE
5	Dk. Green/Orange	Alternator Field Control
6	Yellow	Dwell Control
7	Dk. Blue	Fused J2
8	Dk. Blue	Fused J2
9	NONE	NONE
10	Gray	Distributor Reference
11	Pink	SCI Interface Transmit
12	Gray/Lt. Blue	Tachometer Signal
13	Lt. Green	SCI Interface Transmit
14	Lt. Blue	Fuel Monitor
15	Orange/Lt. Green	Shift Indicator Light
16	Violet/Black	AIS Motor
17	Dk. Blue/Yellow	ASD Relay Control
18	Gray/Red	AIS Motor Open
19	NONE	NONE
20	Violet	AIS Motor
21	Dk. Blue/Pink	Radiator Fan Relay
22	Brown	AIS Motor Close
23	Orange	7.5 Volts Input
24	Black	Signal Ground
25	Black/Lt. Blue	Sensor Return

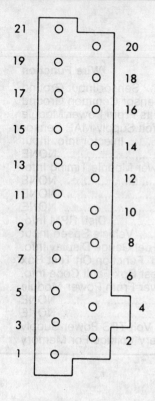

LOGIC MODULE PIN IDENTIFICATION
CONNECTOR 2 FOR 1984 AND EARLIER MODELS

Pin No.	Wire Color	Wire Function
1	Brown/Yellow	Park/Neutral Switch Input
2	Lt. Blue/Yellow	Rear Defrost Input
3	Dk. Green/Red	NONE
4	Dk. Green/Red	MAP Sensor Input
5	Orange/Dk. Blue	TPS Input
6	Dk. Green	Battery Voltage From ASD Relay
7	Brown	A/C Clutch Input
8	Tan	CTS Input
9	Dk. Blue/Yellow	A/C Switch Input
10	Black/Red	Charge Temp Sensor Input
11	Dk. Blue/Orange	WOT Relay Output
12	Gray/Red	AIS Control Output
13	White/Tan	Brake Switch Input
14	Brown	AIS Control Output
15	Lt. Blue/Red	Ground For Logic Module
16	Lt. Blue/Red	Back-Up Ground For Logic Module
17	Black/Orange	Ground For Power Loss Light
18	Gray/Yellow	EGR Ground
19	Pink	Ground For Purge Solenoid
20	NONE	NONE
21	Black	O₂ Sensor Input

LOGIC MODULE PIN IDENTIFICATION
CONNECTOR 2 FOR 1985-87 MODELS

Pin No.	Wire Color	Wire Function
1	Orange/White	TPS 5 Volts
2	Red/White	Battery Standby
3	Dk. Blue/Orange	A/C Cut-Out Relay
4	Black/Orange	Power Loss Light
5	Pink	Surge Solenoid
6	NONE	NONE
7	Lt. Blue/Red	Power Ground
8	Lt. Blue/Red	Power Ground
9	NONE	NONE
10	NONE	NONE
11	Brown	A/C Clutch
12	Brown/Yellow	Park/Neutral Switch
13	White/Tan	Brake Switch
14	White/Orange	Distance Sensor
15	NONE	NONE
16	NONE	NONE
17	NONE	NONE
18	Black	Oxygen Sensor
19		MAP Sensor Signal
20	Red/Black	Battery Temp. Signal
21	Orange/Dk.Blue	TPS
22	Red/White	Battery Voltage Sensor
23	Tan	Coolant Sensor
24	NONE	NONE
25	Black/Red	Injector Temp. Sensor

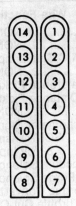

SMEC 14-PIN IDENTIFICATION CONNECTOR FOR 1988 MODELS

Pin No.	Wire Color	Wire Funtion
1	Orange	8-Volt Output
2	Black/White	Ground
3	Dark Blue/White	FJ2 Output
4	Dark Blue	12 Volt
5	Gray/White	Injector Control No. 2 (RWD Only)
6	Black	Ground
7	Black	Ground
8	Violet/Yellow	Injector Control
9	White	Injector Driver No. 1
10	Tan	Injector Driver No. 2 (RWD Only)
11	Dark Green/Orange	Voltage Regulator Signal
12	Black/Yellow	Ignition Coil Driver
13	Yellow	Anti-Dwell Signal
14	Dark Green	Regulator Control

SMEC 60-PIN IDENTIFICATION CONNECTOR FOR 1988 FWD MODELS

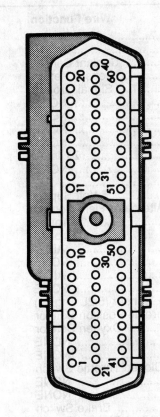

Pin No.	Wire Color	Wire Function
1	Dark Green/Red	Map Sensor
2	NONE	NONE
3	Tan/White	Coolant Sensor
4	Black/Lt. Blue	Sensor Return
5	Black/White	Signal Ground
6	NONE	NONE
7	White/Lt. Green	Speed Control Resume
8	Yellow/Red	Speed Control ON/OFF
9	Brown/Red	Speed Control Set
10	White	B1 Input
11	NONE	NONE
12	Dark Blue/White	FJ2
13	Violet/White	5-Volt Supply
14	Dark Green/Orange	Alternator Field Control
15	Lt. Blue/Red	Ground
16	Lt. Blue/Red	Ground
17	Brown/White	AIS-1
18	Yellow/Black	AIS-2
19	Gray/Red	AIS-3
20	Violet/Black	AIS-4
21	Black/Red	Throttle Body Temperature Sensor
22	Orange/Dark Blue	Throttle Position Sensor
23	Black/Dark Green	Oxygen Sensor
24	NONE	NONE
25	NONE	NONE
26	NONE	NONE
27	NONE	NONE
28	NONE	NONE
29	White/Pink	Brake Switch
30	Brown/Yellow	Park/Neutral Switch
31	Lt. Green	SCI Receive
32	NONE	NONE
33	Violet/Yellow	Injector Control No. 1
34	Yellow	Dwell Control
35	NONE	NONE
36	Lt. Blue/Black	Fuel Monitor
37	NONE	NONE
38	NONE	NONE
39	NONE	NONE
40	Gray/Yellow	EGR Solenoid
41	Red	Direct Battery
42	NONE	NONE

SMEC 60-PIN IDENTIFICATION
CONNECTOR FOR 1988 FWD MODELS (Cont.)

Pin No.	Wire Color	Wire Function
43	NONE	NONE
44	NONE	NONE
45	Brown	A/C Clutch Input
46	NONE	NONE
47	Gray/Black	Reference Pickup
48	White/Orange	Vehicle Distance Pickup
49	NONE	NONE
50	Gray/Lt. Blue	Tachometer Signal
51	Pink	SCI Transmit
52	Orange	8-Volt Input
53	Tan/Red	Speed Control Vacuum Solenoid
54	Pink/Black	Purge Solenoid
55	Orange/Black	Lock-Up Torque Converter
56	Dark Blue/Orange	A/C Wide Open Throttle Cut-Out
57	Dark Blue/Pink	Radiator Fan Relay
58	Dark Blue/Yellow	Auto Shut Down Relay
59	Black/Pink	Check Engine Light
60	Lt. Green/Red	Speed Control Vent Solenoid

SMEC 60-PIN IDENTIFICATION
CONNECTOR FOR 1988 RWD PICKUP & VAN

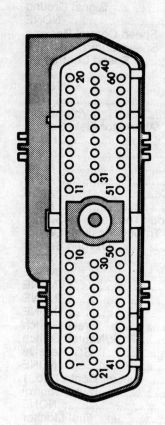

Pin No.	Wire Color	Wire Function
1	Dark Green/Red	Map Sensor
2	NONE	NONE
3	Tan/White	Coolant Sensor
4	Black/Lt. Blue	Sensor Return
5	Black/White	Signal Ground
6	NONE	NONE
7	NONE	NONE
8	NONE	NONE
9	NONE	NONE
10	Dark Green/Black	Z1 Input
11	NONE	NONE
12	Dark Blue/White	FJ2
13	Violet/White	5-Volt Supply
14	Dark Green/Orange	Alternator Field Control
15	Lt. Blue/Red	Ground
16	Lt. Blue/Red	Ground
17	Brown/White	AIS-1
18	NONE	NONE
19	Gray/Red	AIS-3
20	NONE	NONE
21	Black/Red	Throttle Temp. Sensor (5.2L Engine)
22	Orange/Dark Blue	Throttle Position Sensor
23	Black/Dark Green	Oxygen Sensor
24	NONE	NONE
25	NONE	NONE
26	Violet	Closed Throttle Switch
27	NONE	NONE
28	NONE	NONE
29	White/Pink	Brake Switch
30	Brown/Yellow	Park/Neutral Switch
31	Lt. Green	SCI Receive
32	Gray/White	Injector Control No. 2
33	Violet/Yellow	Injector Control No. 1
34	Yellow	Dwell Control
35	NONE	NONE
36	NONE	NONE
54	Pink/Black	Purge Solenoid

176

SMEC 60-PIN IDENTIFICATION
CONNECTOR FOR 1988 RWD PICKUP & VAN (Cont.)

Pin No.	Wire Color	Wire Function
55	Orange/Black	Lock-Up Torq. Conv. or Shift Light
56	Dark Blue/Orange	A/C Wide Open Throttle Cut-Out
57	NONE	NONE
58	Dark Blue/Yellow	Auto Shut Down Relay
59	Black/Pink	Check Engine Light
60	NONE	NONE

POWER MODULE PIN IDENTIFICATION
12-PIN CONNECTOR FOR 1984 AND EARLIER MODELS

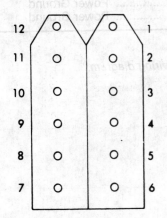

Pin No.	Wire Color	Wire Function
1	Violet/Yellow	Injector Input From Logic Module
2	Black/Lt. Blue	Sensor "Noise" Ground
3	NONE	NONE
4	NONE	NONE
5	NONE	NONE
6	NONE	NONE
7	Gray	Distributor Signal Input
8	NONE	NONE
9	Yellow/Red	NONE
10	Yellow	Advance Input From Logic Module
11	Violet	NONE
12	Orange	8-Volt Output For Hall Pick-Up & Logic Module

POWER MODULE PIN IDENTIFICATION
12-PIN CONNECTOR FOR 1985-87 MODELS

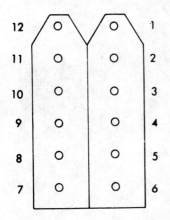

Pin No.	Wire Color	Wire Function
1	Violet/Yellow	Injector Control Unit
2	Black/Lt. Blue	Signal Ground
3	Red/Black	Battery Temp. Sensor
4	NONE	NONE
5	Dk. Blue/Yellow	ASD Relay Control
6	Red/White	Battery Sense & Standby
7	NONE	NONE
8	NONE	NONE
9	NONE	NONE
10	Yellow	Dwell Control
11	Dk. Green/Orange	Alternator Field Control
12	Orange	8-Volt Output

POWER MODULE PIN IDENTIFICATION
10-PIN CONNECTOR FOR 1984 AND EARLIER MODELS

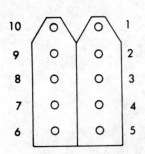

Pin No.	Wire Color	Wire Function
1	Black/Yellow	Coil Ground
2	Dk. Blue	J2 Feed From Ignition Switch
3	Dk. Blue	Fused J2 Power
4	Tan	Pulse Width Signal To Injector
5	White	Injector Feedback Signal
6	NONE	NONE
7	Dk. Blue/Yellow	ASD Relay Ground
8	Dk. Green	Spark & Fuel Power Supply
9	Black	Power Module Ground
10	Black	Back-Up Ground For Power Module

POWER MODULE PIN IDENTIFICATION
10-PIN CONNECTOR FOR 1985-87 MODELS

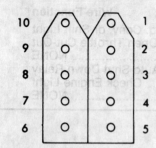

Pin No.	Wire Color	Wire Function
1	Black/Yellow	Coil Neg. Term.
2	Dk. Blue	J2 Feed From Ignition Switch
3	Dk. Blue	Fused J2 Power
4	Tan	Pulse Width Signal To Injector
5	White	Injector Feedback Signal
6	Dk. Green/Black	Switched Battery
7	White	Injector
8	Dk. Green	Alternator Field
9	Black	Power Ground
10	Black	Power Ground

NOTE

- For 1989 connector identification, refer to appropriate wiring diagram.

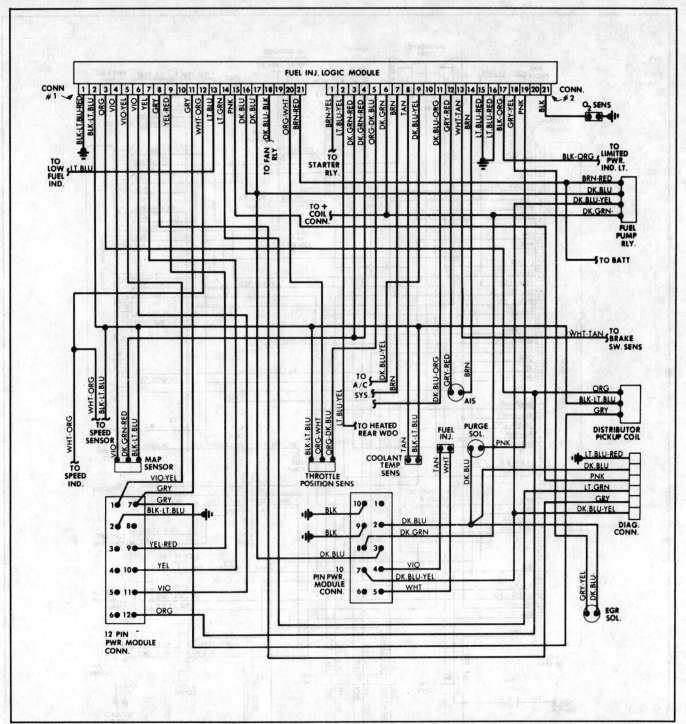

Fig. 14: 1984 & Earlier Throttle Body Fuel Injection Wiring Diagram

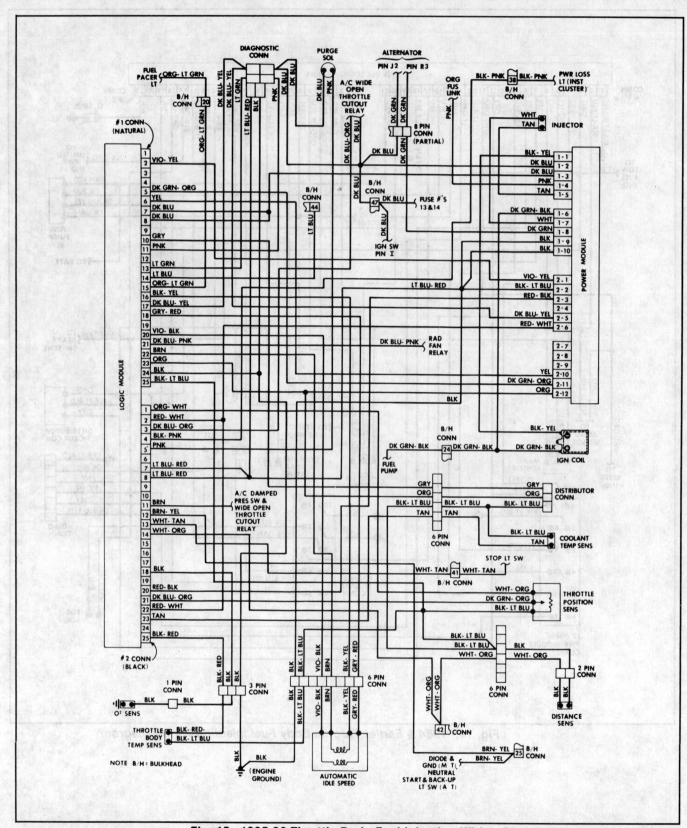

Fig. 15: 1985-86 Throttle Body Fuel Injection Wiring Diagram

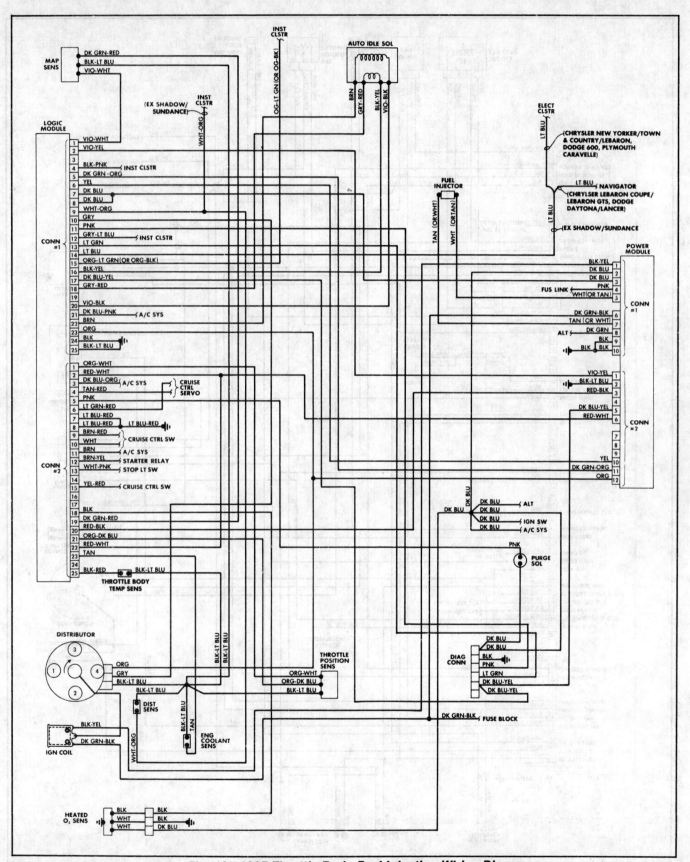

Fig. 16: 1987 Throttle Body Fuel Injection Wiring Diagram

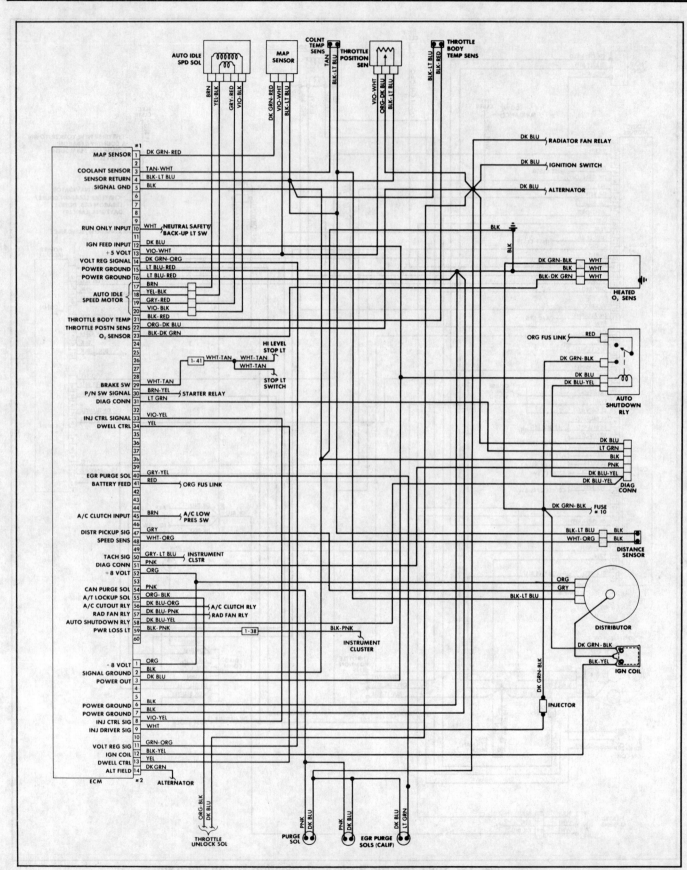

Fig. 17: 1988 Horizon & Omni Throttle Body Fuel Injection System Wiring Diagram

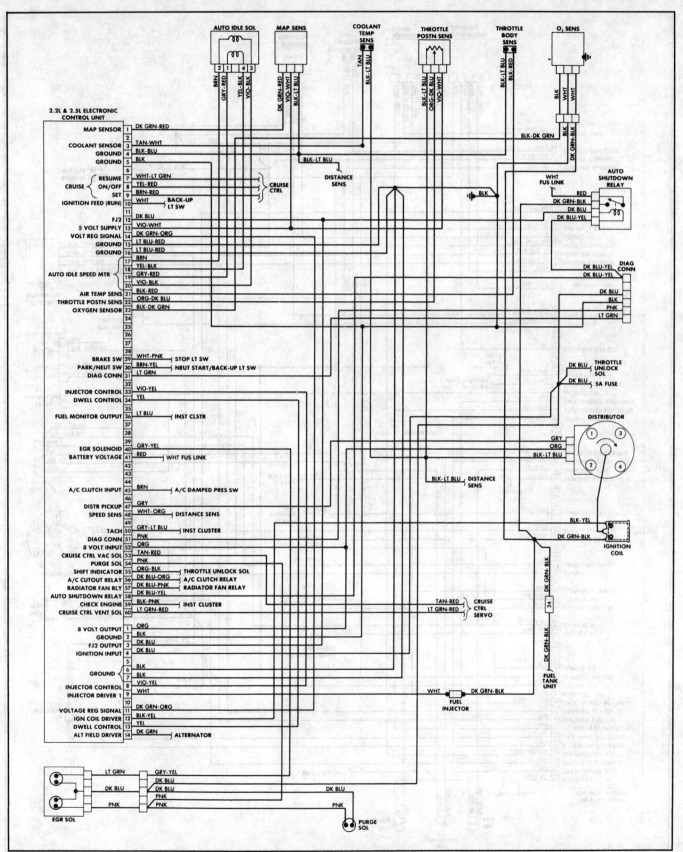

Fig. 18: 1988 Non-Turbo Throttle Body Fuel Injection System Wiring Diagram For FWD Models (Except Horizon & Omni)

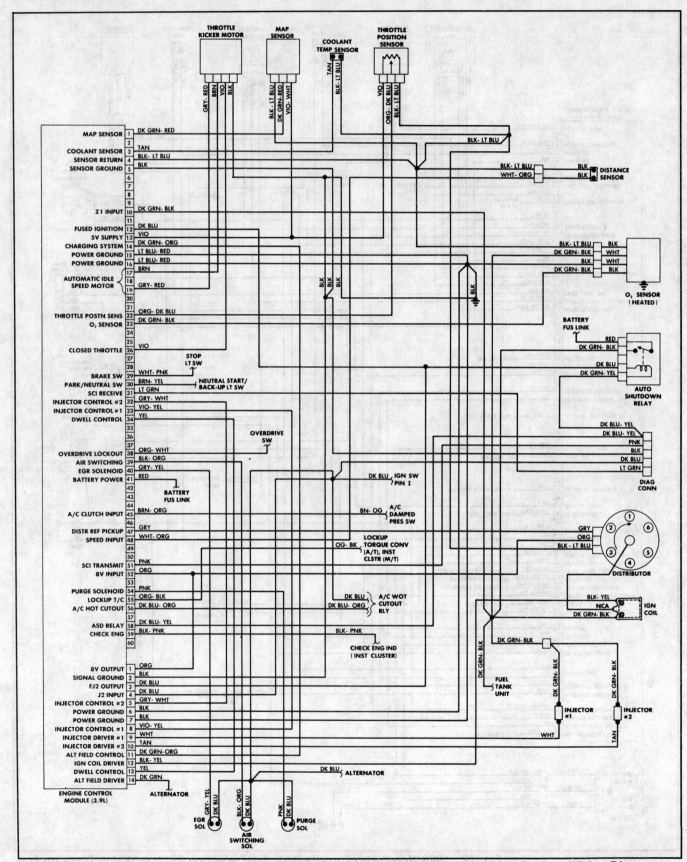

Fig. 19: 1988 Dakota 3.9L Throttle Body Fuel Injection System Wiring Diagram

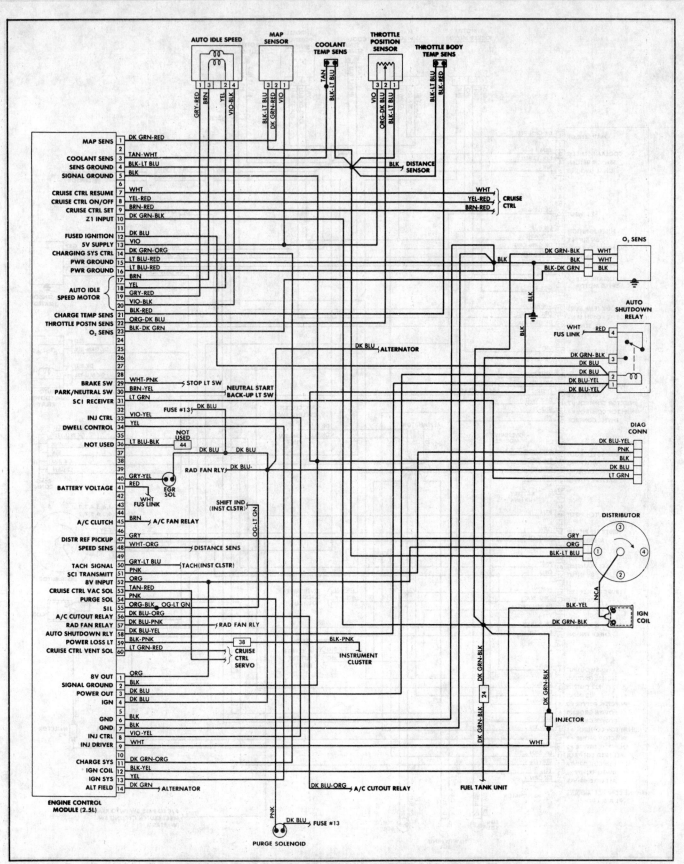

Fig. 20: 1988 FWD Van/Wagon 2.5L Throttle Body Fuel Injection System Wiring Diagram

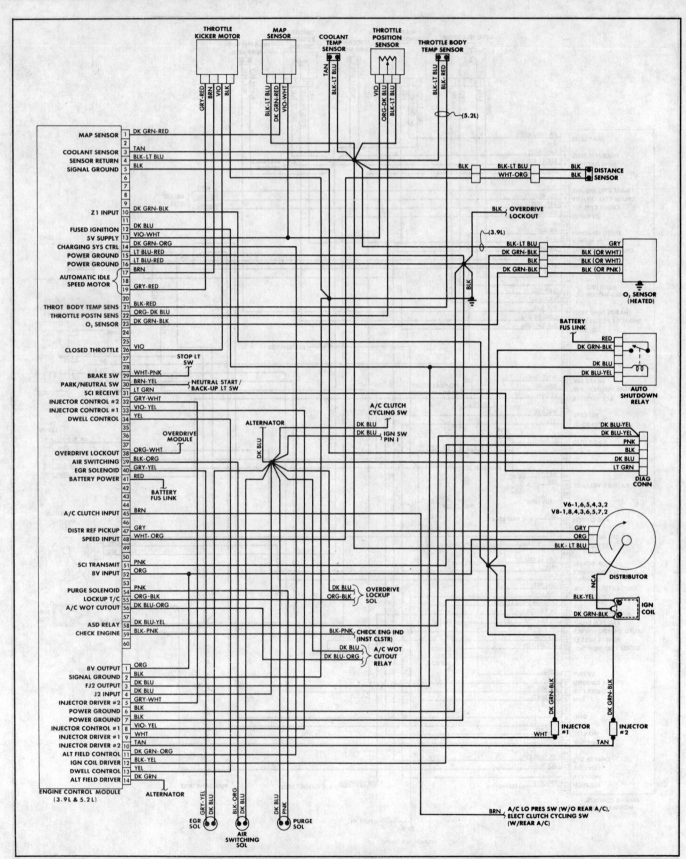

Fig. 21: 1988 RWD Van/Wagon 3.9L & 5.2L Throttle Body Fuel Injection System Wiring Diagram

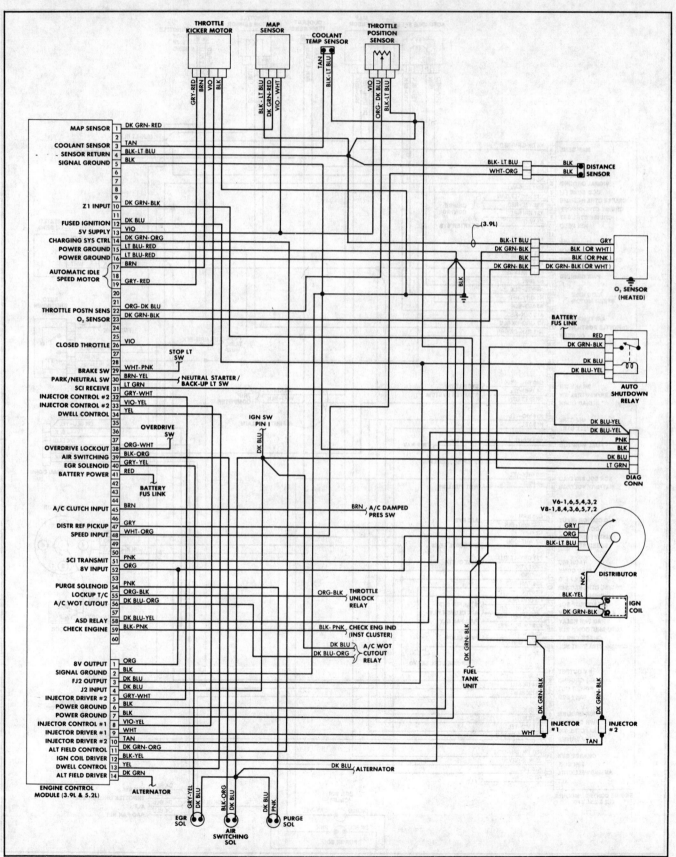

Fig. 22: 1988 RWD Pickup 3.9L & 5.2L Throttle Body Fuel Injection System Wiring Diagram

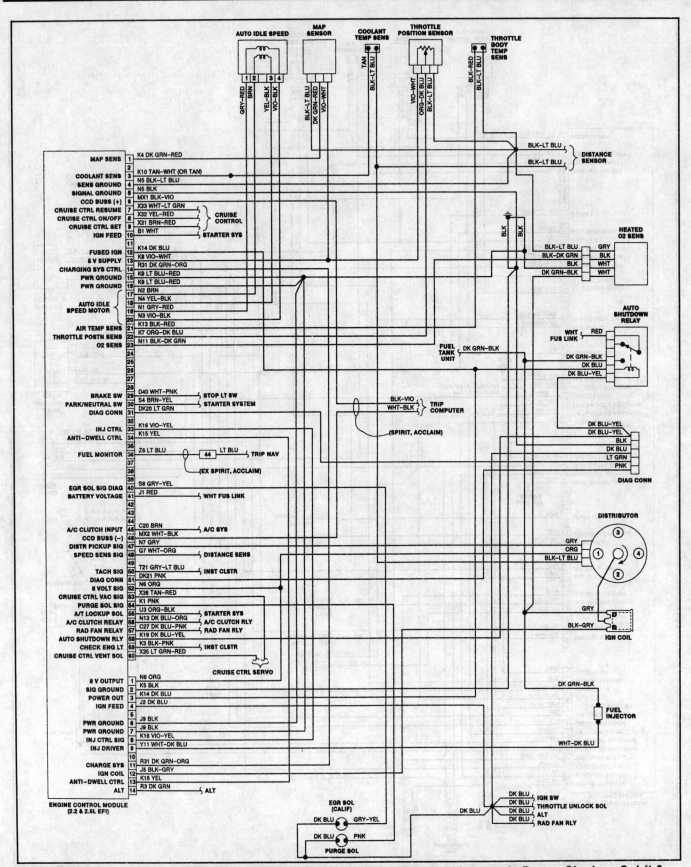

Fig. 23: 1989 2.2L & 2.5L Acclaim, Daytona, Lancer, LeBaron, Shadow, Spirit &
Sundance Throttle Body Fuel Injection System Wiring Diagram

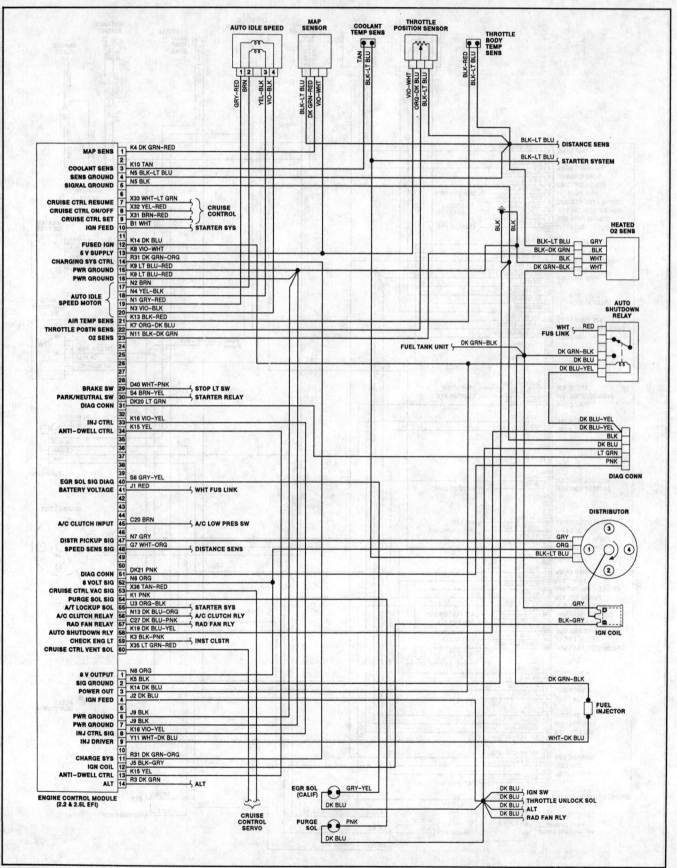

Fig. 24: 1989 2.2L & 2.5L Aries & Reliant Throttle Body Fuel Injection System Wiring Diagram

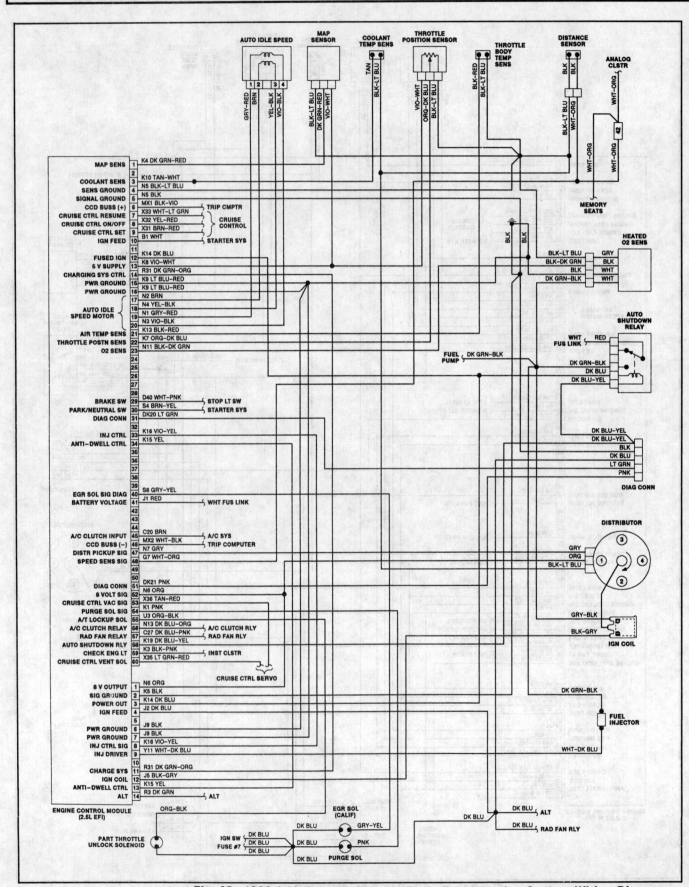

Fig. 25: 1989 2.5L Dynasty Throttle Body Fuel Injection System Wiring Diagram

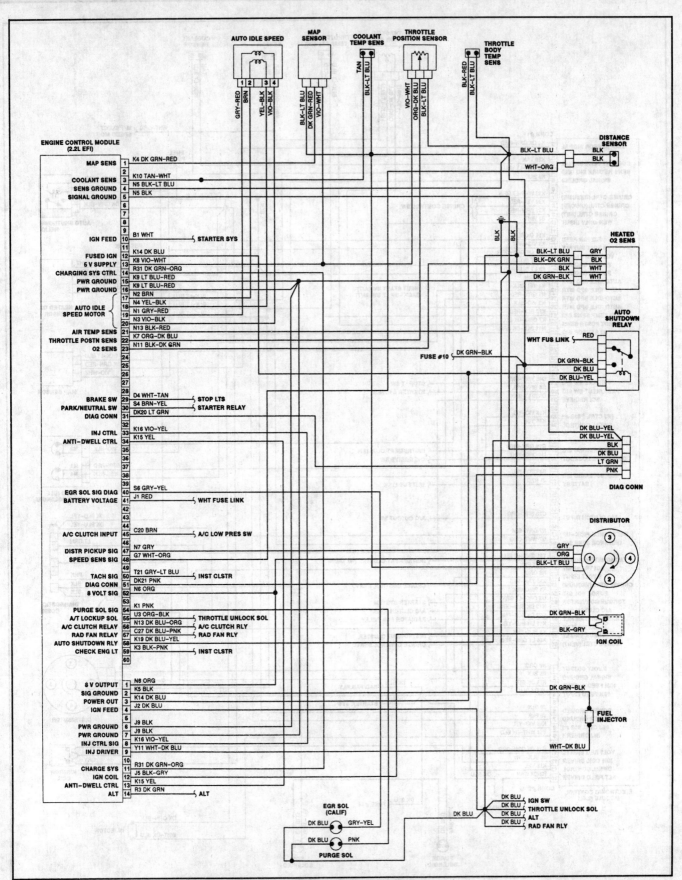

Fig. 26: 1989 2.2L Horizon & Omni Throttle Body Fuel Injection System Wiring Diagram

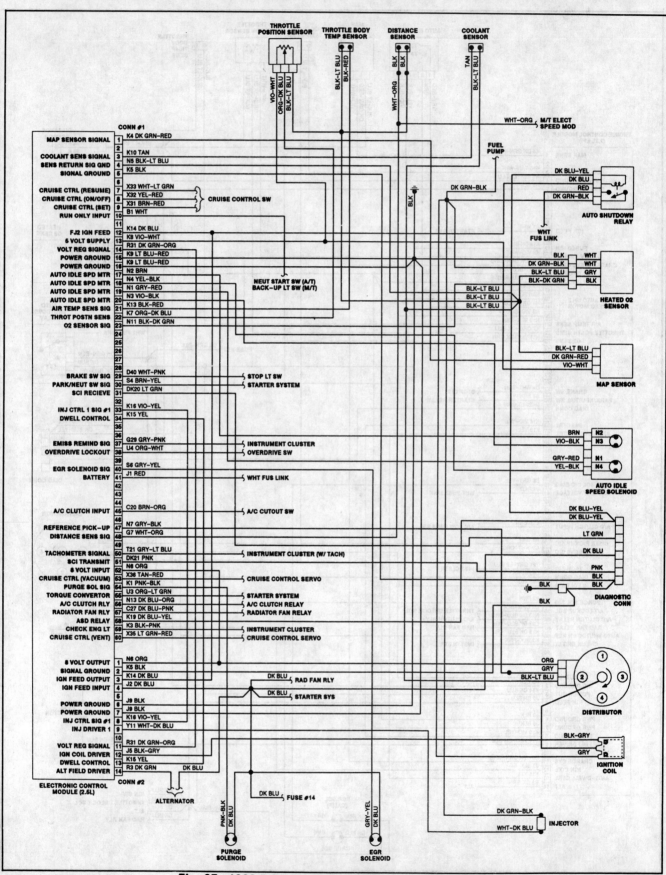

Fig. 27: 1989 2.5L Caravan & Voyager Throttle Body Fuel Injection System Wiring Diagram

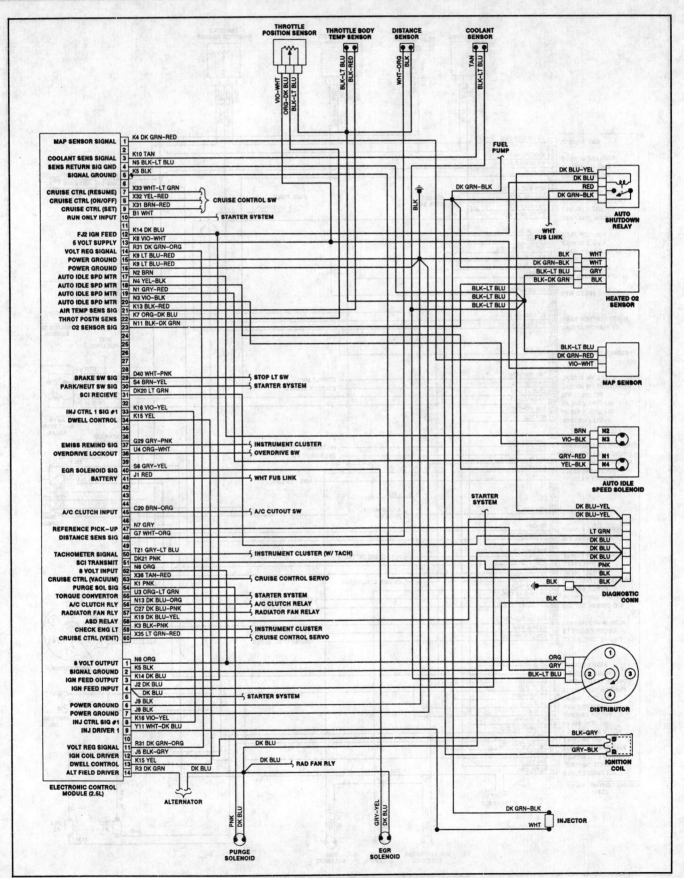

Fig. 28: 1989 Dakota 2.5L TBI Wiring Diagram

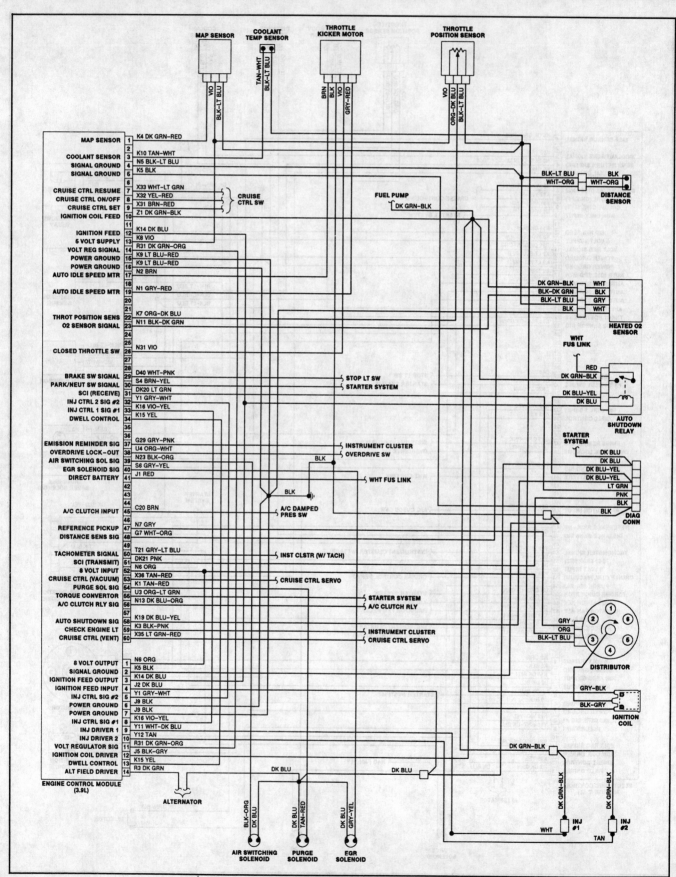

Fig. 29: 1989 3.9L Dakota Throttle Body Fuel Injection System Wiring Diagram

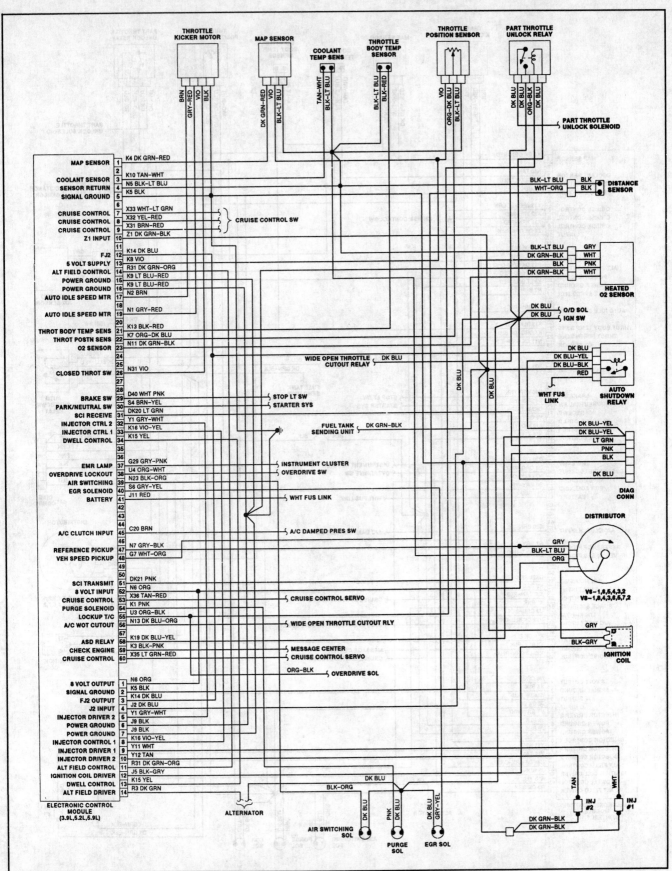

Fig. 30: 1989 3.9L, 5.2L & 5.9L Pickup Throttle Body Fuel Injection System Wiring Diagram

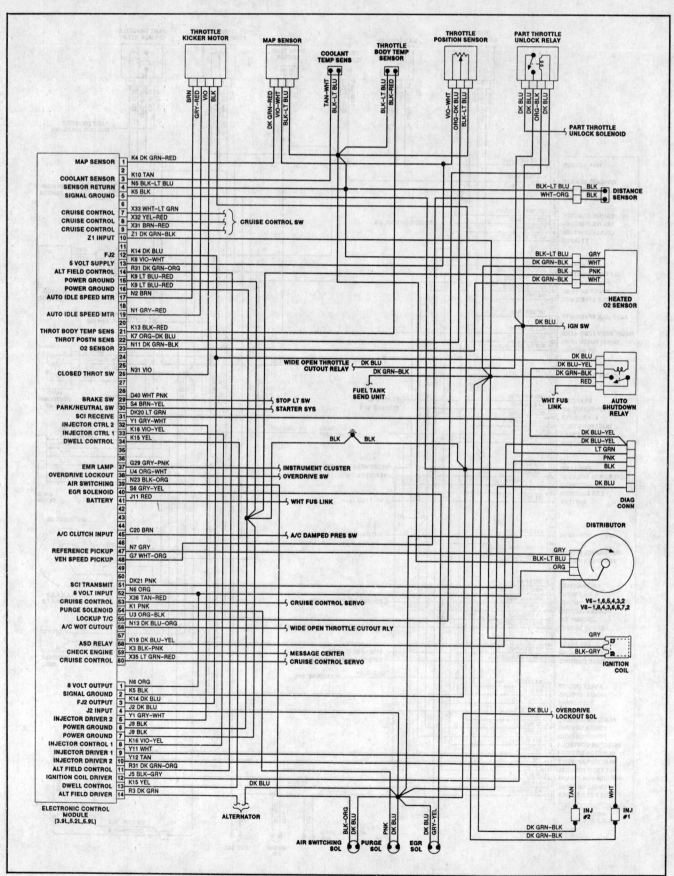

Fig. 31: 1989 3.9L, 5.2L & 5.9L Van Throttle Body Fuel Injection System Wiring Diagram

Premier 3.0L

DESCRIPTION

The 3.0L Eagle Premier Multi-Point Fuel Injection (MPFI) system is a pressure-speed type system. The amount of fuel injected into the engine is dependent upon intake manifold pressure and engine speed.

The fuel delivery system includes an in-tank fuel pump, fuel filter, fuel injectors and fuel pressure regulator. The electronic control system consists of an electronic control unit (ECU), input sensors and engine controls which receive output commands.

OPERATION

ELECTRONIC CONTROL UNIT (ECU)

The ECU is located under the instrument panel on the passenger side of the vehicle. Inputs from various engine sensors to ECU are used to determine engine operating conditions and needs.

FUEL INJECTORS

The multi-point fuel injection system has a fuel injector for each cylinder. When voltage is supplied to injector solenoid, armature and plunger move upward against spring. Check ball above injector nozzle moves off seat and opens small orifice at end of injector, resulting in fine spray of fuel.

FUEL PRESSURE REGULATOR

The fuel pressure regulator is located in-line with the fuel return tube. *See Fig. 1.*

Courtesy of Chrysler Motors.

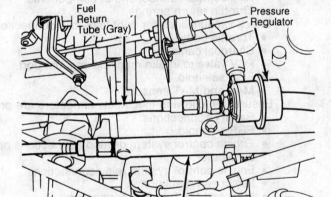

Fuel Return Tube (Gray)

Pressure Regulator

Fuel Rail

Fig. 1: Fuel Pressure Regulator Location

COOLANT TEMPERATURE SENSOR (CTS)

The coolant temperature sensor is located in the thermostat housing, above water pump. *See Fig. 2.*

Courtesy of Chrysler Motors.

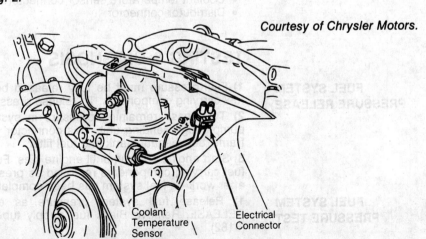

Coolant Temperature Sensor

Electrical Connector

Fig. 2: Coolant Temperature Sensor (CTS)

OXYGEN SENSOR

The amount of oxygen in exhaust gases varies according to the air/fuel ratio of the intake charge. The oxygen sensor, located in the exhaust pipe, detects this content and transmits a low-votage signal to the ECU.

KNOCK SENSOR

The knock sensor is located above right motor mount and provides input to ECU which retards ignition advance to eliminate knock.

MANIFOLD ABSOLUTE PRESSURE (MAP) SENSOR

The MAP sensor detects absolute pressure in the intake manifold as well as ambient atmospheric pressure. This information is supplied to the ECU, through voltage signals, as an indication of engine load. The sensor is located on the firewall, next to the brake power booster.

MANIFOLD AIR/FUEL TEMPERATURE (MAT) SENSOR

The MAT sensor provides signal to ECU that changes depending upon temperature of air/fuel mixture in intake manifold. During high temperature conditions, ECU will compensate for changes in density of air.

ENGINE SPEED SENSOR

The engine speed sensor is attached to bellhousing. It senses and counts teeth on flywheel gear ring as they pass during engine operation. Signal from speed sensor provides ECU with engine speed and crankshaft angle.

TROUBLE SHOOTING

PRELIMINARY CHECKS

Most driveability problems in the throttle body fuel injection system result from faulty wiring and/or leaking vacuum hoses and hose connections. To avoid unnecessary component testing, visual check should be performed before beginning trouble shooting procedures. A preliminary visual check should include:
- Air ducts at air cleaner and throttle body.
- Air filter element.
- All electrical connections at components.
- Throttle return springs.

Check vacuum lines for tight leak-free connections in these areas:
- Throttle body.
- Charcoal canister.
- PCV valve to intake manifold vacuum port.
- EGR solenoid.
- MAP and MAT sensors.

Ensure the following electrical connectors are properly attached:
- Battery connections.
- EGR solenoid.
- Engine control system ground connections on side of frame rail between battery and air cleaner.
- Engine control system relay connections.
- Idle speed regulator.
- MAP and MAT sensor connectors.
- Speed sensor connector.
- Radiator fan relay connector.
- TPS connector.
- Fuel injector connectors.
- Oxygen sensor connector.
- Coolant temperature sensor connector.
- Distributor connector.

TESTING & DIAGNOSIS

FUEL SYSTEM PRESSURE RELEASE

1) Fuel pressure must be fully released before opening fuel system or removing any fuel carrying components. To release pressure in tank, open fuel tank cap slowly.

2) To release remaining pressure in system, disconnect 4-pin connector from fuel pump/gauge sending unit. This connector is located under vehicle, attached to tab on frame, between fuel tank and fuel filter.

3) Start engine and run until engine dies. Fuel injection system can now be opened and fuel carrying components removed as pressure is fully released. Reconnect connector after work on fuel system has been completed.

FUEL SYSTEM PRESSURE TEST

1) Release fuel system pressure as described in FUEL SYSTEM PRESSURE RELEASE. Remove Black fuel supply tube from fuel rail using Fuel Line Disconnect (6182).

EAGLE   MULTI-POINT FUEL INJECTION

2) Install Fuel Tube Adapter (6175) between Black fuel supply line and fuel rail. Attach 0-60 psi Fuel Pressure Gauge (5069) on port of fuel tube adapter. Start engine and check fuel pressure. Fuel pressure should be 28-30 psi (19.3-20.7 kg/cm²).

SYSTEM DIAGNOSIS

The self-diagnostic capabilities of this system, if properly utilized, can greatly simplify testing. The ECU has been programmed to monitor several different circuits of the engine control system. If a problem is sensed with a monitored circuit, a fault code is stored in the ECU.

The setting of a specific fault code is the result of a particular system failure, NOT the reason for that failure, such as failure of a specific component. The existence of a particular code denotes the probable area of the malfunction, not necessarily the failed component itself.

RETRIEVING FAULT CODES

The use of a DRB-II diagnostic scan tool is need to retrieve fault codes. Fualt codes indicate a problem area the ECU has denoted.

FAULT CODES

FAULT CODE IDENTIFICATION

Fault Code	Fault Condition
1000	Ignition Line Low
1001	Ignition Line High
1004	Battery Voltage Low
1005	Sensor Ground Line Out Of Limits
1010	Diagnostic Enable Line Low
1011	Diagnostic Enable Line High
1012	Ignition Control Module Line Low
1013	Ignition Control Module Line High
1014	Fuel Pump Line Low
1015	Fuel Pump Line High
1016	MAT Sensor Line Low
1017	MAT Sensor Line High
1018	No Serial Data From ECU
1021	Engine Failed To Start Due To Mechanical, Fuel or Ignition Problem
1022	Start Line Low
1024	ECU Does Not See Start Signal
1027	ECU Sees Wide-Open Throttle (WOT)
1028	ECU Does Not See Wide-Open Throttle (WOT)
1031	ECU Sees Closed Throttle
1032	ECU Does Not See Closed Throttle
1033	Idle Speed Increase Line Low
1034	Idle Speed Increase Line High
1035	Idle Speed Decrease Line Low
1036	Idle Speed Decrease Line High
1037	TPS Reads Low
1038	Park/Neutral Line High
1040	Latched B+ Line Low
1041	Latched B+ Line High
1042	No Latched B+ .5 Volt Drop
1047	Wrong ECU
1048	Manual Vehicle Equipped With Automatic ECU
1049	Automatic Vehicle Equipped With Manual ECU
1050	Idle Speed Less Than 500 RPM
1051	Idle Speed Greater Than 2000 RPM
1052	MAP Sensor Out Of Limits
1053	Change In MAP Reading Out Of Limits
1054	Coolant Temperature Sensor Line Low
1055	Coolant Temperature Sensor Line High
1056	Inactive Coolant Temperature Sensor
1057	Knock Sensor Shorted
1058	Knock Valve Out Of Limits
1059	A/C Request Line Low
1060	A/C Request Line High
1061	A/C Select Line Low
1062	A/C Select Line High
1063	A/C Clutch Line Low
1064	A/C Clutch Line High

FAULT CODE IDENTIFICATION (Cont.)

Fault Code	Fault Condition
1065	O_2 Sensor Reads Rich
1066	O_2 Sensor Reads Lean
1067	Latch Relay Line Low
1068	Latch Relay Line High
1069	No Tach
1070	A/C Cut-Out Line Low
1071	A/C Cut-Out Line High
1073	ECU Does Not See Crankshaft Position/Engine Speed Sensor Signal
1200	ECU Defective
1202	Injector Shorted To Ground
1209	Injector Open
1218	No Voltage At ECU From Power Latch Relay
1219	No Voltage At ECU From Shift Light
1220	No Voltage At ECU From EGR Solenoid
1221	No Injector Voltage
1222	Ignition Control Module Not Grounded
1223	No ECU Tests Run

IDLE SPEED REGULATOR

1) Unplug idle speed regulator connector. Connect Motor Tester (7088) and Adapter (7195) to idle speed regulator connector. Connect tester to battery. Start engine. Place tester switch in "RETRACT" position.

2) Idle speed regulator should retract and engine idle speed should decrease. Place tester switch in "EXTEND" position. Regulator should extend and engine speed should increase. If regulator does not retract and extand, and idle speed does not change, disconnect air hoses. Watch for valve to open and close as idle speed regulator is moved. If valve does not move, replace regulator.

Courtesy of Chrysler Motors.

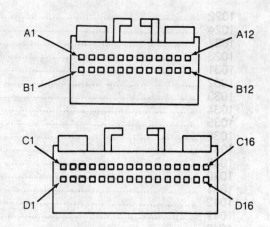

Fig. 3: ECU Connector Terminal Identification

COOLANT TEMPERATURE SENSOR (CTS)

Disconnect electrical connector from coolant temperature sensor. Using DVOM, check the resistance between ECU connector terminal "C10" to sensor connector and ECU connector terminal "D3" to sensor connector. If resistance is more than one ohm, repair wiring harness. Measure resistance between CTS terminals. See CTS/MAT SENSOR RESISTANCE table.

CTS/MAT SENSOR RESISTANCE [1]

°F (°C)	Ohms
212 (100)	185
160 (70)	450
100 (38)	1600
70 (20)	3400
40 (4)	7500
20 (-7)	13,500
0 (-18)	25,500
-40 (-40)	100,700

[1] – Values given are approximate.

MAP SENSOR

1) Ensure MAP sensor vacuum hose connections are secure. Connect a voltmeter between MAP sensor terminal "B" (marked on sensor connector) and ground. Turn ignition on. Voltmeter reading should be 4-5 volts. Start engine and warm to normal operating temperature. Voltmeter reading should be 1.5-2.1 volts.

2) Check ECU terminal "C6" for same voltmeter reading. If voltage readings vary between MAP sensor and ECU, check wiring. Turn ignition off. Connect a voltmeter between MAP sensor terminal "C" and ground. Turn ignition on. Voltage should be 4.5-5.5 volts. Also check voltage at ECU terminal "C14". Turn ignition off. Using an ohmmeter, check for continuity between MAP sensor terminal "A" and ECU terminal "D3". If there is not continuity, repair wiring as necessary.

MAT SENSOR

Unplug connector at MAT sensor. With engine at normal operating temperature, resistance between MAT sensor terminals should be as specified. See CTS/MAT SENSOR RESISTANCE table. Also measure resistance between ECU terminals "C8" and "D3", and MAT sensor connector. If resistance between ECU and connector is greater than one ohm, repair wiring.

THROTTLE POSITION SENSOR (TPS)

1) Turn ignition off. Unplug TPS connector. Turn ignition on. Using a DVOM, measure voltage between connector terminals "B" and "C". See Fig. 4. If voltage is not 5 volts, go to step **3)**.

2) If voltage is 5 volts, turn ignition off. Reconnect TPS connector. Turn ignition on. Measure voltage between TPS terminals "A" and "B" with throttle wide open. If voltage is not 4.6-4.7 volts, adjust TPS. If specified voltage cannot be obtained, replace TPS.

3) Turn ignition off. Using an ohmmeter, chewck for continuity between ECU connector terminal "D3" and TPS harness terminal "B". If continuity is not present, go to next step. If there is continuity, repair TPS wiring ground circuit. Check for continuity between ECU terminal "C7" and TPS harness terminal "A". If there is continuity, replace ECU and retest. If there is not continuity, repair wiring and retest.

Courtesy of Chrysler Motors.

IGNITION MODULE (C506),
OXYGEN SENSOR (C502),
& TPS SENSOR (C514)

Fig. 4: TPS Connector Terminal Identification

ADJUSTMENTS

THROTTLE POSITION SENSOR (TPS)

Turn ignition on, engine off. Do not unplug TPS connector. Insert DVOM leads through back of connector so negative lead is at terminal "B" and positive lead is at terminal "C" (connector is marked). At wide open throttle, voltage should be 5 volts. Measure voltage with positive lead at terminal "A" and throttle closed. Voltage should be .4 volt. If voltage at either point is not as specified, loosen TPS retaining screws and pivot sensor until proper readings are obtained.

HOT (SLOW IDLE SPEED

Hot (slow) idle speed is non-adjustable.

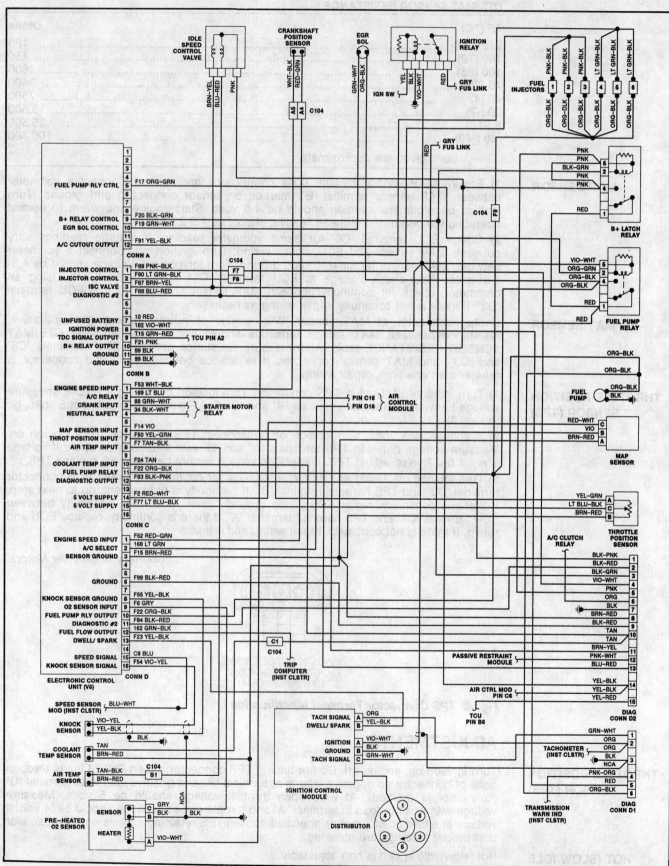

Fig. 5: Eagle Premier Multi-Point Fuel Injection Wiring Diagram

Premier 2.5L

DESCRIPTION

The 2.5L Eagle Premier Throttle Body Injection (TBI) system injects a metered amount of fuel above throttle blade inside throttle body. The throttle body system has a fuel delivery and electronic control system.

The fuel delivery system includes in-tank fuel pump, fuel filter, fuel injector and fuel pressure regulator. The electronic control system consists of Electronic Control Unit (ECU), input sensors and engine controls which receive output commands.

OPERATION

ELECTRONIC CONTROL UNIT (ECU)

The ECU is located under the instrument panel on the passenger side of the vehicle. Inputs from various engine sensors to ECU are used to determine engine operating conditions and needs.

FUEL INJECTOR

A single fuel injector is mounted in the throttle body. When voltage is supplied to injector solenoid, armature and plunger move upward against spring. Check ball above injector nozzle moves off seat and opens small orifice at end of injector, resulting in fine spray of fuel.

FUEL PRESSURE REGULATOR

The fuel pressure regulator is an integral part of throttle body. The pressure regulator has a spring chamber that is vented to same pressure as tip of injector. Because differential pressure between injector nozzle and spring chamber is same, only length of time that injector is energized controls volume of fuel injected.

FUEL PUMP

An electric roller type fuel pump is located in fuel tank. An integral check valve is used to maintain pressure in fuel delivery system after pump stops running. Fuel pump operation is controlled by ECU.

IDLE SPEED ACUATOR /IDLE SPEED CONTROL (ISA/ISC) MOTOR

The ISA/ISC motor acts as a moveable idle stop to change throttle stop angle. Both engine idle speed and deceleration throttle stop are set by ISC. ECU sends varying voltage outputs to control ISA/ISC motor, depending upon engine operating condition.

OXYGEN SENSOR

The oxygen sensor is located in exhaust pipe. ECU receives sensor voltage signal which varies with oxygen content in exhaust gas. Signal is used by ECU as reference for setting air/fuel mixture ratio.

MANIFOLD AIR TEMPERATURE (MAT) SENSOR

The MAT sensor provides signal to ECU that changes depending upon temperature of air/fuel mixture in intake manifold. During high temperature conditions, ECU will compensate for changes in density of air.

MANIFOLD ABSOLUTE PRESSURE (MAP) SENSOR

The MAP sensor measures absolute pressure in intake manifold. Both mixture density and ambient barometric pressure are supplied to the ECU by MAP sensor. MAP sensor is mounted in plenum chamber at middle of firewall in engine compartment. MAP sensor receives manifold pressure information through vacuum line from throttle body. See Fig. 1.

Courtesy of Chrysler Motors.

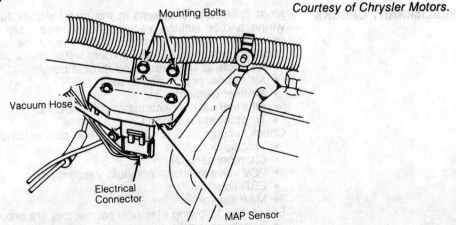

Fig. 1: Manifold Absolute Pressure (MAP) Sensor

203

COOLANT TEMPERATURE SENSOR (CTS)

The CTS is installed in thermostat housing to provide coolant temperature input signal for ECU. *See Fig. 2.*

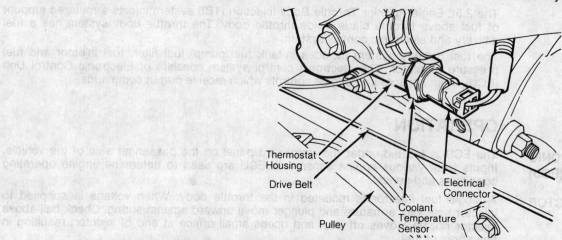

Courtesy of Chrysler Motors.

Thermostat Housing

Drive Belt

Pulley

Electrical Connector

Coolant Temperature Sensor

Fig. 2: Coolant Temperature Sensor (CTS)

ENGINE SPEED SENSOR

The engine speed sensor is attached to bellhousing. It senses and counts teeth on flywheel gear ring as they pass during engine operation. Signal from speed sensor provides ECU with engine speed and crankshaft angle. *See Fig. 3.*

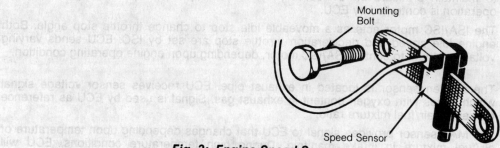

Courtesy of Chrysler Motors.

Mounting Bolt

Speed Sensor

Fig. 3: Engine Speed Sensor

KNOCK SENSOR

The knock sensor is located above right motor mount and provides input to ECU which retards ignition timing to eliminate knock.

TROUBLE SHOOTING

PRELIMINARY CHECKS

Most driveability problems in the throttle body fuel injection system result from faulty wiring and/or leaking vacuum hoses and hose connections. To avoid unnecessary component testing, visual check should be performed before beginning trouble shooting procedures. A preliminary visual check should include:
- Air ducts at air cleaner and throttle body.
- Air cleaner pre-heater hose.
- Air filter element.
- All electrical connections at components.
- Throttle return springs.

Check vacuum lines for tight leak-free connections in these areas:
- Throttle body.
- Charcoal canister.
- PCV valve to intake manifold vacuum port.
- EGR solenoid.
- MAP sensor.

Ensure the following electrical connectors are properly attached:
- Battery connections.
- EGR solenoid.

- Engine control system ground connections on side of frame rail between battery and air cleaner.
- Engine control system relays connections.
- Idle Speed Control (ISC) motor.
- MAP sensor connector.
- Speed sensor connector.
- Throttle body temperature sensor connector.
- Radiator fan relay connector.
- TPS connector.
- Fuel injector connector.
- Oxygen sensor connector.
- Coolant temperature sensor connector.
- Distributor connector.

TESTING & DIAGNOSIS

FUEL SYSTEM PRESSURE RELEASE

1) Fuel pressure must be fully released before opening fuel system or removing any fuel carrying component. To release pressure in tank, open fuel tank cap slowly.

2) To release remaining pressure in system, disconnect 4-pin connector from fuel pump/gauge sending unit. This connector is located under vehicle, attached to tab on frame, between fuel tank and fuel filter.

3) Start engine and run until engine dies. Fuel injection system can now be opened and fuel carrying components removed as pressure is fully released. Reconnect connector after work on fuel system has been completed.

FUEL SYSTEM PRESSURE TEST & REGULATOR ADJUSTMENT

1) Allow engine to cool before performing pressure test. Release fuel system pressure as described in FUEL SYSTEM PRESSURE RELEASE. Remove air filter cover from throttle body. Slowly remove test port plug from side of pressure regulator, using shop towels to catch any spilled fuel. Test port plug is located near fuel return tube connection.

2) Install Fuel Pressure Test Adapter (6173) into test port. Connect Fuel Pressure Gauge (5069) and start engine. Check fuel pressure with engine at idle. Correct pressure reading should be 14-15 psi (.98-1.09 kg/cm²).

3) If pressure is not within specification, adjust fuel pressure regulator until proper pressure is obtained. Remove plug concealing fuel pressure regulator adjusting screw from nose of fuel pressure regulator using small drift. Tap lightly until plug pops out. With engine at idle speed, adjust screw until fuel pressure reading is 14-15 psi (.98-1.09 kg/cm²).

SYSTEM DIAGNOSIS

The self-diagnostic capabilities of this system, if properly utilized, can greatly simplify testing. The ECU has been programmed to monitor several different circuits of the engine control system. If a problem is sensed with a monitored circuit, a fault code is stored in the ECU.

The setting of a specific fault code is the result of a particular system failure, NOT the reason for that failure, such as failure of a specific component. The existence of a particular code denotes the probable area of the malfunction, not necessarily the failed component itself.

RETRIEVING FAULT CODES

The use of a DRB-II diagnostic scan tool is need to retrieve fault codes. Fualt codes indicate a problem area the ECU has denoted.

FAULT CODES

FAULT CODE IDENTIFICATION

Fault Code	Fault Condition
1000	Ignition Line Low
1001	Ignition Line High
1004	Battery Voltage Low
1005	Sensor Ground Line Out Of Limits
1010	Diagnostic Enable Line Low
1011	Diagnostic Enable Line High
1012	Ignition Control Module Line Low
1013	Ignition Control Module Line High
1014	Fuel Pump Line Low
1015	Fuel Pump Line High
1016	MAT Sensor Line Low
1017	MAT Sensor Line High

FAULT CODE IDENTIFICATION (Cont.)

Fault Code	Fault Condition
1018	No Serial Data From ECU
1021	Engine Failed To Start Due To Mechanical, Fuel or Ignition Problem
1022	Start Line Low
1024	ECU Does Not See Start Signal
1027	ECU Sees Wide-Open Throttle (WOT)
1028	ECU Does Not See Wide-Open Throttle (WOT)
1031	ECU Sees Closed Throttle
1032	ECU Does Not See Closed Throttle
1033	Idle Speed Increase Line Low
1034	Idle Speed Increase Line High
1035	Idle Speed Decrease Line Low
1036	Idle Speed Decrease Line High
1037	TPS Reads Low
1038	Park/Neutral Line High
1040	Latched B+ Line Low
1041	Latched B+ Line High
1042	No Latched B+ .5 Volt Drop
1047	Wrong ECU
1048	Manual Vehicle Equipped With Automatic ECU
1049	Automatic Vehicle Equipped With Manual ECU
1050	Idle Speed Less Than 500 RPM
1051	Idle Speed Greater Than 2000 RPM
1052	MAP Sensor Out Of Limits
1053	Change In MAP Reading Out Of Limits
1054	Coolant Temperature Sensor Line Low
1055	Coolant Temperature Sensor Line High
1056	Inactive Coolant Temperature Sensor
1057	Knock Sensor Shorted
1058	Knock Valve Out Of Limits
1059	A/C Request Line Low
1060	A/C Request Line High
1061	A/C Select Line Low
1062	A/C Select Line High
1063	A/C Clutch Line Low
1064	A/C Clutch Line High
1065	O_2 Sensor Reads Rich
1066	O_2 Sensor Reads Lean
1067	Latch Relay Line Low
1068	Latch Relay Line High
1069	No Tach
1070	A/C Cut-Out Line Low
1071	A/C Cut-Out Line High
1073	ECU Does Not See Crankshaft Position/ Engine Speed Sensor Signal
1200	ECU Defective
1202	Injector Shorted To Ground
1209	Injector Open
1218	No Voltage At ECU From Power Latch Relay
1219	No Voltage At ECU From Shift Light
1220	No Voltage At ECU From EGR Solenoid
1221	No Injector Voltage
1222	Ignition Control Module Not Grounded
1223	No ECU Tests Run

ISA/ISC ASSEMBLY

Closed Throttle Switch

1) Ensure all testing is done with ISA/ISC plunger fully extended (as it would be after engine shutdown). If it is necessary to extend ISA/ISC plunger with a test switch, an ISA/ISC motor failure can be suspected. Turn ignition on. Connect a voltmeter between terminals "A" and "B" of ISA/ISC motor. *See Fig. 4.*

2) Voltmeter reading should be zero at closed throttle and more than 2 volts with throttle partially open. If voltage is not present in either position, check for short circuit to ground in wiring harness or switch.

3) Check for an open circuit between ECU connector terminal "D5" and closed throttle switch. *See Fig. 5.* If voltage is always more than 2 volts, check for an open circuit between ECU and ISA/ISC connector.

Courtesy of Chrysler Motors.

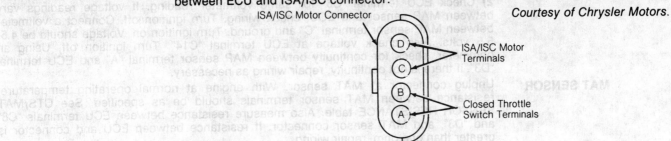

Fig. 4: ISA/ISC Connector Terminal Identification

ISA/ISC Motor

1) Unplug ISA/ISC motor connector. Connect Motor Tester (7088) and Adapter (7195) to ISA/ISC motor. For proper connection, it will be necessary to modify Idle Motor Tester (7088). Cut a groove in terminal "A" the same size and shape as grooves in terminals "B" and "D". Connect tester to battery. Start engine. Place tester switch in "RETRACT" position.

2) ISA/ISC motor should retract and engine idle speed should decrease. If motor does not retract and idle speed is not reduced, replace ISA/ISC motor. Place tester switch in "EXTEND" position. ISA/ISC motor should extend and engine speed should increase. If not, replace ISA/ISC motor. If ISA/ISC motor extends, motor is okay.

Courtesy of Chrysler Motors.

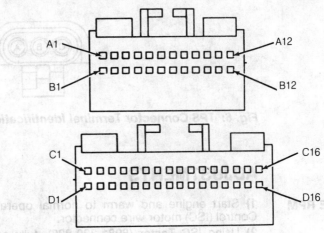

Fig. 5: ECU Connector Terminal Identification

COOLANT TEMPERATURE SENSOR (CTS)

Disconnect electrical connector from coolant temperature sensor. Using DVOM, check the resistance between ECU connector terminal "C10" to sensor connector and ECU connector terminal "D3" to sensor connector. If resistance is more than one ohm, repair wiring harness. Measure resistance between CTS terminals. See CTS/MAT SENSOR RESISTANCE table.

CTS/MAT SENSOR RESISTANCE [1]

°F (°C)	Ohms
212 (100)	185
160 (70)	450
100 (38)	1600
70 (20)	3400
40 (4)	7500
20 (-7)	13,500
0 (-18)	25,500
-40 (-40)	100,700

[1] – Values given are approximate.

MAP SENSOR

1) Ensure MAP sensor vacuum hose connections are secure. Connect a voltmeter between MAP sensor terminal "B" (marked on sensor connector) and ground. Turn ignition on. Voltmeter reading should be 4-5 volts. Start engine and warm to normal operating temperature. Voltmeter reading should be 1.5-2.1 volts.

2) Check ECU terminal "C6" for same voltmeter reading. If voltage readings vary between MAP sensor and ECU, check wiring. Turn ignition off. Connect a voltmeter between MAP sensor terminal "C" and ground. Turn ignition on. Voltage should be 4.5-5.5 volts. Also check voltage at ECU terminal "C14". Turn ignition off. Using an ohmmeter, check for continuity between MAP sensor terminal "A" and ECU terminal "D3". If there is not continuity, repair wiring as necessary.

MAT SENSOR

Unplug connector at MAT sensor. With engine at normal operating temperature, resistance between MAT sensor terminals should be as specified. See CTS/MAT SENSOR RESISTANCE table. Also measure resistance between ECU terminals "C8" and "D3", and MAT sensor connector. If resistance between ECU and connector is greater than one ohm, repair wiring.

THROTTLE POSITION SENSOR (TPS)

1) Turn ignition off. Unplug TPS connector. Turn ignition on. Using a DVOM, measure voltage between connector terminals "B" and "C". *See Fig. 6*. If voltage is not 5 volts, go to step **3)**.

2) If voltage is 5 volts, turn ignition off. Reconnect TPS connector. Turn ignition on. Measure voltage between TPS terminals "A" and "B" with throttle wide open. If voltage is not 4.6-4.7 volts, adjust TPS. If specified voltage cannot be obtained, replace TPS.

3) Turn ignition off. Using an ohmmeter, check for continuity between ECU connector terminal "D3" and TPS harness terminal "B". If continuity is not present, go to next step. If there is continuity, repair TPS wiring ground circuit. Check for continuity between ECU terminal "C7" and TPS harness terminal "A". If there is continuity, replace ECU and retest. If there is not continuity, repair wiring and retest.

Courtesy of Chrysler Motors.

Fig. 6: TPS Connector Terminal Identification

ADJUSTMENTS

HOT (SLOW) IDLE RPM

1) Start engine and warm to normal operating temperature. Disconnect Idle Speed Control (ISC) motor wire connector.

2) Using ISC Tester (8981-320-828), fully extend ISC motor plunger. Adjust plunger screw until engine is running at 3500 RPM.

3) Remove ISC tester and connect idle speed motor electrical connector. Idle speed should automatically return to normal within a few seconds.

THROTTLE POSITION SENSOR (TPS)

1) The use of Diagnostic Tester (DRB-II) is preferred when adjusting throttle position sensor. If digital voltmeter is used, disconnect Idle Speed Control (ISC) motor electrical connector. Connect ISC Tester (7088) and retract ISC plunger until throttle lever contacts the idle stop screw and plunger does not contact the throttle lever.

2) Turn ignition on. On the TPS connector, insert negative lead of voltmeter in terminal "D" and positive lead to terminal "A". Reading should be 5 volts. Insert voltmeter positive lead to terminal "B" and open throttle to wide open position. Reading should be 4.6-4.7 volts. To adjust, loosen TPS retaining screws and pivot sensor to obtain correct reading.

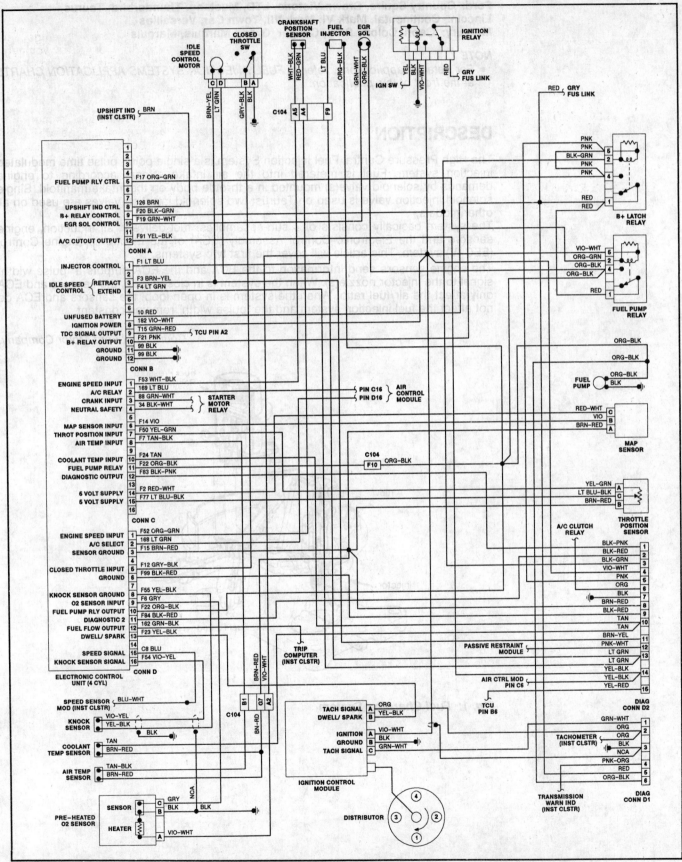

Fig. 7: Eagle Premier Throttle Body Fuel Injection Wiring Diagram

Ford: Country Squire, Crown Victoria, LTD, Mustang, Thunderbird, Taurus
Lincoln: Continental, Mark VI, Mark VII, Town Car, Versailles
Mercury: Capri, Colony Park, Cougar, Grand Marquis, Marquis

NOTE
- *For detailed applications, refer to FUEL INJECTION SYSTEMS APPLICATION CHARTS at the front of this publication.*

DESCRIPTION

The High Pressure Central Fuel Injection System is a single point, pulse time modulated injection system. Fuel is metered into the air intake stream according to engine demands by solenoid valve(s) mounted in a throttle body on the intake manifold. Single solenoid injection valve is used on Taurus, two solenoid injection valves are used on all other models,

The system basically consists of 4 sub-assemblies: fuel delivery, air induction, engine sensors and the Electronic Control Assembly (ECA) of the Electronic Engine Control (EEC-IV) system. This article will cover the first two systems.

The engine sensors send information to the ECA and the ECA outputs a "pulse width" signal to the injector nozzle(s). When the system is in closed loop, the sensors and ECA only affect the air/fuel ratio. When the system is in open loop, the sensors and ECA do not affect the fuel injection system and the "pulse width" remains constant.

Courtesy of Ford Motor Company.

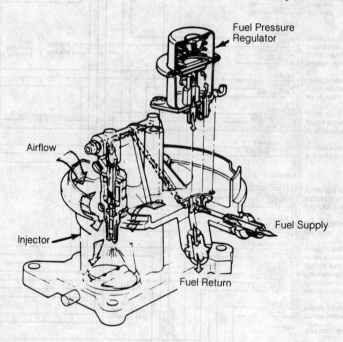

Fig. 1: Fuel Charging System

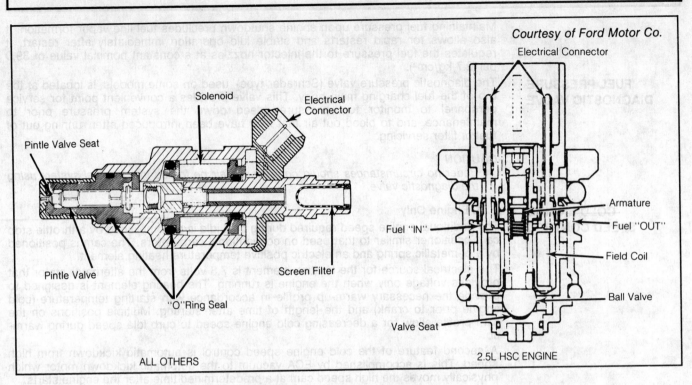

Courtesy of Ford Motor Co.

Fig. 2: Cutaway View of Injector Valve

OPERATION

FUEL DELIVERY

Fuel is supplied from the fuel tank by a high pressure electric fuel pump. Pump is mounted in the fuel tank. The fuel is filtered and sent to the air throttle body, where a regulator keeps the fuel delivery pressure at a constant 39.0 psi (2.7 kg/cm²).

The injector nozzle(s) are mounted vertically above the throttle plates. Taurus models are equipped with a single throttle plate. They are connected in parallel with the fuel pressure regulator. Excess fuel supplied by the pump, but not needed by the engine, is returned to the fuel tank by a steel fuel return line.

The fuel charging assembly is comprised of six individual components that perform the fuel and air metering function to the engine. The throttle body assembly mounts to the conventional carburetor pad of the intake manifold and provides for packaging of air control, fuel injector nozzles, fuel pressure regulator, fuel pressure diagnostic valve, engine idle speed control, and throttle position sensor.

AIR CONTROL

Airflow to the engine is controlled by a single butterfly valve on Taurus or two butterflys on all others, mounted in a two-piece, die cast aluminum housing called the throttle body. The butterfly valves are identical in configuration to the throttle plates of a conventional carburetor and are actuated by similar linkage and pedal cable arrangement.

FUEL INJECTOR NOZZLES

The fuel injector nozzles are mounted vertically above the throttle plates and are electro-mechanical devices that meter and atomize the fuel delivered to the engine. The injector valve bodies consist of a solenoid actuated pintle and needle valve assembly.

An electrical control signal "pulse width" from the ECA activates the solenoid, causing the pintle to move inward off its seat, thus allowing fuel to flow. The injector flow orifice is fixed and the fuel supply pressure is constant, therefore, fuel flow to the engine is controlled by how long the solenoid is energized.

FUEL PRESSURE REGULATOR

The pressure regulator mounts to the fuel charging main body near the rear of the air horn surface. The regulator is located strategically to avoid the affects of the supply line pressure drops. Also, it is not affected by back pressure in the return line to the tank.

A second function of the pressure regulator is to maintain fuel supply pressure upon engine and fuel pump shutdown. The regulator functions as a downstream check valve and traps the fuel between itself and the fuel pump.

Maintaining fuel pressure upon engine shutdown precludes fuel line vapor formation. It also allows for rapid restarts and stable idle operation immediately after restart. It regulates the fuel pressure to the injector nozzles at a constant nominal value of 39.0 psi (2.7 kg/cm²).

FUEL PRESSURE DIAGNOSTIC VALVE

The diagnostic pressure valve (Schrader type), used on some models, is located at the top of the fuel charging main body. This valve provides a convenient point for service personnel to monitor fuel pressure, bleed down the system pressure prior to maintenance, and to bleed out air that may have been introduced after running out of fuel or filter servicing.

CAUTION
- *Under no circumstances should compressed air be forced into the fuel system using the diagnostic valve.*

COLD ENGINE SPEED CONTROL

5.0L Engine Only

The additional engine speed required during cold idle is accomplished by a throttle stop cam positioner similar to that used on conventional carburetors. The cam is positioned by a bi-metallic spring and an electric positive temperature heating element.

The electrical source for the heating element is 7.3 volts from the alternator stator that provides voltage only when the engine is running. The heating element is designed to provide the necessary warm-up profile in accordance with starting temperature (cold engine prior to crank) and the length of time after starting. Multiple positions on the cam profile allow for a decreasing cold engine speed to curb idle speed during warm-up.

A second feature of the cold engine speed control is automatic kickdown from high speed. This is accomplished by ECA vacuum to the automatic kickdown motor which physically moves the high speed cam at a predetermined time after the engine starts.

IDLE SPEED CONTROL MOTOR

2.5L & 3.8L Engine Only

The Idle Speed Control (ISC) motor controls idle speed by modulating the throttle lever. This regulates airflow to maintain the desired engine RPM for both "Warm" engine and the additional engine speed required during "Cold" engine idle.

An Idle Tracking System (ITS), integral to the ISC motor, is utilized to determine when the throttle lever has contacted the actuator. This signals the need to control engine RPM. The ISC motor extends or retracts a linear shaft through a gear reduction system. The motor direction is determined by the polarity of the applied voltage.

THROTTLE POSITION SENSOR

This nonadjustable sensor is mounted to the throttle shaft on the fuel charging assembly. It is used to supply a voltage output proportional to the change in the throttle plate position. The sensor is used by the ECA to determine the operation mode (closed throttle, part throttle, and wide open throttle) for selection of the proper fuel mixture, spark and EGR at selected driving ranges.

INERTIA SWITCH

In the event of a collision, the electrical contacts in the inertia switch open and the fuel pump automatically shuts off. The fuel pump will shut off even if the engine does not stop running. The engine, however, will stop a few seconds after the fuel pump stops. It is not possible to restart the engine until the inertia switch is manually reset.

TROUBLE SHOOTING

PRELIMINARY CHECKS

Following systems and components must be in good condition and operating properly before beginning diagnosis of fuel injection system:
- All Support Systems and Wiring
- Battery Connections and Specific Gravity
- Ignition System
- Compression Pressure
- Fuel Supply System Pressure and Flow
- All Electrical Connections and Terminals
- Vacuum Line, Fuel Hose and Pipe Connections
- Air Cleaner and Air Ducts
- Engine Coolant Level
- Clear Exhaust System

TROUBLE SHOOTING

NOTE
- *Some vehicles may not include all components listed in this section.*

Engine Does Not Crank

Check for hydrostatic lock (liquid or carbon in cylinder) and repair as needed. Check for starting and charging system problems.

Engine Cranks But Does Not Start

1) Ensure fuel tank contains an adequate amount of fuel. Do not assume that fuel gauge is correct. Check fuel for dirt, water or other contamination.

2) Check ignition system for strong secondary current at spark plugs. If none exists or current is weak, repair ignition system problem before continuing with fuel injection diagnosis.

3) Check fuel lines and fittings for leaks. If no leaks are found, check fuel delivery system for proper operation, pressure and volume.

4) Reset inertia switch if necessary.

5) Check for defective or plugged injector, TPS stuck at wide open throttle or bad coolant temperature switch.

6) Check for inoperative Idle Speed Control (ISC) motor.

Hard to Start – Cold

1) Check for defective TPS or coolant temperature sensor.

2) Ensure ISC motor is operating properly.

3) Throttle body or intake manifold gaskets leaking. Replace as required.

Rough Idle – Cold

1) Cold enrichment cycle not functioning. Check fast idle system for proper operation and adjustment. Adjust or replace as required.

2) Air cleaner duct vacuum motor damaged or stuck open. Replace or service as required. Check for vacuum leaks. Check for leaking or defective injector or injector "O" ring seals. Check operation of injector.

Stall, Stumble, Hesitation – Hot or Cold

1) Cold enrichment system may not be functioning correctly. Check for proper operation and adjustment. Adjust or replace as required.

2) Fuel pump output low. Fuel filter possibly clogged. Fuel pump possibly defective. Clean or replaced as required after cause is determined.

3) Fuel injector internal filter clogged. Clean or replace injectors as required.

4) Air cleaner vacuum motor damaged. Service or replaced as required.

5) Check for defective TPS or coolant temperature sensor. Replace as required.

Hard Start – Hot

1) Cold enrichment system may not be functioning correctly. Check for proper operation and adjustment. Adjust or replace as required. Incorrect fuel pressure (too high). Check regulator and fuel return line.

2) Fuel pressure regulator valve and seat may be dirty or clogged. Clean and service regulator. Check fuel system bleed down after stopping engine.

3) Leaking or defective injector or injector "O" ring seal. Check operation of injector.

4) ISC motor not operating properly. Check ISC operation.

5) Throttle body or intake manifold gaskets leaking. Replace defective gaskets as required. Check for defective TPS or coolant temperature sensor.

Rough Idle – Hot

1) Leaking or defective injector or injector "O" ring seal. Check operation of injector.

2) Check ignition system. Check for possible intake system vacuum leaks.

Stalls on Deceleration or Quick Stop

1) Throttle position sensor defectve. Service as required.

2) Throttle body or intake manifold gaskets leaking. Replace gaskets as required.

3) ISC motor not operating properly. Check ISC operation.

Reduced Top Speed/Power

1) Throttle linkage binding. Clean and service as required. Fuel pump delivery volume low. Test fuel delivery system.

2) Fuel filter clogged. Fuel injector internal filter clogged. Injectors may be plugged. Clean, repair, or replace as needed. Pressure regulator may be damaged. Repair or replace as needed.

Surge at Cruise

1) Fuel filter clogged. Fuel pump pressure or delivery volume low. Test fuel delivery system.

2) Injectors may be plugged or leaking. Fuel injector internal filter clogged. Repair or replace as needed.

3) Fuel contaminated by dirt or water. Drain fuel and clean system.

Poor Mileage, But Drives Okay

Leaking injector "O" ring seal. Perform injector leak test.

Flooding

1) Excessive fuel pressure. Test and service fuel pressure regulator as required.

2) Injector stuck open, or injector "O" ring seal leaking. Perform leakage test. Repair or replace as needed.

Engine Diesels or Idles Too Fast

1) Incorrect idle speed adjustment. Adjust idle speed.

2) Vacuum leaks. Check all vacuum hoses and connections.

3) Inoperative ISC motor. Check operation of ISC motor.

4) Sticking throttle plate or linkage. Check and clean as needed.

TESTING & DIAGNOSIS

INERTIA SWITCH

The inertia switch is located in the luggage compartment on the left hinge support on Capri (2-door models), Cougar, Crown Victoria (sedan only), Grand Marquis (sedan only), Mustang (2-door models), Thunderbird, and Town Car. On 3-door Capri and Mustang models, the switch is located near left taillight area. The inertia switch is located in the left side storage compartment on Crown Victoria and Grand Marquis station wagons. On Sable and Taurus station wagons the inertia switch is located behind the access door on the right side trim panel in the rear cargo space of these models. All other models, the switch is located on the left side of the trunk. To reset the inertia switch, depress button on the switch.

CAUTION

- *Do not reset the inertia switch until the complete fuel system has been inspected for leaks.*

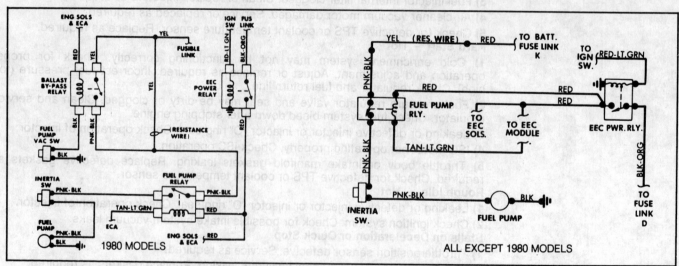

Fig. 3: Fuel Pump Wiring Diagrams - High Pressure In-Tank Pump

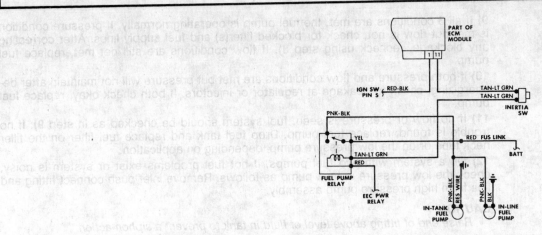

Fig. 4: Fuel Pump Wiring Diagrams – High Pressure In-Line & Low Pressure In-Tank Pumps

FUEL DELIVERY SYSTEM

NOTE
- On 1986-89 models, fuel delivery system is controlled by ECM module (EEC-IV), and entire system must be checked.

1) Check fuel tank for adequate fuel supply. Check for fuel leakage at all fittings and lines.

2) Check for electrical continuity to the fuel pump(s). On a single pump system, disconnect electrical connector just forward of the fuel pump inlet. On dual pumps, disconnect electrical connector to fuel pumps at high pressure pump module. Connect voltmeter to body wiring harness connector. Turn ignition key on while watching the voltmeter.

3) Voltage should rise to battery voltage and then return to zero voltage after approximately one second. If voltage is not as specified, check inertia switch and electrical system. Connect ohmmeter to pump wiring harness connector. If no continuity is present, check continuity directly at pump terminals. If no continuity is at pump terminals, replace pump(s). If continuity is present at pump terminals, service and replace wiring harness.

4) For dual pump systems only, connect the ohmmeter across the body wiring harness connector. If continuity is present (about 5 ohms), low pressure pump circuit is electrically okay.

5) For dual pump systems only, if no continuity is present in step **4)**, it is necessary to remove fuel tank and check for continuity at fuel pump/pump sender flange terminals. If no continuity is present at pump flange terminals, replace the assembly. If continuity is present at pump but not in step **4)**, service or replace the wiring harness in the low pressure pump.

6) Check operation of fuel pump(s). Disconnect return line at fuel rail. Use care to avoid spillage. Connect hose from fuel rail fitting to a calibrated container of at least one quart. Connect pressure gauge to fuel diagnostic valve on fuel rail. Disconnect the electrical connector to electric fuel pump(s) located just forward of the pump outlet.

7) Connect auxiliary wiring harness to fuel pump(s) electrical connector. Energize fuel pump(s) for 10 seconds by connecting auxiliary wiring harness to fully charged 12-volt battery. Observe pressure while energized. If no pressure, check polarity of wiring harness and also check terminal connections at fuel pump(s).

8) Allow fuel to drain from hose into container and observe volume. The fuel pump(s) are operating properly if: fuel pressure reaches 35-45 psi (2.5-3.2 kg/cm²), fuel flow is a minimum of 7.5 oz. (220 cc) for a single pump system; 9.5 oz. (280 cc) for a dual pump system, in 10 seconds, and fuel pressure remains at a minimum of 30 psi (2.1 kg/cm²) immediately after de-energizing.

9) If all 3 conditions are met, the fuel pump is operating normally. If pressure condition is met but flow is not, check for blocked filter(s) and fuel supply lines. After correcting any blockage, recheck using step **8)**. If flow conditions are still not met, replace fuel pump.

10) If both pressure and flow conditions are met but pressure will not maintain after de-energizing, check for leakage at regulator or injectors. If both check okay, replace fuel pump.

11) If no flow or pressure is seen, fuel system should be checked as in step **9)**. If no trouble is found, replace fuel pump. Drop fuel tank and replace fuel filter on the filler neck tube or on the low pressure pump, depending on application.

12) On a system with two fuel pumps, if hot fuel problems exist or system is noisy, check the low pressure in-tank pump as follows. Remove inlet push connect fitting and line from high pressure pump assembly.

CAUTION

- *Raise end of fitting above level of fluid in tank to prevent a siphon action.*

13) Connect hose from fuel tank to a calibrated container of at least one quart. Place ignition key in the "RUN" position until fuel pump times out (about one second). Check container for presence of fuel (if necessary, lift container to prevent siphon action). The amount of fuel should be a minimum of 1.5 oz. (50 cc) for one second of operation.

14) If no fuel is present, recycle ignition key. If no fuel is present, check for a pinched line between fuel tank and fitting, and then proceed to drop the fuel tank.

15) Connect a voltmeter to the chassis electrical connector from the fuel pump and turn ignition key to "ON" position. Voltage should rise to battery voltage for one second and then return to zero volts. If voltage is as specified and electrical connector was okay at pump, replace fuel pump assembly and repeat low pressure fuel pump test. If no voltage is present, service electrical circuit in vehicle.

ADJUSTMENTS

HOT (SLOW) IDLE SPEED

1981-82 5.0L

NOTE

- *Relocate air cleaner while making adjustments, but leave all hoses connected. Install air cleaner while making speed checks. Do not apply foot brake on hydro-boost systems as engine RPM will be affected during speed checks.*

1) Warm engine to normal operating temperature and connect tachometer. Turn A/C off. Stop engine, then restart and run at 2000 RPM for 60 seconds. Let engine speed stabilize at idle for 15 seconds, then place transmission selector in "D".

NOTE

- *Idle stabilizing time must not exceed 60 seconds before curb idle speed is checked.*

2) Check idle speed. If adjustment is required, stop engine. If idle speed is low, turn saddle bracket adjusting screw clockwise one turn. If idle speed is high, turn screw counterclockwise. See *Fig. 5*.

3) Restart engine and run at 2000 RPM for 60 seconds. Let engine speed stabilize at idle for 15 seconds, then place transmission selector in "D" and check idle speed. Repeat procedure until speed is correct.

1983-85 5.0L

1) Place transmission in "P" (auto. trans.) or Neutral (man. trans.) and firmly set parking brake. Start engine and let it warm to normal operating temperature. Place A/C-heater selector in "OFF" position.

2) On Capri and Mustang models, run engine at 2000 RPM for more than 10 seconds in Neutral. Allow engine to return to idle and stabilize for 10 seconds. Place transmission in "R" (auto. trans.). Adjust curb idle using saddle bracket adjusting screw. See *Fig. 5*. If RPM is too low, turn screw clockwise. If too high, turn screw counterclockwise. Repeat steps **1)** and **2)** until correct RPM is obtained.

3) On all other models, turn engine off. Restart engine and run at 2000 RPM for 60 seconds. Allow engine to return to idle and stabilize for 15 seconds. Place transmission in "R" (auto. trans.). Adjust curb idle to specification using saddle bracket adjusting screw.

4) If RPM is too low, turn engine off and turn adjusting screw one full turn clockwise. If RPM is too high, turn engine off and turn adjusting screw one full turn counterclockwise. Repeat steps **1)** and **3)** until correct RPM is obtained.

Courtesy of Ford Motor Co.

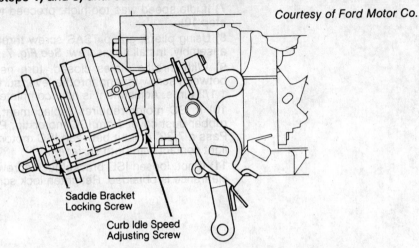

Saddle Bracket
Locking Screw

Curb Idle Speed
Adjusting Screw

Fig. 5: Curb Idle Speed Adjustment Location on 5.0L Engine

1984-87 3.8L

NOTE
- *Curb idle is controlled by EEC-IV and idle speed control motor. If control system is operating properly, these speeds are fixed and cannot be changed by standard adjustments. If these operating limit adjustments do not correct curb idle speed, perform appropriate 3.8L EEC-IV diagnosis.*

1) Set parking brake and block drive wheels. If equipped with automatic parking brake on all except 1987 models, always place transmission in Reverse when checking curb idle speed. On 1987 models, disconnect and plug vacuum hose when checking curb idle speed. Make all adjustments with engine at normal operating temperature and all accessories off.

2) On 1986 models, disconnect A/C clutch wire and turn A/C on. After engaging transmission in Drive (or Reverse) and allowing engine to idle for 60 seconds, check idle speed.

3) If idle speed is incorrect and throttle lever is not in contact with ISC motor, but held open by Throttle Stop Adjustment Screw (TSAS), this screw must be adjusted. This is the only reason for adjusting the TSAS. It is strictly a throttle stop and does not affect idle speed, unless it is holding throttle open through misadjustment. Perform the following procedure in its exact sequence.

4) To adjust TSAS, the ISC motor must be fully retracted. Shut engine off and remove air cleaner. In engine compartment, locate self-test connector and self-test input (STI) connector (both located next to each other on driver's side, behind strut tower on EEC harness).

5) Connect jumper wire between STI connector and signal return pin on self-test connector. *See Fig. 6.* Turn ignition to "RUN" position. DO NOT start engine.

Courtesy of Ford Motor Co.

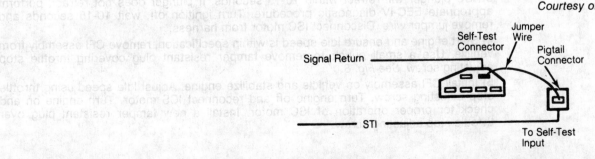

Signal Return

Self-Test
Connector

Jumper
Wire

Pigtail
Connector

STI

To Self-Test
Input

Fig. 6: Inserting Jumper Wire to Retract ISC Plunger on 2.5 & 3.8L Engine

6) The ISC plunger should retract. If not, EEC-IV service is required. Wait until plunger is fully retracted (about 10 seconds). Turn ignition to "OFF" position. Remove jumper wire.

7) If idle speed was too high, proceed to step **8)**. If idle speed was too low, proceed to step **10)**.

8) Using pliers, grasp TSAS screw threads and turn screw until it is removed from CFI assembly. Install new screw. *See Fig. 7.*

9) With throttle plates closed, turn new screw in until there is .005" (.13 mm) gap between screw tip and throttle lever surface that it contacts. Turn screw in an additional 1 1/2 turns. Adjustment is now complete.

10) If ISC motor requires replacement, adjust by using following procedure: remove rubber dust cover from ISC motor tip. Push tip back toward motor to remove any lash. Pass a 7/32" drill bit between ISC motor tip and throttle lever, making sure there is only light contact with both surfaces.

11) If not, loosen ISC bracket lock screw. Turn ISC bracket adjusting screw until proper clearance is obtained. Retighten lock screw and replace rubber dust cover.

Courtesy of Ford Motor Co.

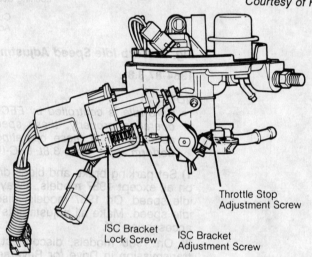

Throttle Stop
Adjustment Screw

ISC Bracket
Lock Screw

ISC Bracket
Adjustment Screw

Fig. 7: ISC Adjustment Location on 3.8L Engine

1989 2.5L CFI

NOTE

- If idle RPM is not correct after performing this procedure, testing and repair of EEC-IV system will be required.

1) Set parking brake and block wheels. Start engine and warm to operating temperature. Remove air cleaner. Turn all accessories off.

2) Place transmission in Drive (auto. trans.) or Neutral (man. trans.). Connect jumper wire between self-test input connector and signal return pin on self-test connector. *See Fig. 6.* Turn ignition on with engine off.

3) ISC plunger will retract within 10-15 seconds. If plunger does not retract, perform appropriate EEC-IV diagnostic procedure. Turn ignition off, wait 10-15 seconds and remove jumper wire. Disconnect ISC motor from harness.

4) Start engine and ensure idle speed is within specification, remove CFI assembly from vehicle. Use a small punch to remove tamper resistant plug covering throttle stop adjusting screw. *See Fig. 8.*

5) Install CFI assembly on vehicle and stabilize engine. Adjust idle speed using throttle stop adjusting screw. Turn engine off and reconnect ICS motor. Turn engine on and check for proper operation of ISC motor. Install a new tamper resistant plug over throttle stop adjusting screw.

Courtsey of Ford Motor Co.

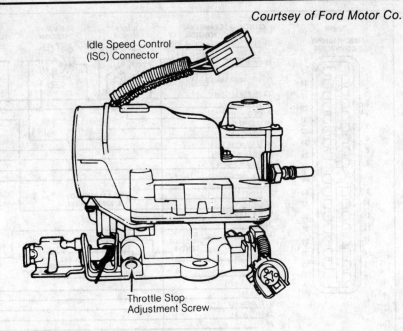

Idle Speed Control (ISC) Connector

Throttle Stop Adjustment Screw

Fig. 8: 2.5L CFI Throttle Adjustment Screw Location

IDLE SPEED (RPM)

Application	Curb Idle
1984-85 3.8L	
Capri & Mustang	475-575
All Others	500-600
1986-87 3.8L	500-600
1980-85 5.0L	550
1989	[1]

[1] – Refer to underhood decal for specifications.

COLD (FAST) IDLE SPEED

NOTE
- *Fast idle adjustment is not possible on 2.5L and 3.8L engine.*

1980-85 5.0L

1) Place transmission in Neutral (man. trans.) or "P" (auto. trans.). Start and warm engine to normal operating temperature. Disconnect and plug vacuum hose at EGR valve and fast idle pull-down motor. On all models except 1985 Capri and Mustang, stop engine.

2) On 1985 Capri and Mustang, set fast idle lever on specified step of fast idle cam. Start engine and check idle speed within 20-60 seconds after engine is started. Adjust if necessary.

NOTE
- *If speed is not checked between 20-60 seconds after engine starts, stop engine and repeat procedure.*

3) Reconnect vacuum line at EGR valve and fast idle pull-down motor. Remove test equipment.

FAST IDLE SPEED (RPM)

Application	Fast Idle	Cam Step
1980-81 5.0L		
Federal	2200	1st High
Calif.	2150	1st High
1982-85 5.0L		
High Output	2400	Highest Step
All Others	2200	Highest Step

HIGH PRESSURE CENTRAL FUEL INJECTION (CFI)

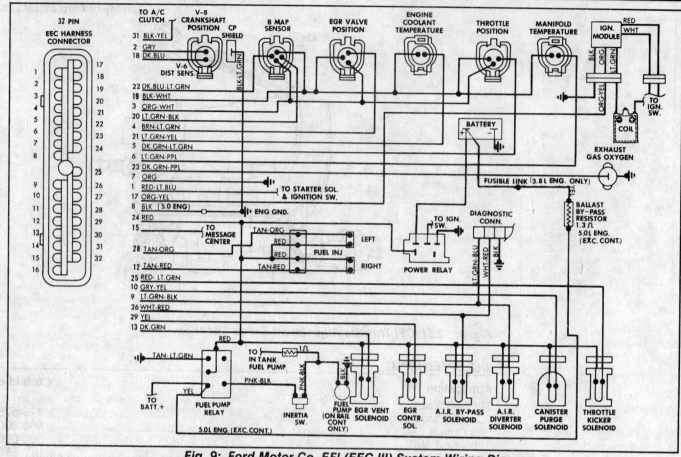

Fig. 9: Ford Motor Co. EFI (EEC-III) System Wiring Diagram

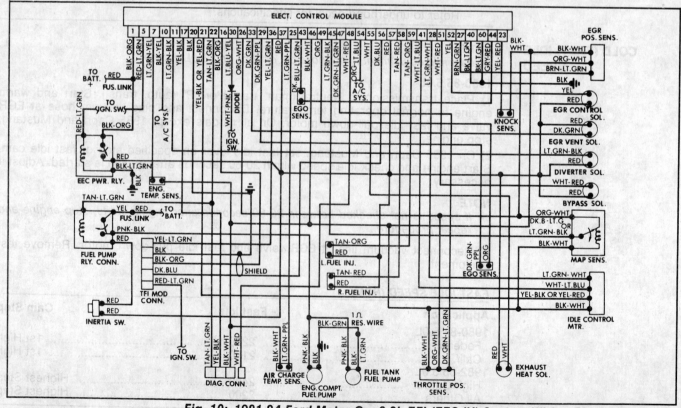

Fig. 10: 1981-84 Ford Motor Co. 3.8L EFI (EEC-IV) System Wiring Diagram

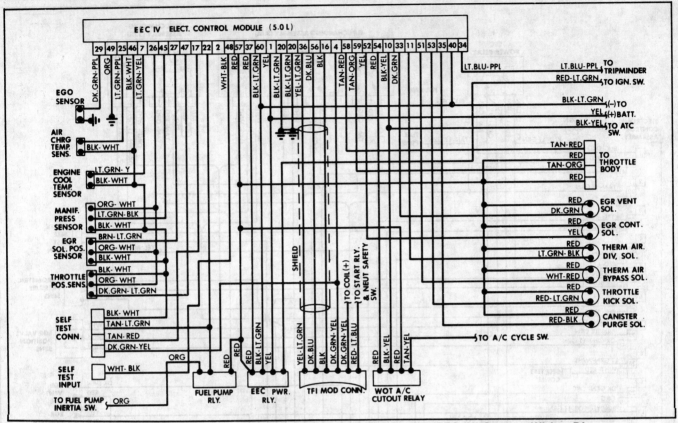

Fig. 11: 1981-84 Ford Motor Co. 5.0L EFI (EEC-IV) System Wiring Diagram

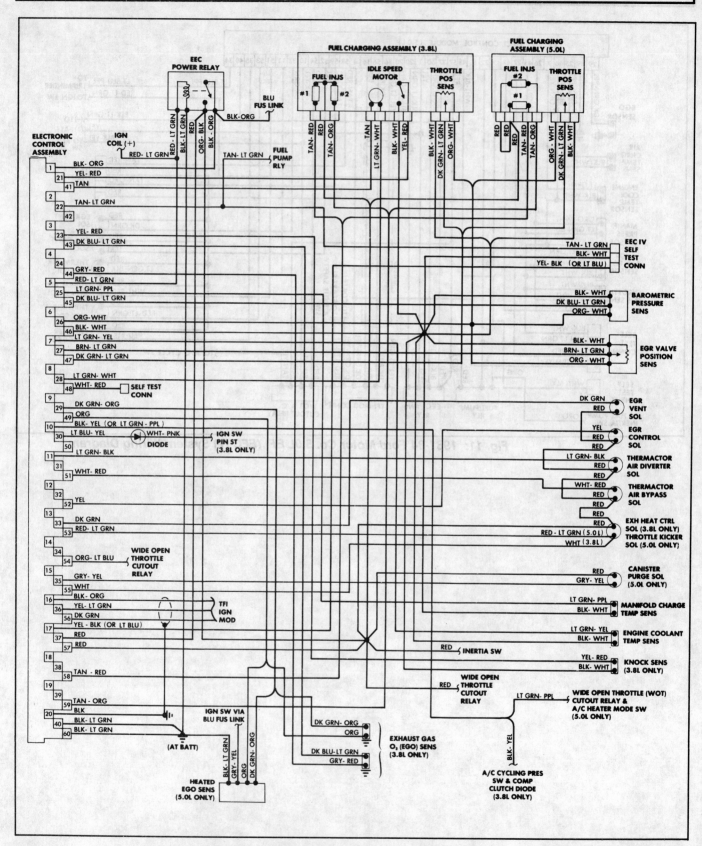

Fig. 12: 1985 Ford Motor Co. 3.8L/5.0L & 1986 3.8L EFI (EEC-IV) System Wiring Diagram

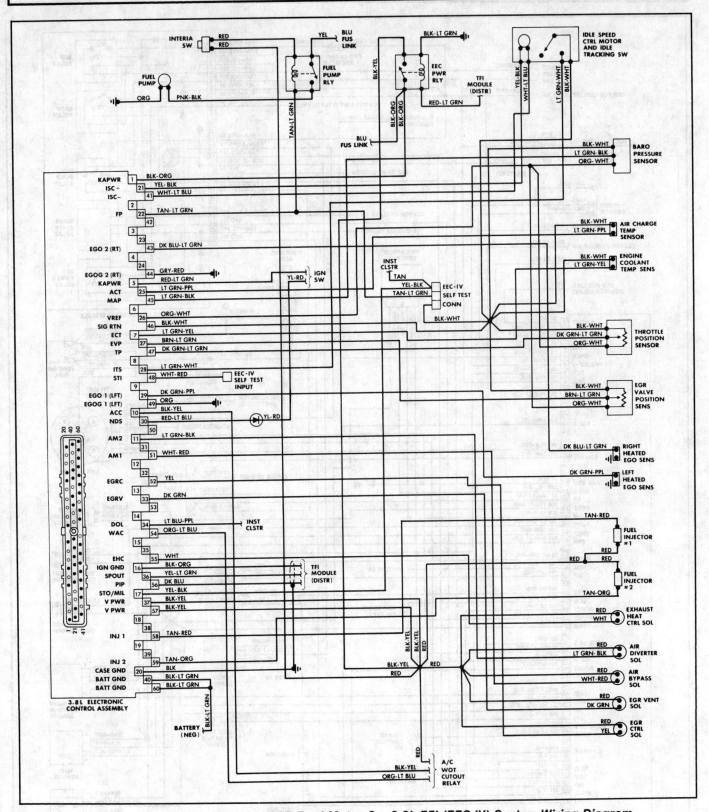

Fig. 13: 1987 Ford Motor Co. 3.8L EFI (EEC-IV) System Wiring Diagram

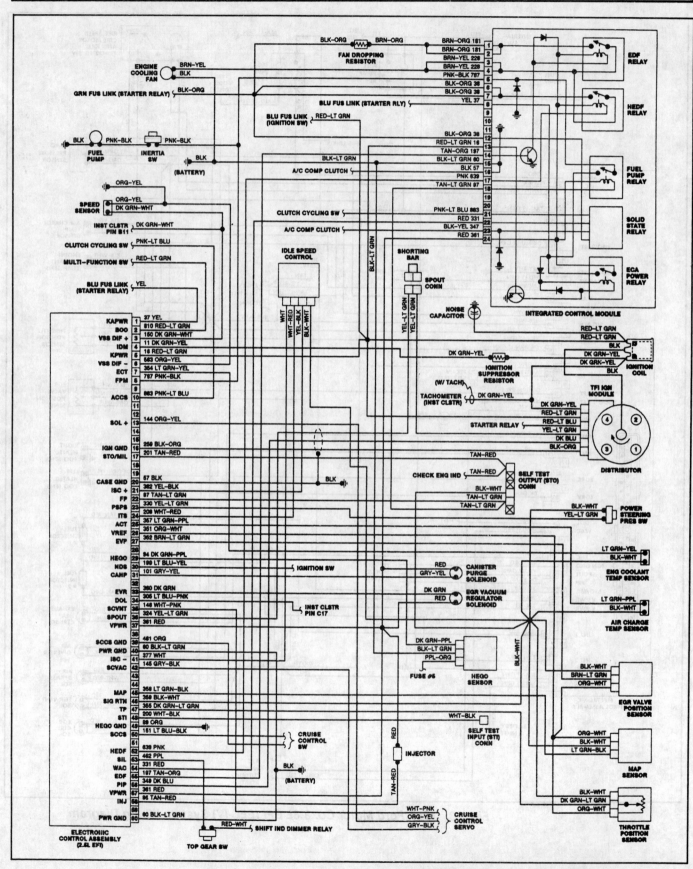

Fig. 14: 1989 Ford Motor Co. 2.5L CFI (EEC-IV) System Wiring Diagram

Ford: Escort & EXP 1.9L, Taurus 2.5L HSC, Tempo 2.3L HSC
Mercury: Lynx 1.9L, Topaz 2.3L HSC

NOTE
- *For detailed applications, refer to FUEL INJECTION SYSTEMS APPLICATION CHARTS at the front of this publication.*

DESCRIPTION

The Central Fuel Injection (CFI) System is a single point, pulse time modulated injection system. Fuel is metered into the air intake stream according to engine demands by a single solenoid injection valve, mounted in a throttle body on the intake manifold.

Fuel is supplied by a low pressure, electric fuel pump mounted in the fuel tank. The fuel is filtered and sent to the fuel charging assembly injector fuel cavity and then to the regulator where the fuel delivery pressure is maintained at a nominal value of 14.5 psi. (1.0 kg/cm²).

A single injector nozzle is mounted vertically above the throttle plate and is connected in series with the fuel pressure regulator. Excess fuel is returned to the fuel tank by a fuel return line.

Courtesy of Ford Motor Co.

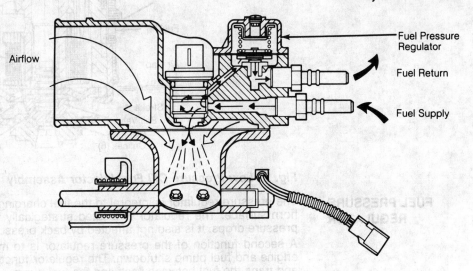

Fig. 1: Low Pressure CFI Fuel Charging Assembly

OPERATION

AIR & FUEL CONTROL

The fuel charging assembly is comprised of 5 individual components that perform the fuel and air metering function to the engine. The throttle body assembly mounts to the conventional carburetor pad of the intake manifold and provides for packaging of:
- Air Control
- Fuel Injector Nozzles
- Fuel Pressure Regulator
- Idle Speed Control
- Throttle Position Sensor

AIR CONTROL

Airflow to the engine is controlled by a single butterfly valve mounted in a two-piece, die cast aluminum housing called the throttle body. The butterfly valve is identical in configuration to the throttle plate of a conventional carburetor and is actuated by similar linkage.

FUEL INJECTOR NOZZLE

The fuel injector nozzle is mounted vertically above the throttle plate. It is an electromechanical device that meters and atomizes the fuel delivered to the engine. The injector valve body consists of a solenoid actuated ball and seat valve assembly.

An electrical control signal from the Electronic Control Assembly (ECA) activates the solenoid, causing the ball to move inward off its seat and allow fuel to flow. Injector flow orifice is fixed and fuel supply pressure is constant. Fuel flow to the engine is controlled by how long the solenoid is energized.

Courtesy of Ford Motor Co.

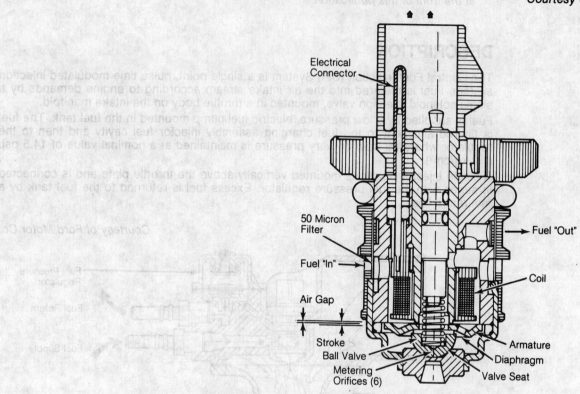

Fig. 2: Low Pressure CFI Fuel Injector Assembly

FUEL PRESSURE REGULATOR	The pressure regulator is integral to the fuel charging main body near the rear of the air horn surface. The regulator is located strategically to avoid the affects of supply line pressure drops. It is also not affected by back pressure in the return line to the tank.

A second function of the pressure regulator is to maintain fuel supply pressure upon engine and fuel pump shutdown. The regulator functions as a downstream check valve and traps the fuel between itself and the fuel pump.

The maintenance of fuel pressure upon engine shutdown precludes fuel line vapor formation and allows for rapid restarts and stable idle operation immediately thereafter. It regulates the fuel pressure to the injector nozzles at a constant nominal value of 14.5 psi. ($1.0 kg/cm^2$).

IDLE SPEED CONTROL (ISC)

Idle speed is controlled by the throttle actuator DC motor. Speed is regulated by modulating the throttle lever for the required airflow to maintain the desired engine RPM with a warm or cold engine.

An Idle Tracking Switch (ITS), integral with the DC motor, is utilized to determine when the throttle lever has contacted the actuator, thereby signaling the need to control engine RPM. The DC motor extends or retracts a linear shaft through a gear reduction system. The motor direction is determined by the polarity of the applied voltage.

THROTTLE POSITION SENSOR (TPS)

This nonadjustable sensor is mounted to the throttle shaft of the fuel charging assembly. It is used to supply a voltage output proportional to the change in throttle position. The sensor is used by the computer to determine the operation mode (closed throttle, part throttle, and wide open throttle) for selection of the proper fuel mixture, spark and EGR at selected driving modes.

TROUBLE SHOOTING

PRELIMINARY CHECKS

Following systems and components must be in good condition and operating properly before beginning diagnosis of fuel injection system:
- All Support Systems and Wiring
- Battery Connections and Specific Gravity
- Ignition System
- Compression Pressure
- Fuel Supply System Pressure and Flow
- All Electrical Connections and Terminals
- Vacuum Line, Fuel Hose and Pipe Connections
- Air Cleaner and Air Ducts
- Clear Exhaust System
- Cooling System

NOTE
- *Some vehicles may not include all components listed in this section.*

TROUBLE SHOOTING

Engine Does Not Crank

Check for hydrostatic lock (liquid or carbon in cylinder) and repair as necessary. Check for starting and charging system problems.

Engine Cranks But Does Not Start

1) Ensure fuel tank contains an adequate amount of fuel. Do not assume fuel gauge is correct. Check fuel for dirt, water, or other contamination.

2) Check ignition system for strong secondary current at spark plugs. If none exists or current is weak, repair ignition system problem before continuing with fuel injection diagnosis.

3) Check fuel lines and fittings for leaks. If no leaks are found, check fuel delivery system for proper operation, pressure, and volume.

4) Reset inertia switch if necessary.

5) Check for defective or plugged injector, TPS stuck at wide open throttle or bad coolant temperature switch.

6) Check for inoperative Idle Speed Control (ISC) motor.

Hard to Start – Cold

1) Check for defective TPS or coolant temperature sensor.

2) Ensure ISC motor is operating properly.

3) Cold enrichment system may not be functioning correctly. Check TPS and DC motor for proper operation and adjustment. Adjust or replace as required.

4) Throttle body or intake manifold gaskets leaking. Replace leaking gaskets as required.

Rough Idle – Cold

1) Cold enrichment cycle not functioning. Check fast idle system for proper operation and adjustment. Adjust or replace as required.

2) Air cleaner duct vacuum motor damaged or stuck open. Service or replace as required. Check for vacuum leaks. Check for leaking or defective injector or injector "O" ring seals. Check operation of injector.

Stall, Stumble, Hesitation – Hot or Cold

1) Cold enrichment system may not be functioning properly. Check for proper operation and adjustment. Adjust or replace as required.

2) Fuel pump output low. Possible clogged fuel filter or defective fuel pump. Clean, service, or replace as required after cause is determined.

3) Fuel injector internal filter clogged. Check for leaking "O" ring seals. Clean or replace injector as required.

4) Air cleaner vacuum motor damaged. Service or replace as required.

5) Check for defective TPS or coolant temperature sensor.

Hard Start – Hot

1) Cold enrichment system may not be functioning properly. Check for proper operation and adjustment. Adjust or replace as required.

2) Incorrect fuel pressure (too high). Check regulator and fuel return line. Fuel pressure regulator valve and seat may be dirty or clogged. Clean and service regulator. Check fuel system bleed down after stopping engine.

3) Leaking or defective injector or injector "O" ring seal. Check operation of injector.

4) ISC motor not operating properly. Check ISC operation.

5) Throttle body or intake manifold gaskets leaking. Replace leaking gaskets as required.

6) Check for defective TPS or coolant temperature sensor.

Rough Idle – Hot

1) Check ignition system. Check for possible intake vacuum leaks.

2) Leaking or defective injector or injector "O" ring seal. Check operation of injector.

Stalls on Deceleration or Quick Stop

1) Throttle position sensor defective. Service as required.

2) Throttle body or intake manifold gaskets leaking. Replace leaking gaskets as necessary.

3) ISC motor not operating properly. Check ISC operation.

Reduced Top Speed/Power

1) Throttle linkage binding. Clean and service as required.

2) Fuel pump delivery low. Check for clogged fuel filter. Test fuel delivery system.

3) Injectors may be plugged. Clean, repair, or replace as needed.

4) Pressure regulator may be damaged. Repair or replace as needed.

Lack of Power

1) Fuel filter clogged. Fuel injector internal filter clogged. Clean or replace as required.

2) Check fuel delivery and repair as required.

Surge at Cruise

1) Fuel filter clogged. Fuel pump delivery pressure or volume low. Test fuel delivery system. Service, repair, or replace as required.

2) Fuel contaminated by dirt or water. Drain and clean system as required.

3) Injectors may be plugged or leaking. Clean, repair, or replace as needed.

Poor Mileage, But Drives Okay

1) Check air filter. Replace as required.

2) Leaking injector "O" ring seal. Perform injector leak test.

Flooding

1) Excessive fuel pressure. Test and service fuel pressure regulator as required.

2) Injector stuck open, or injector "O" ring seal leaking. Perform leakage test. Repair or replace as needed.

Engine Diesels or Idles Too Fast

1) Incorrect idle speed adjustment. Adjust idle speed.

2) Vacuum leaks. Check all vacuum hoses and connections.

3) Inoperative ISC motor. Check operation of ISC motor.

4) Sticking throttle plate or linkage. Check and clean as needed.

DIAGNOSIS & TESTING

PRESSURE CHECK

CAUTION

- *Use care when opening fuel lines and testing fuel pressure due to danger of fire in case of fuel spillage. System is under residual pressure which MUST be discharged before opening fuel lines. Exercerise care to prevent combustion from fuel spillage.*

1) Disconnect electrical connector from inertia switch. On all models, except Sable and Taurus station wagons, the inertia switch is located on the left side of the luggage compartment. On Taurus and Sable wagons the inertia switch is located behind the right access panel in the rear compartment area. Crank engine for 15 seconds to reduce system pressure. Remove clip on fuel supply line at fuel charging assembly.

NOTE

- *On Taurus and Sable models use In-Line Adapter (D85L-9974-B) and Fuel Pressure Gauge (T87L-9974A)*

2) Install Fuel Line Adapter (T85L-9974-C). Connect Fuel Pressure Gauge (T80L-9974-B) to adapter. Reconnect electrical connector to inertia switch in trunk. Start engine and check fuel system pressure. Accelerate engine. Pressure should remain stable throughout acceleration.

3) The fuel pump is operating properly if the pressure reaches 13-16 psi (.9-1.1 kg/cm²) on 1.9L and 2.5L engines, or 14.5-17.5 psi (1.0-1.2 kg/cm²) on 2.3L engine. Fuel flow should be a minimum of 6 oz. (.18L) in ten seconds and fuel pressure must remain at a minimum of 11.6 psi (.8 kg/cm²) immediately after shutdown.

4) If pressure condition is met, but flow is not, check for blocked filters and fuel supply lines. After correcting any blockages, recheck. If flow conditions are still not met, replace fuel pump-sender assembly.

5) If both pressure and flow conditions are met, but pressure will not maintain after de-energization, check for leaking injectors or regulator. If both check okay, replace fuel pump-sender assembly. If no flow or pressure is seen, the fuel system should be rechecked for blockages. If no trouble is found, remove the fuel tank and replace the fuel pump-sender assembly.

6) Disconnect electrical connector at inertia switch. Crank engine 15 seconds. Remove gauge and adapter. Reconnect fuel line. Reconnect inertia switch. Crank engine and check for leaks.

INERTIA SWITCH

In the event of a collision, the electrical contacts in the inertia switch open and fuel pump automatically shuts off. The fuel pump will shut off even if the engine does not stop running. The engine will stop a few seconds after the fuel pump stops.

It is not possible to restart the engine until the inertia switch is manually reset. On all models except Sable and Taurus station wagons, the inertia switch is located on the left side of the luggage compartment. On Taurus and Sable station wagons the inertia switch is located behind the right access panel in the rear compartment area. To reset the inertia switch, depress the button on the switch.

CAUTION
• *Do not reset inertia switch until fuel system has been inspected for leaks.*

Courtesy of Ford Motor Co.

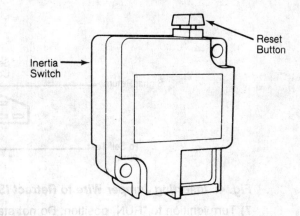

Fig. 3: *Inertia Switch*

INJECTOR RESISTANCE

To measure injector resistance, disconnect injector harness connector. Set ohmmeter on 10-ohm scale. Injector resistance should be within specification. See INJECTOR RESISTANCE table.

INJECTOR RESISTANCE

Engine	Ohms
All Models	1.0-2.0

ADJUSTMENTS

HOT (SLOW) IDLE SPEED

NOTE

- *Curb idle is controlled by EEC-IV and idle speed control motor on all fuel injected engines. If control system is operating properly, these speeds are fixed and cannot be changed by standard adjustments. If these operating limit adjustments do not correct curb idle speed, perform appropriate EEC-IV diagnosis. If battery is disconnected for any reason, or vehicle has to be jump-started, perform idle speed check.*

1.9L (1985-88)

1) Set parking brake and block drive wheels. Start engine and warm to normal operating temperature. Check for vacuum leaks. Turn all accessories off.

2) Place transmission in "D" (auto. trans.) or Neutral (man. trans.). Run engine at idle speed for 2 minutes and check idle speed. Idle speed should be within specifications.

3) Place transmission in "P" (auto. trans.). Engine idle speed should increase by about 75 RPM. Lightly step on and off accelerator. Engine RPM should return to specified idle RPM.

4) If idle RPM remains high, repeat procedure as it may take the EEC-IV system 2 minutes to reset. If idle speed is not correct, perform appropriate EEC-IV diagnosis.

5) If curb idle speed is within specification, ensure throttle linkage is free and cruise control (if equipped) is not holding throttle open. If throttle lever is not in contact with ISC motor when engine is running, but held open by throttle stop screw, this screw must be adjusted using following procedure.

6) Ensure no vacuum leaks are present. Turn engine off and remove air cleaner. Find self-test connector and self-test input connector in engine compartment. Connect jumper wire between self-test input connector and signal return pin on self-test connector. *See Fig. 4.*

Courtesy of Ford Motor Co.

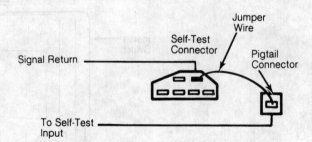

Fig. 4: Inserting Jumper Wire to Retract ISC Plunger

7) Turn ignition to "RUN" position. Do not start engine. ISC plunger should retract. Wait until plunger is fully retracted (about 10-15 seconds). If plunger does not retract, perform appropriate EEC-IV diagnosis.

8) Return ignition to "OFF" position, wait 10-15 seconds and remove jumper wire from self-test connector. Unplug ISC motor electrical connector from vehicle wiring harness.

9) Start engine and allow engine speed to stabilize. Check and adjust idle RPM to 550-650 RPM by turning throttle stop adjusting screw. *See Fig. 5.* Stop engine and reconnect ISC motor connector to wiring harness. Check for proper operation of ISC system.

1.9L (1989)

1) Set parking brake and block wheels. Start engine and warm to normal operating temperature. Remove air cleaner. Turn all accessories off.

2) Place transmission in Drive (auto. trans.) or Neutral (man. trans.). Connect a jumper wire between self-test input connector and signal return pin self-test connector. *See Fig. 4.* Turn ignition on with engine off.

3) ISC plunger will retract within 10-15 seconds. If plunger does not retract, perform appropriate EEC-IV diagnostic procedure. Turn ignition off, wait 10-15 seconds and remove jumper wire. Disconnect ISC motor from harness.

Courtesy of Ford Motor Co.

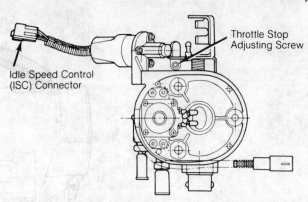

Fig. 5: Location of Throttle Stop Adjusting Screw on 1.9L Engine

4) Start engine and ensure that idle speed is within specification. If idle speed is not within specification, remove CFI assembly from vehicle. Use a small punch to remove tamper resistant plug covering throttle stop adjusting screw. Remove and replace screw. *See Figs. 5 & 6.*

5) Install CFI assembly on vehicle start engine and stabilize idle. Adjust idle speed using throttle stop adjusting screw. Turn engine off and reconnect ISC motor. Turn engine on and check for proper operation of ISC motor. Install a new tamper resistant plug over throttle stop adjusting screw.

2.3L HSC & 2.5L HSC

1) Set parking brake and block drive wheels. If vehicle is equipped with automatic parking brake, disconnect and plug vacuum hose during idle speed check. Start engine and warm to normal operating temperature. Check for vacuum leaks. Turn all accessories off.

2) Place transmission in "D" (auto. trans.) or Neutral (man. trans.). Run engine at idle speed for 2 minutes and check idle speed. Idle speed should be within specifications.

3) Place transmission in "P" (auto. trans.). Engine idle speed should increase by about 100 RPM. Lightly step on and off accelerator. Engine RPM should return to specified idle RPM.

4) If idle RPM remains high, repeat procedure as it may take the EEC-IV system 2 minutes to reset. If idle speed is not correct, perform appropriate EEC-IV diagnosis.

5) If curb idle speed is within specification, ensure throttle linkage is free and cruise control (if equipped) is not holding throttle open. If throttle lever is not in contact with ISC motor when engine is running, but held open by throttle stop screw, this screw must be adjusted by using the following procedure.

6) Ensure no vacuum leaks are present. Turn engine off and remove air cleaner. Find self-test connector and self-test input connector in engine compartment. Connect jumper wire between self-test input connector and signal return pin on self-test connector. *See Fig. 4.*

7) Turn ignition to "ON" position. Do not start engine. ISC plunger should retract. Wait until plunger is fully retracted (about 10-15 seconds). If plunger does not retract, perform appropriate EEC-IV diagnosis.

8) Return ignition to "OFF" position, wait 10-15 seconds and remove jumper wire from self-test connector. Remove CFI assembly from vehicle. Use a small punch to remove welsh plug covering throttle stop adjusting screw. Remove and replace screw. *See Fig. 6.*

9) With throttle plates closed, turn screw inward until there is a .005" (.13 mm) gap between screw tip and throttle lever surface it contacts. Turn screw one additional turn. Install new welsh plug. Install CFI assembly on vehicle. Start and stabilize engine. Set idle speed 100 RPM less (50 RPM on auto. trans.) than specified by turning throttle stop adjusting screw. Check for proper operation of ISC system.

Courtesy of Ford Motor Co.

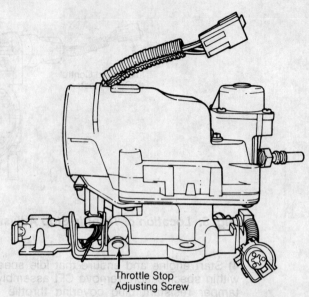

Throttle Stop
Adjusting Screw

Fig. 6: Location of Throttle Stop Adjusting Screw on 2.3L & 2.5L Engines

IDLE SPEED (RPM)

Application	Curb Idle
1985 2.3L HSC	
Man. Trans.	
High Output	775-825
All Others	725-775
Auto. Trans.	570-630
1986-88 2.3L HSC & 2.5L HSC	
Man. Trans.	725-775
Auto. Trans.	625-675
1986-87 2.3L High Output	775-825
1987-88 1.9L	
Man. Trans.	900
Auto. Trans.	800
1989	[1]

[1] – Refer to specification on underhood decal.

COLD (FAST) IDLE SPEED

NOTE
- *Fast idle adjustment is not possible on fuel injected engines.*

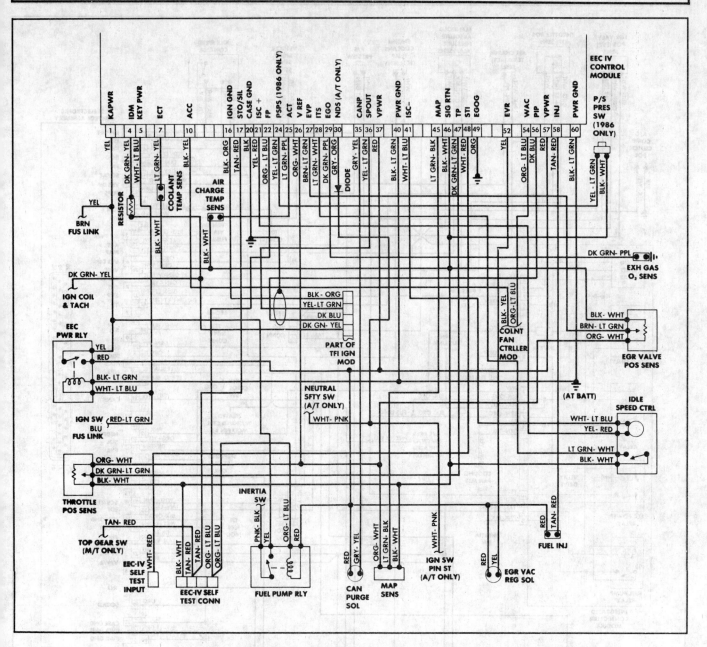

Fig. 7: 1985-86 2.3L HSC Low Pressure CFI System Wiring Diagram

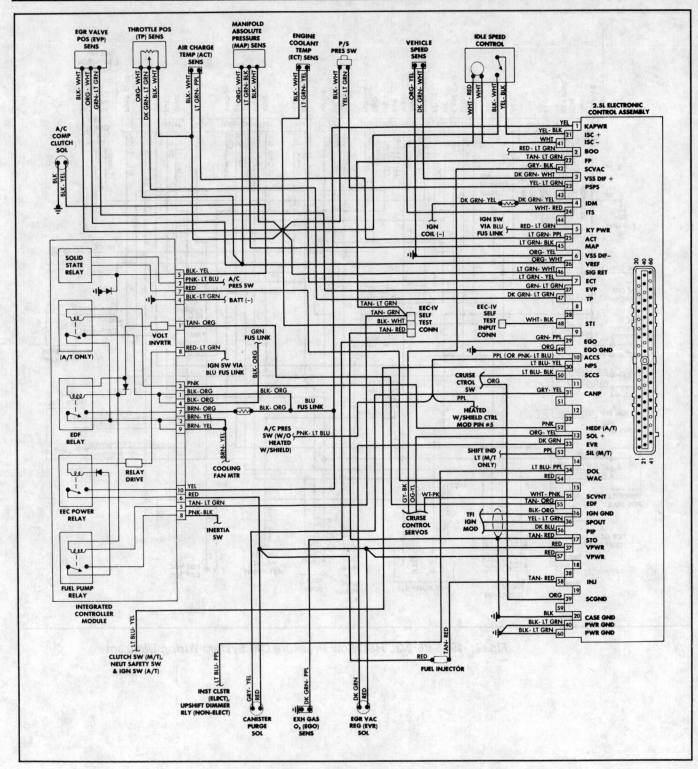

Fig. 8: 1986 2.5L HSC Low Pressure CFI System Wiring Diagram

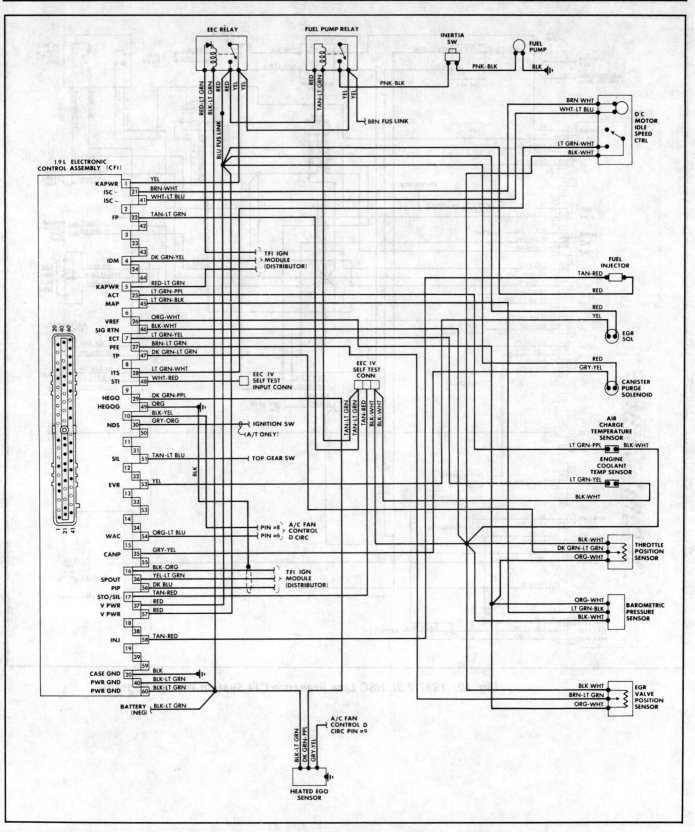

Fig. 9: 1987 1.9L Low Pressure CFI System Wiring Diagram

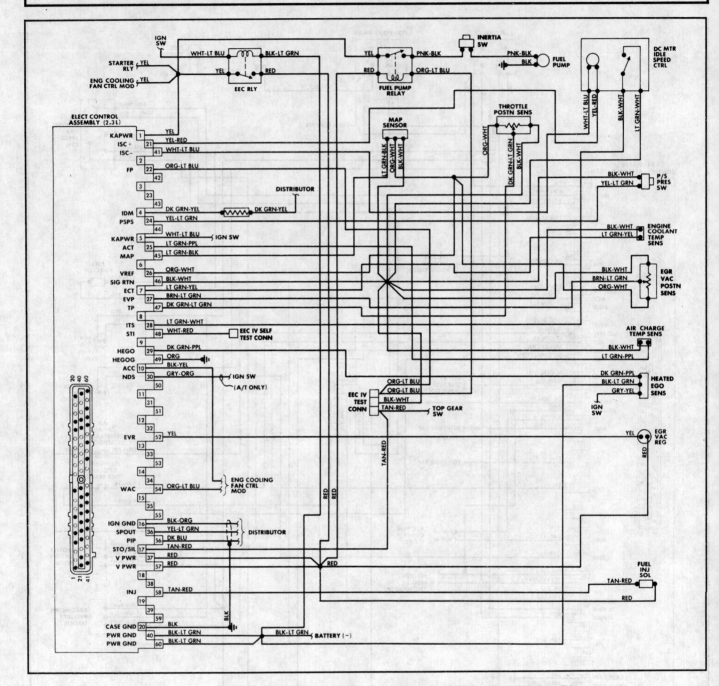

Fig. 10: 1987 2.3L HSC Low Pressure CFI System Wiring Diagram

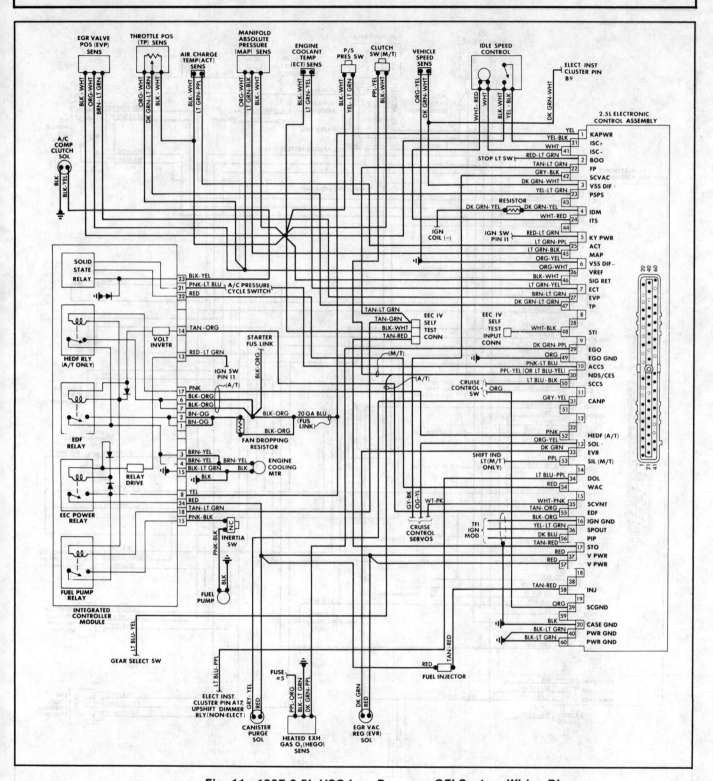

Fig. 11: 1987 2.5L HSC Low Pressure CFI System Wiring Diagram

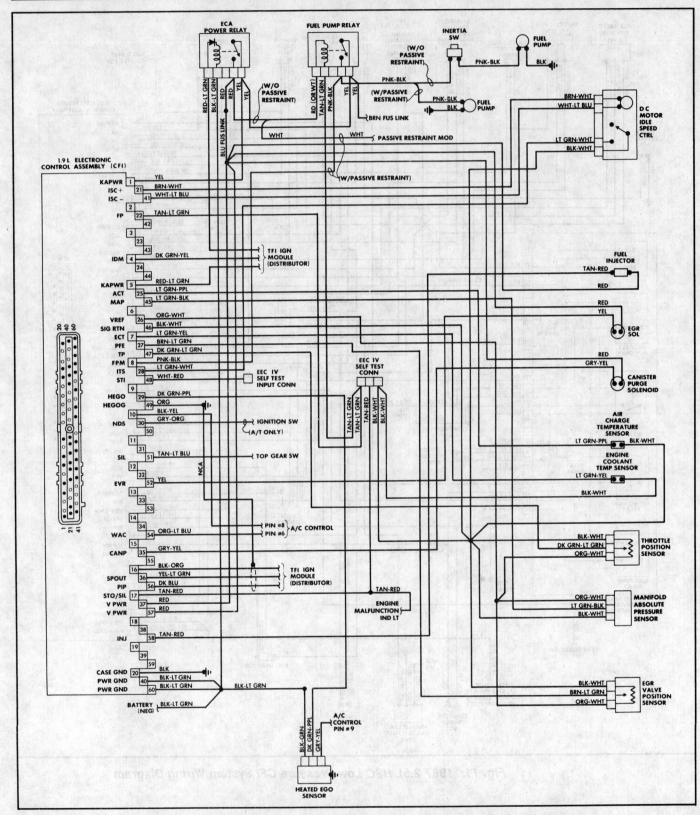

Fig. 12: 1988 1.9L Low Pressure CFI System Wiring Diagram

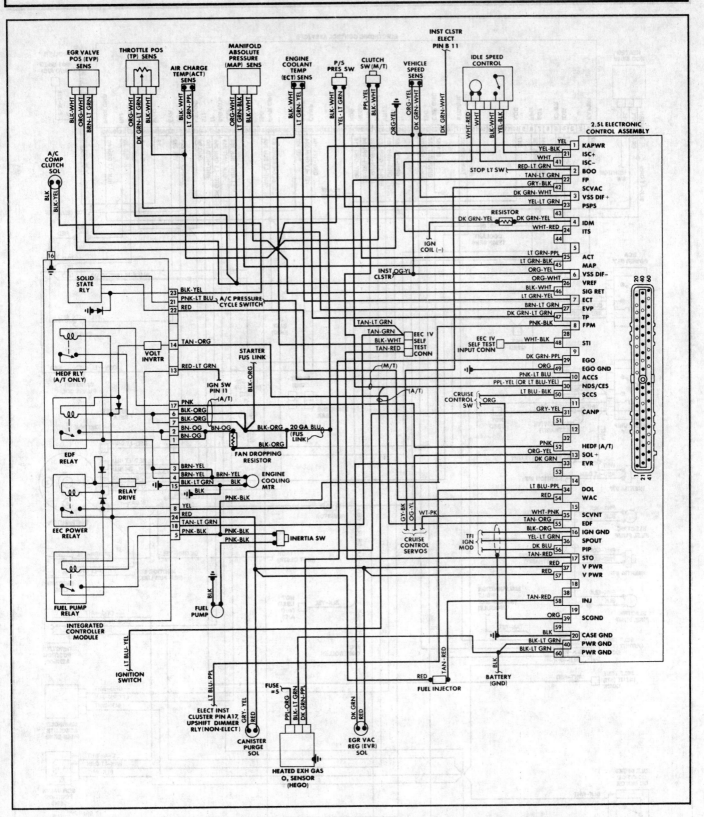

Fig. 13: 1988 2.5L Low Pressure CFI System Wiring Diagram

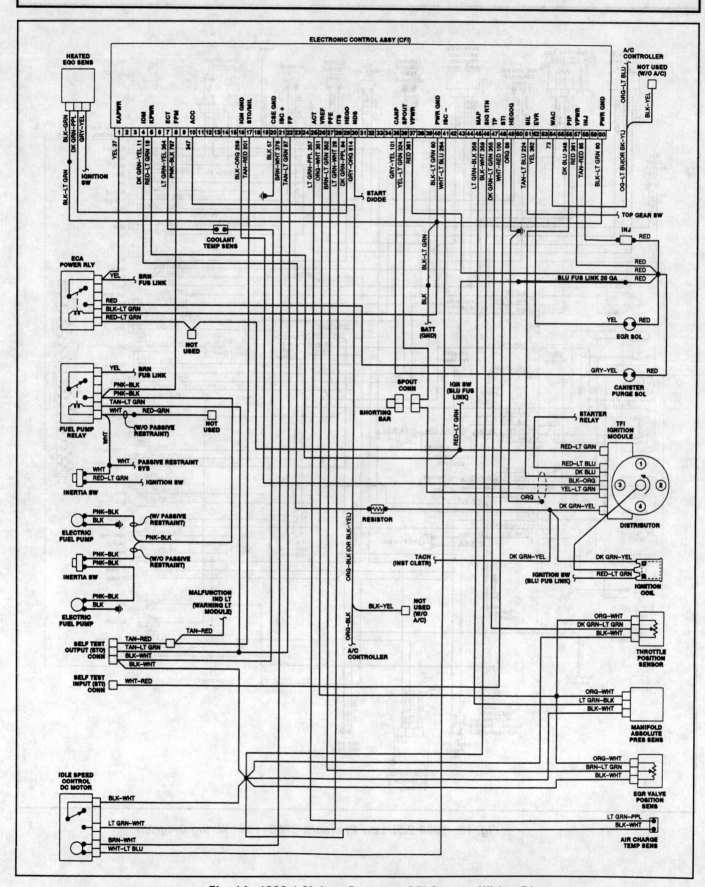

Fig. 14: 1989 1.9L Low Pressure CFI System Wiring Diagram

Ford: Aerostar, Bronco, Bronco II, Country Squire, Crown Victoria,
 "E" Series Vans, "F" Series Pickups, Escort, EXP,
 Mustang, Ranger, Tempo, Taurus, Thunderbird, Probe
Lincoln: Continental, Mark VII, Town Car
Mercury: Capri, Colony Park, Cougar, Grand Marquis,
 LN7, Lynx, Marquis, Merkur XR4Ti, Sable, Topaz

NOTE
- *For detailed applications, refer to FUEL INJECTION SYSTEMS APPLICATION CHARTS at the front of this publication.*

DESCRIPTION

The Electronic Fuel Injection System is a multi-point, pulse time system using a mass airflow sensor on 1.6L, 1.9L and 2.3L Turbo engines or Barometric Manifold Atmosphere Pressure sensor (BMAP) and separate air temperature sensor on all other engines. Fuel is metered into the intake air stream in accordance with engine demand through injectors mounted on a tuned intake manifold.

An on-board Electronic Engine Control (EEC-IV) system has an Electronic Control Assembly (ECA) which uses inputs from various engine sensors to compute the required fuel flow rate necessary to maintain the prescribed air-fuel ratio throughout the entire engine operational range. The computer then outputs a command to the fuel injectors to meter the precise quantity of fuel. The ECA also controls injector "ON" time is the only controlled variable in the fuel delivery system.

Courtesy of Ford Motor Co.

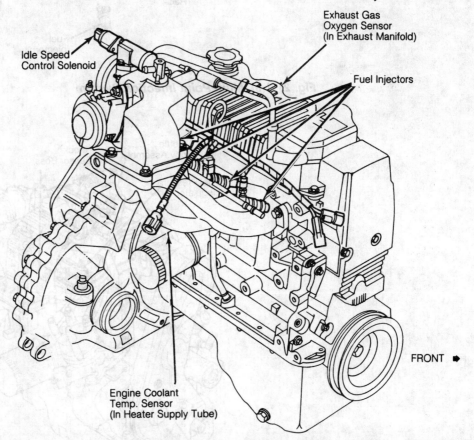

Fig. 1: *1.6L Engine With Multi-Point Injection System*

Additional air flow metering is provided on 1.9L, 2.2L and 2.2L Turbo engines by a Vane Airflow (VAF) meter. The 5.0L Mass Airflow (MA) engine used in Mustangs uses a Mass Airflow (MA) sensor to measure quantity of air entering the engine.

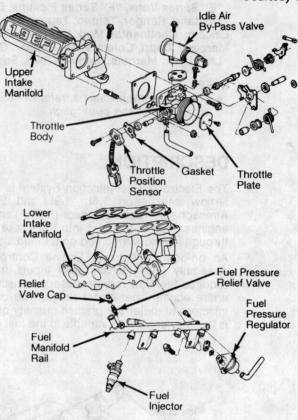

Courtesy of Ford Motor Co.

Fig. 2: 1.9L Multi-Point Injection System

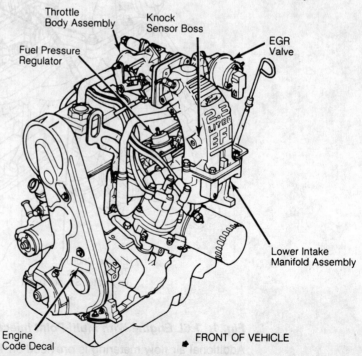

Courtesy of Ford Motor Co.

Fig. 3: 2.3L Non-Turbo With Multi-Point Injection System

Courtesy of Ford Motor Co.

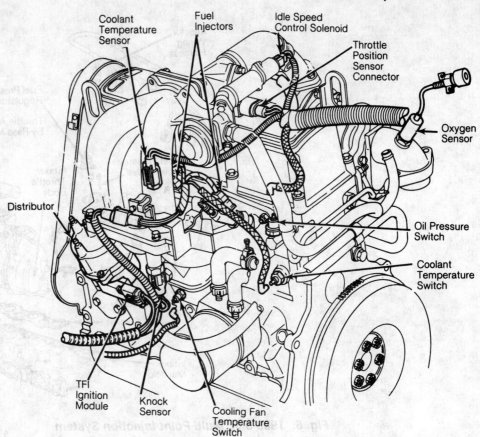

Fig. 4: 2.3L Turbo Engine With Multi-Point Injection System

Courtesy of Ford Motor Co.

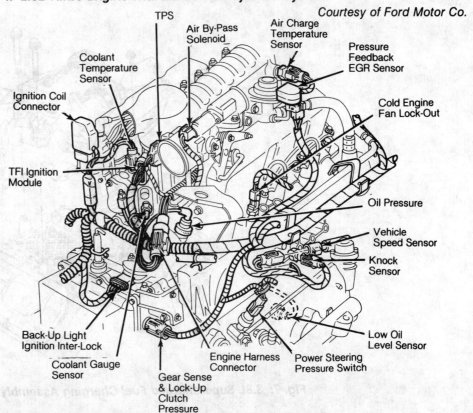

Fig. 5: 3.0L Engine With Multi-Point Injection System

243

Courtesy of Ford Motor Co.

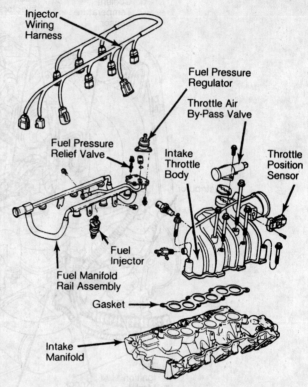

Fig. 6: 1989 3.0L Multi-Point Injection System

Courtesy of Ford Motor Co.

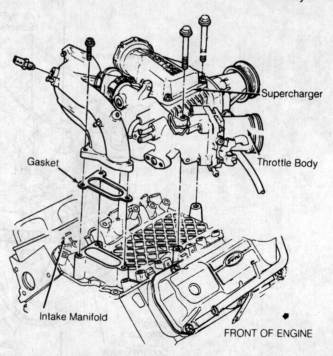

Fig. 7: 3.8L Supercharged Fuel Charging Assembly

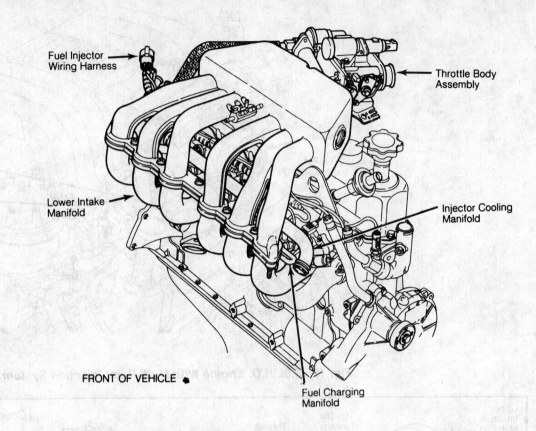

FRONT OF VEHICLE

Fig. 8: 4.9L Engine With Multi-Point Injection System

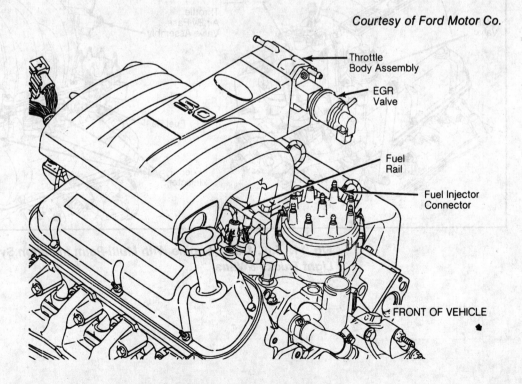

FRONT OF VEHICLE

Fig. 9: 5.0L Engine With Multi-Point Injection System

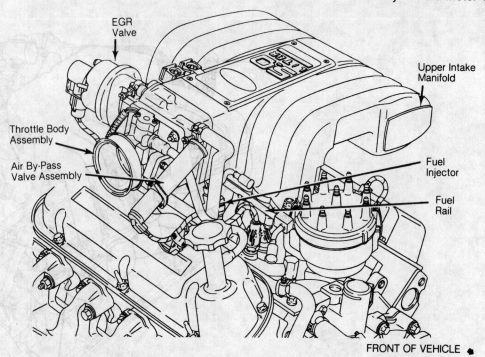

Fig. 10: 5.0L H.O. Engine With Multi-Point Injection System

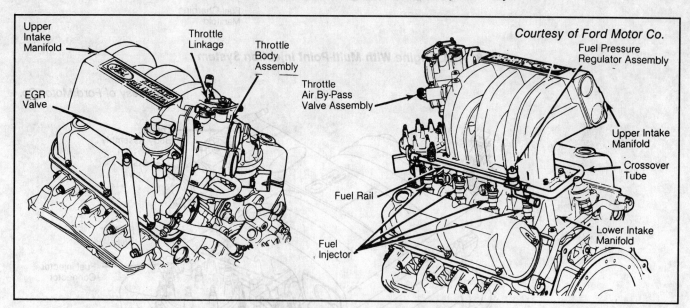

Fig. 11: 5.0L & 5.8L Engine With Multi-Point Injection System Used on 1988-89 Light Trucks & Vans

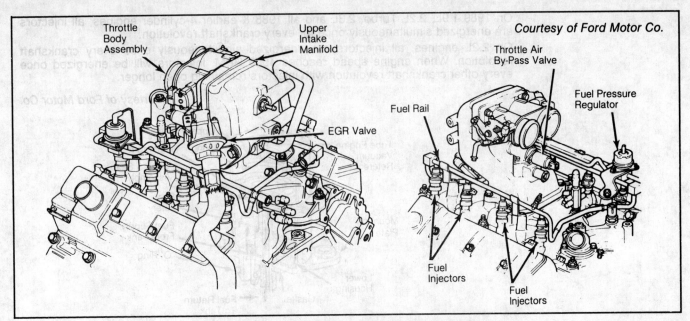

Courtesy of Ford Motor Co.

Fig. 12: 7.5L Engine With Multi-Point Injection System

OPERATION

FUEL DELIVERY

The fuel delivery sub-system consists of a high pressure, chassis or in-tank mounted electric fuel pump. Some models also use one low pressure in-tank pump and one high pressure pump. Pump delivers fuel from the fuel tank through a 20 micron fuel filter to a fuel charging manifold assembly. The fuel charging manifold assembly incorporates electrically actuated fuel injectors directly above each intake port. The injectors spray a metered quantity of fuel into the intake air stream.

A constant fuel pressure drop is maintained across the injector nozzles by a pressure regulator. The regulator is connected in series and positioned downstream from the fuel injectors. Excess fuel not required by the engine, passes through the regulator and returns to the fuel tank through a fuel return line.

Courtesy of Ford Motor Co.

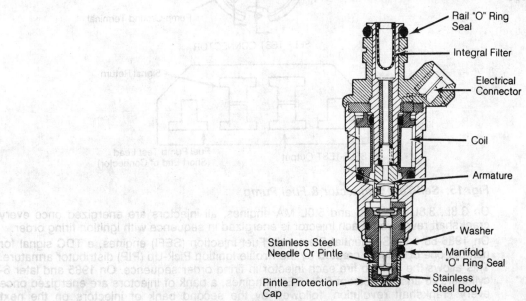

Fig. 13: Cutaway View of Fuel Injector

On 1989 1.9L, 2.2L Turbo, 2.3L and all 1988 & earlier 4-cylinder engines, all injectors are energized simultaneously once for every crankshaft revolution.

On 2.2L engines, all injectors are energized simultaneously once every crankshaft revolution. When engine speed reaches 4500 RPM, injectors will be energized once every other crankshaft revolution with injectors remaining open longer.

Courtesy of Ford Motor Co.

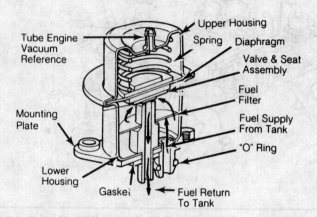

Fig. 14: *Fuel Pressure Regulator*

On 1989 3.0L engines, injectors are energized in 3 pairs, in sequence of 1-2, 3-4, 5-6. The injectors are energized once every other crankshaft revolution.

On 1989 3.0L SHO and 5.0L HO engines, injectors are energized once every other crankshaft revolution in sequence with engine firing order.

Courtesy of Ford Motor Co.

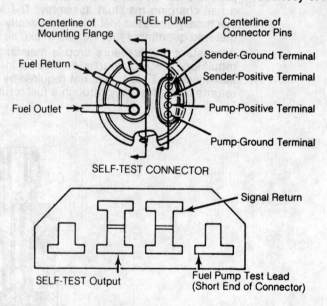

Fig. 15: *Self-Test Connector & Fuel Pump*

On 3.8L, 3.8L SC, 5.0L and 5.0L MA engines, all injectors are energized once every crankshaft revolution. Each injector is energized in sequence with ignition firing order.

On 1986-88 5.0L Sequential Electronic Fuel Injection (SEFI) engines, a TDC signal for number one cylinder is created by the Profile Ignition Pick-Up (PIP) distributor armature. This allows the ECA to fire each injector in firing order sequence. On 1985 and later 6-cylinder, V6 and V8 (except 5.0L SEFI) engines, a bank of injectors are energized once every crankshaft revolution, followed by the second bank of injectors on the next crankshaft revolution.

The period of time the injectors are energized (injector "on" time or "pulse width") is controlled by the ECA. Air entering the engine is measured by a vane airflow meter or BMAP and separate air temperature sensor.

This airflow or BMAP information and input from various other engine sensors is used to compute the required fuel flow rate necessary to maintain the prescribed air-fuel ratio for the given engine operation. The computer determines the needed injector pulse width and outputs a command to the injector to meter the exact quantity of fuel.

VAF system measures intake air quantity with vane airflow meter and integral air charge temperature. Speed/density control system uses Throttle Position Sensor (TPS), Manifold Absolute Pressure (MAP) sensor and Air Charge Temperature (ACT) sensor to determine intake air quantity.

FUEL INJECTORS

The fuel injector nozzles are electromechanical devices that both meter and atomize fuel delivered to the engine. The injectors are mounted in the lower intake manifold and are positioned so their tips direct fuel just ahead of the intake valves.

The injector bodies consist of a solenoid actuated pintle and needle valve assembly. An electrical control signal from the ECA unit activates the injector solenoid, causing the pintle to move inward off the seat, allowing fuel to flow.

Because injector flow orifice is fixed and fuel pressure drop across the injector tip is constant, fuel flow to the engine is regulated by how long the solenoid is energized. Atomization is obtained by contouring the pintle at the point where the fuel separates.

FUEL PRESSURE REGULATOR

The fuel pressure regulator is attached to the fuel supply manifold assembly downstream of the fuel injectors. It regulates the fuel pressure supplied to the injectors. The regulator is a diaphragm operated relief valve in which one side of the diaphragm senses fuel pressure and the other side is subjected to intake manifold pressure.

The nominal fuel pressure is established by spring preload applied to the diaphragm. Balancing one side of the diaphragm with manifold pressure maintains a constant fuel pressure drop across the injectors. Excess fuel is by-passed through the regulator and returned to the fuel tank.

VANE AIR METER ASSEMBLY

The vane airflow meter assembly is used on 1.6L, 1.9L, 2.2L, 2.2L Turbo and 2.3L Turbo engines. VAF contains two sensors that furnish input to the ECA, an airflow sensor and a Vane Air Temperature (VAT) sensor. VAT measures temperature of incoming air. The vane air meter measures the volume of airflow to the engine. Airflow through the body moves a vane mounted on a pivot pin. The vane is connected to a variable resistor (potentiometer) that in turn is connected to 5-volt reference voltage.

The output of this potentiometer varies depending on the volume of air flowing through the sensor. The temperature sensor in the vane air meter measures the incoming air temperature. The air volume and temperature inputs are used by the ECA to compute the fuel flow (injector pulse duration) necessary for the optimum air/fuel ratio.

BAROMETRIC PRESSURE (BP) SENSOR

The BP sensor is used on 2.2L and 2.2L Turbo engines. Mounted on right side of engine compartment firewall, this sensor senses atmospheric pressure then sends the information to the ECA.

MASS AIRFLOW (MA) SENSOR

The mass airflow sensor, used on 5.0L Mustangs, measures airflow to intake system. Sensor output is used by ECA to calculate injector impulse width. This sensor has no moving parts. A heated, glass covered, platinium wire is used to heat air flowing through a fixed orifice to a temperature sensing wire. The temperature of air reaching sensor wire indicates amount of air which has flowed past the heated wire and through MA sensor.

MAP & ACT SENSORS

Manifold Absolute Pressure (MAP) sensor compares manifold vacuum and manifold pressures to obtain manifold absolute pressures. Resulting input signal from sensor provides ECA with engine load and air density information. This sensor is located on firewall in engine compartment or on air cleaner.

Air Charge Temperature (ACT) sensor measures air/fuel mixture temperature. Resulting input signal from sensor provides ECA with density correction factor, which is used to calculate airflow and to proportion cold enrichment fuel flow. This sensor is located in intake manifold runner or side of throttle body.

BMAP & AIR TEMPERATURE SENSOR

The BMAP and air temperature sensor is used on all other engines. The BMAP sensor measures both atmospheric and manifold absolute pressure. This sensor provides the ECA with engine air intake information. The air temperature sensor provides the ECA with air/fuel mixture temperature information. This sensor may be located in the intake manifold runner or air cleaner assembly.

FUEL SUPPLY MANIFOLD ASSEMBLY

The fuel supply manifold assembly is the component that delivers high pressure fuel from the vehicle fuel supply line to the fuel injectors. The assembly consists of preformed tube(s) or stamping with injector connectors, a mounting flange to the fuel pressure regulator, and mounting attachments that locate the fuel manifold assembly and provide fuel injector retention. A fuel pressure relief valve is installed on fuel supply manifold assembly to release fuel system pressure.

AIR THROTTLE BODY ASSEMBLY

The throttle body assembly controls air flow to the engine through a double butterfly (4.9L and 5.0L non-SEFI engines) or single butterfly (all other engines) valve. The throttle position is controlled by cable/cam mechanism or multiple-link, throttle linkage that opens progressively. The body is a single piece die casting made of aluminum. It has a single or double bore with an air by-pass channel around the throttle plate.

This by-pass channel controls both cold and warm engine idle air flow as regulated by an air by-pass valve assembly mounted to the upper intake manifold (2.9L engine), or throttle body on all other engines. The valve assembly is an electromechanical device controlled by the ECA. It incorporates a linear actuator that positions a variable area metering valve.

AIR INTAKE SYSTEM

Air intake system on 1989 3.0L engine incorporates throttle body and air intake manifold in one-piece unit. Runner lengths are tuned for optimium engine torque output. On 1989 truck & van models, the intake manifold is a 2-piece (upper and lower) aluminum casting. Manifold provides mounting flanges for throttle body assembly, fuel supply manifold, accelerator controls, EGR valve and supply tube.

Vacuum ports are provided to support various engine accessories. Mounting sockets for fuel injectors are machined to prevent both air and fuel leakage. Pockets in which injectors are mounted are placed so injectors spray fuel directly in front of each intake valve.

TROUBLE SHOOTING

PRELIMINARY CHECKS

Following systems and components must be in good condition and operating properly before beginning diagnosis of fuel injection system:
- All Support Systems and Wiring.
- Battery Connections and Specific Gravity.
- Ignition System.
- Compression Pressure.
- Fuel Supply System Pressure and Flow.
- All Electrical Connections and Terminals.
- Vacuum Line, Fuel Hose, and Pipe Connections.
- Air Cleaner and Air Ducts.
- Engine Coolant Level.
- Clear Exhaust System.
- State of Tune.

TROUBLE SHOOTING

NOTE

- *Some vehicles may not include all components listed in this section.*

Engine Does Not Crank

Check for hydrostatic lock (liquid or carbon in cylinder) and repair as necessary. Check for starting and charging system problems.

Engine Cranks But Does Not Start

1) Ensure fuel tank contains an adequate amount of fuel. Do not assume fuel gauge is correct. Check fuel for dirt, water or other contamination.

2) Check ignition system for strong secondary current at spark plugs. If none exists or current is weak, repair ignition system problem before continuing with fuel injection diagnosis.

3) Check fuel lines and fittings for leaks. If no leaks are found, check fuel delivery system for proper operation, pressure and volume.

4) Reset inertia switch if necessary.

5) Check for defective or plugged injector, TPS stuck at wide open throttle or bad coolant temperature switch.

6) Check for inoperative Idle Speed Control (ISC) motor.

Hard to Start – Cold

1) Check for TPS sticking at Wide Open Throttle (WOT), defective TPS or coolant temperature sensor.

2) Crank engine with TPS disconnected. Check for discharging injector.

3) Ensure ISC motor is operating properly. EEC-IV diagnosis may be required.

Rough Idle – Cold

1) Check for possible vacuum leaks.

2) Check for leaking or defective injector or injector "O" ring seals.

3) Perform injector/regulator leakage test. Check operation of injector.

Stall, Stumble, Hesitation – Hot or Cold

Check for defective TPS or coolant temperature sensor. EEC-IV diagnosis may be required.

Hard Start – Hot

1) Incorrect fuel pressure (too high). Check regulator and fuel return line.

2) Fuel pressure regulator valve and seat may be dirty or clogged. Clean and service regulator. Check fuel system bleed down after stopping engine.

3) Leaking or defective injector or injector "O" ring seal. Check operation of injector.

4) ISC motor not operating properly. Check ISC operation.

Rough Idle – Hot

Leaking or defective injector or injector "O" ring seal. Check operation of injector. Perform injector leakage test.

Stalls on Deceleration or Quick Stop

ISC motor not operating properly. Check ISC operation. EEC-IV diagnosis may be required.

Reduced Top Speed/Power

1) Injectors may be plugged. Clean, repair, or replace as needed.

2) Pressure regulator may be damaged. Repair or replace as needed.

Surge at Cruise

Injectors may be plugged or leaking. Check fuel filter. Service or replace as needed.

Poor Mileage, But Drives Okay

Leaking injector "O" ring seal. Perform injector leak test.

Flooding

1) Excessive fuel pressure. Test and service fuel pressure regulator as required.

2) Injector stuck open, or injector "O" ring seal leaking. Perform leakage test. Repair or replace as needed.

Engine Diesels or Idles Too Fast

1) Incorrect idle speed adjustment. Adjust idle speed.

2) Vacuum leaks. Check all vacuum hoses and connections.

3) Inoperative ISC motor. Check operation of ISC motor.

4) Sticking throttle plate or linkage. Check and clean as needed.

TESTING & DIAGNOSIS

FUEL PUMP CONTROL

When the ignition switch is first turned to the "ON" position, the EEC power relay is energized, closing its contacts. Power is provided to both the fuel pump relay and a timing device in the EEC module. The fuel pump runs through the contacts of the fuel pump relay. If the ignition switch is not turned to the "START" position, the timing device in the EEC module will open the ground circuit after about one second. EEC senses engine speed and shuts off fuel pump by opening ground circuit to fuel pump relay when engine stops, or when engine speed drops to less than 120 RPM.

Opening the ground circuit de-energizes the fuel pump relay (opening its contacts), which in turn de-energizes the fuel pump. This circuitry provides for pre-pressurization of the fuel system. When the ignition switch is turned to the "START" position, the EEC module operates the fuel pump relay to provide fuel for starting the engine while cranking.

FUEL PUMP

The fuel pump may be activiated by grounding the fuel pump lead at SELF-TEST connector. Use a jumper lead, and ground the "FP" terminial with ignition on. This activates the fuel pump. *See Fig. 15.*

CAUTION
* *Inspect fuel system for leaks or damage before resetting inertia switch or testing fuel pump.*

INERTIA SWITCH

In the event of a collision, the electrical contacts in the inertia switch open and the fuel pump automatically shuts off. The fuel pump will shut off even if the engine does not stop running. The engine, however, will stop a few seconds after the fuel pump stops. It is not possible to restart the engine until the inertia switch is manually reset. See INERTIA SWITCH LOCATION table for location of the inertia switch. To reset the inertia switch, depress the button on top of the switch.

INERTIA SWITCH LOCATION

Application	Location
Capri & Mustang	Floor of trunk, to left of spare tire.
Continental & Mark VII	Inboard side of left trunk hinge support.
Cougar & Thunderbird	On left or right wheelwell in trunk.
Escort, EXP, Lynx & LN7	
Sedan	Behind left rear quarter panel on wheelwell housing.
Station Wagon	Behind left rear quarter panel, at rear light housing.
Pickup & Van	
1985-87	
Models	Behind left side of instrument panel, near steering column.
1988-89 Models	
F-Series	On left floorboard near parking brake assembly.
E-Series	On right cowl panel forward of right front door.
Sable & Taurus	
Sedan	Left rear side of trunk, behind trunk liner.
Station Wagon	Behind access panel at right rear corner.
Crown Victoria & Grand Marquis	
Station Wagon	Left rear side storage compartment.
Mustang 3-Door	
RWD Model	Near left taillight
FWD Model	Behind trim panel in left side of trunk.
All Others	On left rear wheelwell in trunk.

MODIFYING FUEL CUT-OFF RELAY (ALL MODELS)

1) Using relay (E3EB-9345-BA, CA, DA or E3TF-9345-AA), drill a 1/8" hole in-line with pins and as close to relay base as possible.

NOTE
* *Part of relay skirt may be cut away to provide easier access to pins.*

2) Solder a 16-18 gauge jumper wire between pins 2 and 4. Feed one end of an 8-10 ft. flexible wire through hole drilled in relay skirt and solder to pin No. 1. *See Fig. 16.*

NOTE
* *Leads should be soldered as close to relay base as possible to permit insertion of relay into socket with minimum interference.*

3) Solder an alligator clip to other end of 8-10 ft. flexible wire.

Courtesy of Ford Motor Co.

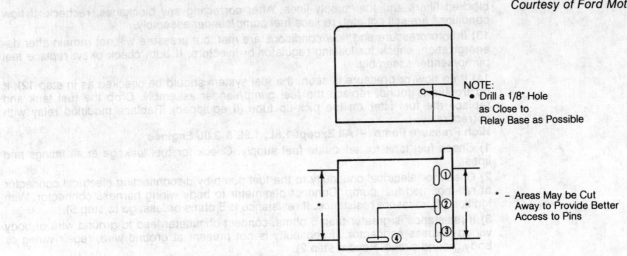

NOTE:
• Drill a 1/8" Hole as Close to Relay Base as Possible

* – Areas May be Cut Away to Provide Better Access to Pins

Fig. 16: Modifying Fuel Pump Cut-Off Relay

CAUTION
• *Fuel system is under pressure. Pressure will remain in fuel lines for long period of time, even after engine shut down.*

FUEL DELIVERY SYSTEM

1.6L, 1.9L & 3.0L Engines

1) Check fuel tank for adequate fuel supply. Check for fuel leakage at all fittings and lines.

2) Check for electrical continuity to fuel pump by disconnecting electrical connector at inertia switch. Connect ohmmeter to one lead of wiring harness. Check for continuity between wiring harness and ground.

3) If no continuity is present, check continuity of other wire in connector. If continuity is not present at either wire, remove fuel tank and test for continuity between wiring harness and switch.

4) If continuity is present on at least one wire, check continuity across in-tank fuel pump terminals. If continuity is not present across pump terminals, replace fuel pump/sender assembly. If continuity is present across pump terminals, check ground circuit or connections to pump from body connector.

5) Reconnect electrical connector to inertia switch. Connect voltmeter to wiring harness on pump side of inertia switch (side showing continuity). Turn ignition to "RUN" position while watching voltage to fuel pump. Voltage should rise to 10 volts or more for about one second, then return to zero.

6) If voltage is not as specified, check inertia switch to ensure it has been reset. Also check electrical system and continuity of inertia switch.

7) Check fuel pump operation. Disconnect electrical connector at inertia switch. Crank vehicle for at least 15 seconds to reduce fuel system pressure. Disconnect return line at fuel rail, using care to avoid fuel spillage. Connect hose from fuel rail fitting to a calibrated container of at least one quart.

8) If vehicle is equipped with fuel diagnostic valve on fuel rail, connect pressure gauge to diagnostic valve. If vehicle is not equipped with fuel diagnostic valve, disconnect hose between throttle body and fuel filter. Attach an in-line pressure gauge.

9) Remove fuel pump relay. Replace relay with modified relay shown in *Fig. 16*. Ensure ground lead is placed within reach outside vehicle. Energize the fuel pump for 10 seconds by grounding the ground lead from modified relay. Observe pressure while energized.

10) If no pressure is present, check voltage past inertia switch to determine if proper voltage is being received by fuel pump. Correct electrical problems as required. Also check terminal connections at fuel pump.

11) The fuel pump is operating properly if the fuel pressure reaches 35-45 psi (2.5-3.2 kg/cm²), fuel flow is a minimum of 5.6 oz. (.17L) in 10 seconds, and fuel pressure remains at a minimum of 30 psi (2.1 kg/cm²) immediately after de-energization.

12) If all three conditions are met, the fuel pump is operating normally. Check for engine and electrical problems. If pressure condition is met, but flow is not, check for blocked filters and fuel supply lines. After correcting any blockages, recheck. If flow conditions are still not met, replace fuel pump/sender assembly.

13) If both pressure and flow conditions are met, but pressure will not remain after de-energization, check for leaking regulator or injectors. If both check okay, replace fuel pump/sender assembly.

14) If no flow or pressure is seen, the fuel system should be checked as in step **12)**. If no trouble is found, replace the fuel pump/sender assembly. Drop the fuel tank and replace the fuel filter on the pick-up tube (if equipped). Replace modified relay with correct relay.

High Pressure Pump – All Except 1.6L, 1.9L & 3.0L Engines

1) Check fuel tank for adequate fuel supply. Check for fuel leakage at all fittings and lines.

2) Check for electrical continuity to the fuel pump by disconnecting electrical connector at rail-mounted fuel pump. Connect ohmmeter to body wiring harness connector. With ignition off, measure resistance. If resistance is 5 ohms or less, go to step **5)**.

3) If resistance is greater than 5 ohms, connect ohmmeter lead to ground wire of body wiring harness connector. If continuity is not present at ground wire, repair wiring or body ground circuit. Repeat step **2)**.

4) Connect ohmmeter to Pink/Black wire of body wiring harness connector. If continuity is not present, check wiring at fuel pump/sender assembly. If wiring checks okay, test continuity across fuel pump terminals. Replace pump if continuity is not present.

5) Connect voltmeter to body wiring harness connector. Turn ignition to "RUN" position while watching the voltmeter. Voltage should rise to battery voltage for about one second, and then return to zero. If voltage is not as specified, check electrical system and repair as required.

6) Connect ohmmeter across pump wiring harness connector. If no continuity is present, check continuity directly at pump terminals. If no continuity is present at pump terminals, replace pump/sender assembly. If continuity is present, go to next step.

7) Check fuel pump pressure and flow. If vehicle is equipped with fuel diagnostic valve, connect Pressure Gauge (T80L-9974-A) to diagnostic valve. If vehicle is not equipped with diagnostic valve, attach pressure gauge between fuel rail and filter (or pump outlet). Perform preliminary check by turning ignition to "RUN" position and observing pressure gauge. If reading is 30-40 psi (2.1-2.8 kg/cm²), pump is operating correctly and all other system components are okay.

8) Disconnect fuel return line at fuel rail, using care to avoid spillage. Connect hose from fuel rail fitting to calibrated container of at least one quart. Remove fuel pump relay. Replace relay with modified relay shown in *Fig. 16*. Ensure ground lead is placed within reach outside vehicle. Energize the fuel pump for 10 seconds by grounding the ground lead from modified relay. Observe pressure while energized.

9) The fuel pump is operating properly if the fuel pressure reaches 35-45 psi (2.5-3.2 kg/cm²), fuel flow is a minimum of 8.1 oz. (.24L) for 2.3L Turbo and 5.0L SEFI engines or 5.6 oz. (.17L) for all other engines in 10 seconds, and fuel pressure remains at a minimum of 30 psi (2.1 kg/cm²) immediately after de-energization.

10) If all three conditions are met, the fuel pump is operating normally. Check for engine and electrical problems. If pressure condition is met, but flow is not, check for blocked filters, fuel supply lines, and/or tank selector valves (trucks with dual tanks). After correcting any blockages, recheck. If flow conditions are still not met, replace fuel pump/sender assembly.

11) If both pressure and flow conditions are met, but pressure will not remain after de-energization, check for leaking regulator or injectors. If both check okay, replace fuel pump/sender assembly.

12) If no flow or pressure is seen, the fuel system should be checked as in step **10)**. If no trouble is found, replace the fuel pump/sender assembly. Replace modified relay with correct relay.

Low Pressure Pump – All Except 1.6L, 1.9L & 3.0L Engines

1) The low pressure pump should be tested if there is a problem with hot fuel handling and/or excessive noise. Check pump by opening fuel line at high pressure pump inlet. Connect hose to line from fuel tank and place hose in calibrated container of at least one quart. Disconnect high pressure fuel pump electrical connector from body wiring harness.

2) Remove fuel pump relay. Replace relay with modified relay shown in *Fig. 16*. Ensure ground lead is placed within reach outside vehicle. Energize the fuel pump for 10 seconds by grounding the ground lead from modified relay. Observe flow while energized.

NOTE
- *The fuel hose may have to be blocked momentarily to prime low pressure pump. This is normal when outlet is open with no back pressure on outlet.*

3) The pump is operating properly if volume is 16 oz. (.47L) in 10 seconds. If there is no flow from low pressure pump, check pump inlet for restriction, and check electrical circuit. Replace fuel pump, if required.

4) If fuel pump noise comes from low pressure pump, high pressure pump may be damaged. To check, compare vibration of inlet and outlet fuel lines at high pressure pump. If there is a big difference in vibration level between the two lines, replace high pressure pump. Recheck vibration level. Reconnect fuel and electrical lines. Install fuel pump relay.

INJECTORS

1) With engine running at operating temperature, disconnect one injector. Engine speed should decrease about 100 RPM. Repeat procedure for each injector. If engine RPM does not change, check and if necessary replace injector.

2) Disconnect all injector harness connectors. Using a digital ohmmeter, check resistance across injector terminals. Ohmmeter should read 2.0-2.7 ohms on 1.6L, 1.9L, and 2.3L Turbo engines, 12.0-16.0 ohms on 2.2L Non-Turbo, 11.0-15.0 ohms on 2.2L Turbo, 13.5-16.0 ohms on 1989 2.3L HSC, 2.3L HSO, 3.0L SHO and 3.8L engine, 15.0-18.0 ohms on 1989 3.0L, 1.5-19.0 ohms on 1989 5.0L Mustang, 13.5-19.0 ohms on all other 1989 models or 15.0-19.0 ohms on all other engines. If not within specification, replace injector.

ADJUSTMENTS

NOTE
- *Curb idle is controlled by EEC-IV and idle speed control air by-pass valve on all engines. If control system is operating properly, these speeds are fixed and cannot be changed by standard adjustments. If these operating limit adjustments do not correct curb idle speed, perform appropriate EEC-IV diagnosis. Perform this adjustment procedure only if curb idle speed is not within specifications. Fast idle adjustment is not possible on fuel injected engines.*

HOT (SLOW) IDLE SPEED

1.6L & 1.6L Turbo

1) Place transmission in Neutral (man. trans.) or "P" (auto. trans.). Start engine and warm to normal operating temperature. Turn engine off.

2) Disconnect and plug both vacuum hoses at EGR solenoid. Disconnect idle speed control (ISC) air by-pass valve power lead.

NOTE
- *Ensure cooling fan is operating when idle adjustment is made.*

3) Start engine and operate at 2000 RPM for 60 seconds. Place transmission in Neutral (man. trans.) or "D" (auto. trans.). Adjust curb idle speed to specified RPM using throttle plate adjusting screw. See Fig. 17.

4) Adjustment must be made within 2 minutes after restarting engine. Remove all test equipment. Reconnect all vacuum hoses and electrical connectors to original locations.

Courtesy of Ford Motor Co.

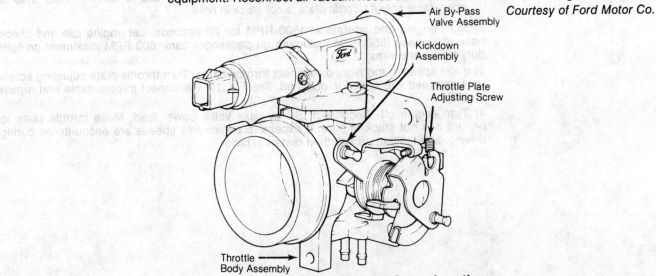

Fig. 17: 1.6L Throttle Plate Adjusting Screw Location

1.9L

1) Place transmission in Neutral. Apply parking brake and block drive wheels. Start engine and warm to normal operating temperature. Stop engine. Disconnect and plug both vacuum hoses at EGR solenoid. Ensure throttle plate is not stuck in bore, or that linkage is not preventing throttle plate from closing. Disconnect ISC connector.

2) Start engine and operate at 2000 RPM for 60 seconds. Let engine idle and check base idle speed. Ensure cooling fan is not operating when base idle speed check/adjustment is made. Check of base idle must be made within 2 minutes after returning to idle. Adjust curb idle speed to specified RPM using throttle plate adjusting screw. *See Fig. 18.*

3) If idle speed is not checked or adjusted within 2 minutes, stop engine. Restart engine and repeat step **2)**.

4) Reconnect ISC connector and vacuum lines. Start engine and operate at 2000 RPM for 60 seconds. Return to idle and recheck idle speed within 2 minutes. If idle speed is not checked within 2 minutes, stop engine. Restart engine and repeat step **4)**. If curb idle is still not within specifications, perform EEC-IV diagnosis.

Courtesy of Ford Motor Co.

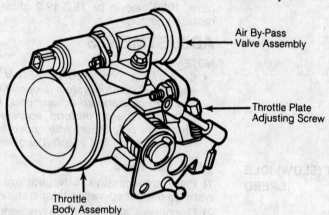

Fig. 18: 1.9L Throttle Plate Adjusting Screw Location

2.3L Non-Turbo OHC

1) Place transmission in Neutral (man. trans.) or "P" (auto. trans.). Turn A/C-heater switch off. Start engine and warm to normal operating temperature. Stop engine. Disconnect power lead to idle speed control (ISC) air by-pass valve.

NOTE
- *Engine may stall when ISC air by-pass valve power lead is disconnected. This is acceptable only if throttle plate is not stuck in bore.*

2) Start engine and operate at 1500 RPM for 20 seconds. Let engine idle and check base idle speed (550 RPM maximum on passenger cars; 600 RPM maximum on light duty trucks and vans).

3) If idle speed is too high, disconnect throttle cable. Turn throttle plate adjusting screw until specified idle speed is obtained. *See Fig. 19.* Reconnect throttle cable and repeat step **3)**.

4) Turn engine off. Reconnect ISC by-pass valve power lead. Move throttle plate to ensure it is not stuck in bore. If excessive engine idle speeds are encountered during driving, turn ignition off and then restart vehicle.

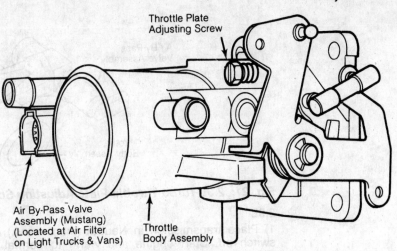

Courtesy of Ford Motor Co.

Fig. 19: 2.3L Non-Turbo Throttle Plate Adjusting Screw Location

2.3L Non-Turbo High Swirl Combustion (HSC)

1) Disconnect Spark Output (SPOUT) connector, near distributor. Ensure ignition timing is base plus 2 degrees BTDC. Remove PCV hose at PCV valve. Install Orfice (T86P-9600-A).

2) Disconnect idle speed control/air bypass solenoid. Place transmission lever in "D" (automatic transmission) or Neutral (manual transmission). Adjust idle RPM to specification. *See Fig. 20.* Start and operate engine at 2500 RPM for 30 seconds.

3) Turn engine off and reconnect spout line. Remove orifice from PCV hose and reconnect to PCV valve. Reconnect idle speed control/air bypass solenoid. Ensure throttle is not stuck in bore and linkage is not preventing throttle from closing.

4) Start and operate engine at 2500 RPM for 30 seconds. Recheck idle speed.

Courtesy of Ford Motor Co.

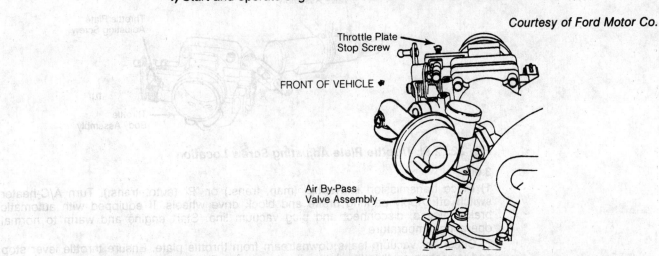

Fig. 20: 2.3L High Swirl Combustion Throttle Plate Adjusting Screw Location

2.3L Turbo

1) Place transmission in Neutral and apply parking brake. Turn A/C-heater switch off. Start engine and warm to normal operating temperature. Stop engine. Disconnect power lead to Idle Speed Control (ISC) air by-pass valve.

2) Start engine and operate at 2000 RPM for 2 minutes. If electric cooling fan comes on, disconnect power lead to fan. Let engine idle and check base idle speed (800 RPM maximum). Adjust speed by turning throttle plate adjusting screw. *See Fig. 21.*

3) Turn engine off. Reconnect by-pass valve and electric cooling fan power leads. Move throttle plate to ensure it is not stuck in bore. If idle speed is too high, turn ignition off and then restart vehicle.

Courtesy of Ford Motor Co.

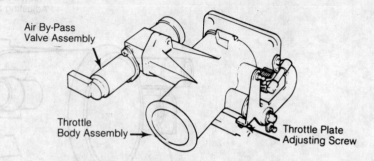

Fig. 21: 2.3L Turbo Throttle Plate Adjusting Screw Location

2.9L

1) Place transmission in Neutral (man. trans.) or "P" (auto. trans.). Turn A/C-heater switch off. Start engine and warm to normal operating temperature. Stop engine. Disconnect power lead to idle speed control (ISC) air by-pass valve.

2) Start engine and operate at 2000 RPM for 2 minutes. Let engine idle and check base idle speed (600-700 RPM maximum on man. trans.; 675-725 RPM maximum on auto. trans.).

3) If idle speed is too high, turn throttle plate adjusting screw until specified idle speed is obtained. *See Fig. 22.* Turn engine off. Restart engine and recheck idle speed. Readjust idle speed if necessary.

4) Turn engine off. Reconnect ISC by-pass valve power lead. Move throttle plate to ensure it is not stuck in bore. If excessive engine idle speeds are encountered during driving, turn ignition off and then restart vehicle.

Courtesy of Ford Motor Co.

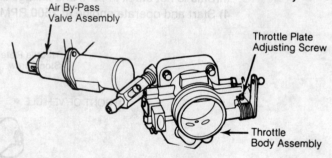

Fig. 22: 2.9L Throttle Plate Adjusting Screw Location

3.0L

1) Place transmission in Neutral (man. trans.) or "P" (auto. trans.). Turn A/C-heater switch off. Apply parking brake and block drive wheels. If equipped with automatic brake release, disconnect and plug vacuum line. Start engine and warm to normal operating temperature.

2) Check for vacuum leaks downstream from throttle plate. Ensure throttle lever stop pad is resting on throttle plate adjusting screw. Ensure throttle linkage is not binding. Perform EEC-IV diagnosis and repair any faults.

3) Stop engine. Disconnect negative battery terminal for minimum of 5 minutes. Reconnect negative battery terminal.

4) Start engine and check curb idle speed (700 RPM maximum on passenger cars; 650-750 RPM maximum on auto. trans. light duty trucks and vans; 750-850 RPM maximum on man. trans. light duty trucks and vans). If idle speed is not as specified, unplug "SPOUT" line and ensure ignition timing is 8-12 degrees BTDC. Reset ignition timing if required.

5) With "SPOUT" line unplugged, remove PCV entry line at PCV valve and install Orifice (T86P-9600-A) in PCV entry line. Disconnect power lead to idle speed control (ISC) air by-pass valve.

6) Start engine. With transmission in "D" (auto. trans.) or Neutral (man. trans.), check idle speed.

NOTE

- *If electric cooling fan engages during idle speed check, unplug fan electrical connector, or wait until fan stops.*

7) If idle speed is not within specifications, turn throttle plate adjusting screw until specified idle speed is obtained. *See Fig. 23.* Turn engine off and wait 3-5 minutes. Repeat step **6)**. Stop engine and repeat step **3)**.

8) Remove orifice and reconnect PCV entry line at PCV valve. Reconnect "SPOUT" line. Reconnect ISC by-pass valve power lead and electric cooling fan (if disconnected). Start engine and run for 1-2 minutes. Recheck curb idle speed. On models with auto. trans., check throttle valve pressure adjustment.

Courtesy of Ford Motor Co.

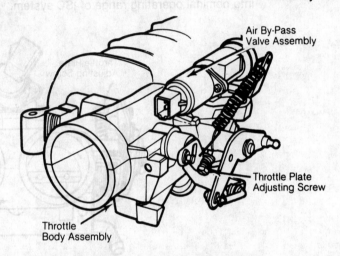

Fig. 23: 3.0L Throttle Plate Adjusting Screw Location

3.8L RWD & AXOD Transmission

1) Place transmission in Neutral (man. tans.) or "P" (auto. trans.). Turn A/C-heater switch off. Apply parking brake and block drive wheels. If equipped with automatic brake release, disconnect and plug vacuum line. Start engine and warm to normal operating temperature.

2) Operate engine at 2500 RPM for 30 seconds. Set idle speed according to specifications. *See Fig. 24.*

Courtesy of Ford Motor Co.

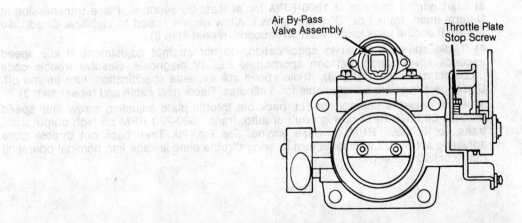

Fig. 24: 3.8L RWD & AXOD Throttle Plate Adjusting Screw Location

4.9L & 5.0L Light Trucks & Vans

1) Set parking brake and block drive wheels. Start engine and warm to normal operating temperature. Stop engine. Place transmission in Neutral (man. trans.) or "P" (auto. trans.). Turn A/C-heater switch off.

2) Ensure throttle linkage is free and travel is unobstructed. Ensure cruise control cable (if equipped) is not holding throttle open. Check for vacuum leaks.

3) Start engine and run at 1800 RPM for at least 30 seconds. Place transmission in Neutral (man. trans.) or "D" (auto. trans.). Allow engine speed to stabilize. Check idle speed. If engine idles longer than 30 seconds, repeat step **3)**.

4) If idle speed is less than or equal to specification, do not attempt adjustment. If idle speed exceeds specification, turn engine off. Disconnect negative battery cable for 5 minutes, then reconnect cable. If speed still exceeds specification, perform EEC-IV diagnosis and repeat step **3)**.

5) If idle speed is still not correct, back out throttle plate adjusting screw until speed reaches 580-620 RPM (auto. trans.) or 680-720 RPM (man. trans.). *See Fig. 25.* Then back out throttle plate adjusting screw 1/2 additional turn to bring throttle plate linkage into nominal operating range of ISC system.

Courtesy of Ford Motor Co.

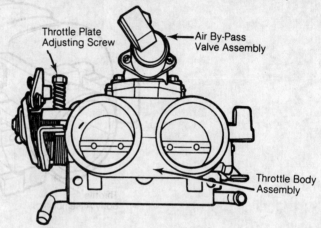

Throttle Plate Adjusting Screw

Air By-Pass Valve Assembly

Throttle Body Assembly

Fig. 25 : Throttle Plate Adjusting Screw Location for 4.9L, 5.0L & 5.8L Light Trucks & Vans

5.0L Passenger Cars

1) Set parking brake and block drive wheels. Start engine and warm to normal operating temperature. Stop engine. Place transmission in Neutral (man. trans.) or "P" (auto. trans.). Turn A/C-heater switch off.

2) Ensure throttle linkage is free and travel is unobstructed. Ensure cruise control cable (if equipped) is not holding throttle open. Check for vacuum leaks.

3) Start engine and run at 1800 RPM for at least 30 seconds. Place transmission in Neutral (man. trans.) or "D" (auto. trans.). Allow engine speed to stabilize. Check idle speed. If engine idles longer than 30 seconds, repeat step **3)**.

4) If idle speed is equal to specification, do not attempt adjustment. If idle speed exceeds specification, perform appropriate EEC-IV diagnosis. Resolve trouble code problems and repeat step **3)**. If idle speed still exceeds specification, turn engine off. Disconnect negative battery cable for 5 minutes. Reconnect cable and repeat step **3)**.

5) If idle speed is still not correct, back out throttle plate adjusting screw until speed reaches 605-645 RPM on high output auto. trans., 680-720 RPM on high output man. trans., or 555-595 RPM on base engine. *See Fig. 26.* Then back out throttle plate adjusting screw 1/2 additional turn to bring throttle plate linkage into nominal operating range of ISC system.

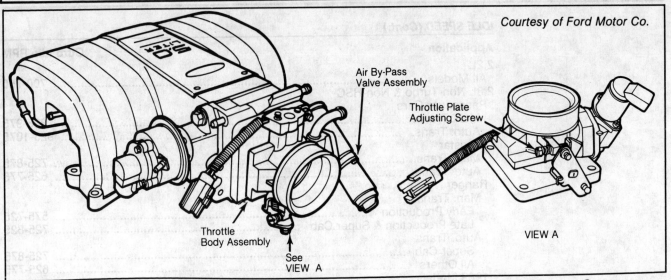

Courtesy of Ford Motor Co.

Fig. 26: Throttle Plate Adjusting Screw Location for 5.0L Passenger Cars

5.8L & 7.5L Light Trucks & Vans

1) Place transmission in Neutral (man. trans.) or "P" (auto trans.). Turn A/C-heater switch off. Apply parking brake and block drive wheels. Start engine and warm to normal operating termperature.

2) On 7.5L engines, disconnect idle speed control/air by-pass solenoid. Idle engine at 2500 RPM for 30 seconds. *See Fig. 27.* On 5.8L engine, idle at 1800 RPM for one minute. *See Fig. 25.* If engine stalls turn throttle stop screw clockwise and repeat test.

3) Adjust idle speed to specification. On 5.8L engine, idle speed must be adjusted within 30 seconds. On 7.5L engine, idle speed must be adjusted within 40 seconds. With engine off, reconnect idle speed control/air by-pass solenoid. Ensure throttle is not stuck in bore and linkage is not preventing throttle from opening.

Courtesy of Ford Motor Co.

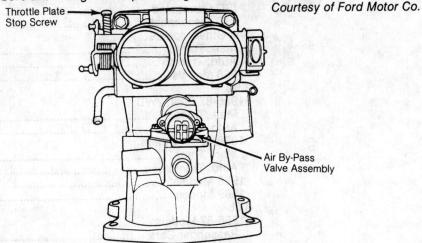

Fig. 27: 7.5L Throttle Plate Adjusting Screw Location

IDLE SPEED

Application	Curb Idle RPM
1.6L & 1.6L Turbo	
Man. Trans.	800
Auto. Trans.	[1] 750
1.9L	
1986-87	
Man. Trans.	900
Auto. Trans.	800
1988-89 Models	900-1000

[1] – Set Turbo models to 800 RPM.

IDLE SPEED (Cont.)

Application	Curb Idle RPM
2.2L	
All Models	700-750
2.3L Non-Turbo & Non-HSC	
Passenger Cars	
Man. Trans.	825-975
Auto. Trans.	925-1075
Aerostar	
Man. Trans.	725-875
Auto. Trans.	625-775
Ranger	
Man. Trans.	
Early Production	575-725
Late Production & Super Cab	725-825
Auto. Trans.	
Super Cab	725-875
All Others	625-775
1989 Models	
All Trucks	550-600
2.3L OHC	
Passenger Car	500-550
1989 Models	
Auto. Trans.	625-675
Man. Trans.	575-625
Light Truck or Van	600-650
2.3L HSC	
Man. Trans.	1500-1600
Auto. Trans.	975-1075
2.3L Turbo	
1984 Models	700-800
1985-86 Merkur XR4Ti	
ISC Connected	825-975
ISC Disconnected	700-800
All Other Models	
Man. Trans.	825-975
Auto. Trans.	925-1075
1988 Models	700-800
2.9L	
1986-87 Calif. 2WD	
Man. Trans.	900
Auto. Trans.	800
1986-87 All Others	
Man. Trans.	850
Auto. Trans.	800
1988 Models	700
1989 Models	725
3.0L	
1986-87 Models	
Passenger Cars	595-655
Light Trucks & Vans	
Man. Trans.	875-925
Auto. Trans.	990-1090
1988-89 Models	
Passenger Cars	740-780
Light Trucks & Vans	
Man. Trans.	700-750
Auto. Trans.	600-650
3.8L	
Front and Rear Wheel Drive	
Auto. & Man. Trans.	620-720
Supercharged	[1]

IDLE SPEED (Cont.)

Application	Curb Idle RPM
4.9L	
1987 Models	
Man. Trans.	775 Max.
Auto. Trans.	675 Max.
1988 Models	
Man. Trans.	670-750
Auto. Trans.	570-650
1989 Models	
All Trucks	650-750
5.0L	
1985-87 Light Trucks & Vans	
Man. Trans.	775 Max.
Auto. Trans.	675 Max.
1986-87 Passenger Cars	
Base Engine	595-655
High Output Engine	
Man. Trans.	625-775
Auto. Trans.	550-700
1988 Light Trucks & Vans	
Man. Trans.	675-725
Auto. Trans.	600-650
1988 Passenger Cars	
Base Engine	555-595
High Output Engine	
Man. Trans	680-720
Auto. Trans.	605-645
1989 Light Trucks & Vans	
Auto. Trans.	675-725
Man. Trans.	700-750
1989 Passenger Cars	
Base Engine	530-570
High Output Engine	
Auto. Trans.	675-725
Man. Trans.	700-750
5.8L	
Man. Trans.	625-675
Auto. Trans.	600-650
1989 Models	
Auto. Trans.	
C-6	780-830
E4OD	730-780
Man. Trans.	730-780
7.5L	
All Models	650
1989 Models	
Auto. Trans.	650-700
Man. Trans.	650-675

[1] – Information not available from manufacturer.

COLD (FAST) IDLE SPEED	*NOTE* • *Fast idle adjustment is not possible on fuel injected engines.*

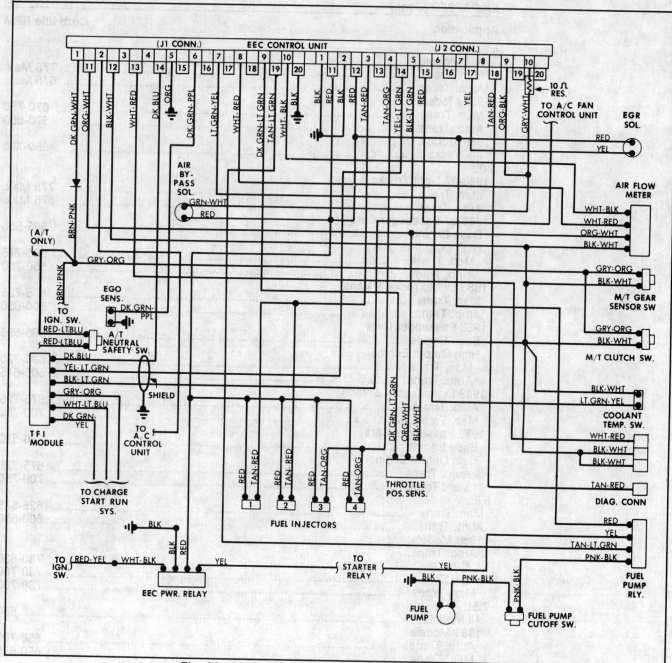

Fig. 28: 1983 1.6L Passenger Car Fuel Injection System Wiring Diagram

MULTI-POINT FUEL INJECTION

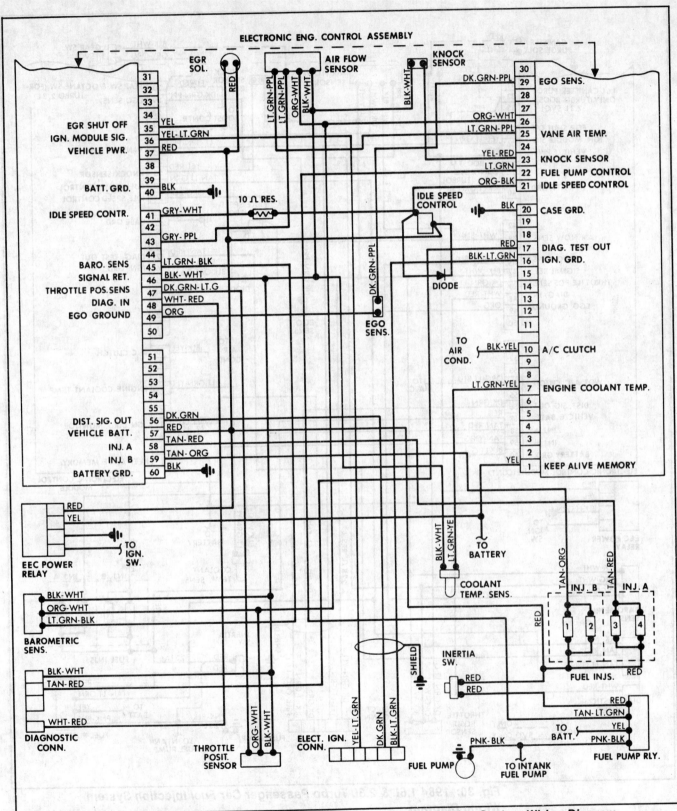

Fig. 29: 1983 2.3L Passenger Car Fuel Injection System Wiring Diagram

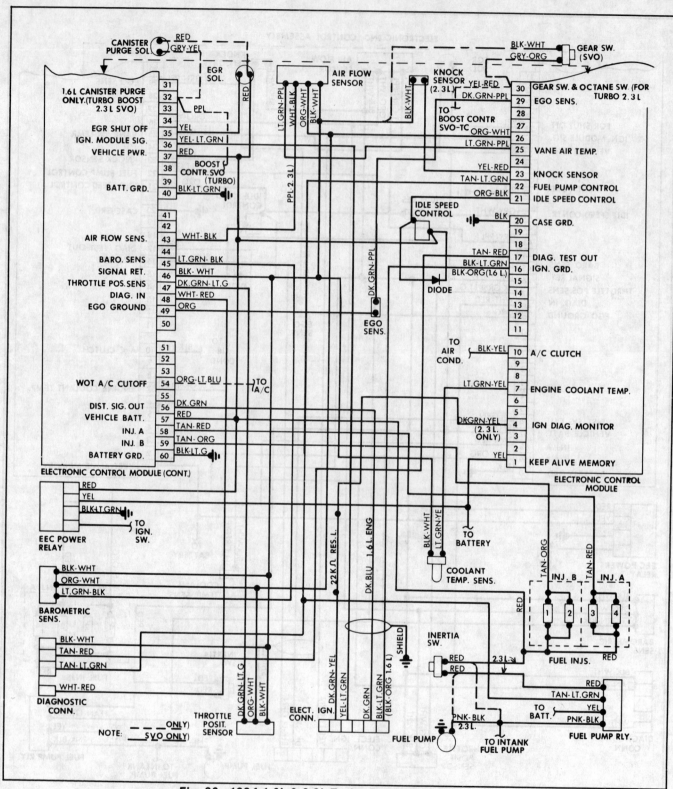

Fig. 30: 1984 1.6L & 2.3L Turbo Passenger Car Fuel Injection System Wiring Diagram

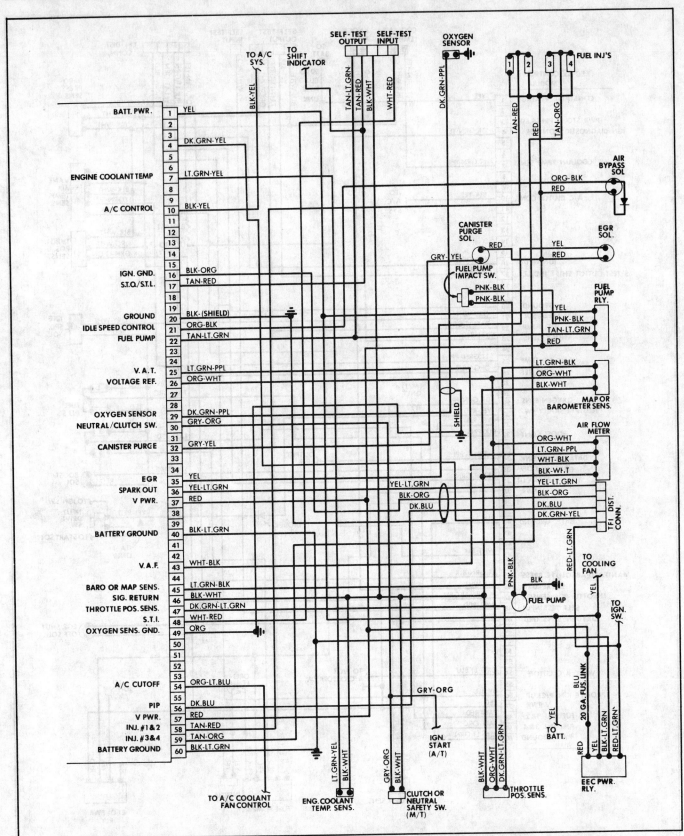

Fig. 31: 1985 1.6L & 1.6L Turbo Passenger Car Fuel Injection System Wiring Diagram

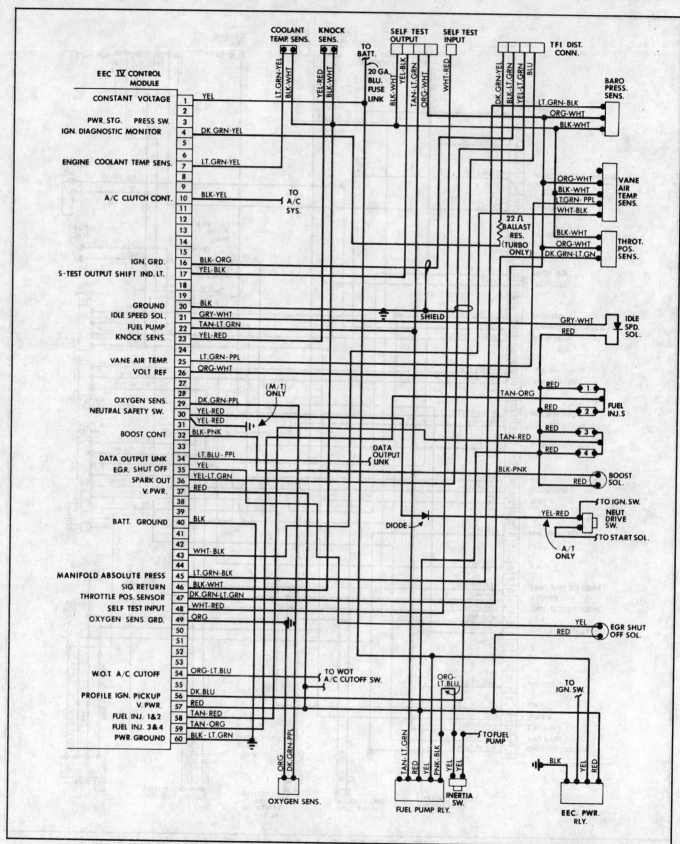

Fig. 32: 1985 2.3L Turbo Passenger Car Fuel Injection System Wiring Diagram

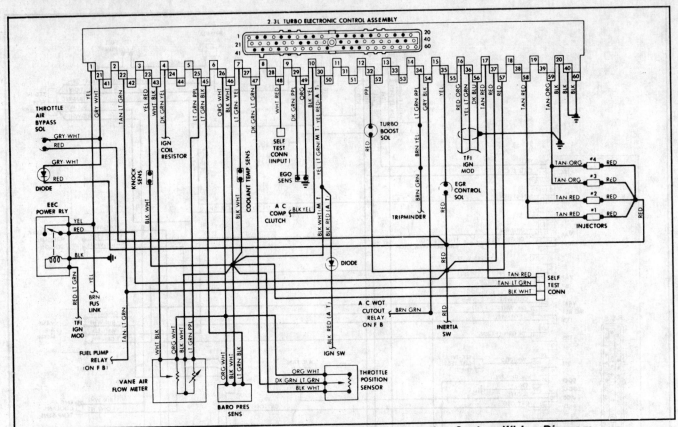

Fig. 33: 1985/86 Merkur XR4Ti Fuel Injection System Wiring Diagram

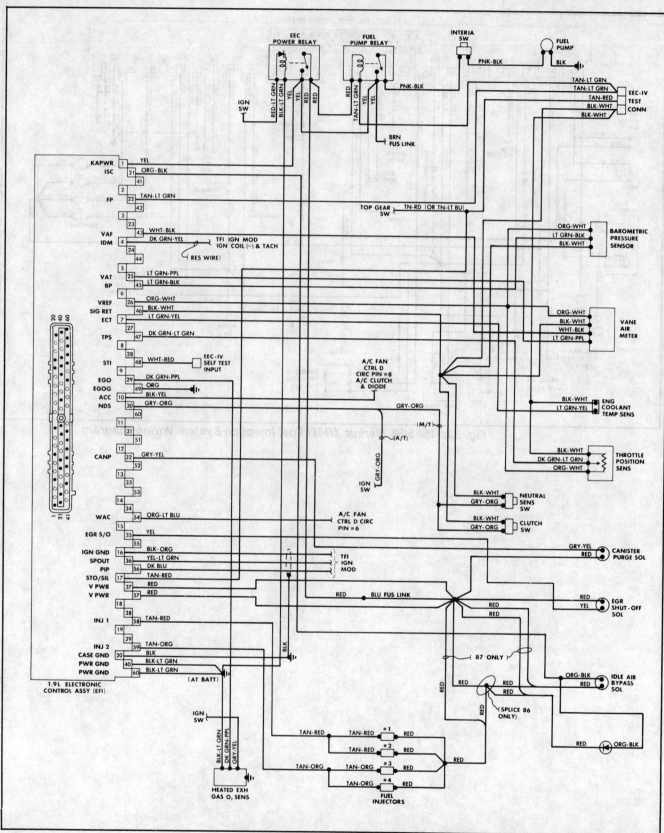

Fig. 34: 1986/87 1.9L Passenger Car Fuel Injection System Wiring Diagram

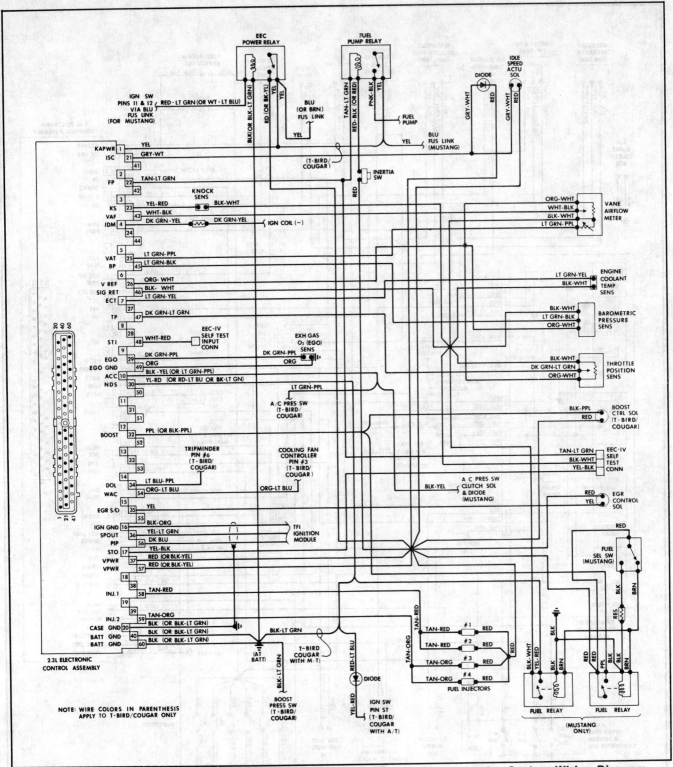

Fig. 35: 1986 2.3L Turbo Passenger Car Fuel Injection System Wiring Diagram

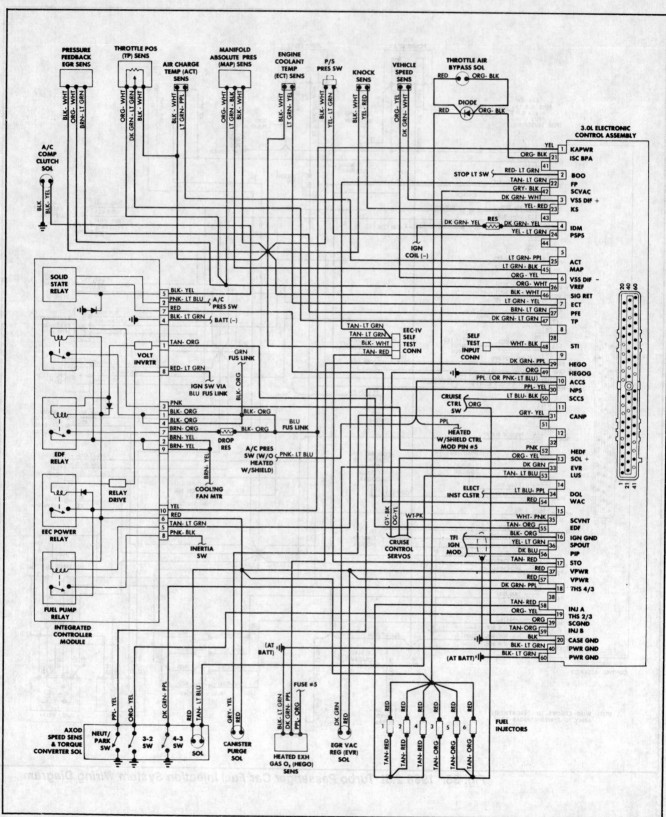

Fig. 36: 1986 3.0L Passenger Car Fuel Injection System Wiring Diagram

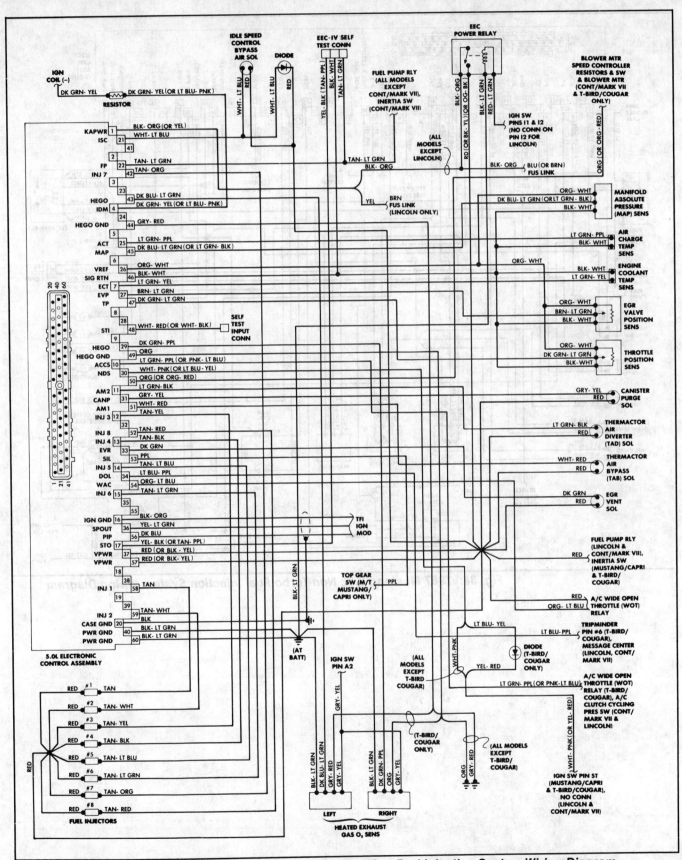

Fig. 37: 1986 5.0L Passenger Car Fuel Injection System Wiring Diagram

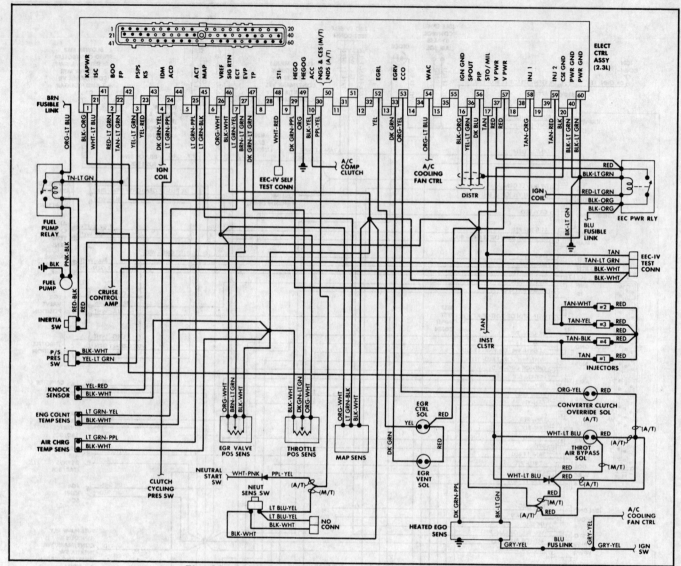

Fig. 38: 1987 Mustang 2.3L Non-Turbo Fuel Injection System Wiring Diagram

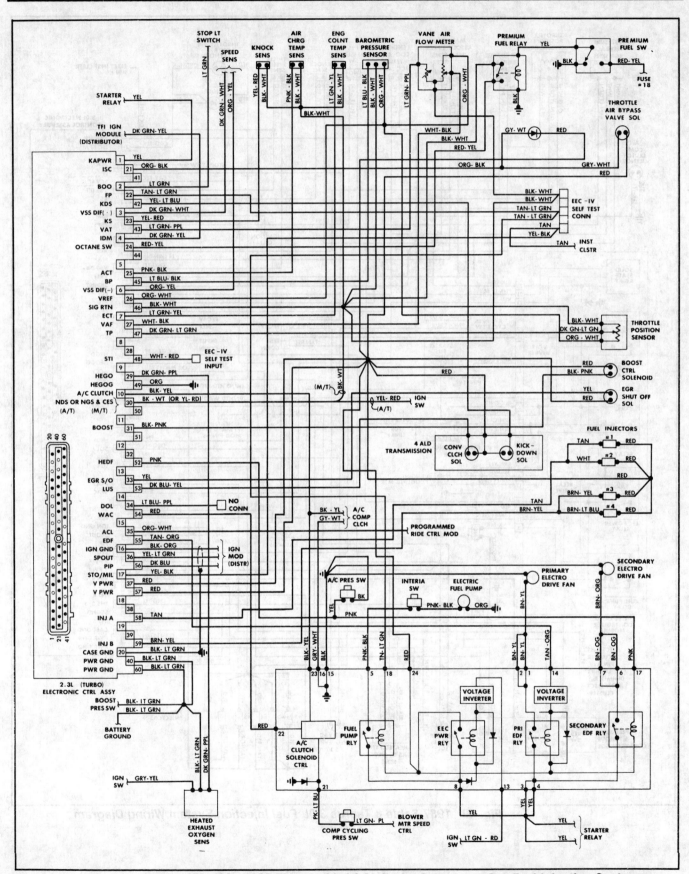

Fig. 39: 1987 Thunderbird 2.3L Turbo Passenger Car Fuel Injection System Wiring Diagram

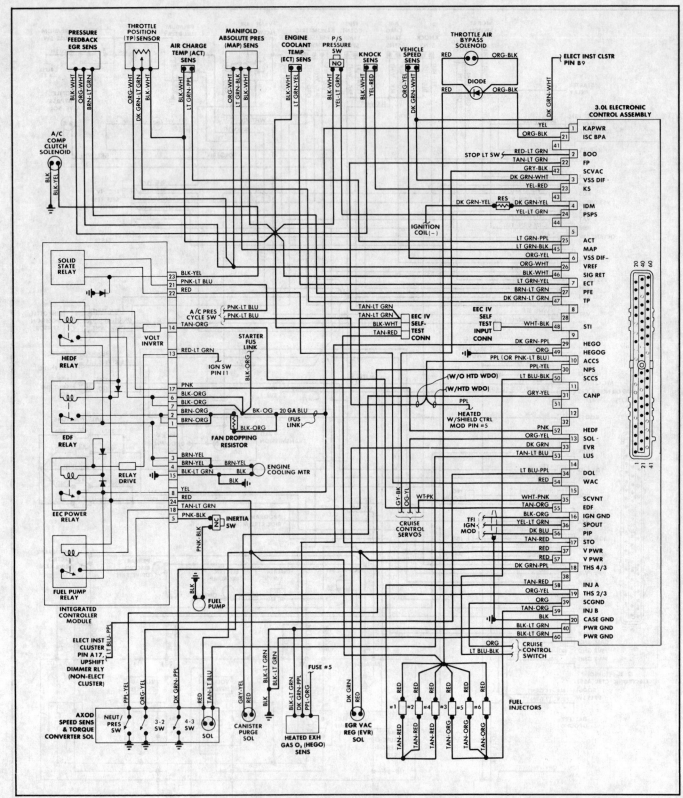

Fig. 40: 1987 Sable & Taurus 3.0L Fuel Injection System Wiring Diagram

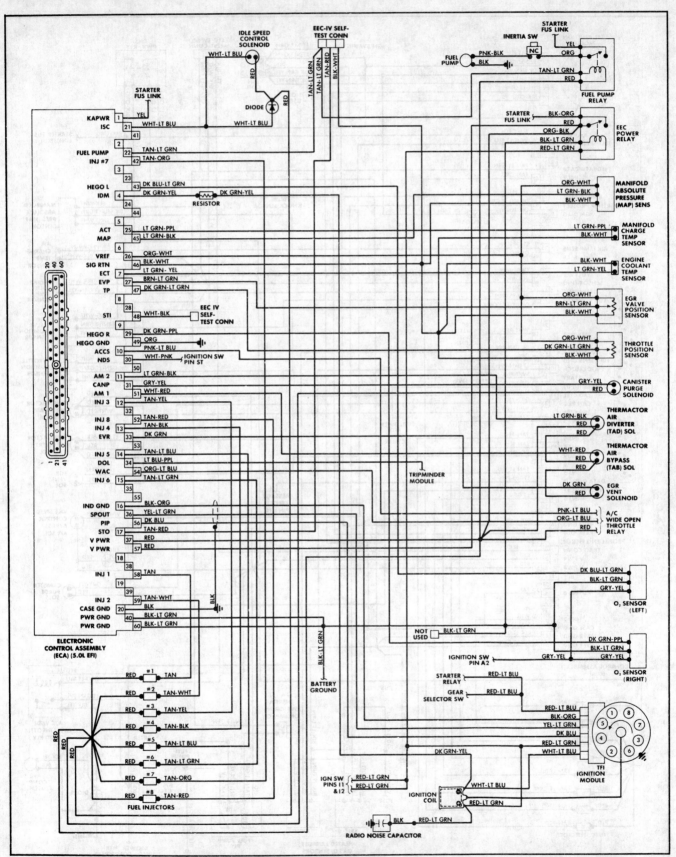

Fig. 41: 1987 Crown Victoria & Grand Marquis 5.0L Fuel Injection System Wiring Diagram

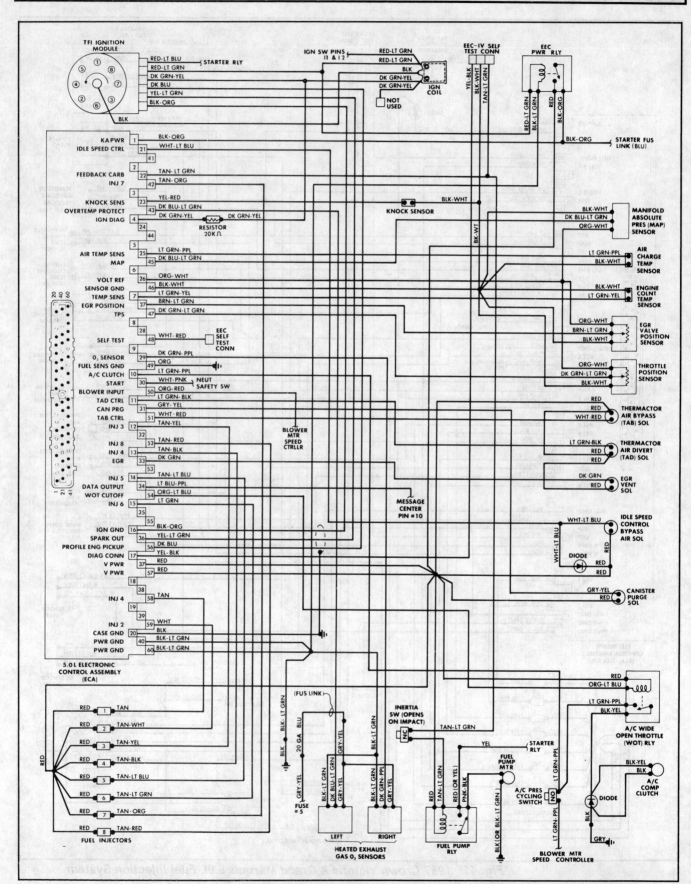

Fig. 42: 1987 Continental & Mark VII 5.0L Fuel Injection System Wiring Diagram

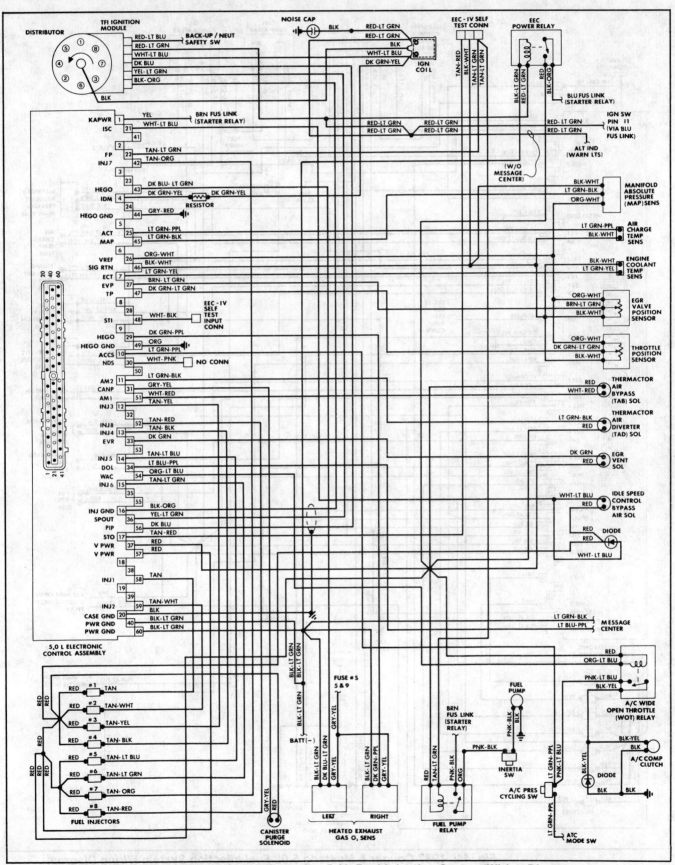

Fig. 43: 1987 Town Car 5.0L Fuel Injection System Wiring Diagram

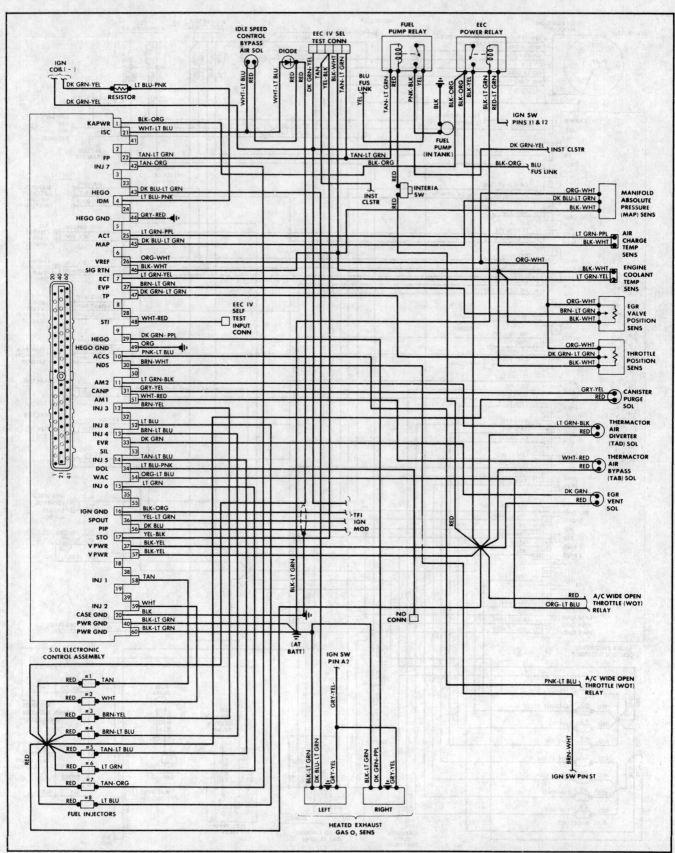

Fig. 44: *1987 Cougar & Mustang 5.0L Fuel Injection System Wiring Diagram*

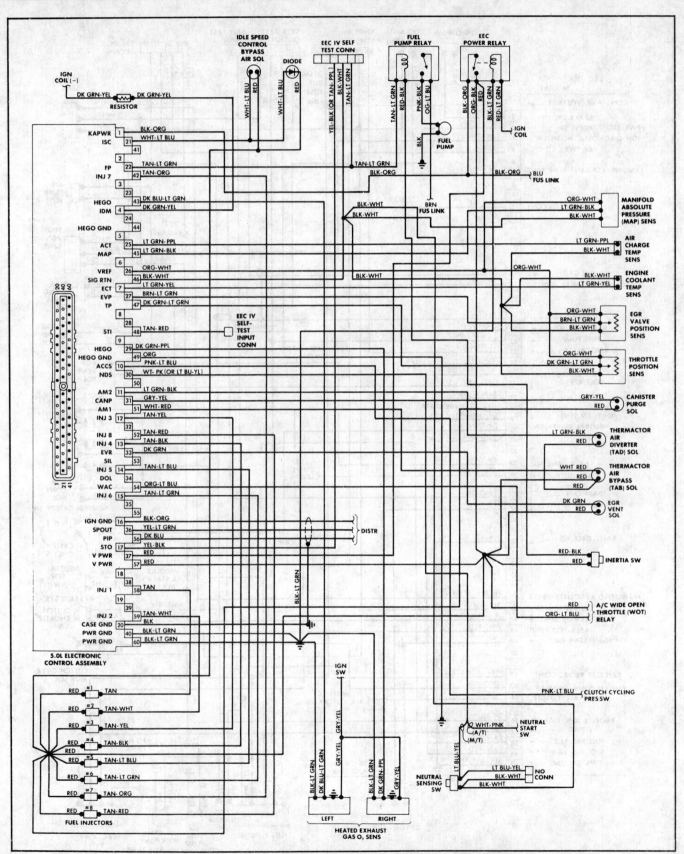

Fig. 45: 1987 Mustang 5.0L Fuel Injection System Wiring Diagram

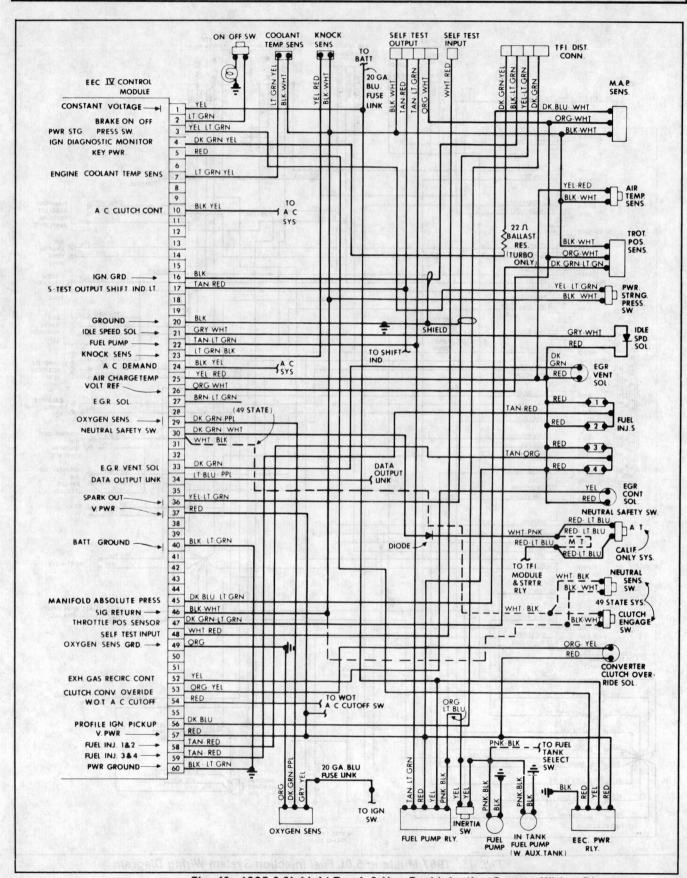

Fig. 46: 1985 2.3L Light Truck & Van Fuel Injection System Wiring Diagram

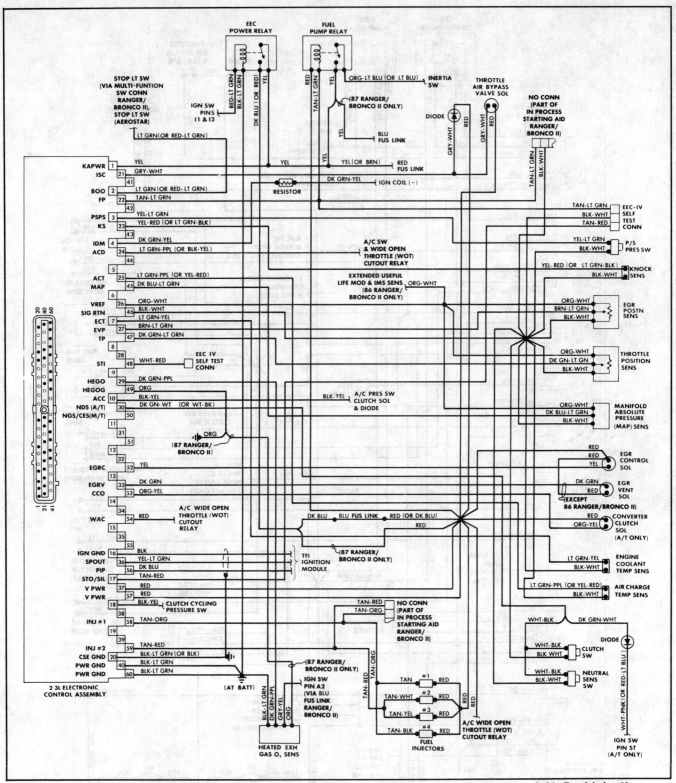

Fig. 47: 1986-87 Bronco II & Ranger and 1986-89 Aerostar 2.3L Fuel Injection System Wiring Diagram

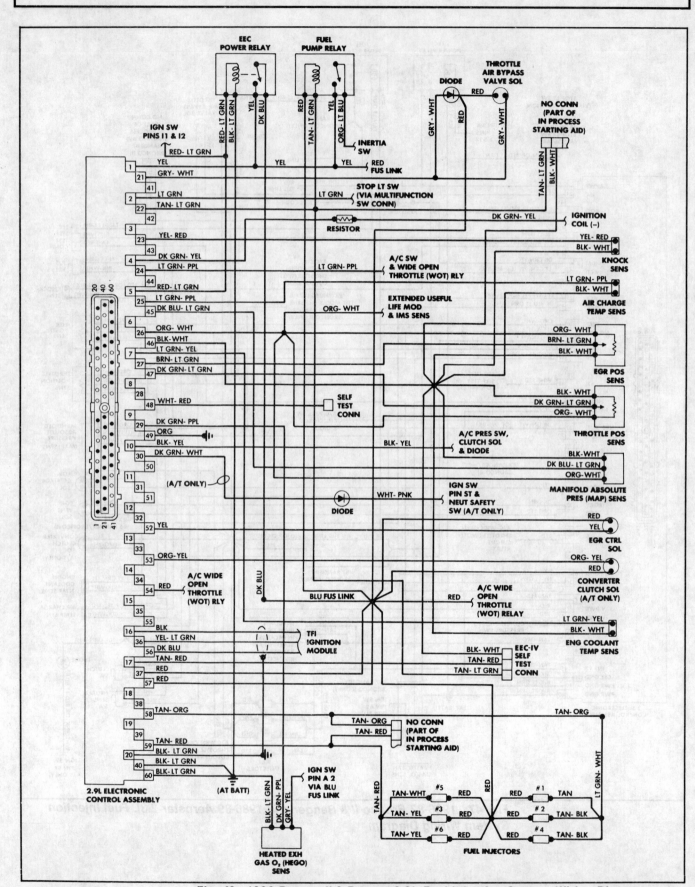

Fig. 48: 1986 Bronco II & Ranger 2.9L Fuel Injection System Wiring Diagram

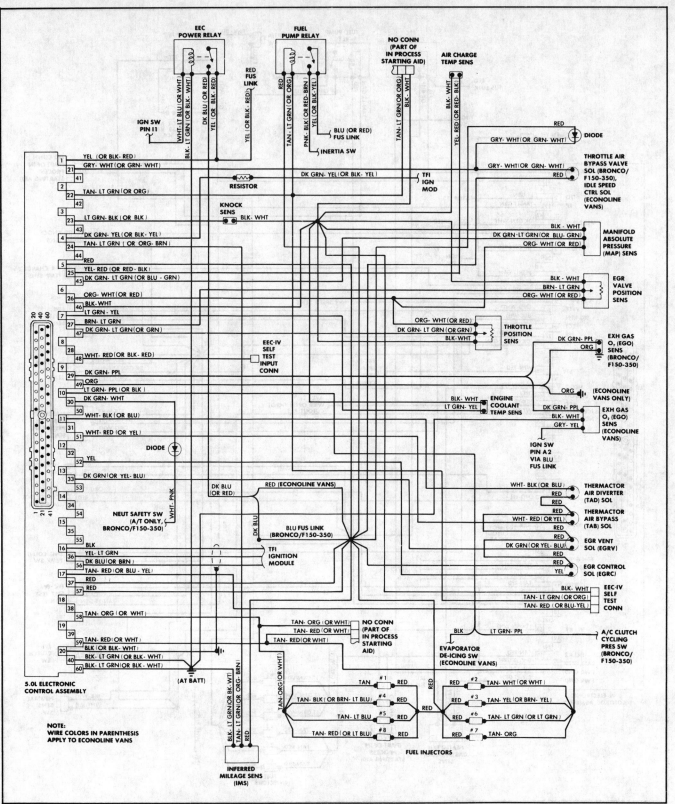

Fig. 49: 1985-86 Bronco, "E" Series & "F" Series 5.0L Fuel Injection System Wiring Diagram

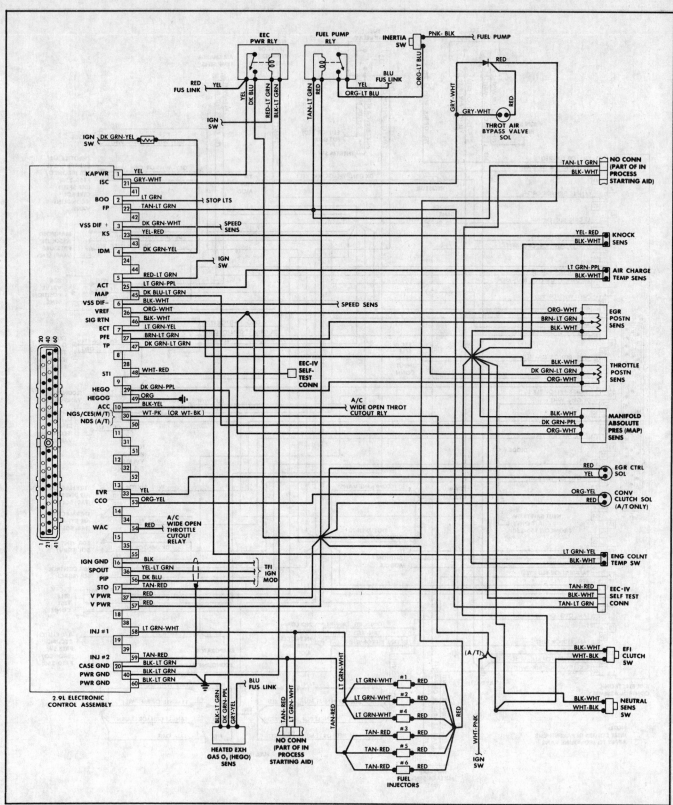

Fig. 50: 1987 Bronco II & Ranger 2.9L Fuel Injection System Wiring Diagram

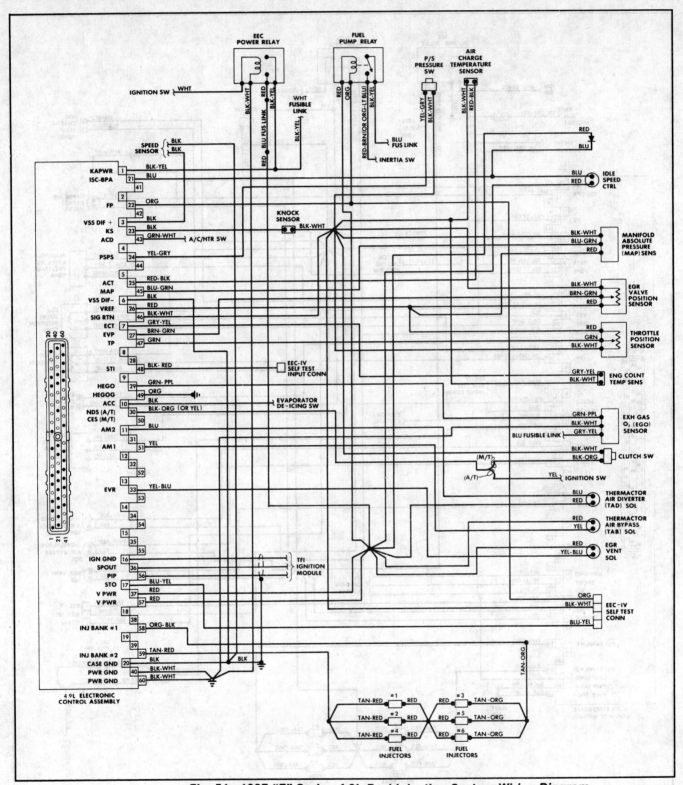

Fig. 51: 1987 "E" Series 4.9L Fuel Injection System Wiring Diagram

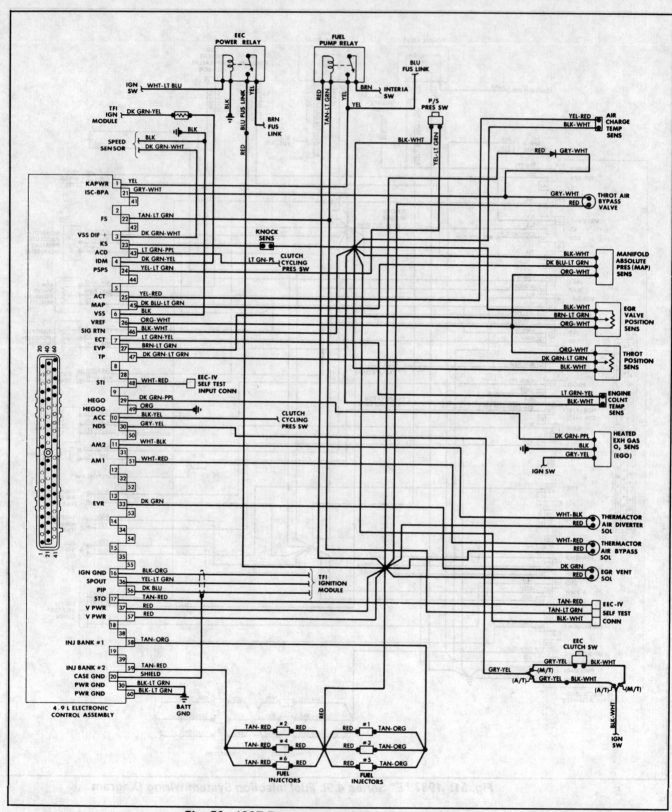

Fig. 52: *1987 Bronco & "F" Series 4.9L Fuel Injection System Wiring Diagram*

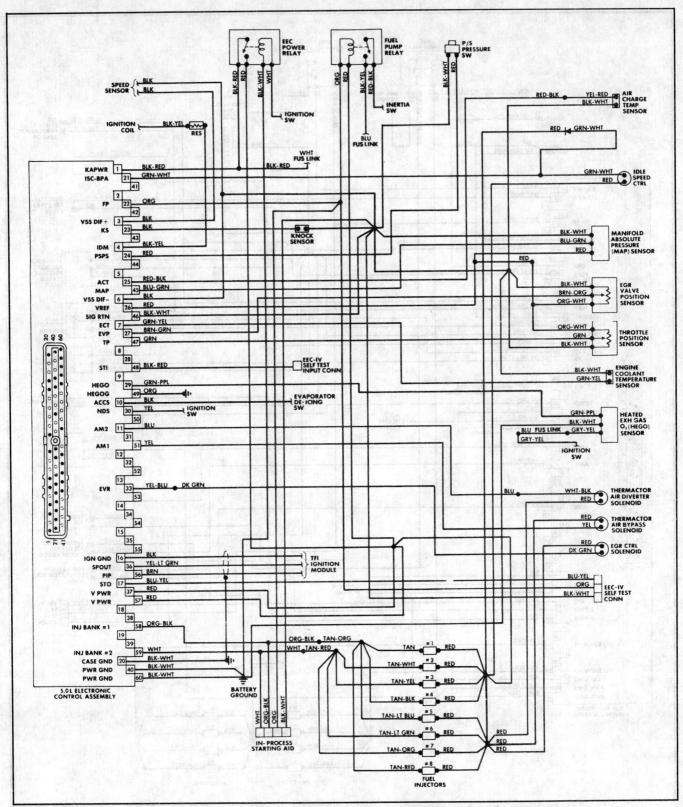

Fig. 53: 1987 "E" Series 5.0L Fuel Injection System Wiring Diagram

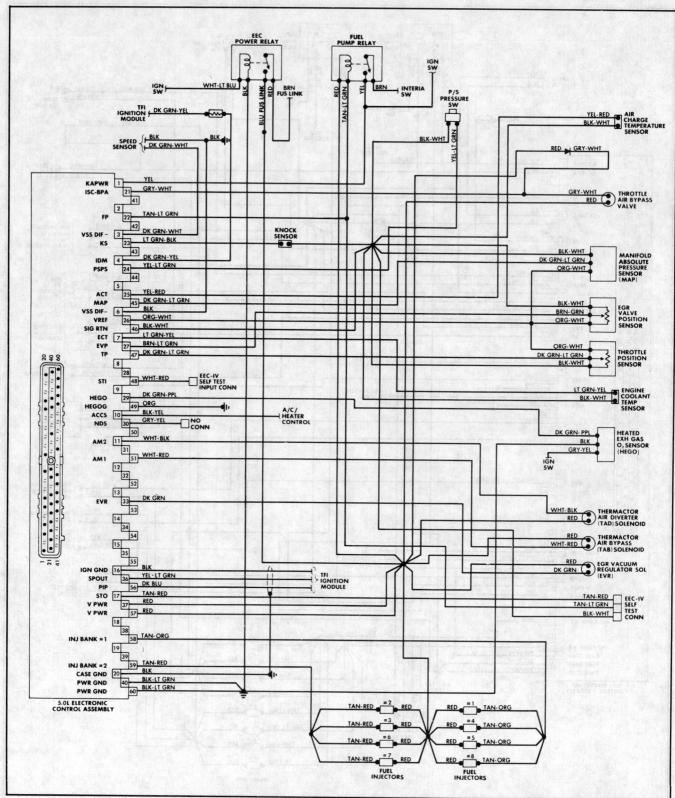

Fig. 54: 1987 Bronco & "F" Series 5.0L Fuel Injection System Wiring Diagram

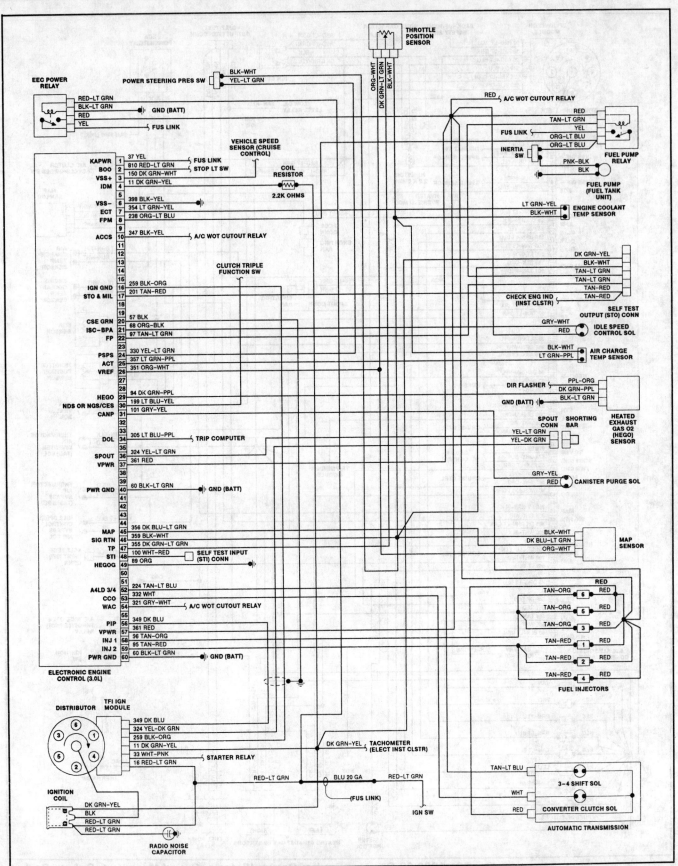

Fig. 55: *1987-89 Aerostar 3.0L Electronic Fuel Injection System Wiring Diagram*

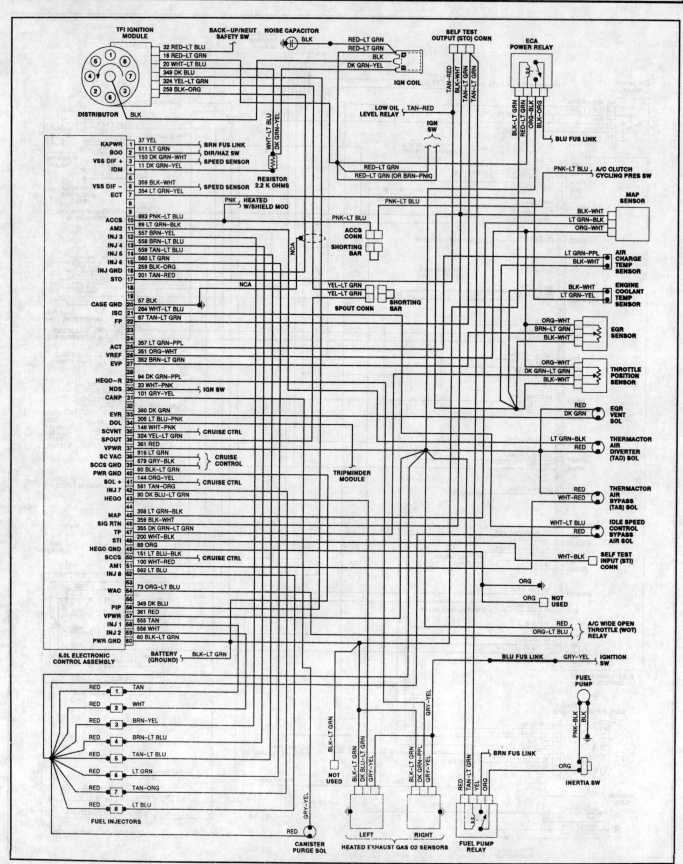

Fig. 56: 1988-89 Country Squire, Crown Victoria, Grand Marquis & Colony Park 5.0L Electronic Fuel Injection Wiring Diagram

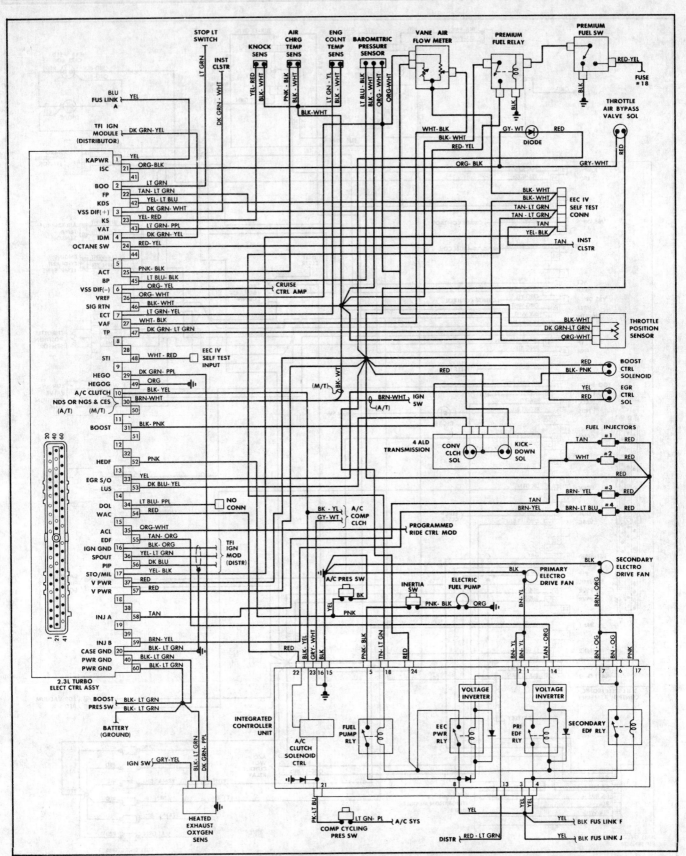

Fig. 57: 1988 Cougar/Thunderbird 2.3L Turbo Electronic Fuel Injection System Wiring Diagram

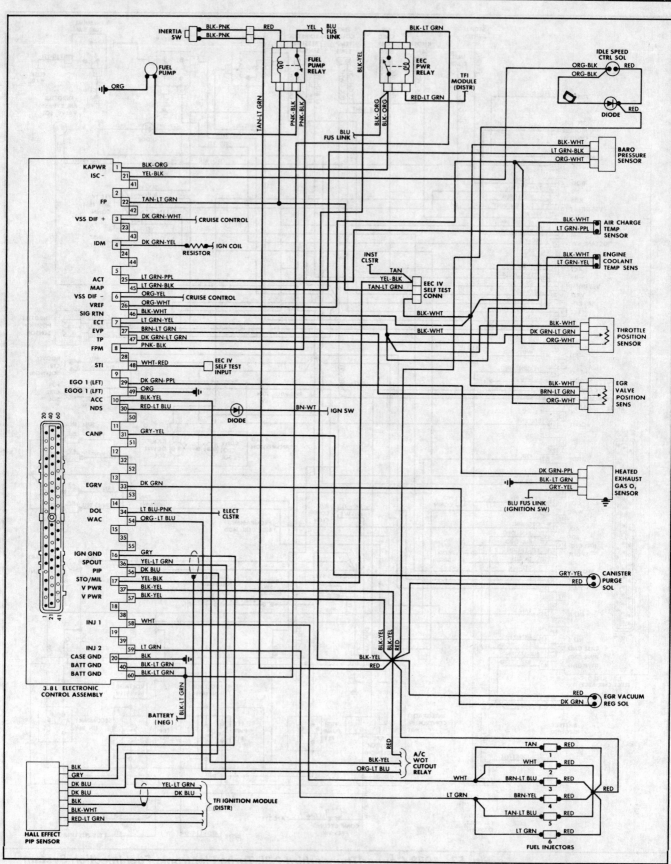

Fig. 58: 1988 Cougar/Thunderbird 3.8L Electronic Fuel Injection System Wiring Diagram

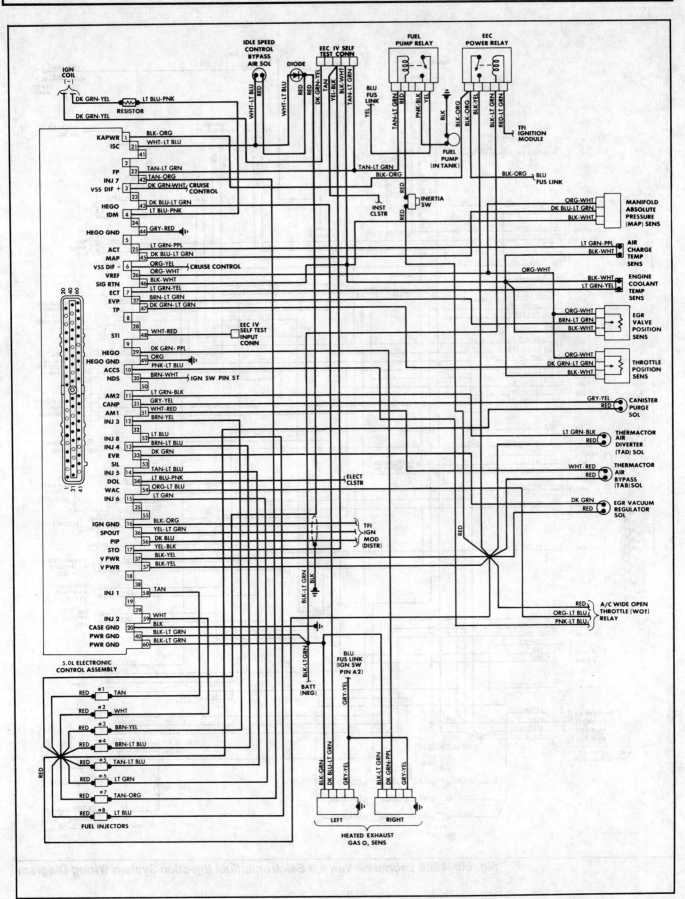

***Fig. 59:** 1988 Cougar/Thunderbird 5.0L Electronic Fuel Injection System*
Wiring Diagram

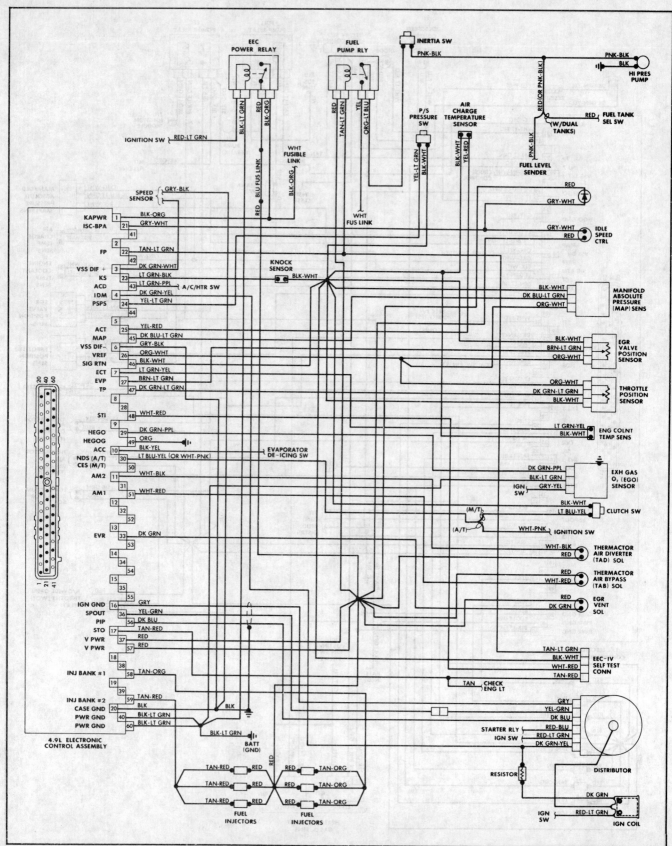

Fig. 60: 1988 Econoline Van 4.9 Eelctronic Fuel Injection System Wiring Diagram

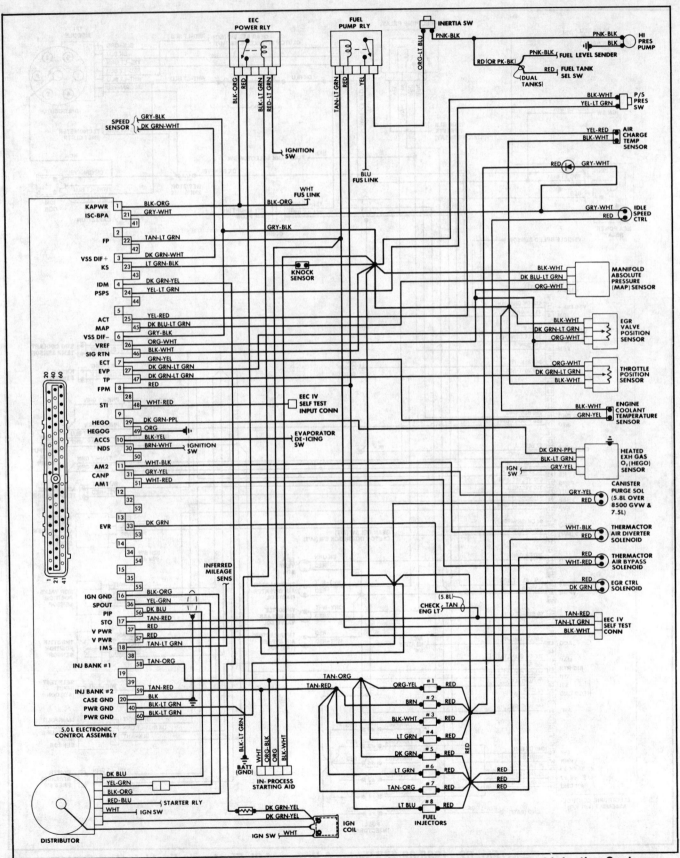

Fig. 61: 1988 Econoline Van 5.0L, 5.8L, 7.5L Electronic Fuel Injection System Wiring Diagram

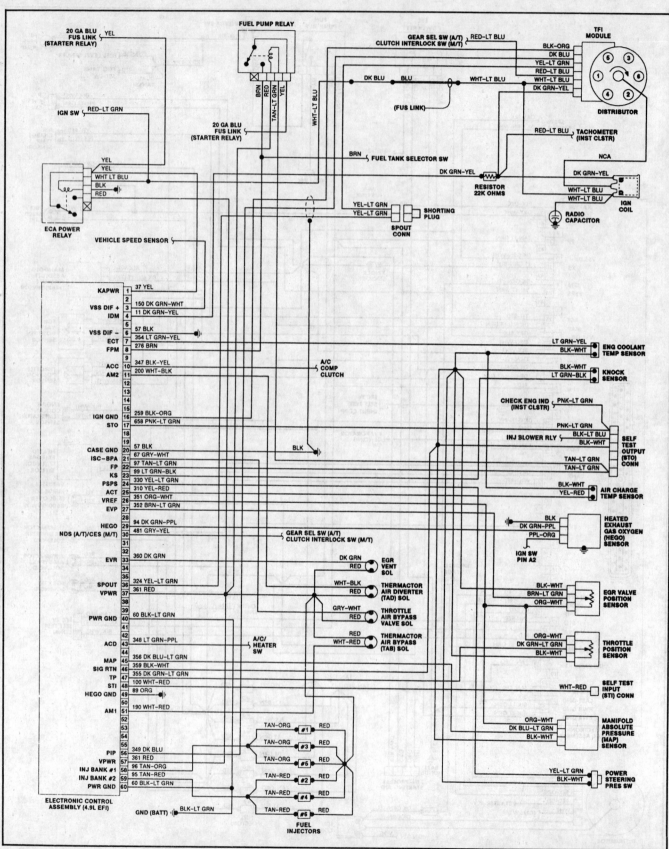

Fig. 62: 1988-89 "F" Series & Bronco 4.9L Electronic Fuel Injection System Wiring Diagram

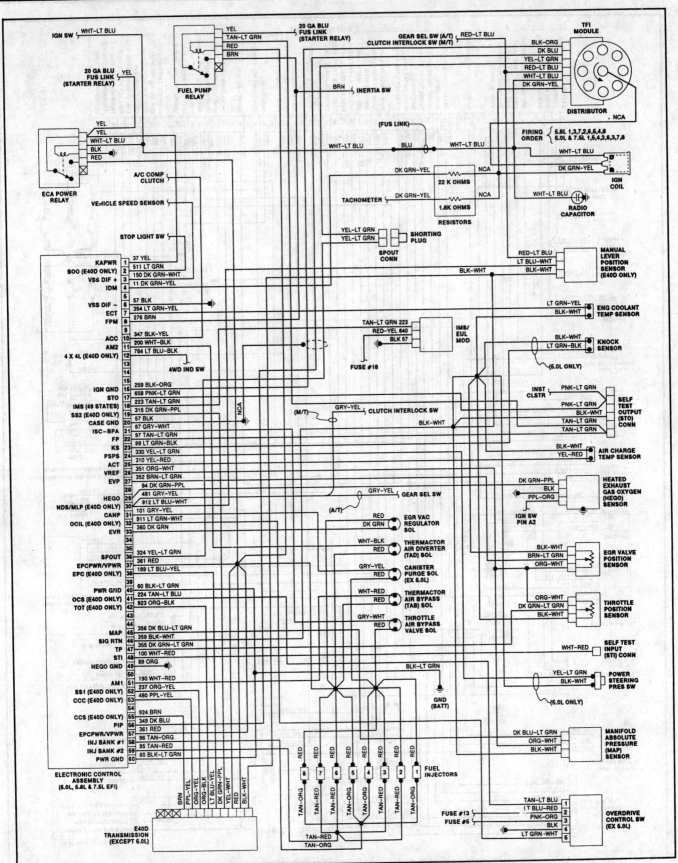

Fig. 63: 1988-89 "F" Series & Bronco 5.0L, 5.8L, & 7.5L Electronic Fuel Injection System Wiring Diagram

299

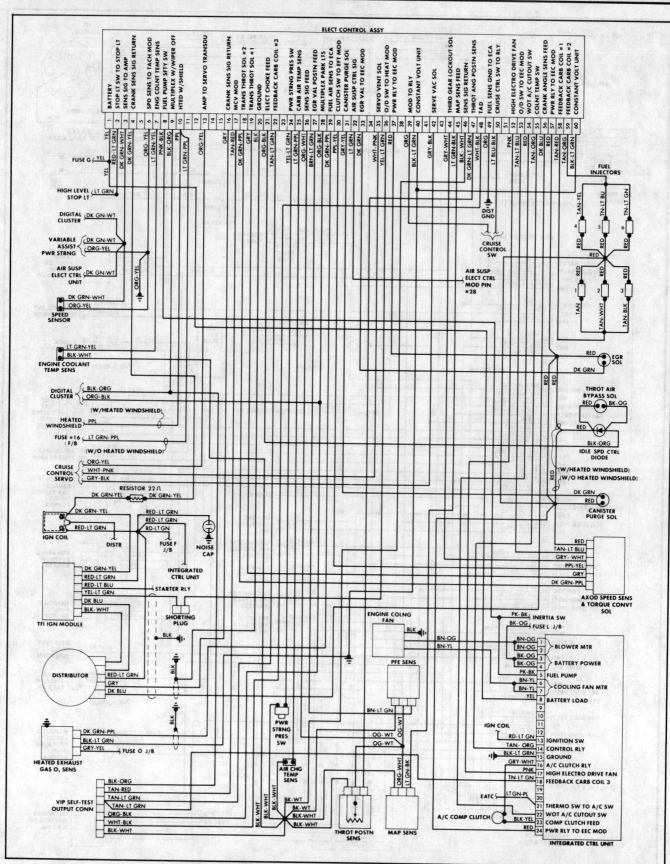

Fig. 64: 1988 Lincoln Continental 3.8L Electronic Fuel Injection System Wiring Diagram

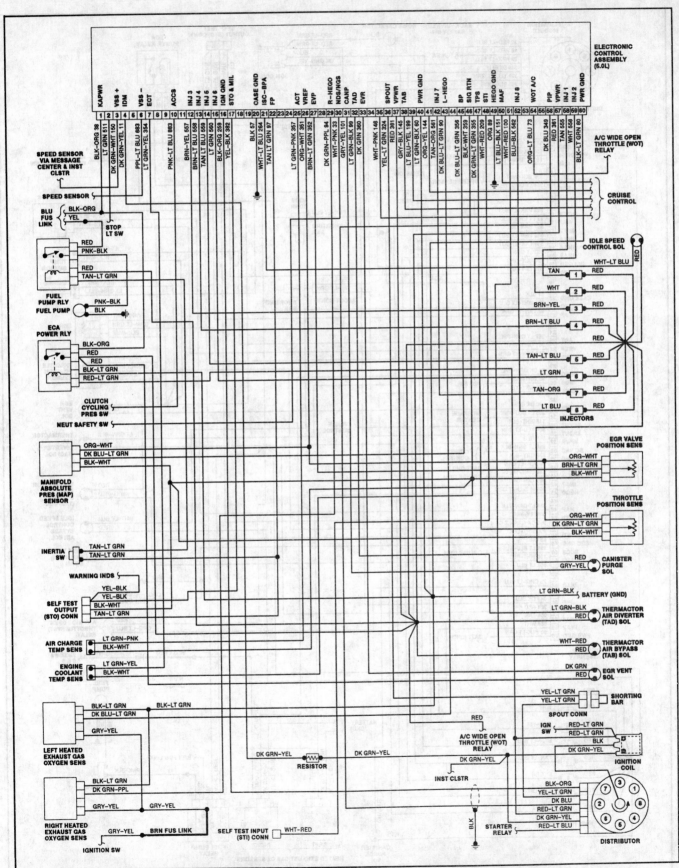

Fig. 65: 1988-89 Mark VII 5.0L Electronic Fuel Injection System Wiring Diagram

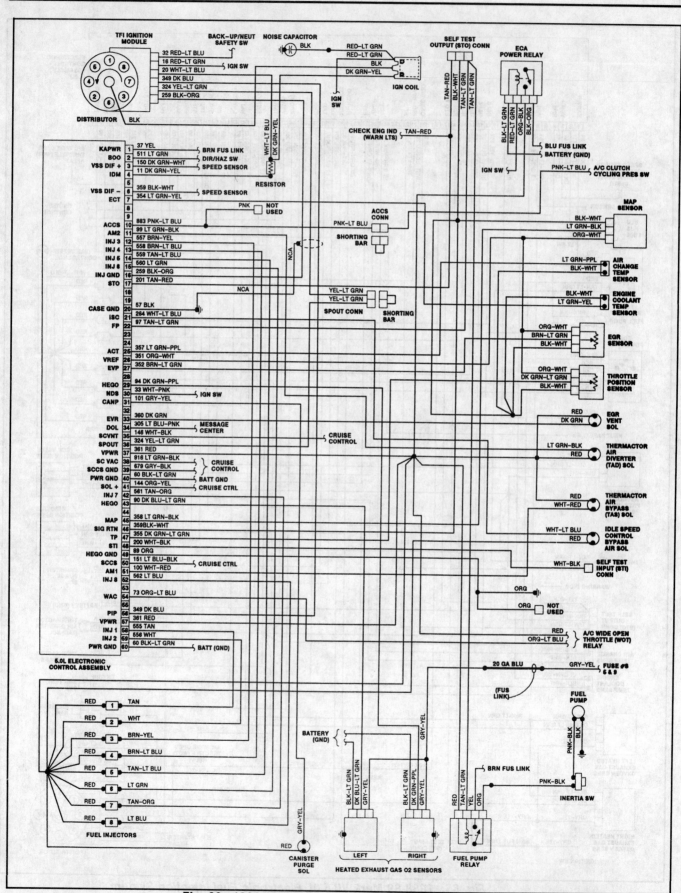

Fig. 66: 1988-89 Lincoln Town Car 5.0L Electronic Fuel Injection System Wiring Diagram

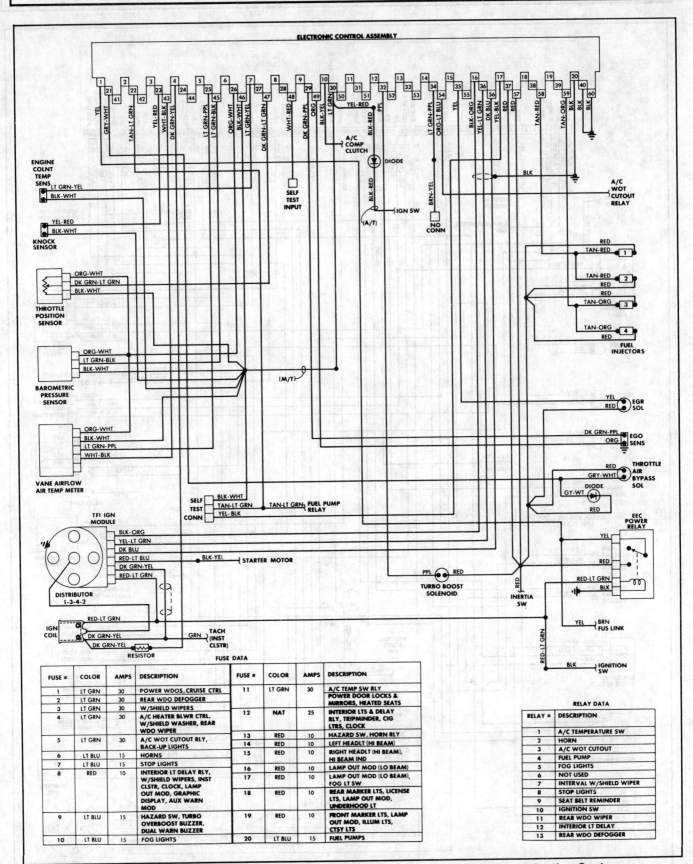

Fig. 67: 1988 Merkur XR4Ti 2.3L Turbo Electronic Fuel Injection System Wiring Diagram

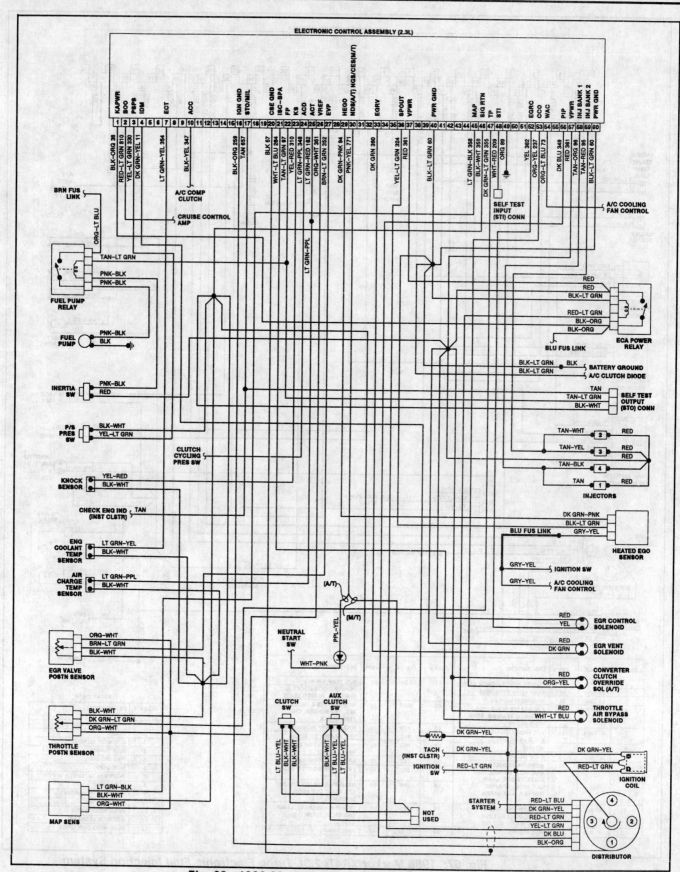

Fig. 68: 1988-89 Mustang 2.3L OHC Electronic Fuel Injection System
Wiring Diagram

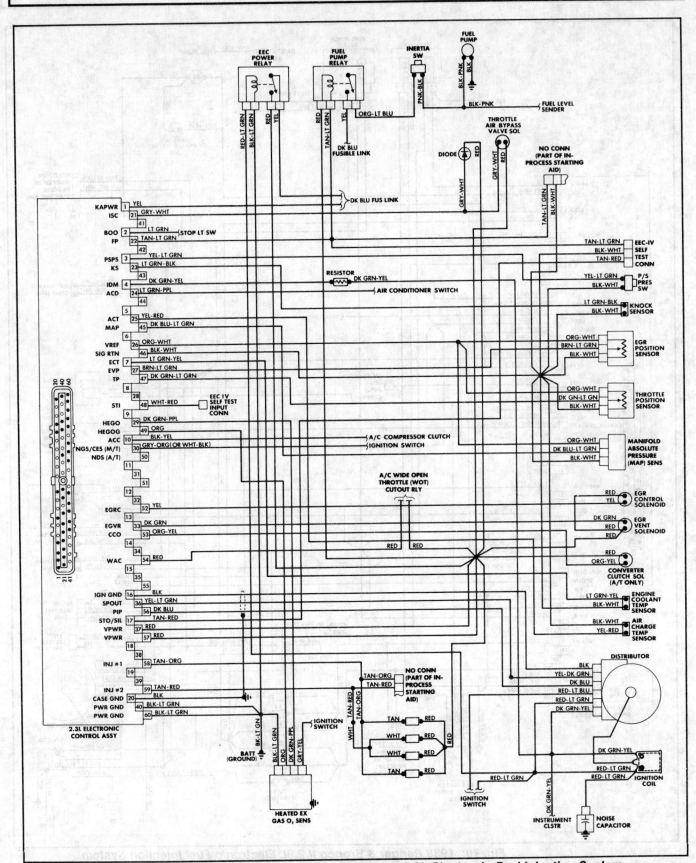

Fig. 69: 1988 Ranger & Bronco II 2.3L Electronic Fuel Injection System Wiring Diagram

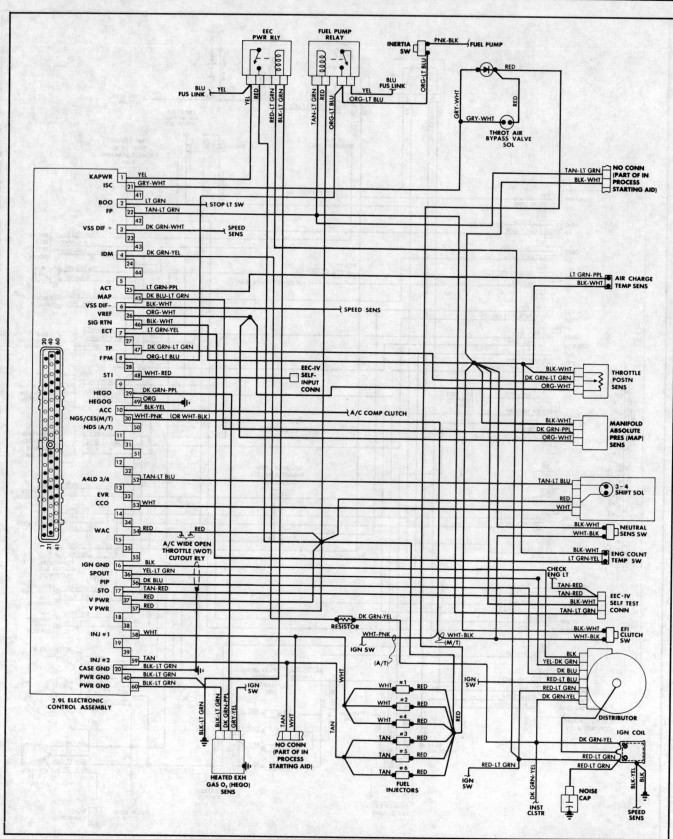

Fig. 70: 1988 Ranger & Bronco II 2.9L Electronic Fuel Injection System Wiring Diagram

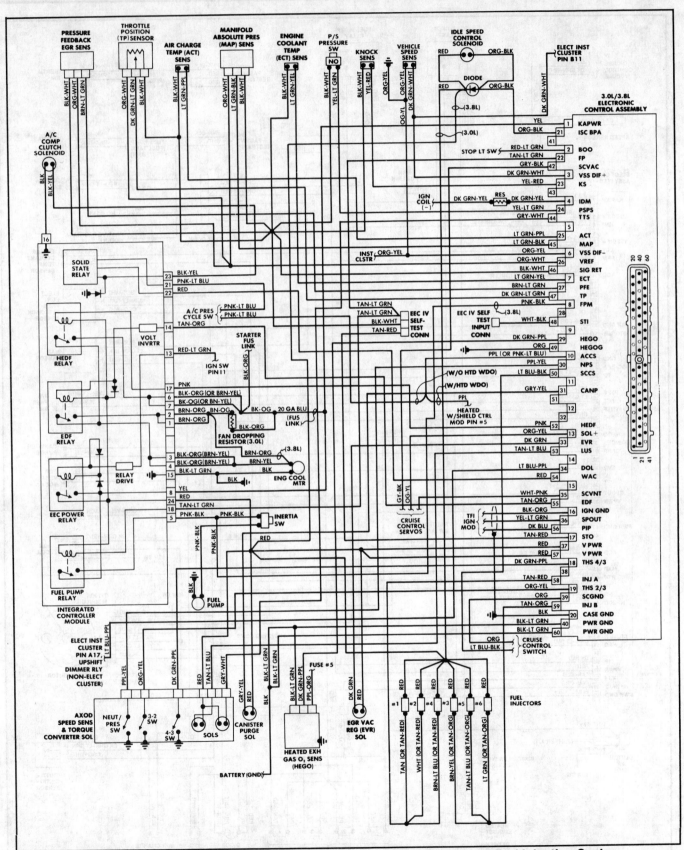

**Fig. 71: 1988 Taurus/Sable 3.0L & 3.8L Electronic Fuel Injection System
Wiring Diagram**

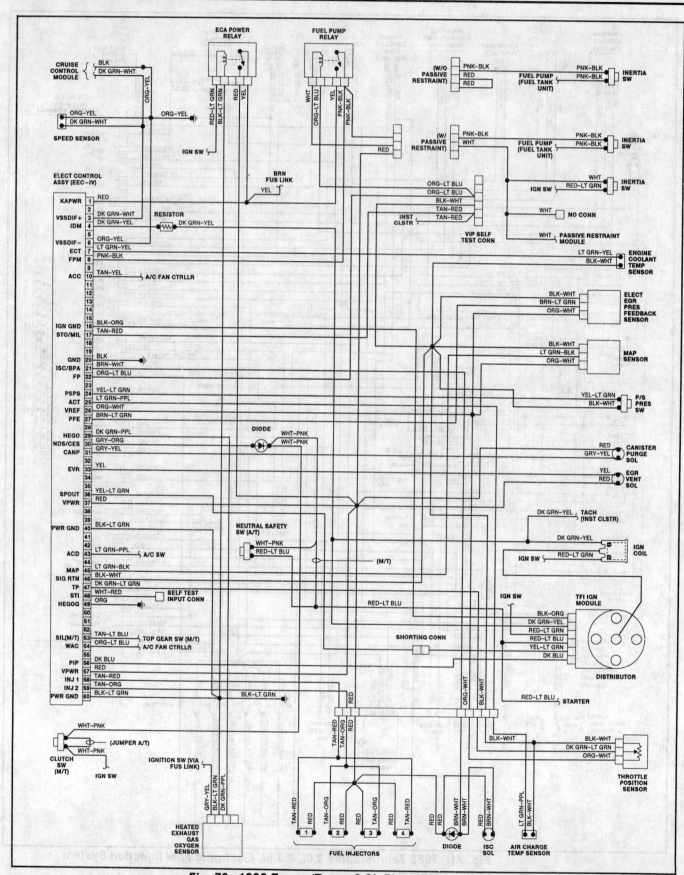

Fig. 72: 1988 Tempo/Topaz 2.3L Electronic Fuel Injection Wiring Diagram

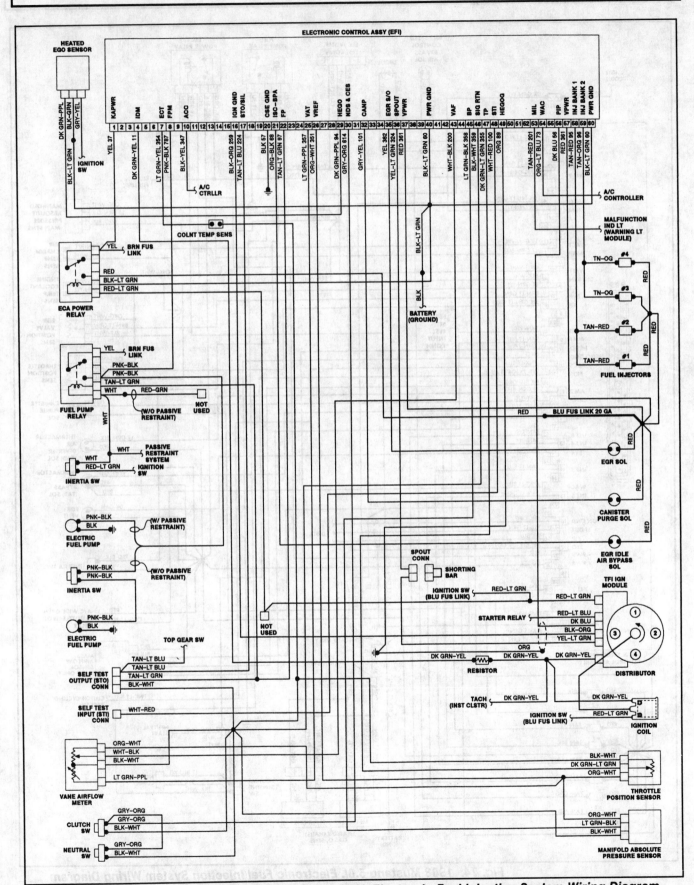

Fig. 73: 1988-89 Escort 1.9L Electronic Fuel Injection System Wiring Diagram

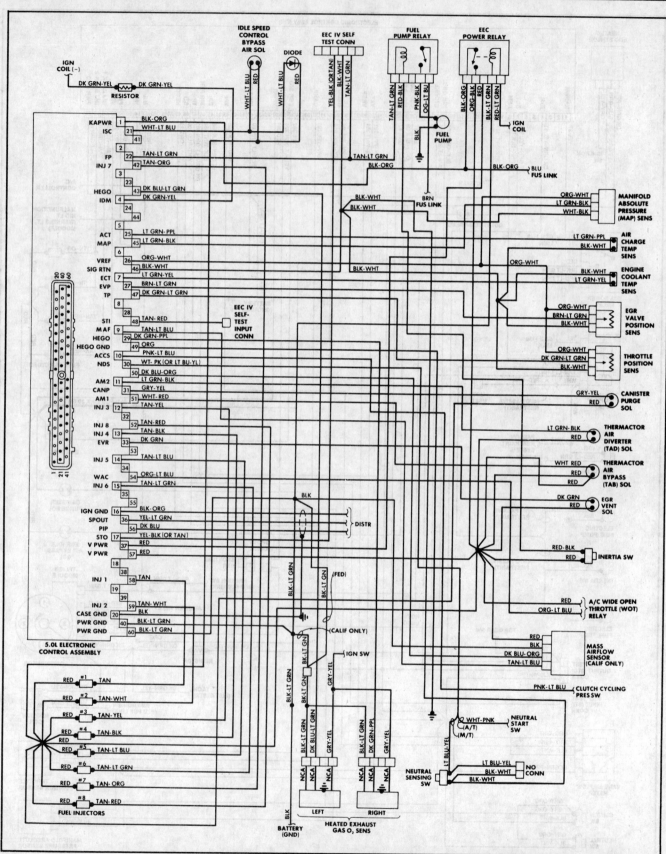

Fig. 74: 1988 Mustang 5.0L Electronic Fuel Injection System Wiring Diagram

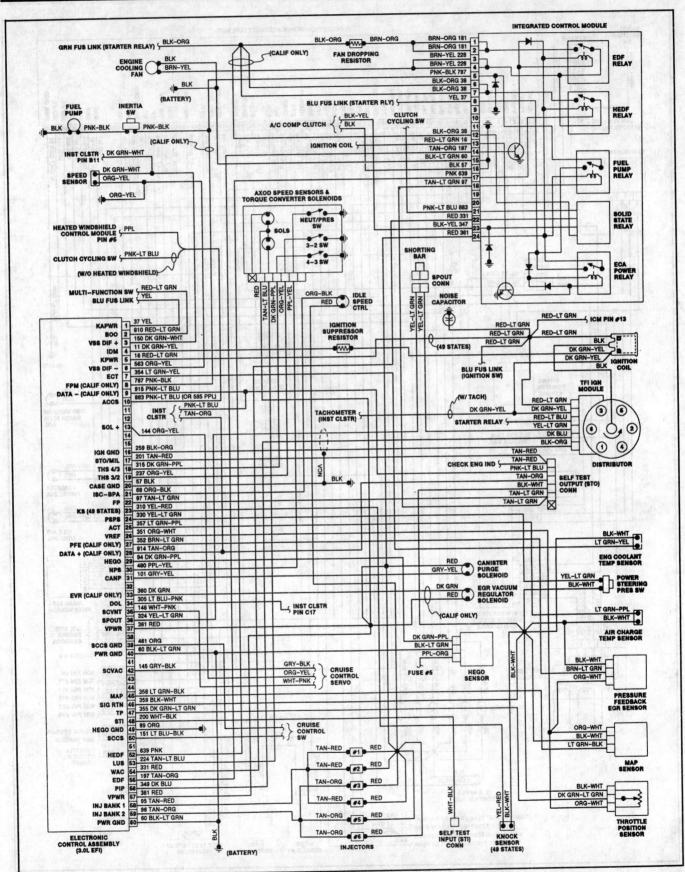

Fig. 75: *1989 Sable & Taurus 3.0L Electronic Fuel Injection System Wiring Diagram*

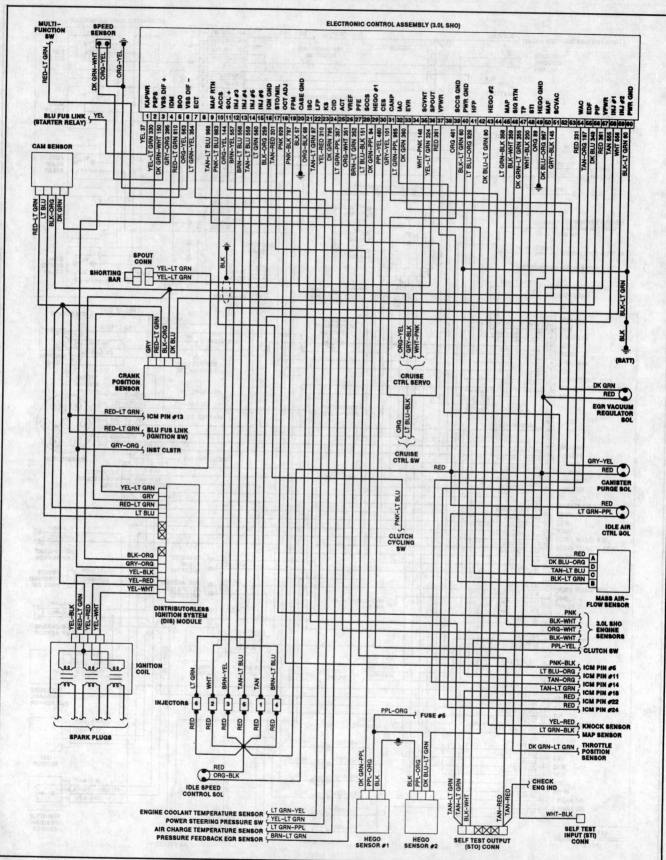

Fig. 76: 1989 Taurus SHO 3.0L Electronic Fuel Injection System Wiring Diagram (Part 1 of 2)

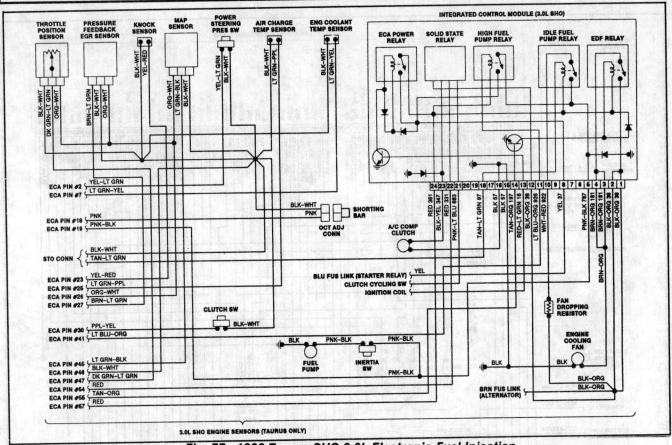

Fig. 77: 1989 Taurus SHO 3.0L Electronic Fuel Injection System Wiring Diagram (Part 2 of 2)

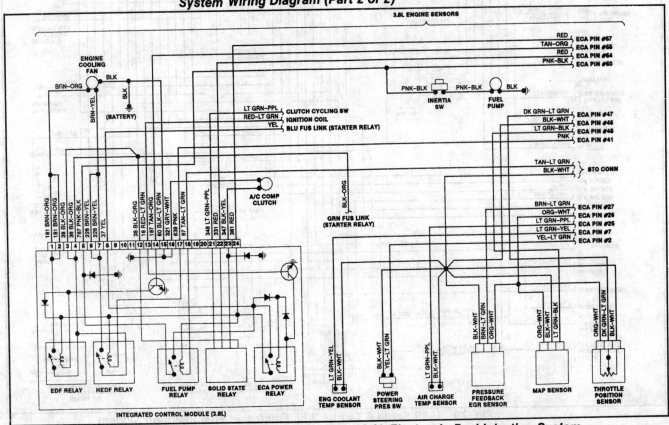

Fig. 78: 1989 Sable & Taurus 3.8L Electronic Fuel Injection System Wiring Diagram (Part 1 of 2) 313

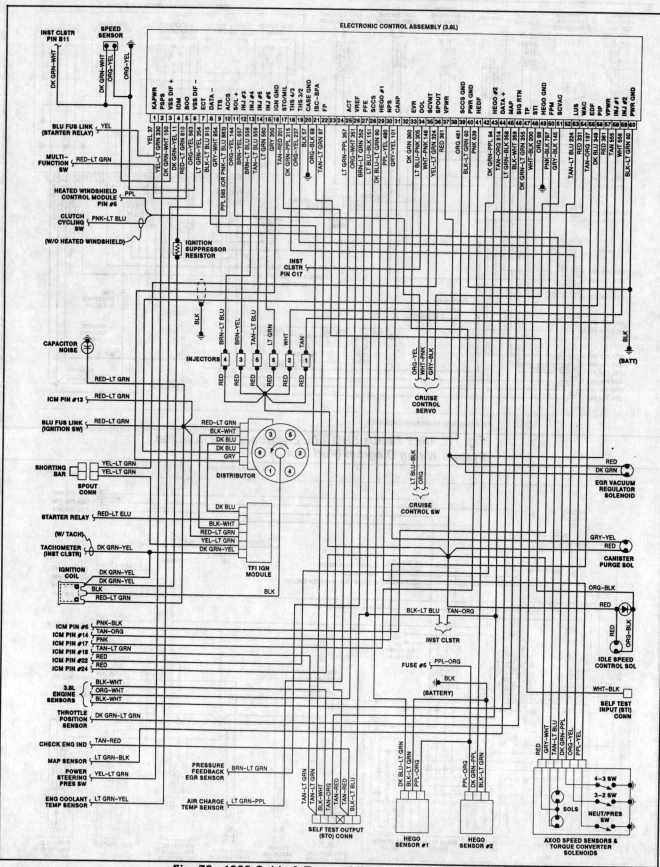

Fig. 79: 1989 Sable & Taurus 3.8L Electronic Fuel Injection System Wiring Diagram (Part 2 of 2)

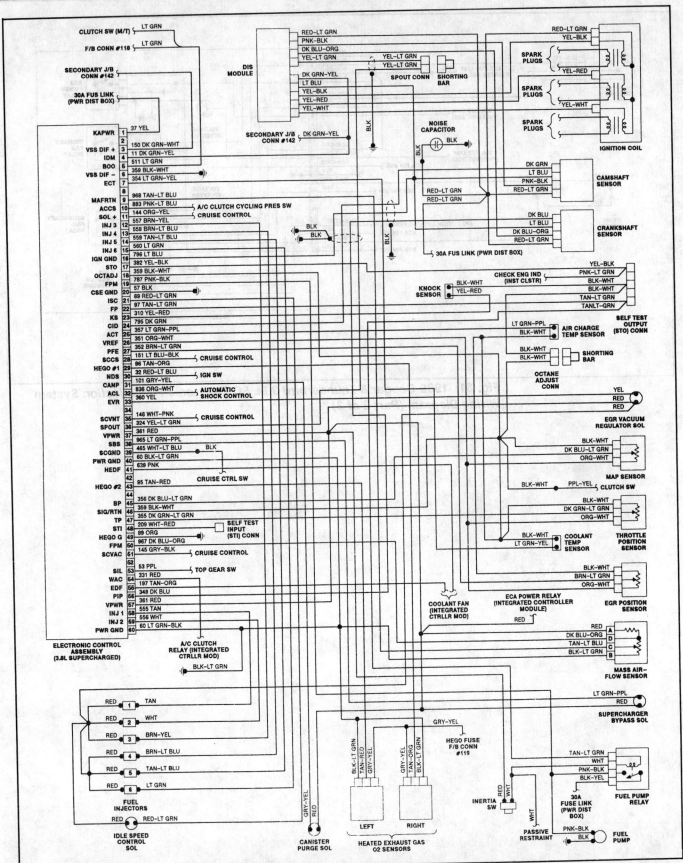

Fig. 80: 1989 Cougar & Thunderbird 3.8L SC Electronic Fuel Injection System Wiring Diagram (Part 1 of 2)

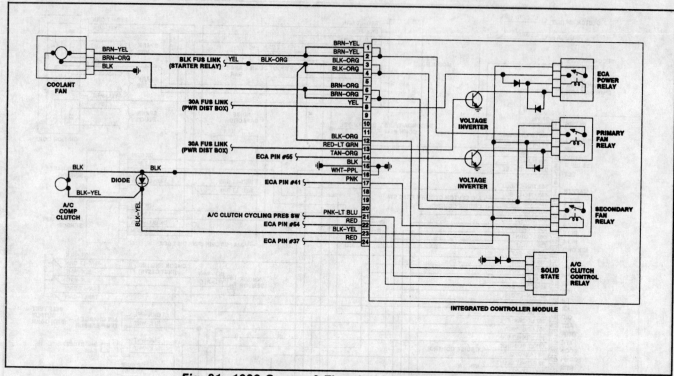

Fig. 81: 1989 Cougar & Thunderbird 3.8L SC Electronic Fuel Injection System Wiring Diagram (Part 2 of 2)

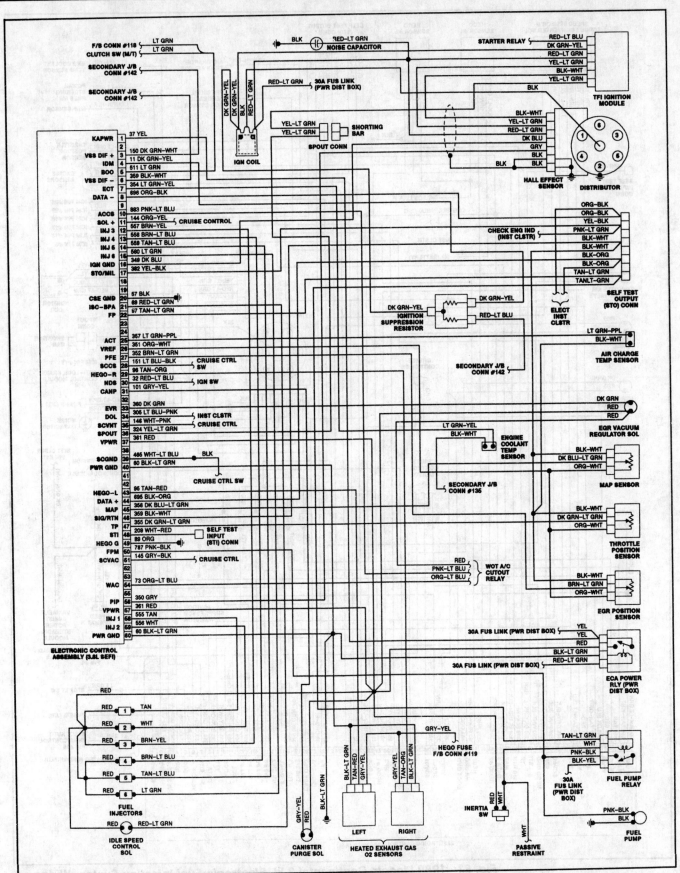

Fig. 82: 1989 Cougar & Thunderbird 3.8L Electronic Fuel Injection System Wiring Diagram

317

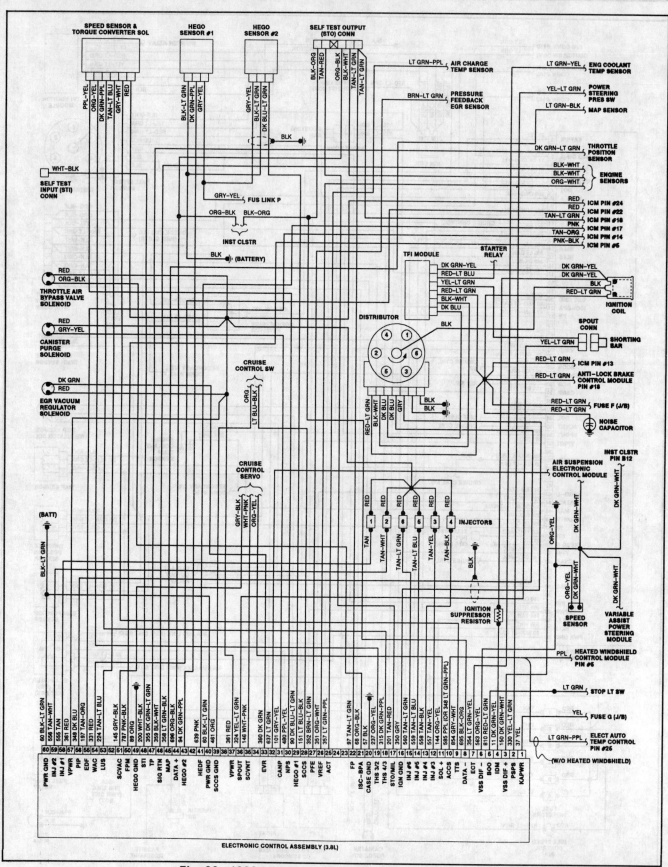

Fig. 83: *1989 Lincoln Continental 3.8L Electronic Fuel Injection System Wiring Diagram (Part 1 of 2)*

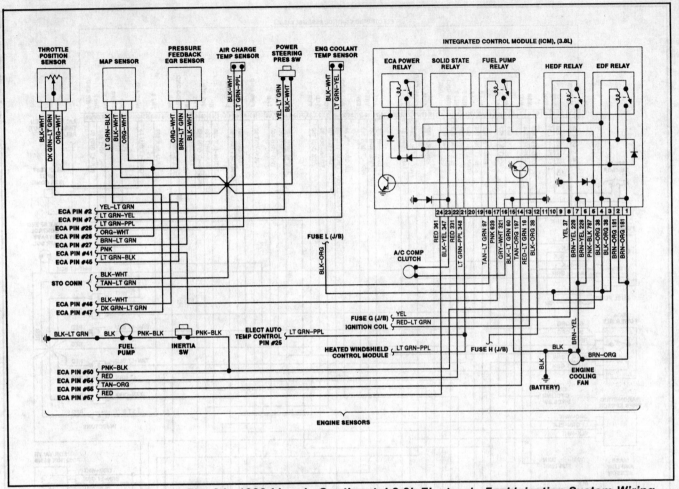

Fig. 84: 1989 Lincoln Continental 3.8L Electronic Fuel Injection System Wiring Diagram (Part 2 of 2)

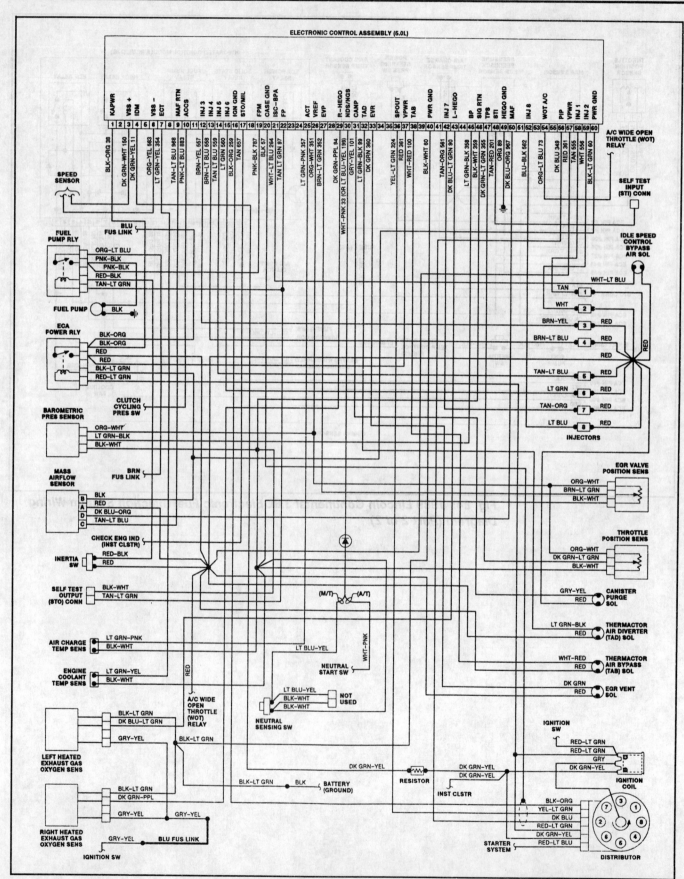

Fig. 85: 1989 Mustang 5.0L Electronic Fuel Injection System Wiring Diagram

Ford Motor Co.

MUTI-POINT FUEL INJECTION

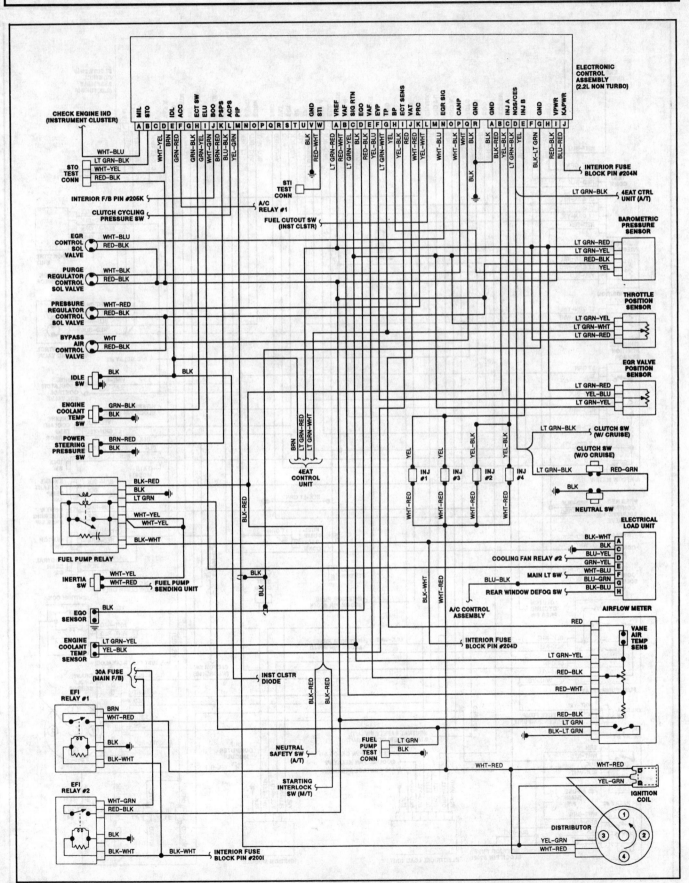

Fig. 86: 1989 Probe Electronic Fuel Injection System Wiring Diagram (Non-Turbo)

321

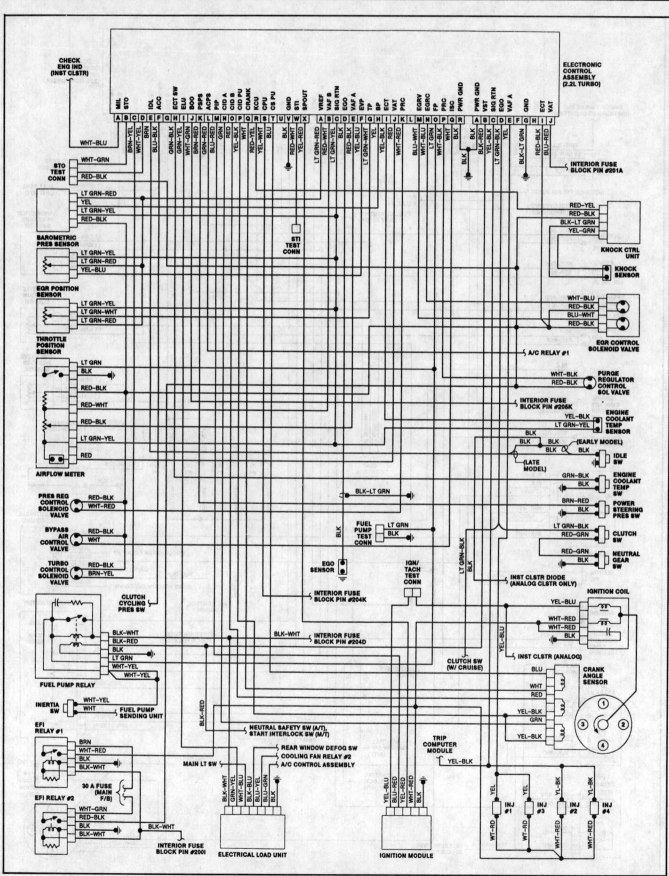

Fig. 87: 1989 Probe Electronic Fuel Injection System Wiring Diagram (Turbo)

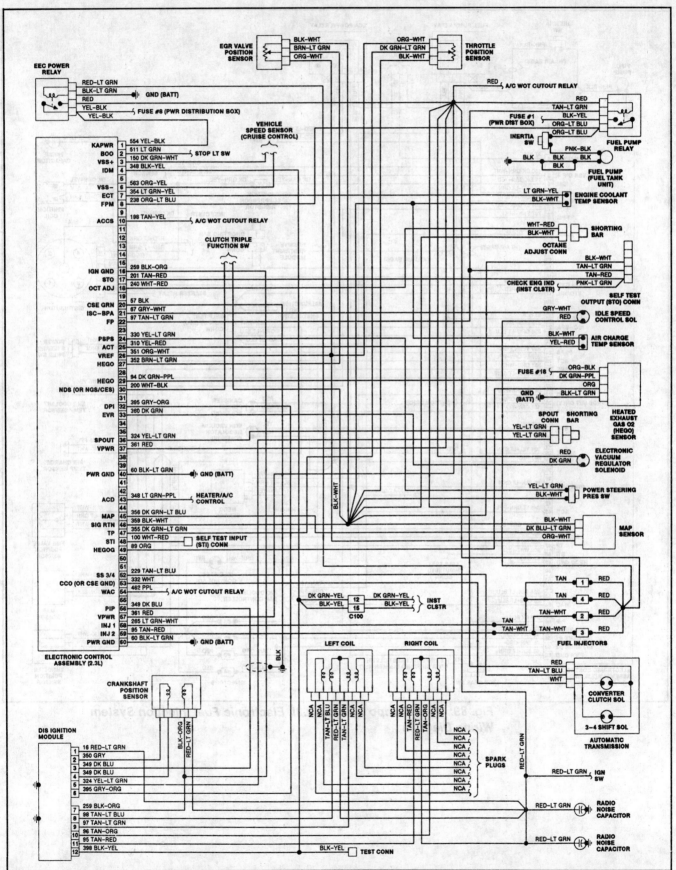

Fig. 88: 1989 Ranger 2.3L Electronic Fuel Injection System Wiring Diagram

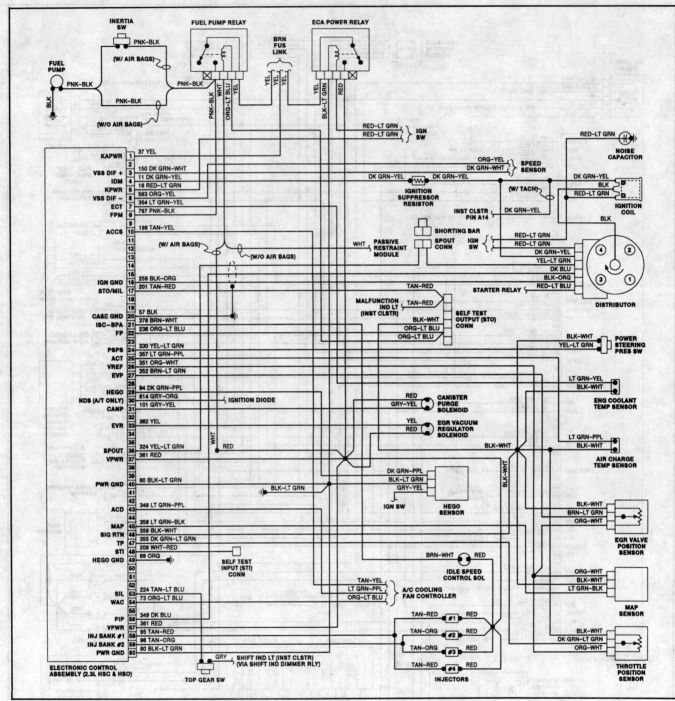

Fig. 89: 1989 Tempo & Topaz 2.3L Electronic Fuel Injection System Wiring Diagram

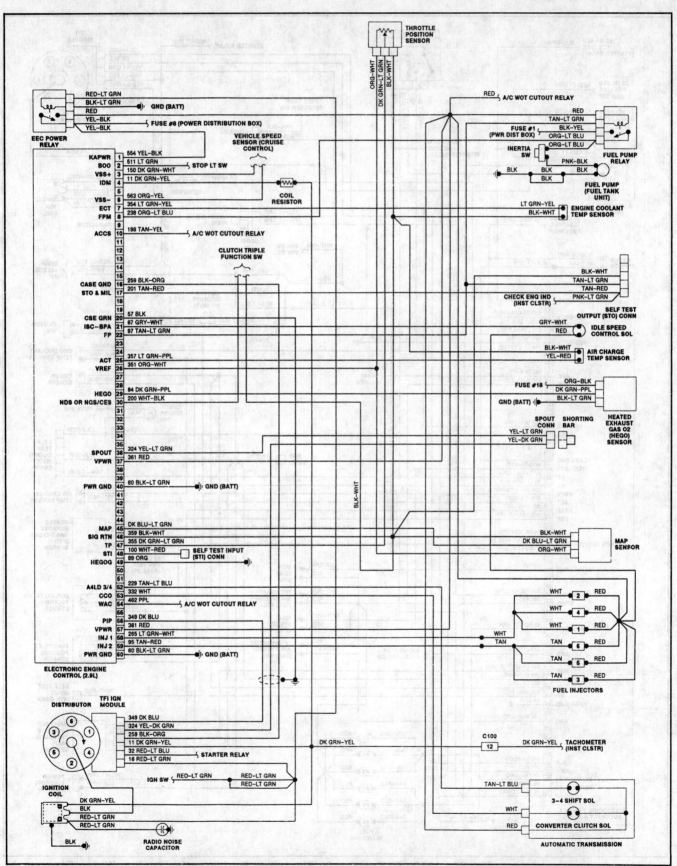

Fig. 90: 1989 Bronco II & Ranger 2.9L Electronic Fuel Injection System Wiring Diagram

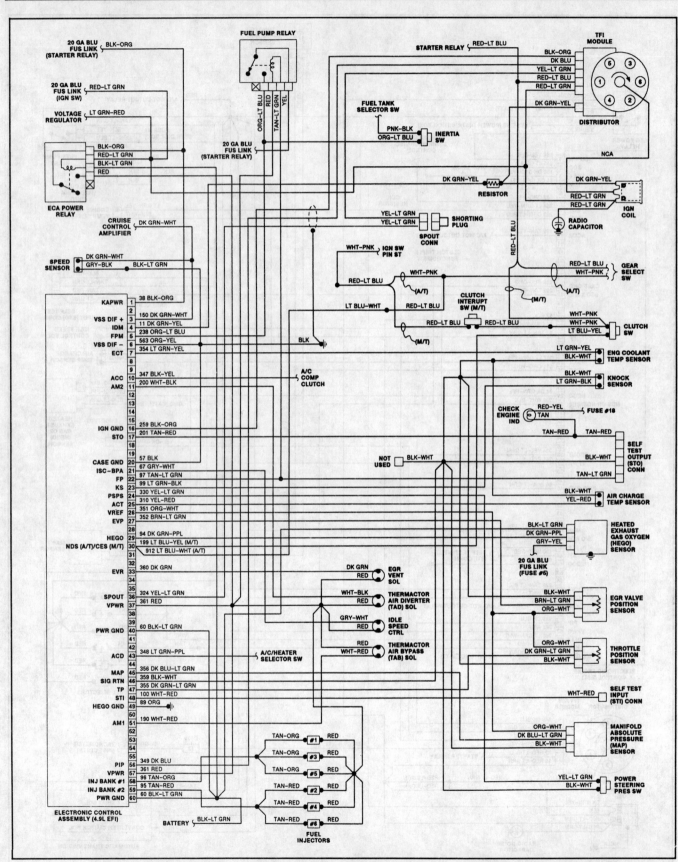

Fig. 91: 1989 "E" Series 4.9L Electronic Fuel Injection System Wiring Diagram

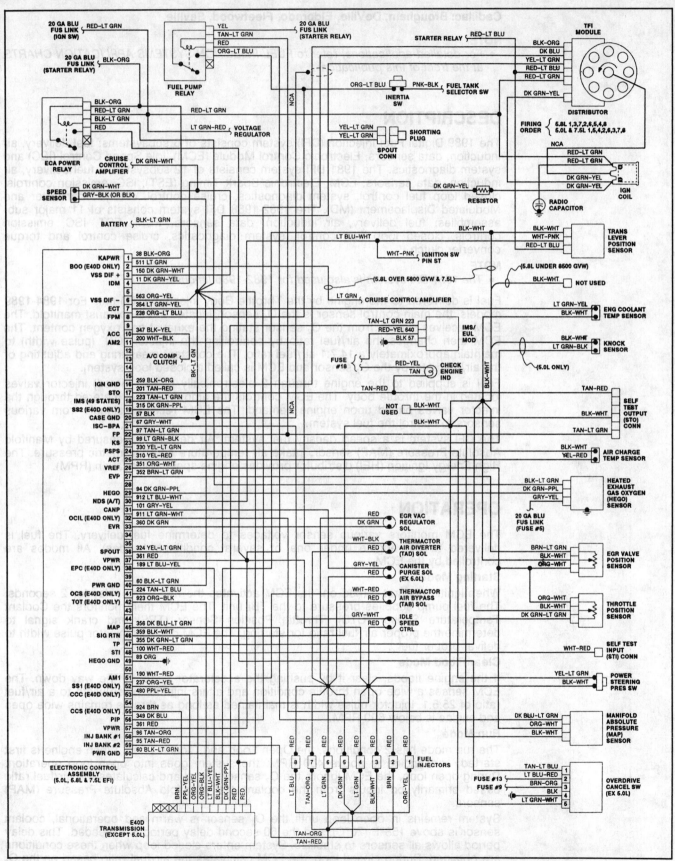

Fig. 92: 1989 "E" Series 5.0L, 5.8L & 7.5L Electronic Fuel Injection System Wiring Diagram

Cadillac: Brougham, DeVille, Eldorado, Fleetwood, Seville

NOTE
- *For detailed applications, refer to FUEL INJECTION SYSTEMS APPLICATION CHARTS at the front of this publication.*

DESCRIPTION

The 1989 Digital Fuel Injection (DFI) system consists of 6 subsystems: fuel delivery, air induction, data sensors, Electronic Control Module (ECM), Idle Speed Control (ISC) and system diagnostics. The 1981 DFI system consists of 12 subsystems: fuel delivery, air induction, data sensors, ECM, Electronic Spark Timing (EST), ISC, emission controls, closed loop fuel control, system diagnostics, cruise control, catalytic converter and Modulated Displacement (MD). The 1982-1988 DFI system consists of 11 major sub-assemblies: fuel delivery, air induction, data sensors, ECM, EST, ISC, emission controls, closed loop fuel control, system diagnostics, cruise control and torque converter clutch.

NOTE
- *The 1981 DFI system is also used for 1982-1986 limousines.*

Fuel is delivered to the engine by the Throttle Body Injection (TBI) unit. For 1984-1989 models, the main control sensor is the O_2 sensor, located in the exhaust manifold. The ECM receives signals from the O_2 sensor stating the exhaust gas oxygen content. The ECM then changes the air/fuel ratio by controlling the injector "ON" (pulse width) to maintain approximately a 14.7:1 air/fuel ratio. The constant measuring and adjusting of the air/fuel ratio by the O_2 sensor and ECM is called a closed loop system.

Fuel is supplied to the engine through 2 electronically pulsed (timed) injector valves located in the throttle body. The ECM controls the amount of fuel metered through the injector valves based upon engine demand. The ECM receives signals from various sensors to control the fuel system.

The DFI system is a speed density fuel system. Air density is measured by Manifold Absolute Pressure (MAP) sensor, intake air temperature and barometric pressure. The High Energy Ignition (HEI) distributor provides engine speed information (RPM).

OPERATION

The ECM monitors various sensor voltages to determine fuel delivery. The fuel is delivered to the engine under one of several conditions (modes). All modes are controlled by the ECM.

Starting Mode

When ignition is first turned on, the ECM activates the fuel pump relay for 2 seconds. The fuel pump supplies pressure to the TBI unit. The ECM then monitors the Coolant Temperature Sensor (CTS), Throttle Position Sensor (TPS) and crank signal to determine the proper air/fuel ratio for starting. The ECM controls injector pulse width to deliver proper fuel.

Clear Flood Mode

If the engine floods, clear it by pushing the accelerator pedal all the way down. The ECM senses a wide open throttle condition and alters injector pulse width to a air/fuel ratio of 25.5:1. Injector pulse width is maintained as long as throttle remains wide open and engine is below 600 RPM.

Run Mode

The run mode has 2 conditions, the Open Loop and Closed Loop. When engine is first started, and engine is above 400 RPM, the system goes into open loop operation. During open loop, the ECM ignors the O_2 sensor signal and calculates the air/fuel ratio based primarily on inputs from the coolant and Manifold Absolute Pressure (MAP) sensors.

System remains in open loop until the O_2 sensor is warm and operational, coolant sensor is above 158°F (70°C), and the 60 second delay period is exceeded. This delay period allows all sensors to stabilize. System enters closed loop when these conditions are obtained. During closed loop, the ECM calculates the air/fuel ratio based on the O_2 sensor.

FUEL DELIVERY SYSTEM

Fuel Pump

The in-tank electric fuel pump supplies pressurized fuel to the throttle body. The ECM activates the fuel pump relay to operate the fuel pump when ignition switch is in the "ON" or "START" position. If engine stalls or is not cranked within 2 seconds, the ECM deactivates the fuel pump.

The fuel pump can also be activated by the oil pressure switch. The oil pressure switch contacts close when oil pressure reaches approximately 4 psi (.3 kg/cm²). If the fuel pump relay fails, the oil pressure switch operates the fuel pump. Inoperative fuel pump relay may result in extended cranking times.

Fuel Pressure Regulator

The fuel pressure regulator, contained in the throttle body, maintains a constant pressure of 9-12 psi (.63-.84 kg/cm²) across the fuel injectors. The diaphragm-operated relief valve senses the fuel pressure on one side, while the other side is exposed to atmospheric pressure and a calibrated spring. *See Fig. 1.* The spring preload provides nominal pressure. Excess fuel is returned to the tank.

Courtesy of General Motors Corp.

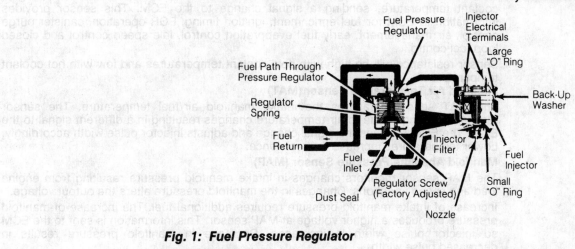

Fig. 1: Fuel Pressure Regulator

Fuel Injectors

The 2 injector valves are electronically actuated solenoids and meter the fuel into the throttle body above the throttle plates. The valve body contains a solenoid with plunger, (core piece) which is pulled upward by the soleniod coil. When solenoid coil is energized, the plunger moves the ball valve away from the valve seat, allowing fuel to flow through the valve. With constant pressure across the injectors, the quantity of fuel injected is determined by the injector pulse width. During normal operation, the 2 fuel injectors are actuated alternately by the ECM. During cranking, both injectors are actuated simultaneously.

Courtesy of General Motors Corp.

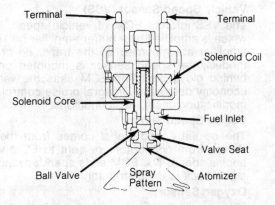

Fig. 2: View of Fuel Injector

IDLE SPEED CONTROL (ISC)

The ISC, mounted to the throttle body, is an electrically driven actuator which changes throttle angle according to demands from the ECM. An internal switch is contained in the ISC. Switch position determines if idle speed is to be controlled by the ISC.

The throttle lever rests against the ISC plunger which extends and retracts to change the throttle blade angle. When switch is closed, the ECM controls the ISC actuator to maintain the programmed idle speed. When the throttle lever moves away from the actuator plunger, the switch opens and the ECM no longer controls the ISC actuator.

When engine is stopped, the ISC fully extends, allowing the plunger to open the throttle blades. This allows the proper idle speed to start the engine. When the engine is cold, the ECM holds the throttle valve open for a longer period of time to allow adequate warm-up time.

AIR INDUCTION

The air induction system consists of throttle body and intake manifold. Airflow rate is controlled by throttle valves. Idle speed is determined by the position of throttle valves and is controlled by the ISC.

ENGINE SENSORS

Coolant Temperature Sensor (CTS)

The CTS is mounted in the coolant stream. The CTS varies resistance according to coolant temperature, sending a signal change to the ECM. This sensor provides information to ECM for fuel enrichment, ignition timing, EGR operation, canister purge control, air management, early fuel evaporation control, idle speed control and closed loop fuel control.

Sensor resistance will be high with cold coolant temperatures and low with hot coolant temperatures.

Manifold Air Temperature Sensor (MAT)

The MAT sensor measures the intake manifold air/fuel temperature. The sensor resistance changes due to air temperature changes resulting in a different signal to the ECM. The ECM receives the signal change and adjusts injector pulse width accordingly. Lower temperatures increase the resistance.

Manifold Absolute Pressure Sensor (MAP)

The MAP sensor monitors changes in intake manifold pressure resulting from engine load and speed variations. Changes in the manifold pressure alters the output voltage.

Increase of intake manifold pressure requires additional fuel. The increase of manifold pressure produces a higher voltage at MAP sensor. This information is sent to the ECM so injector pulse width is increased. Decrease in manifold pressure results in decreased pulse width.

Barometric Pressure Sensor (BARO)

This sensor is used on the Brougham, Eldorado and Seville prior to 1988 models. The BARO sensor measures ambient or barometric pressures and signals the ECM on pressure changes due to altitude and weather.

Throttle Position Sensor (TPS)

The TPS is mounted on the throttle body connected directly to throttle shaft. This sensor monitors throttle movement and position, transmitting electrical signals to the ECM. At closed throttle the output of the TPS is low, approximately .5 volts. At wide open throttle the output voltage is approximately 5 volts. The ECM processes these signals to operate the ISC system and supply fuel enrichment.

Vehicle Speed Sensor (VSS)

The VSS informs ECM of vehicle speed. The speed sensor produces a weak signal which is amplified by a buffer amplifier on 1981-88 models, the speed sensor and buffer amplifier are mounted on the transaxle or behind the speedometer cluster. On 1989 models, the speed sensor is mounted on transaxle and buffer amplifier is located behind glove box. The ECM uses the vehicle speed sensor signals to operate fuel economy data panel, integral cruise control, idle speed control system and fuel delivery modification under load conditions.

Engine Speed Sensor

The engine speed signal comes from the 7 terminal HEI module in the distributor. Pulses from distributor are sent to ECM where the pulses are used to calculate the engine speed. The ECM adds spark advance modifications to the signal and sends this signal back to the distributor.

Oxygen Sensor

The oxygen O_2 sensor is mounted in the exhaust gas stream. The O_2 sensor compares the oxygen content of exhaust gases to the amount present in the atmosphere. The sensor produces a signal which is proportional to the oxygen content of the exhaust gases.

Increase in the oxygen content of exhaust gases indicates a lean fuel mixture by a higher output voltage. The ECM uses the signal and adjusts injector pulse width for proper air/fuel mixture.

NOTE

- *No attempt should be made to measure oxygen sensor voltage output. Current drain of a conventional voltmeter could permanently damage the sensor, shift sensor calibration and/or render sensor unusable. Do not connect a jumper wire, test leads or other electrical connectors to sensor.*

Power Steering Pressure Switch (PSPS)

The PSPS is normally closed and opens when power steering pressure is indicated. Normal voltage is supplied to the ECM through the PSPS. Under pressure the switch opens and voltage loss is read at the ECM. The ECM uses this signal to extend the ISC and stabilize idle speed.

Park/Neutral Switch

The Park/Neutral switch is used to send a signal to the ECM when placed in Park or Neutral. The ECM uses this signal to stabilize the idle speed.

ELECTRONIC CONTROL MODULE (ECM)

The ECM monitors information from various sensors, and controls the systems that affect vehicle performance. Sensor inputs are supplied to the ECM and processed to maintain proper pulse duration, correct idle speed and proper spark advance. The ECM also performs a system diagnostic function. Operational problems are noted and activates the "Engine Control System" light, and stores a code or codes to identify the problem areas.

A device called the PROM (Programmable Read Only Memory) is used inside the ECM. It contains information to define vehicles weight, engine, transmission, axle ratio and other miscellaneous features.

The CALPAK inside the ECM, is used to allow fuel delivery in the event of component failures of the ECM. The CALPAK contains back-up fuel and spark calibrations for use by the ECM when the PROM information is incorrect or unreadable.

DIAGNOSIS & TESTING

PRELIMINARY CHECKS

The following systems and components must be in good operating condition prior to diagnosing the fuel injection system.

- Ensure no hard codes are stored in the memory.
- All support systems and wiring.
- Battery connections and specific gravity.
- Ignition system.
- Compression pressure.
- Fuel supply system pressure and flow.
- All electrical connections and terminals.
- Vacuum line, fuel hose and pipe connections.
- Exhaust system.
- Ensure "Service Engine Soon" light is operable.

FLOODING, ROUGH IDLE

1) If the INJECTOR SYSTEM DIAGNOSIS chart indicated fuel injector(s) continue to spray fuel with electrical connection removed, replace injector(s) as required.

2) After using appropriate diagnostic chart, remove injector(s). Inspect large and small "O" rings for cuts, distortion or other damage. Check that steel back-up washer is located beneath large (upper) "O" ring. Use new "O" rings during installation.

3) Inspect fuel injector fuel filters for cleanliness and damage. Clean or replace as necessary. DO NOT immerse injectors, filters or rubber parts in carburetor cleaner.

NOTE

- *Following symptoms should be checked when no hard codes are stored.*

HARD START

1) Visually check for fuel line or injector leaks with the ignition on and the engine off. Install fuel pressure gauge. Cap the fuel return line at throttle body. Turn ignition on, then off while noting pressure drop.

2) If pressure bleeds off rapidly, check fuel pump check valve, or coupling in fuel tank. If pressure holds with ignition on and return line capped, inspect the fuel pressure regulator for leakage. Check quality of fuel.

3) Check ignition timing. Check HEI voltage output and spark plug wire resistance. Replace spark plug wire if resistance exceeds 50,000 ohms. Remove spark plugs and check for wet plugs, cracks, improper gap, burned electrodes, heavy carbon deposits, type and heat range. Check EST operation.

4) Check for sticking or binding TPS. Check MAP sensor for restrictions in the supply hose or throttle body MAP port. Check for vacuum trapped in the MAP sensor. Perform INJECTOR DIAGNOSIS as outlined in flow charts. Check for accuracy and high resistance of CTS and MAT sensors and circuits.

STALL AFTER START

1) Inspect all vacuum hoses and electrical connections. Ensure hot air tube is connected to air cleaner. Check air cleaner thermostatic valve operation.

2) Check PCV valve and EGR valve operation. Check Electronic Spark Timing (EST) if engine will run with EST disabled. Check fuel pressure. Fuel pressure should be a steady 9-12 psi (.6-.8 kg/cm²) in all speed ranges. If fuel pressure is not within range, see FUEL SYSTEM DIAGNOSIS chart.

3) Check for overcharged A/C system. Check for high A/C head pressure. Check for damaged spark plugs or faulty ignition coil. Check for restricted exhaust or plugged catalytic converter.

BACKFIRE

1) Inspect for loose wiring grounds at engine block. Check ignition system for faulty components. Check ignition timing.

2) Inspect distributor for worn components. Ensure the A.I.R. system is pumping air into the exhaust manifolds at all times.

3) Check intake manifold bolts for tightness. Check for damaged internal engine components.

HESITATION, SAG, STUMBLE

1) Inspect all vacuum hoses and electrical connections. Check air cleaner thermostatic valve operation. Check vacuum hose to MAP sensor for leaks or restrictions, and restrictions at MAP port on throttle body.

2) Check fuel pressure. Pressure should be a steady 9-12 psi (.6-.8 kg/cm²) at all speed ranges. If fuel pressure is not within range see FUEL SYSTEM DIAGNOSIS chart. Check ignition timing.

3) Check for defective spark plugs. Check PROM part number for correct application. Check for sticking or binding of TPS. Check for proper operation of EGR system.

4) Check fuel injectors for fuel leakage. Separate injector connectors and remove injectors. Crank engine while observing injectors. No fuel should be emitted through nozzle opening while engine is cranking. Replace injectors if necessary. Check HEI ground for good contact. Check canister purge system. Check for carboned or sticking valves. Check for restricted exhaust and catalytic converter.

SURGES AND/OR CHUGGLE

1) Verify transmission for proper operation in all ranges. Check air cleaner thermostatic valve for proper operation. Check EGR valve operation.

2) Check for open or short in the following circuits; HEI by-pass number 424, Electronic Spark Timing (EST) number 423 and VCC numbers 420 and 422. See WIRING DIAGRAMS at end of this article.

NOTE
- *The following step will require a vehicle of the same make, model and year to be used for comparison.*

3) Check for exhaust system restrictions. Use Pressure Gauge Tap (J-35314) with a 0-15 psi (0-1.1 kg/cm²) range. Block drive wheels. Apply brakes firmly. Place the vehicle in drive and load the engine against the brakes at a steady 1600 RPM. If there is a 1.0 psi (0.1 kg/cm²) or greater backpressure difference, exhaust system is restricted.

4) Check fuel system. Fuel pressure should be 9-12 psi (.6-.8 kg/cm²) in all speed ranges. If fuel pressure is not within range, see FUEL SYSTEM DIAGNOSIS chart. Check for vacuum leaks or defective ignition system. Decarbon engine using top cylinder cleaner following instructions on can. Remove spark plugs and check for wet plugs, cracks, improper gap, burned electrodes, heavy carbon deposits, type and heat range.

DETONATION, SPARK KNOCK

1) Check for overheating conditions. Check ignition timing. Check EGR valve operation. Inspect MAP sensor for restrictions in MAP hose or throttle body MAP port.

2) Check fuel system pressure. Fuel pressure should be 9-12 psi (.6-.8 kg/cm²) in all speed ranges. If fuel pressure is not within range, see FUEL SYSTEM DIAGNOSIS chart. Check MEM-CAL for correct application.

3) Check transmission for proper operation. Check for loose intake manifold bolts. Decarbon engine using top cylinder cleaner following instructions on can. If condition still exists, vehicle may require use of high octane gasoline.

ROUGH, UNSTABLE OR INCORRECT IDLE, STALLING

1) Inspect all vacuum hoses and electrical connections. Observe ISC operation in output cycling. Watch for ISC sticking or binding while manipulating the four terminal ISC connector.

2) Check for binding throttle linkage or cruise control cable. Check ignition system and ignition timing.

3) Check fuel pressure. Fuel pressure should be 9-12 psi (.6-.8 kg/cm²) at all speed ranges. If incorrect, see FUEL SYSTEM DIAGNOSIS chart.

4) Check for leaking fuel injectors. Inspect for intermittent injector operation. Check PCV and EGR operation. Check A/C clutch for rapid cycling. Check for loose ECM grounds. Check distributor for wear. Check for air leak at throttle body and intake manifold.

LACK OF POWER OR SLUGGISH

1) Inspect air cleaner condition. Check air cleaner thermostatic valve operation. Ensure throttle valve fully opens. Check ignition timing. Check for faulty spark plugs or ignition system.

2) Check for restricted fuel filter. Check fuel pressure. Fuel pressure should be 9-12 psi (.6-.8 kg/cm²) at all speed ranges. If incorrect, see FUEL SYSTEM DIAGNOSIS chart.

3) Check EGR valve operation. Inspect for restricted MAP vacuum fitting. Inspect exhaust system for restrictions. Check for vacuum leaks at intake manifold and throttle body. Ensure transmission operates properly.

DIESELING

Check for leaking fuel injectors. See FUEL SYSTEM DIAGNOSIS chart. Check canister purge operation.

EXCESSIVE EXHAUST EMISSIONS (ODORS)

1) Operate engine to normal operating temperature. If car runs rich, inspect air filter condition. Check for high fuel pressure. Fuel pressure should be 9-12 psi (.6-.8 kg/cm²) at all speed ranges. If incorrect, see FUEL SYSTEM DIAGNOSIS chart.

2) Check engine idle speed. Check for fuel in the crankcase. If fuel exists, check for proper PROM usage.

3) Check PCV operation. Inspect spark plug condition. Inspect catalytic converter for lead contamination. Check for leaking injectors.

4) If car runs lean, check EGR valve operation. Check for vacuum leaks and air cleaner thermostatic valve operation. Inspect cooling system for overheat conditions. Check ignition timing and the amount of advance. Check for air at exhaust manifold ports at all times.

INTERMITTENT ENGINE STALL, NO CODES

1) Check fuel pressure during stall period. Fuel pressure should be 9-12 psi (.6-.8 kg/cm²) at all speed ranges. If fuel pressure is not steady, see FUEL SYSTEM DIAGNOSIS chart.

2) Inspect ICS throttle switch operation. Enter diagnostics and operate throttle from open to closed position. Note throttle switch status light during operation. If light operation is intermittent, check ISC switch or throttle linkage for proper operation.

3) Check TPS operation. False TPS reading may create a stall due to improper acceleration enrichment or if false high throttle angle during coast down is indicated.

4) Check for intermittent operation of coolant, MAT or MAP sensors. Ensure hose to MAP sensor is not restricted. Check EGR valve operation.

5) Inspect fuel injectors for intermittent operation. Check for voltage loss at the time of stall. If voltage loss exists, inspect charging system, battery and wiring.

6) Check for improperly applied Viscous Converter Clutch (VCC). Inspect ignition switch circuits for intermittent open circuit.

7) On stalls existing during low fuel level or deceleration, inspect fuel sending unit and fuel pump strainer.

8) If stall exists during normal closed throttle while coasting to a stop, check TPS setting and ISC to throttle linkage gap.

9) Stalls may exist when power steering pressure is applied or when shifting transmission into gear. Check proper switch condition. Check for proper ISC to throttle linkage gap.

10) If stall exists during A/C operation, check A/C clutch indicator light operation or for high A/C head pressure.

11) On stalls occuring during engine starts with hot ambient air temperatures, check the evaporative emission control caniter and solenoid.

12) If stall exists during cold or warm weather, check air cleaner thermostatic valve operation.

POOR FUEL ECONOMY

1) Verify air cleaner thermostatic valve operation. Inspect air cleaner condition. Check ignition timing and ignition system operation.

2) Check for improperly applied Viscous Converter Clutch (VCC). Inspect ignition switch circuits for intermittent open circuit.

3) Inspect exhaust systems for restrictions. Verify fuel data panel for proper accuracy.

FUEL SYSTEM

Fuel Pressure

To check fuel pressure, connect Fuel Pressure Gauge (J-25400-300) to fuel line service fitting and measure fuel pressure while cranking engine. Use FUEL SYSTEM DIAGNOSIS chart if pressure is not between 9-12 psi (.6-.8 kg/cm²). Use INJECTOR SYSTEM DIAGNOSIS chart if pressure is between 9-12 psi (.6-.8 kg/cm²).

ADJUSTMENTS

Maximum ISC Extension

1) Check TPS adjustment. See Throttle Position Sensor (TPS) adjustment. With ignition on and engine off, fully extend the ISC motor. Using 2 jumper wires, apply positive battery voltage to terminal "D" and ground terminal "C". *See Fig. 3.*

2) Apply battery voltage to ISC motor until plunger reaches maximum travel and ratchets a few times. Observe ISC nose positions to determine maximum travel. Remove jumper wires.

3) With ISC at maximum extension, TPS parameter displayed should be 10.5-11.5 (1.05-1.15 volts). If parameter displayed is incorrect, adjust ISC plunger adjusting screw.

4) Disconnect all test equipment and plug in all harness connectors. Turn off ignition for at least 10 seconds. Start engine and check ISC motor for proper operation. After all connections have been made and system is restored to normal operation, codes must be cleared.

Throttle Position Sensor (TPS) Adjustment

1) Ensure throttle is against idle speed stop screw. With ignition on and engine off, enter diagnostics on climate control panel.

2) Note throttle angle displayed in diagnostics on ECM parameter P.0.1 TPS. Displayed reading should be -1.0°-1.0°. If reading is incorrect, loosen TPS screws and rotate TPS to achieve a parameter display of -.5°-.5° (.45-.55 volts). Tighten TPS screws securely and recheck adjustment.

Courtesy of General Motors Corp.

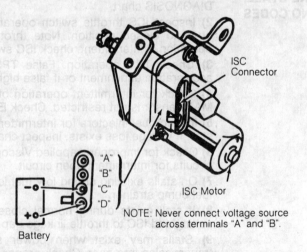

ISC Connector

ISC Motor

NOTE: Never connect voltage source across terminals "A" and "B".

"A"
"B"
"C"
"D"

Battery

Fig. 3: ISC Terminal Identification

Courtesy of General Motors.

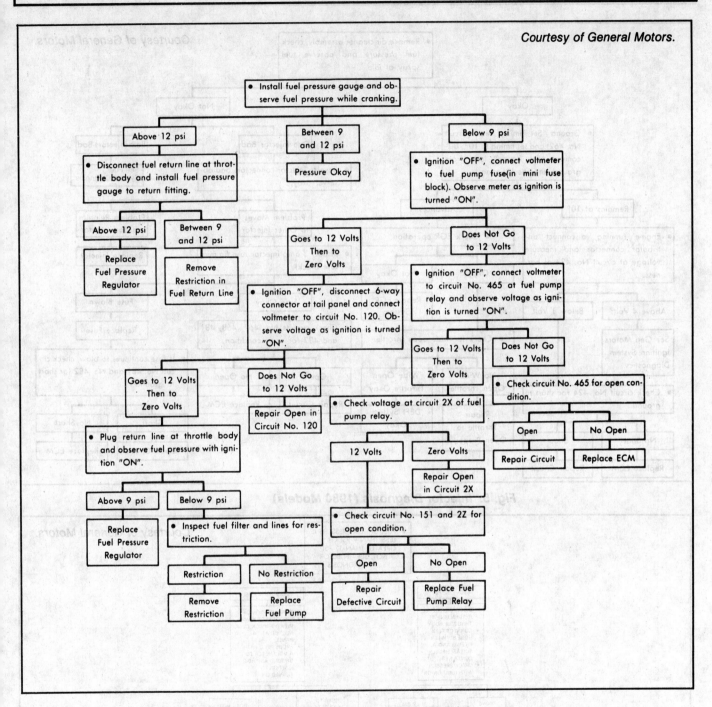

Fig. 4: Fuel System Diagnosis (1980 Models)

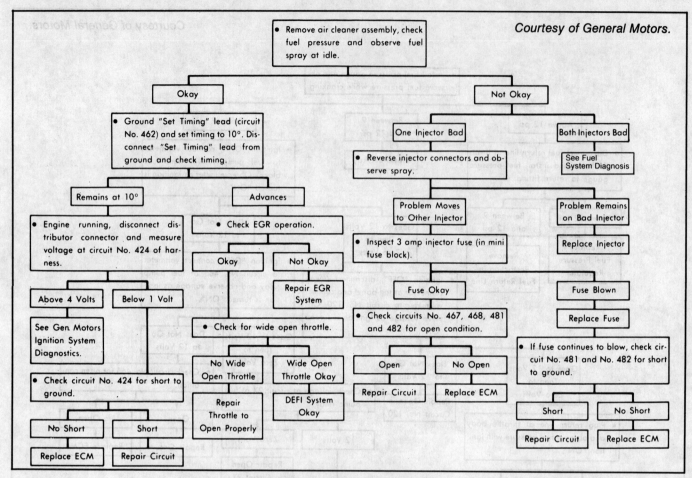

Fig. 5: Injector Diagnosis (1980 Models)

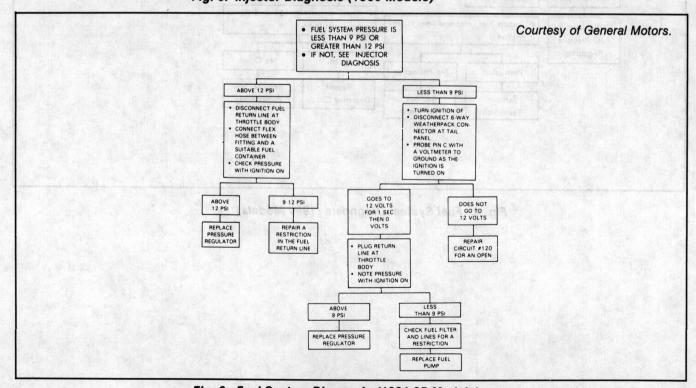

Fig. 6: Fuel System Diagnosis (1981-85 Models)

Courtesy of General Motors.

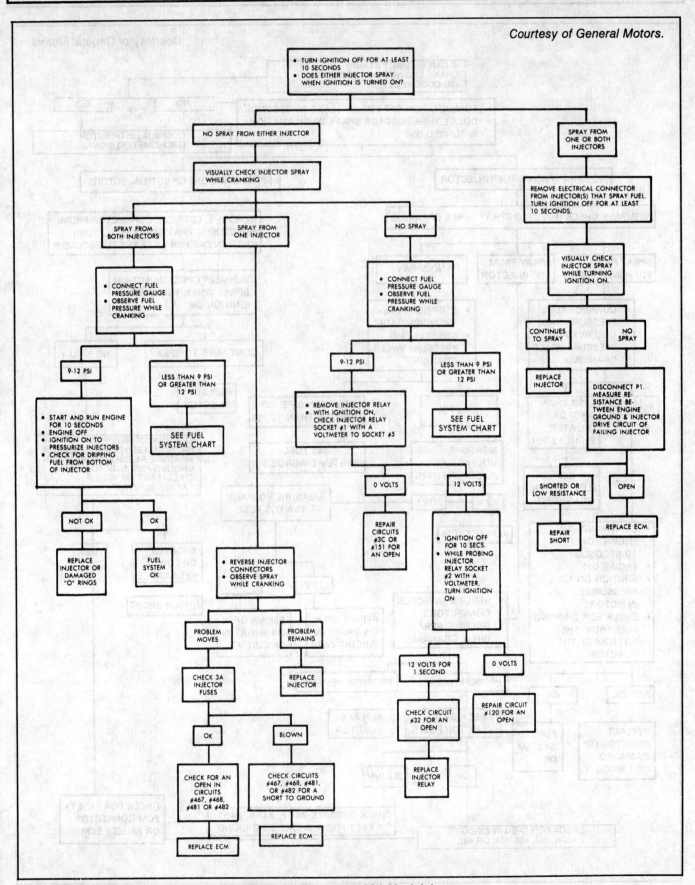

Fig. 7: Injector Diagnosis (1981 Models)

Courtesy of General Motors.

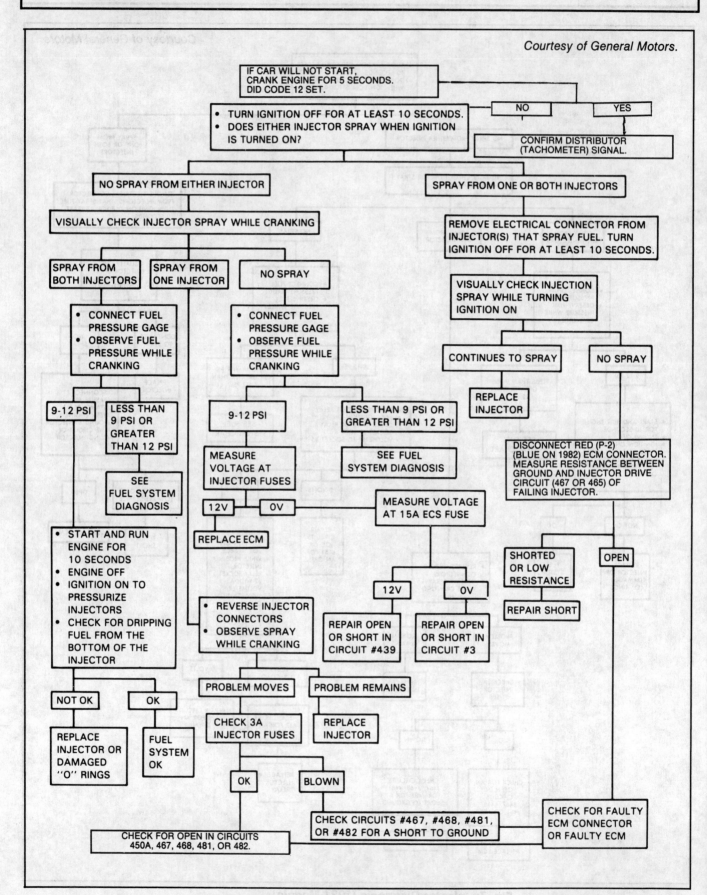

Fig. 8: Injector Diagnosis (1981-82 Models)

Courtesy of General Motors.

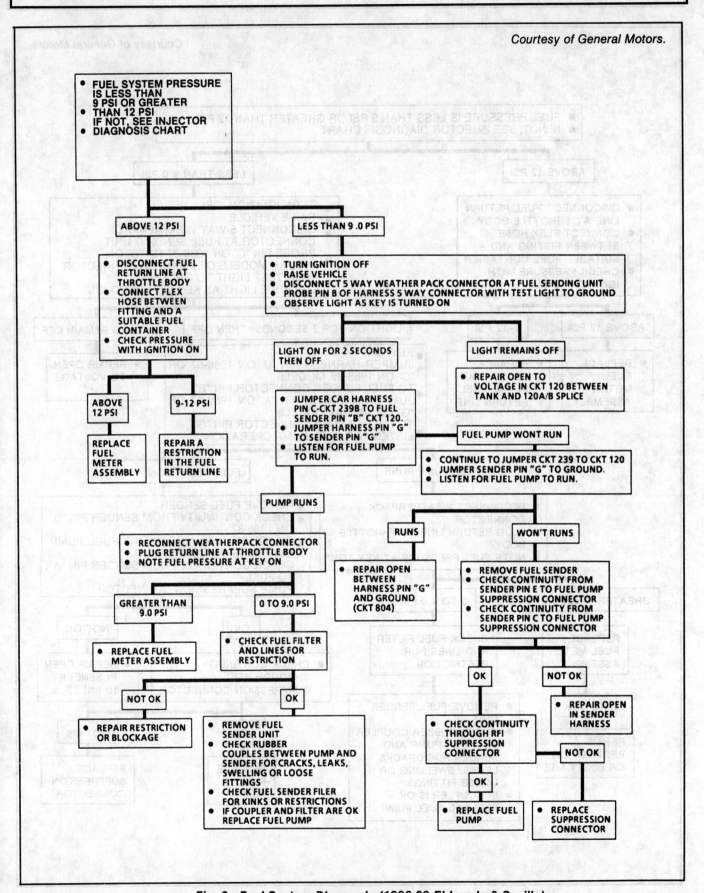

- FUEL SYSTEM PRESSURE IS LESS THAN 9 PSI OR GREATER THAN 12 PSI
- IF NOT, SEE INJECTOR DIAGNOSIS CHART

ABOVE 12 PSI

- DISCONNECT FUEL RETURN LINE AT THROTTLE BODY
- CONNECT FLEX HOSE BETWEEN FITTING AND A SUITABLE FUEL CONTAINER
- CHECK PRESSURE WITH IGNITION ON

ABOVE 12 PSI

- REPLACE FUEL METER ASSEMBLY

9-12 PSI

- REPAIR A RESTRICTION IN THE FUEL RETURN LINE

LESS THAN 9.0 PSI

- TURN IGNITION OFF
- RAISE VEHICLE
- DISCONNECT 5 WAY WEATHER PACK CONNECTOR AT FUEL SENDING UNIT
- PROBE PIN B OF HARNESS 5 WAY CONNECTOR WITH TEST LIGHT TO GROUND
- OBSERVE LIGHT AS KEY IS TURNED ON

LIGHT ON FOR 2 SECONDS THEN OFF

- JUMPER CAR HARNESS PIN C-CKT 239B TO FUEL SENDER PIN "B" CKT 120.
- JUMPER HARNESS PIN "G" TO SENDER PIN "G"
- LISTEN FOR FUEL PUMP TO RUN.

PUMP RUNS

- RECONNECT WEATHERPACK CONNECTOR
- PLUG RETURN LINE AT THROTTLE BODY
- NOTE FUEL PRESSURE AT KEY ON

GREATER THAN 9.0 PSI

- REPLACE FUEL METER ASSEMBLY

NOT OK

- REPAIR RESTRICTION OR BLOCKAGE

0 TO 9.0 PSI

- CHECK FUEL FILTER AND LINES FOR RESTRICTION

OK

- REMOVE FUEL SENDER UNIT
- CHECK RUBBER COUPLES BETWEEN PUMP AND SENDER FOR CRACKS, LEAKS, SWELLING OR LOOSE FITTINGS
- CHECK FUEL SENDER FILER FOR KINKS OR RESTRICTIONS
- IF COUPLER AND FILTER ARE OK REPLACE FUEL PUMP

LIGHT REMAINS OFF

- REPAIR OPEN TO VOLTAGE IN CKT 120 BETWEEN TANK AND 120A/B SPLICE

FUEL PUMP WONT RUN

- CONTINUE TO JUMPER CKT 239 TO CKT 120
- JUMPER SENDER PIN "G" TO GROUND.
- LISTEN FOR FUEL PUMP TO RUN.

RUNS

- REPAIR OPEN BETWEEN HARNESS PIN "G" AND GROUND (CKT 804)

WON'T RUNS

- REMOVE FUEL SENDER
- CHECK CONTINUITY FROM SENDER PIN E TO FUEL PUMP SUPPRESSION CONNECTOR
- CHECK CONTINUITY FROM SENDER PIN C TO FUEL PUMP SUPPRESSION CONNECTOR

OK

- CHECK CONTINUITY THROUGH RFI SUPPRESSION CONNECTOR

OK

- REPLACE FUEL PUMP

NOT OK

- REPAIR OPEN IN SENDER HARNESS

NOT OK

- REPLACE SUPPRESSION CONNECTOR

Fig. 9: Fuel System Diagnosis (1986-89 Eldorado & Seville)

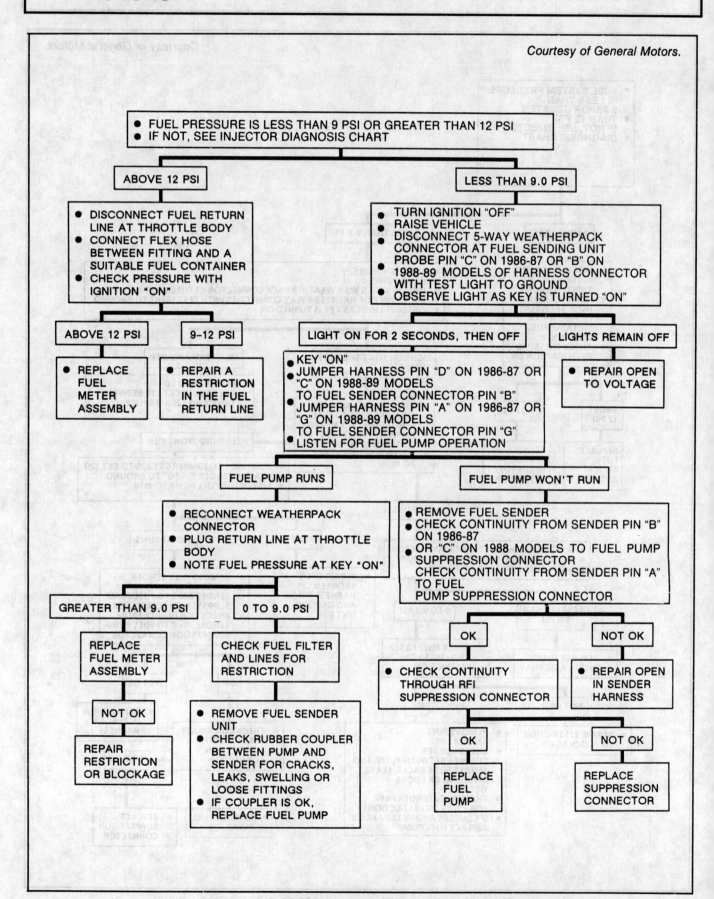

Courtesy of General Motors.

- FUEL PRESSURE IS LESS THAN 9 PSI OR GREATER THAN 12 PSI
- IF NOT, SEE INJECTOR DIAGNOSIS CHART

ABOVE 12 PSI

- DISCONNECT FUEL RETURN LINE AT THROTTLE BODY
- CONNECT FLEX HOSE BETWEEN FITTING AND A SUITABLE FUEL CONTAINER
- CHECK PRESSURE WITH IGNITION "ON"

ABOVE 12 PSI

- REPLACE FUEL METER ASSEMBLY

9–12 PSI

- REPAIR A RESTRICTION IN THE FUEL RETURN LINE

LESS THAN 9.0 PSI

- TURN IGNITION "OFF"
- RAISE VEHICLE
- DISCONNECT 5-WAY WEATHERPACK CONNECTOR AT FUEL SENDING UNIT
- PROBE PIN "C" ON 1986-87 OR "B" ON 1988-89 MODELS OF HARNESS CONNECTOR WITH TEST LIGHT TO GROUND
- OBSERVE LIGHT AS KEY IS TURNED "ON"

LIGHT ON FOR 2 SECONDS, THEN OFF

- KEY "ON"
- JUMPER HARNESS PIN "D" ON 1986-87 OR "C" ON 1988-89 MODELS TO FUEL SENDER CONNECTOR PIN "B"
- JUMPER HARNESS PIN "A" ON 1986-87 OR "G" ON 1988-89 MODELS TO FUEL SENDER CONNECTOR PIN "G"
- LISTEN FOR FUEL PUMP OPERATION

LIGHTS REMAIN OFF

- REPAIR OPEN TO VOLTAGE

FUEL PUMP RUNS

- RECONNECT WEATHERPACK CONNECTOR
- PLUG RETURN LINE AT THROTTLE BODY
- NOTE FUEL PRESSURE AT KEY "ON"

FUEL PUMP WON'T RUN

- REMOVE FUEL SENDER
- CHECK CONTINUITY FROM SENDER PIN "B" ON 1986-87
- OR "C" ON 1988 MODELS TO FUEL PUMP SUPPRESSION CONNECTOR
  CHECK CONTINUITY FROM SENDER PIN "A" TO FUEL PUMP SUPPRESSION CONNECTOR

GREATER THAN 9.0 PSI

- REPLACE FUEL METER ASSEMBLY

NOT OK

- REPAIR RESTRICTION OR BLOCKAGE

0 TO 9.0 PSI

- CHECK FUEL FILTER AND LINES FOR RESTRICTION

- REMOVE FUEL SENDER UNIT
- CHECK RUBBER COUPLER BETWEEN PUMP AND SENDER FOR CRACKS, LEAKS, SWELLING OR LOOSE FITTINGS
- IF COUPLER IS OK, REPLACE FUEL PUMP

OK

- CHECK CONTINUITY THROUGH RFI SUPPRESSION CONNECTOR

NOT OK

- REPAIR OPEN IN SENDER HARNESS

OK

REPLACE FUEL PUMP

NOT OK

REPLACE SUPPRESSION CONNECTOR

Fig. 10: Fuel System Diagnosis (1986-89 DeVille & Fleetwood)

Courtesy of General Motors.

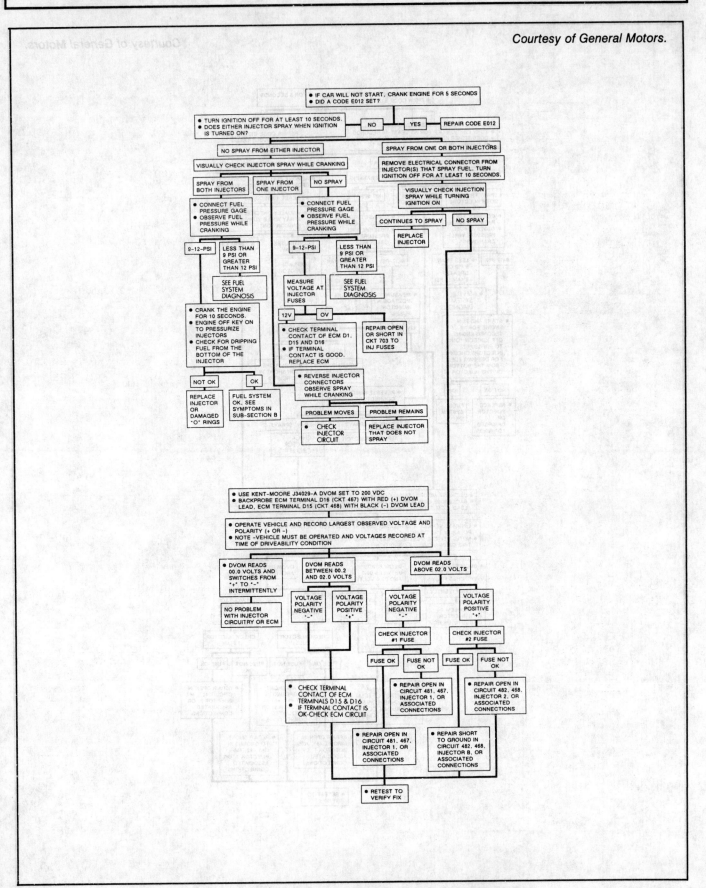

Fig. 11: Injector Diagnosis (1986-89 Eldorado & Seville)

Courtesy of General Motors.

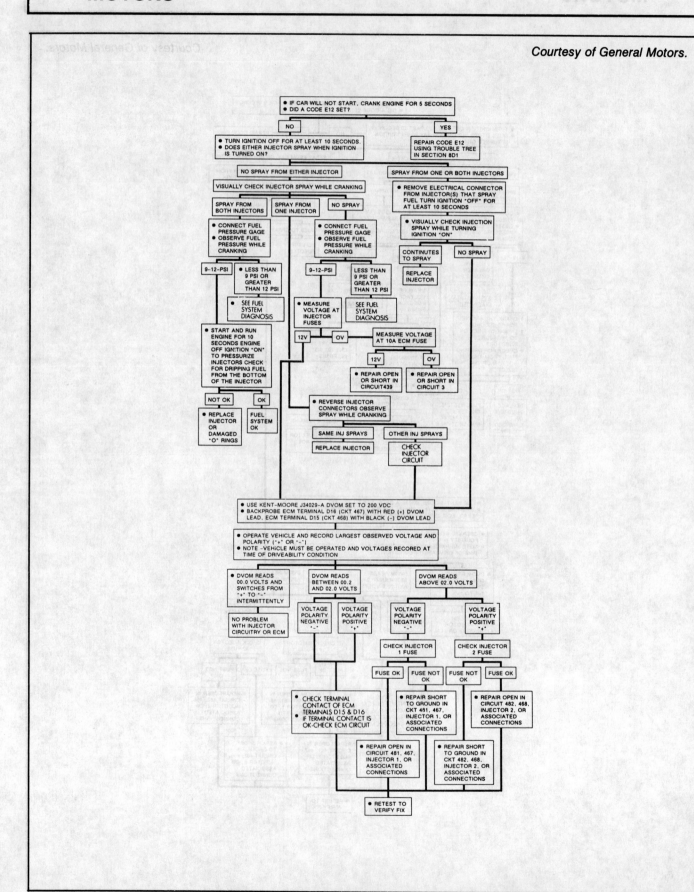

Fig. 12: Injector Diagnosis (1986-89 DeVille & Fleetwood)

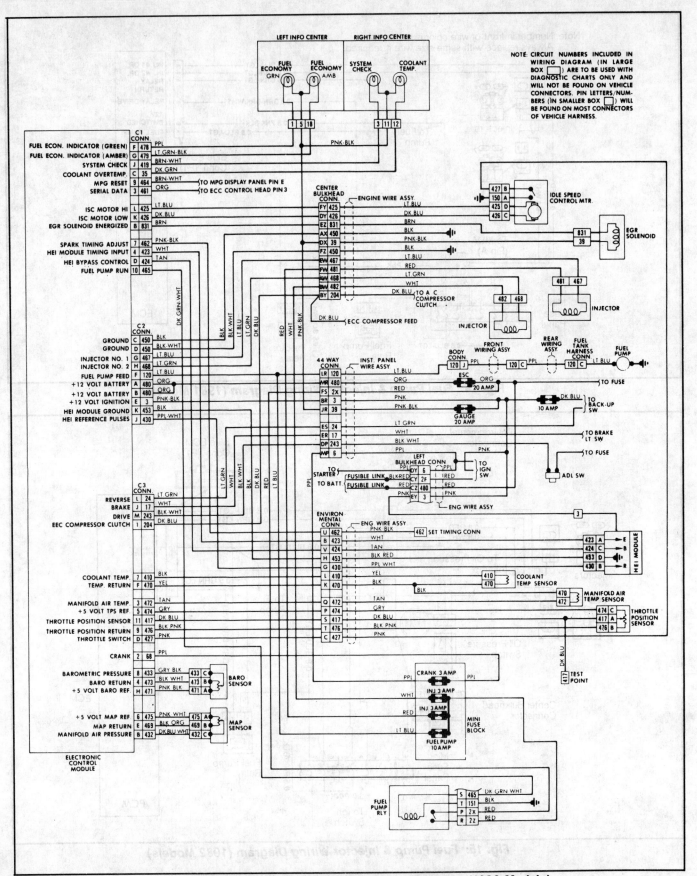

Fig. 13: Digital Fuel Injection Wiring Diagram (1980 Models)

Note: Number in front of wire color is metric size.
Always replace with same size wire if required.

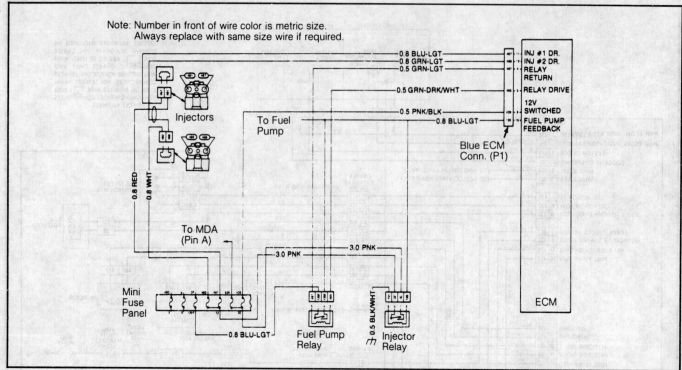

Fig. 14: Fuel Pump & Injector Wiring Diagram (1981 Models)

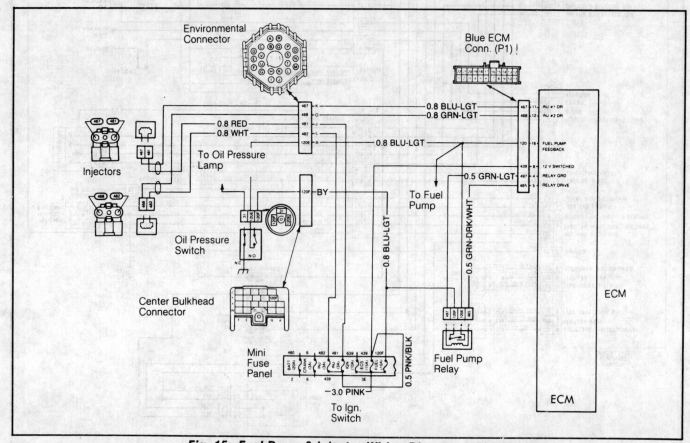

Fig. 15: Fuel Pump & Injector Wiring Diagram (1982 Models)

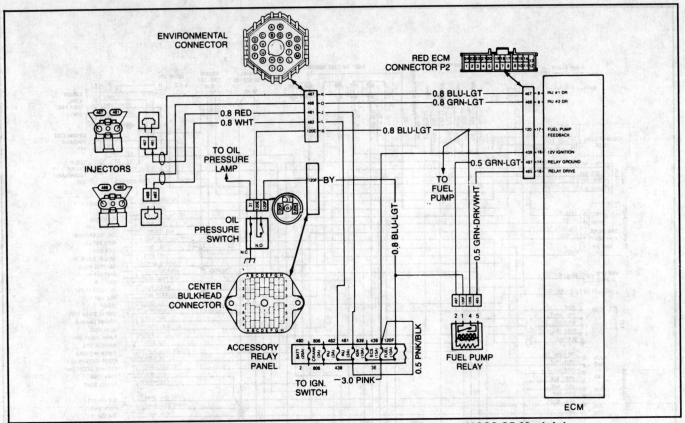

Fig. 16: Fuel Pump & Injector Wiring Diagram (1983-85 Models)

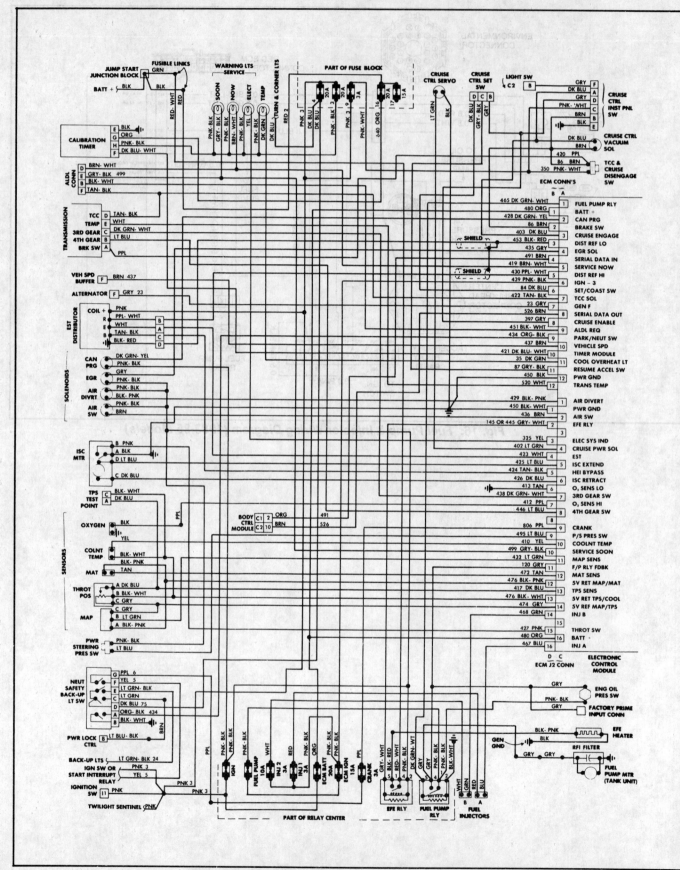

Fig. 17: Digital Fuel Injection Wiring Diagram (1986 DeVille & Fleetwood)

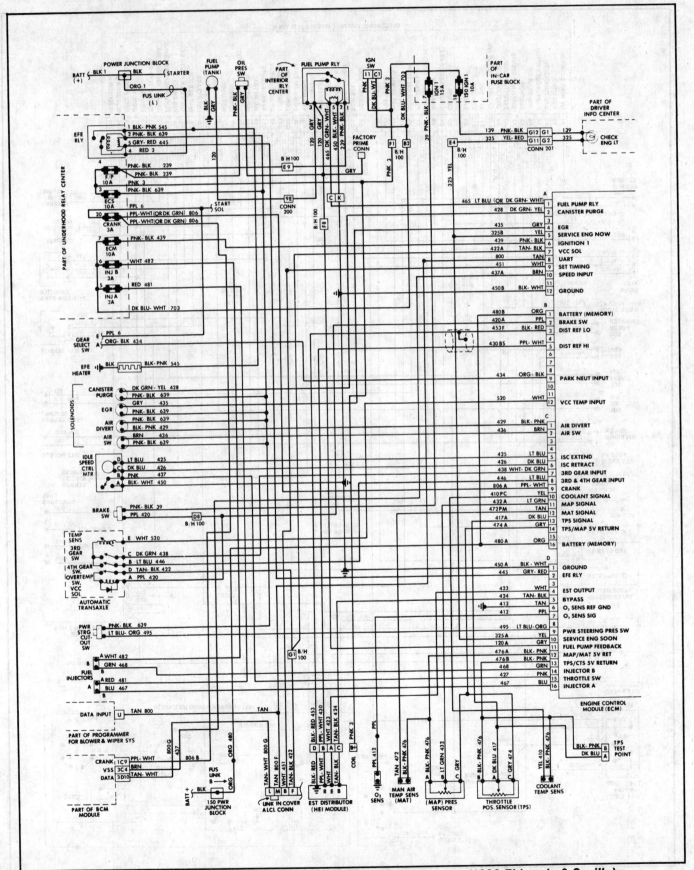

Fig. 18: Digital Fuel Injection Wiring Diagram (1986 Eldorado & Seville)

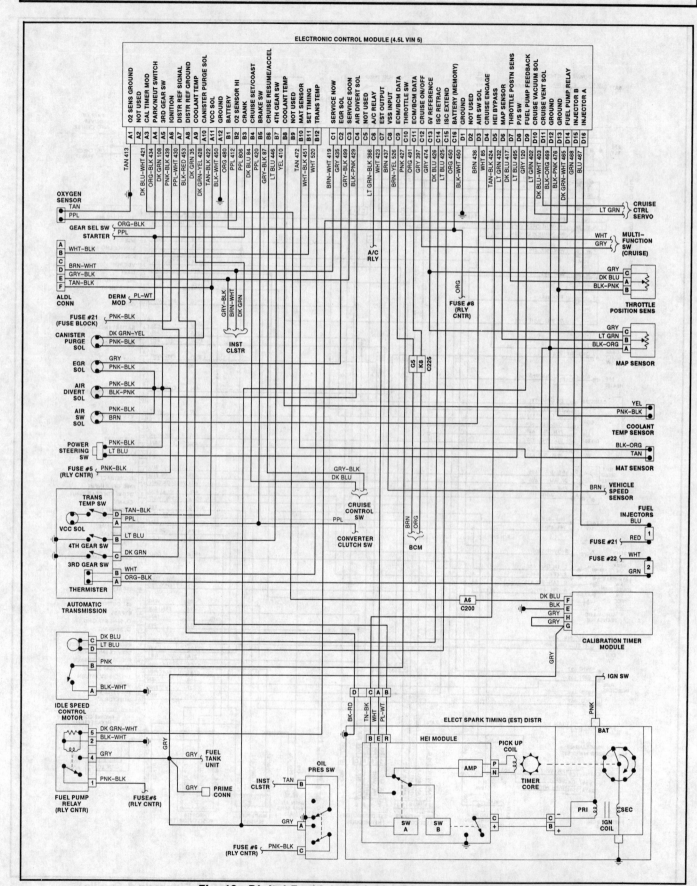

Fig. 19: Digital Fuel Injection Wiring Diagram (1987-89 DeVille & Fleetwood)

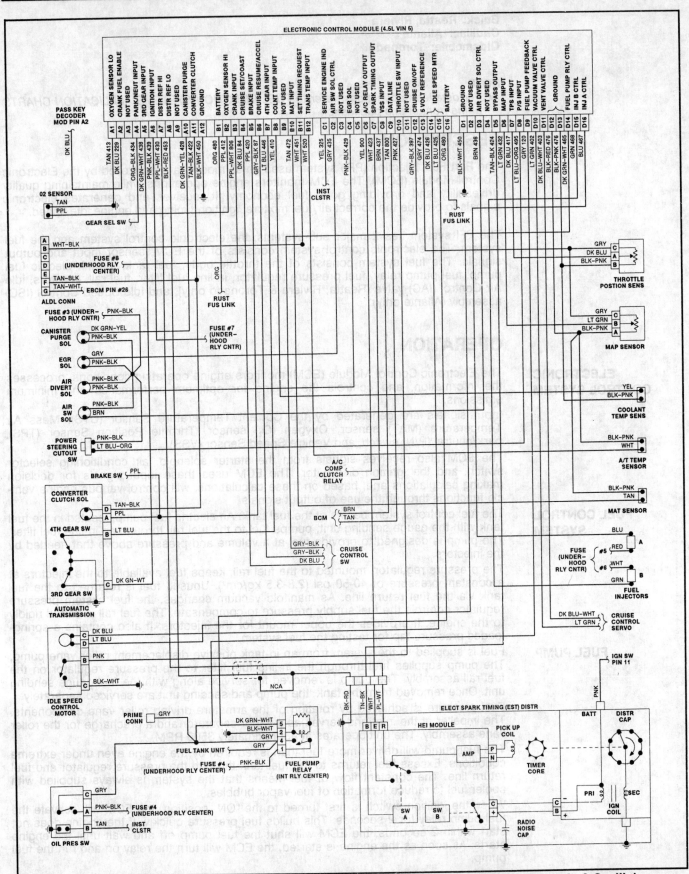

Fig. 20: Digital Fuel Injection Wiring Diagram (1987-89 Eldorado & Seville)

Buick: Reatta, Riviera
Cadillac: Allanté
Oldsmobile: Toronado

NOTE
- *For detailed applications, refer to FUEL INJECTION SYSTEMS APPLICATION CHARTS at the front of this publication.*

DESCRIPTION

The Port Fuel Injection (PFI) system used on all models are controlled by the Electronic Control Module (ECM). The ECM monitors engine operations while maintaining quality driveability and providing good fuel economy. It receives and generates electronic signals to provide the correct air/fuel mixture, ignition timing and engine idle speed.

The PFI system has 2 major sub-systems; the electronic control system and the fuel system. The electronic control system consists of the ECM, sensor input and output signals. The fuel system consists of the throttle body, fuel tank, in-tank electric fuel pump, fuel pump relay, fuel pressure regulator, in-line fuel filter, fuel rail, injectors, Idle Air Control (IAC) valve (Reatta, Riviera & Toronado only), and Idle Speed Control (ISC) assembly (Allanté only).

OPERATION

ELECTRONIC CONTROL SYSTEM

The Electronic Control Module (ECM) monitors engine operating conditions, processes the information, and controls the engine for optimum performance and minimum emissions.

Input signals are generated by the Coolant Temperature Sensor (CTS), Mass Air Temperature (MAT) sensor, Oxygen (O_2) sensor, Throttle Position Sensor (TPS), park/neutral (P/N) switch, and Vehicle Speed Sensor (VSS).

The ECM also receives signals from the starter solenoid, air conditioning selector switch, and the ignition distributor. The ECM uses these input signals for decision making calculations and, based on these calculations, will control various engine/vehicle functions through the use of output signals.

FUEL CONTROL SYSTEM

The fuel control system begins at the fuel tank. An electric fuel pump, located in the fuel tank with the gauge sending unit, pumps fuel to the fuel rail through an in-line fuel filter. The pump is designed to provide fuel at a volume and pressure above that needed by the injectors.

The pressure regulator, mounted to the fuel rail, keeps fuel available to the injectors at a constant pressure of 40-50 psi (2.8-3.5 kg/cm²). Unused fuel is returned to the fuel tank via the fuel return line. As manifold vacuum changes, the fuel system pressure regulator controls the fuel supply pressure to compensate. The fuel rail is bolted rigidly to the engine. It provides the upper mount for the injectors. It also contains a spring-loaded pressure tap for testing the fuel system.

FUEL PUMP

Fuel is supplied to the system from an in-tank positive displacement roller vane pump. The pump supplies fuel through the in-line fuel filter to the pressure regulator on the fuel rail assembly. The pump is removed for service along with the fuel gauge sending unit. Once removed from the tank, the pump and sending unit are serviced separately.

Fuel pressure is achieved by rotation of the armature driving roller vane components. The impeller at the inlet end serves as a vapor separator and a precharge for the roller vane assembly. The unit operates at approximately 3500 RPM.

The fuel pump will deliver more fuel than is required by the engine even under extreme conditions. Excess fuel returns to the fuel tank through the pressure regulator and fuel return line. The constant flow of fuel means that the system is always supplied with cooler fuel to reduce formation of fuel vapor bubbles.

When the ignition switch is first turned to the "ON" position, the ECM will activate the fuel pump relay for 2 seconds. This builds fuel pressure quickly. If the engine does not start within 2 seconds, the ECM will shut the fuel pump off and wait until the engine starts. As soon as the engine is started, the ECM will turn the relay on and run the fuel pump.

As a back-up system to the fuel pump relay, the fuel pump can also be activated by the oil pressure switch. The oil pressure switch is a normally open switch that closes when oil pressure reaches 4 psi (.3 kg/cm²). If the fuel pump relay fails, the oil pressure switch will close and run the fuel pump. An inoperative fuel pump relay can result in long cranking times, particulary if the engine is cold. *See Fig. 1.*

Courtesy of General Motors Corp.

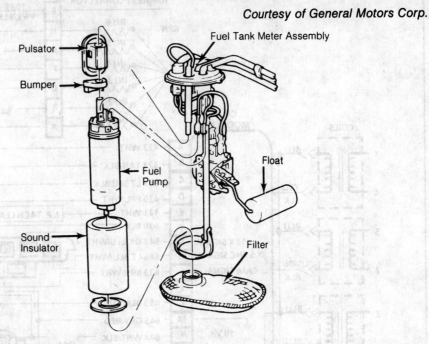

Fig. 1: In-Tank Fuel Pump

Courtesy of General Motors Corp.

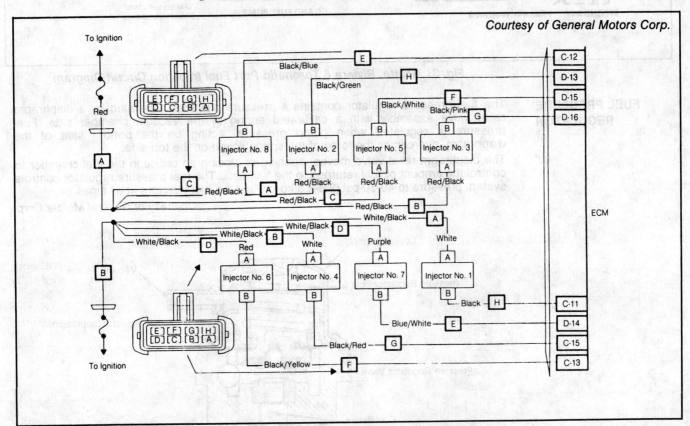

Fig. 2: Allanté Port Fuel Injection Circuit Diagram

As a back up system to the fuel pump relay, the fuel pump can also be activated by the oil pressure switch. The oil pressure switch is a normally open switch that closes when oil pressure reaches 4 psi (3 kg/cm²). If the fuel pump relay fails, the oil pressure switch will close and run the fuel pump. An inoperative fuel pump relay can result in long crank times, particularly when the engine is cold. See Fig. 3.

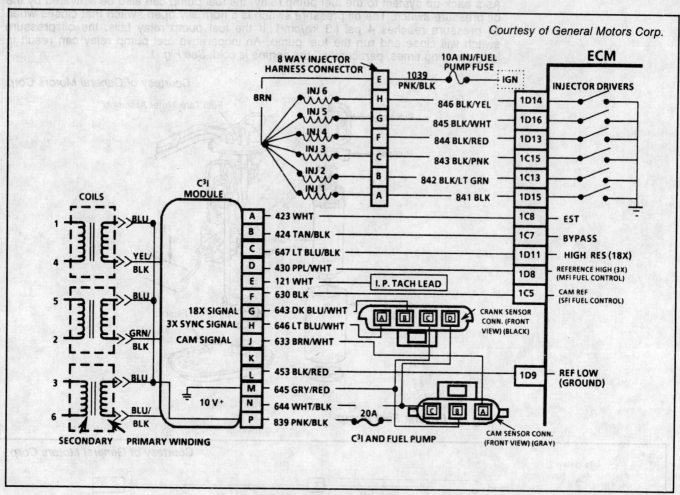

Courtesy of General Motors Corp.

Fig. 3: Reatta, Riviera & Toronado Port Fuel Injection Circuit Diagram

**FUEL PRESSURE
REGULATOR**

The fuel pressure regulator contains a pressure chamber separated by a diaphragm relief valve assembly with a calibrated spring in the vacuum chamber side. Fuel pressure is regulated when pump pressure, acting on the bottom side of the diaphragm, overcomes the force of the spring action on the top side.

The diaphragm relief valve moves, opening or closing an orifice in the fuel chamber to control the amount of fuel returned to the fuel tank. The fuel pressure regulator controls system pressure to 40-50 psi (2.8-3.5 kg/cm²) at the fuel injectors at all times.

Courtesy of General Motors Corp.

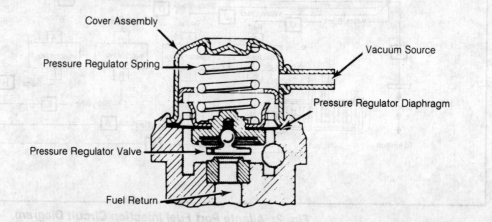

Fig. 4: Sectional View of Fuel Pressure Regulator

Vacuum acting on the top side of the diaphragm along with spring pressure controls the fuel pressure. An increase in vacuum creates a decrease in fuel pressure. Under heavy load conditions, the engine requires more fuel flow. Vacuum decreases under heavy load conditions because of throttle opening. The decrease in vacuum will allow additional pressure to the top side of the pressure relief valve to increase fuel pressure. *See Fig. 4.*

FUEL RAILS

The fuel rail assembly includes a fuel pressure regulator and 8 individual high pressure fuel injectors. *See Fig. 5.* The fuel rail assembly is bolted to the intake manifold assembly and the injectors fit into individual sockets on the base plate. The injectors are force fit into the opening on the rail. The injectors are locked in place with the aid of the injector retainer clip. *See Fig. 7.*

Courtesy of General Motors Corp.

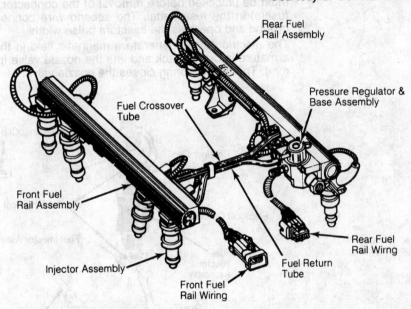

Fig. 5: Fuel Rails Allanté Engines

Courtesy of General Motors Corp.

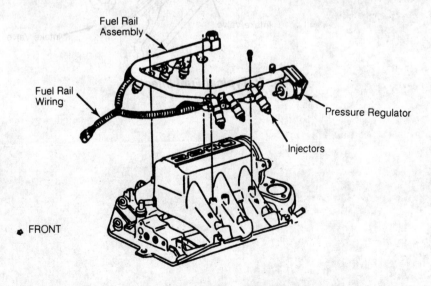

Fig. 6: Fuel Rails Reatta, Riviera & Toronado Engines

FUEL INJECTORS

A fuel injector is installed in the intake manifold at each cylinder. Nozzle spray pattern is on a 25 degree angle. The 2 rubber "O" rings used are lubricated and replaced as necessary whenever the injector is removed from the intake manifold. The "O" rings provide thermal insulation to reduce the formation of vapor bubbles and promote good hot start characteristics. The "O" rings also reduce injector vibration.

The injectors are identified with an ID number cast on the injector near the top side. Injectors manufactured by Rochester Products have an "RP" positioned near the top side in addition to the ID number.

The solenoid-operated injector consists of a valve body and nozzle valve with a special ground pintle. The moveable armature is attached to the nozzle valve which is pressed against the nozzle body sealing seat by a helical spring. At the back of the valve body is the solenoid winding and in the front section is the guide for the nozzle valve.

Each injector has a 2-wire connector. The injector connectors have a spring clip that must be unlocked before removal of the connector. One wire supplies voltage from the fuse(s) in the fuse panel. The second wire connects to the ECM, which controls the ground and operates the injectors pulse width.

The electric pulses generate a magnetic field in the solenoid winding. As a result, the armature is drawn back and lifts the nozzle valve from its seat approximately .038" (.15 mm). The helical spring closes the nozzle valve.

Courtesy of General Motors Corp.

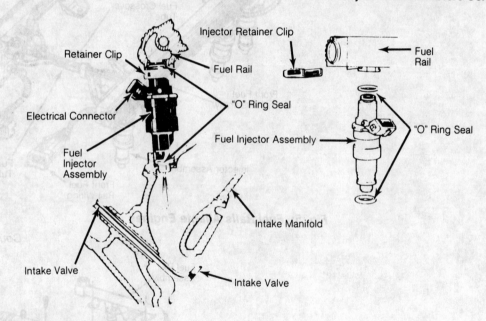

Fig. 7: Typical Fuel Injector

THROTTLE BODY

The throttle body is used to control the amount of air that enters the engine as well as the amount of vacuum in the throttle body vacuum manifold. The throttle body also supports and controls the movement of the Throttle Position Sensor (TPS), enabling the ECM to know the throttle position under all operating conditions. *See Fig. 8 and Fig. 9.*

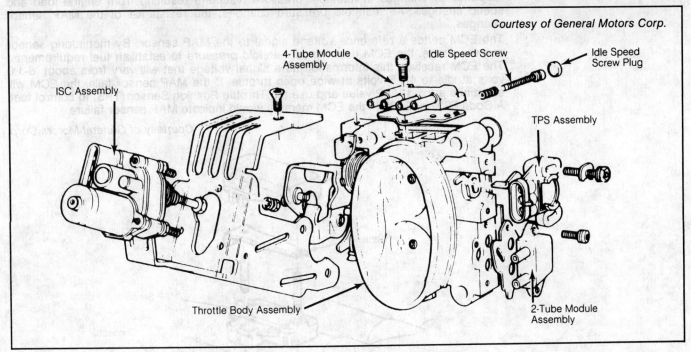

Courtesy of General Motors Corp.

Fig. 8: Exploded View of Allanté Throttle Body

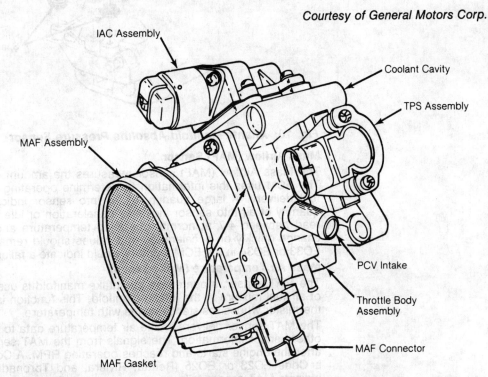

Courtesy of General Motors Corp.

Fig. 9: Exploded View of Reatta, Riviera & Toronado Throttle Body

AIRFLOW SENSING

Manifold Absolute Pressure (MAP) Sensor

Manifold absolute pressure and temperature along with other input engine variables are used to calculate airflow in the ECM. The Manifold Absolute Pressure (MAP) sensor responds to changes in manifold pressure (vacuum) resulting from engine load and speed changes. As manifold pressure changes, the resistance of the MAP sensor changes.

The ECM sends a reference voltage signal to the MAP sensor. By monitoring sensor output voltage, the ECM determines manifold pressure to establish fuel requirements. The ECM receives this information as a signal voltage that will vary from about .5-1.0 volts at idle to 4-4.5 volts at wide open throttle. If the MAP sensor fails, the ECM will substitute a fixed MAP value and use the Throttle Position Sensor (TPS) to control fuel. A Code E031 or E032 in the ECM memory would indicate MAP sensor failure.

Courtesy of General Motors Corp.

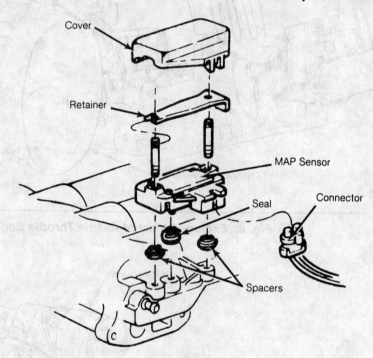

Fig. 10: Allanté Manifold Absolute Pressure Sensor

Mass Airflow (MAF) Sensor

The Mass Airflow (MAF) sensor measures the amount of air which passes through it. The ECM uses this information to determine operating condition of engine, to control fuel delivery. A large quantity of air into sensor indicates acceleration and a small quantity of air into sensor indicates deceleration or idle. MAF parameter values should read between 4-7 at normal operating temperature and engine idling. Values should change quickly on acceleration and values should remain stable at any RPM. A Code EO33 or EO34 in the ECM memory would indicate a failure in MAF sensor or circuit.

Mass Air Temperature (MAT) Sensor

The MAT sensor, mounted on the intake manifold is used to measure the temperature of air/fuel mixture in the intake manifold. This function is made possible by an integral thermistor whose resistance varies with temperature.

The MAT sensor delivers intake air temperature data to the ECM to be processed with other sensor information. The signals from the MAT sensor are accepted by the ECM after the engine starts and reaches operating RPM. A Code E037 or E038 (Allanté) and a Code EO23 or EO25 (Reatta, Riviera, and Toronado in the ECM memory would indicate MAT sensor failure.

Courtesy of General Motors Corp.

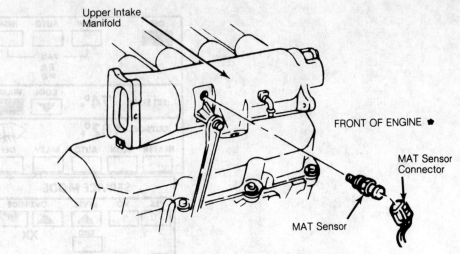

Fig. 11: Allanté Mass Air Temperature Sensor Assembly

TROUBLE SHOOTING

ECM TROUBLE CODES

Entering On-Board Diagnostics

The ECM uses the Electronic Climate Control (ECC) panel to display trouble codes as they are stored in the ECM memory. The following procedure is designed to retreive any trouble codes stored.

1) Turn ignition on, enter into diagnostic mode by depressing the "OFF" and "WARMER" buttons at the same time. *See Fig. 12.* Hold buttons down until all displays illuminate (Allanté and Toronado) or 2 "BEEPS" is heard or see a display "SERVICE MODE" (Reatta and Riviera).

2) The diagnostic code level will display ECM codes followed by BCM codes. Proceed to the next selection, press "LO". To exit diagnostics, press "RESET" on the Driver Information Center (Allanté) or press "BI-LEVEL" on Climate Control panel (Reatta, Riviera, and Toronado).

Courtesy of General Motors Corp.

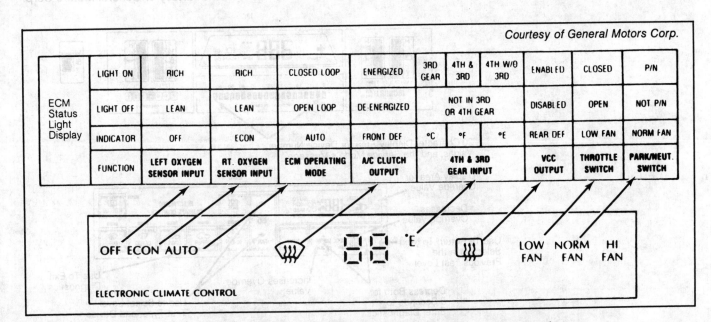

		RICH	RICH	CLOSED LOOP	ENERGIZED	3RD GEAR	4TH & 3RD	4TH W/O 3RD	ENABLED	CLOSED	P/N
ECM Status Light Display	LIGHT ON										
	LIGHT OFF	LEAN	LEAN	OPEN LOOP	DE-ENERGIZED	NOT IN 3RD OR 4TH GEAR			DISABLED	OPEN	NOT P/N
	INDICATOR	OFF	ECON	AUTO	FRONT DEF	°C	°F	°E	REAR DEF	LOW FAN	NORM FAN
	FUNCTION	LEFT OXYGEN SENSOR INPUT	RT. OXYGEN SENSOR INPUT	ECM OPERATING MODE	A/C CLUTCH OUTPUT	4TH & 3RD GEAR INPUT			VCC OUTPUT	THROTTLE SWITCH	PARK/NEUT. SWITCH

OFF ECON AUTO   `88` °E   LOW FAN   NORM FAN   HI FAN

ELECTRONIC CLIMATE CONTROL

Fig. 12: Climate Control Panel & Diagnostic Center (Allanté)

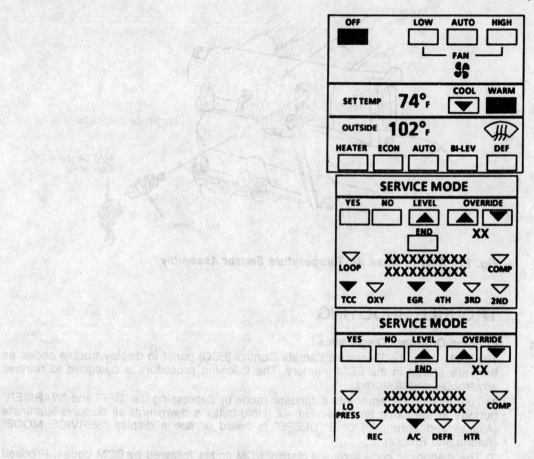

Fig. 13: Climate Control Panel & Diagnostic Center (Reatta, Riviera)

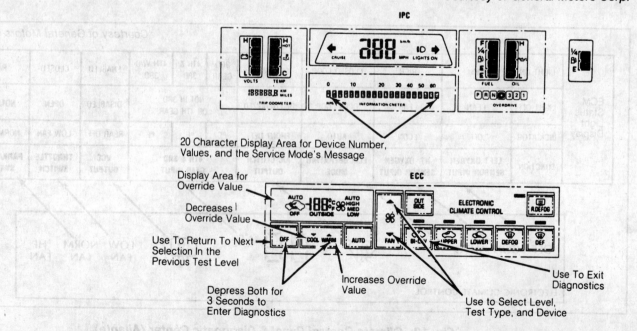

20 Character Display Area for Device Number,
Values, and the Service Mode's Message

Display Area for
Override Value

Decreases
Override Value

Use To Return To Next
Selection In the
Previous Test Level

Depress Both for
3 Seconds to
Enter Diagnostics

Increases Override
Value

Use To Exit
Diagnostics

Use to Select Level,
Test Type, and Device

Fig. 14: Climate Control Panel & Diagnostic Center (Toronado)

The following is a list and description of ECM trouble codes associated with the fuel system:

Code 12
No distributor signal.

Code 13 (Allanté)
Right oxygen sensor inoperative.

Code 13 (Reatta, Riviera, Toronado)
Oxygen sensor cicuit open.

Code 14
Coolant sensor circuit shorted.

Code 15
Coolant sensor circuit open.

Code 16
Alternator voltage circuit failure.

Code 17
Left oxygen sensor inoperative.

Code 19
Shorted fuel pump circuit.

Code 20
Open fuel pump circuit.

Code 21
Throttle position sensor shorted.

Code 22
Throttle position sensor open.

Code 23 (Allanté)
Est/By-pass circuit disrupted.

Code 23 (Reatta, Riviera, Toronado)
MAT sensor circuit open.

Code 24
Vehicle speed sensor disrupted.

Code 25 (Reatta, Riviera, Toronado)
MAT sensor circuit shorted.

Code 26 (Allanté)
Throttle switch circuit shorted.

Code 26 (Reatta, Riviera, Toronado)
Quad driver error.

Code 27 (Allanté)
Throttle switch circuit open.

Code 27 (Reatta, Riviera, Toronado)
Second gear switch circuit disrupted.

Code 28
Third gear or 2-3 shift solenoid circuit shorted.

Code 29 (Allanté)
Third gear or 2-3 shift solenoid circuit open.

Code 29 (Reatta, Riviera, Toronado)
Fourth gear switch circuit open.

Code 30
ISC circuit disrupted.

Code 31 (Allanté)
Shorted MAP sensor circuit.

Code 31
Park/Neutral switch circuit disrupted.

Code 32
Open MAP sensor circuit.

Code 34 (Allanté)
MAP sensor reading too high.

Code 34 (Reatta, Riviera, Toronado)
Open MAF sensor circuit.

Code 37
MAT sensor circuit shorted.

Code 38 (Allanté)
MAT sensor circuit open.

Code 38 (Reatta, Riviera, Toronado)
Brake switch circuit disrupted.

Code 39 (Allanté)
VCC engagement disrupted.

Code 39 (Reatta, Riviera, Toronado)
Torque converter clutch circuit disrupted.

Code 40
Power steering sensor circuit disrupted.

Code 41
Cam sensor circuit disrupted.

Code 42 (Allanté)
Lean left oxygen sensor signal.

Code 42 (Reatta, Riviera, Toronado)
EST or bypass circuit failure.

Code 43 (Allanté)
Rich left oxygen sensor signal.

Code 43 (Reatta, Riviera, Toronado)
EST system failure.

Code 44
Lean right oxygen sensor signal.

Code 45
Rich right oxygen sensor signal.

Code 46 (Allanté)
Left/right fuel rail imbalance.

Code 46 (Reatta, Riviera, Toronado)
Power steering pressure switch circuit open.

Code 47
ECM-BCM data disruption.

Code 48 (Allanté)
EGR system inoperative.

Code 48 (Reatta, Riviera, Toronado)
Misfire.

Code 49
AIR system inoperative.

Code 52
ECM memory reset indicator.

Code 53
Interrupted distributor signal.

Code 55
Misadjusted TPS.

Code 56
Fourth gear or 3-4 shift solenoid circuit open.

Code 57
Fourth gear or 3-4 shift solenoid circuit shorted.

Code 59
Transmission temperature sensor circuit disrupted.

Code 60
Cruise control, transmission not in drive.

Code 61
Cruise control, vent solenoid circuit disrupted.

Code 62

Cruise control, vacuum solenoid circuit disrupted.

Code 63 (Allanté)

Cruise control, set speed and vehicle speed difference.

Code 63 (Reatta, Riviera, Toronado)

EGR flow disrupted (small).

Code 64 (Allanté)

Cruise control, vehicle acceleration too high.

Code 64 (Reatta, Riviera, Toronado)

EGR flow disrupted (medium).

Code 65 (Allanté)

Cruise control, servo position sensor failure.

Code 65 (Reatta, Riviera, Toronado)

EGR flow disrupted (large).

Code 66

Cruise control, engine RPM too high.

Code 67

Shorted cruise control switch.

After the diagnosis and repairs are made, trouble codes should be cleared from the ECM memory.

Clearing ECM Trouble Codes

After all of the trouble codes have been displayed, the ECM will enter the system "service mode". The "service mode" has several service options to choose from: "DATA", "INPUTS", "OUTPUTS", "OVERRIDES" and "CLEAR CODES". If the ECM "CLEAR CODES" option is chosen, the ECM will clear all stored trouble codes.

DIAGNOSIS & TESTING

PRELIMINARY CHECKS

Prior to diagnosing the fuel injection system, the following systems and components must be in good condition and operating properly:

- The ECM "ENGINE CONTROL SYSTEM FAULT" light is operational.
- There are no trouble codes stored.
- ECM grounds tight and clean.
- All support systems and wiring.
- Battery connections and specific gravity.
- Compression pressure.
- Fuel supply system pressure and flow.
- All electrical connections.
- Air filter.
- Vacuum lines, fuel hoses and pipe connections.

FUEL SYSTEM DIAGNOSIS

NOTE

- *The fuel cap on this system must comply to manufacturer's specification. If the incorrect application is used, malfunction of the system could result.*

Pressure Test

1) Connect fuel Pressure Gauge (J-34730-1) to fuel system. Wrap a shop towel around the fuel pressure tap to absorb any fuel leakage that may occur when installing gauge. Turn ignition on. Fuel pump pressure should be 40-50 psi (2.8-3.5 kg/cm²). This pressure is controlled by spring pressure within the regulator assembly.

2) When the engine is idling, manifold pressure is low (high vacuum). This high vacuum is applied to the fuel regulator diaphragm. This will offset the spring and result in a lower fuel pressure of 35-38 psi (2.5-2.7 kg/cm²). The idle pressure will vary some depending on barometric pressure. If the pressure at idle is less than 35 psi (2.5 kg/cm²) a problem with the pressure regulator control is indicated.

3) If fuel is observed in vacuum hose to pressure regulator, the regulator is faulty and must be replaced. Pressure that continues to fall is caused by one of the following: in-tank fuel pump check valve not holding, pump coupling hose leaking, fuel pressure regulator valve leaking or a injector sticking open.

Low Pressure

Normally, a vehicle with a fuel pressure of less than 24 psi (1.7 kg/cm²) at idle will not be driveable. However, if the pressure drop occurs only while driving, the engine will normally surge and then stop as pressure begins to drop rapidly. If fuel pressure is low check fuel pump output pressure between fuel pump and fuel filter. If pressure is within specification, replace fuel filter. If pressure is still low, pinch off fuel return line. Restricting the fuel return line allows the fuel pump to develop its maximum pressure. If pressure is still low, problem is in-tank fuel pump leaks or defective pump.

High Pressure.

If fuel pump pressure is higher than specification, this will result in rich air/fuel mixtures. Test system to determine if the high fuel pressure is due to a restricted fuel return line or a pressure regulator problem. This can be done by removing pressure return line and placing the end of the hose in a suitable container (relieve pressure from system before loosening fuel line). Turn ignition on (engine off) and note pressure gauge reading. If pressure is within specification, blockage in return line is the cause of high pressure. If the pressure is still high, the problem is in the fuel pressure regulator.

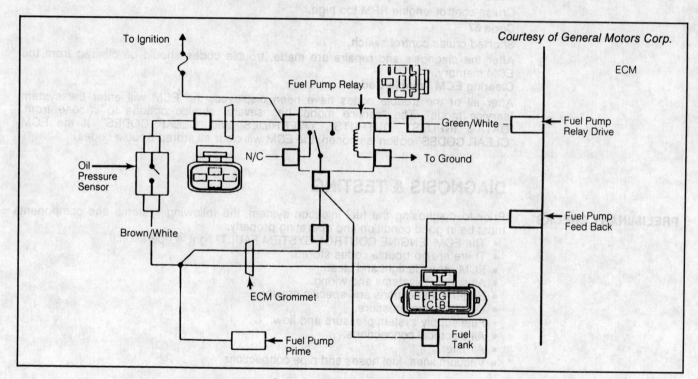

Courtesy of General Motors Corp.

Fig. 15: Allanté PFI Fuel Pump Circuit

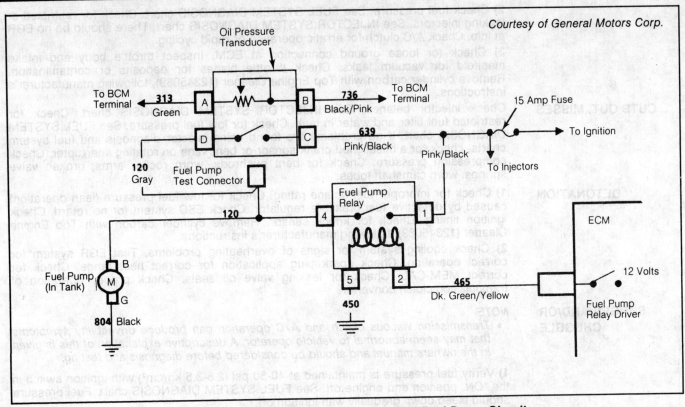

Courtesy of General Motors Corp.

Fig. 16: Reatta, Riviera & Toronado PFI Fuel Pump Circuit

HARD START

1) Verify fuel pressure is maintained at between 40-50 psi (2.8-3.5 kg/cm²) with ignition in the "ON" position and engine off. See FUEL SYSTEM DIAGNOSIS chart. Fuel pressure should bleed down gradually with ignition off.

2) Inspect for contaminated fuel. If condition exists in warm climate or after hot soak, check for improper or high alcohol content fuel. Inspect TPS and throttle linkage for sticking or binding. Check MAP sensor vacuum fitting and manifold port for damage or restrictions.

3) Check exhaust system for restrictions and repair as necessary. Check for high resistance in coolant sensor circuit or sensor. Check fuel pump relay circuit. Check for faulty in-tank fuel pump check valve.

4) Check ignition system for proper spark, loose connections, loose ground mounting screws, crank sensor. If engine starts and stalls, disconnect MAF sensor. If engine then runs and sensor connections are okay, replace the sensor.

STALL AFTER START

1) Verify fuel pressure is maintained at 40-50 psi (2.8-3.5 kg/cm²) with ignition in the "ON" position and engine off. See FUEL SYSTEM DIAGNOSIS chart. Fuel pressure should bleed down gradually when ignition is turned off.

2) Check for proper operation of PCV valve. Check for high A/C pressure or overcharged system. Check to verify no EGR operation with engine cold or at idle.

HESITATION, SAG, STUMBLE

1) Check fuel pressure. Verify fuel pressure is maintained at between 40-50 psi (2.8-3.5 kg cm²) with ignition switch in the "ON" position and engine off. See FUEL SYSTEM DIAGNOSIS chart. Fuel pressure should bleed down gradually with ignition off.

2) Check for water contaminated fuel. Check TPS for sticking or binding. Check vacuum hose to MAP sensor for cuts or restrictions. Check canister purge system for proper operation. Check alternator output voltage. Voltage should be 9-16 volts.

3) Check for proper seal of oil filler cap and tube. Check for air leaks at air duct between MAF sensor and throttle body. Test injector balance by disabling each individual injector while noting RPM drop. Inspect injector if RPM does not decrease. Check high energy ignition ground, and check circuit No. 453.

4) Check ignition ground cicuit. Check canister purge system. Check detonation sensor and ignition operation on acceleration. Check crank sensor for malfunction.

ROUGH, UNSTABLE IDLE

1) Check throttle linkage, cruise control and TV cables for sticking or binding. Check idle speed and observe ISC operation while manipulating connector. There should be no erratic movement.

2) Check fuel pressure. See FUEL SYSTEM DIAGNOSIS chart. Check for restricted or leaking injectors. See INJECTOR SYSTEM DIAGNOSIS chart. There should be no EGR at idle. Check A/C clutch for erratic operation or rapid cycling.

3) Check for loose ground connection at ECM. Inspect throttle body and intake manifold for vacuum leaks. Check throttle blades for deposits or contamination. Remove cylinder carbon with Top Engine Cleaner (12345089), following manufacturer's instructions.

CUTS OUT, MISSES

Check injector balance. See INJECTOR SYSTEM DIAGNOSIS chart. Check for restricted fuel filter and water in tank. Check for low fuel pressure. See FUEL SYSTEM DIAGNOSIS chart. If no problems are found using injector diagnosis and fuel system charts, check for a misaligned crank sensor or bent vane on rotating interrupter. Check compression pressure. Check for bent pushrods, worn rocker arms, broken valve springs, worn camshaft lobes.

DETONATION

1) Check for improper fuel octane rating. Check for low fuel pressure (lean operation) caused by defective fuel pressure regulator. Check ESC system for no retard. Check ignition timing. Check for knock sensor. Remove cylinder carbon with Top Engine Cleaner (12345089), following manufacturer's instructions.

2) Check cooling system for signs of overheating problems. Test EGR system for correct operation. Check spark plug application for correct heat range. Check for correct MEM-CAL. Check for leaking valve oil seals. Check proper operation of transmission clutch converter.

SURGES AND/OR CHUGGLE

NOTE
- *Transmission viscous clutch and A/C operation can produce driveability symptoms that may seem abnormal to vehicle operator. A descriptive explanation of this is given in the owners manual and should be considered before diagnosis and testing.*

1) Verify fuel pressure is maintained at 40-50 psi (2.8-3.5 kg/cm²) with ignition switch in the "ON" position and engine off. See FUEL SYSTEM DIAGNOSIS chart. Fuel pressure should bleed down gradually with ignition off.

2) Inspect for contaminated fuel. Inspect TPS and throttle linkage for sticking or binding. Check exhaust system for restrictions and repair as necessary. Check for faulty in-tank fuel pump check valve. If engine starts and stalls, disconnect MAF sensor.

3) Check EGR valve operation, there should be no EGR at idle. Check all vacuum hoses for kinks or leaks. Check for dirty or plugged in-line fuel filter. Remove cylinder carbon with Top Engine Cleaner (12345089), following manufacturer's instructions. Check alternator output voltage. Voltage should be 9-16 volts.

4) Check spark plug application for correct heat range. Check spark plugs for signs of oil consumption. If necessary, repair cause of oil consumption. Test drive vehicle to isolate symptom. If problem is in the engine and not body or chassis, check balance components such as alternator and air conditioner.

LACK OF POWER, SLUGGISH, OR SPONGY

1) Test drive vehicle to verify complaint. Remove air cleaner to inspect air filter and throttle valve operation. Check spark plugs for damage or wear. Check intake manifold and throttle for vacuum leaks. Check ESC system.

2) Check for contaminated fuel, or restricted fuel filter. Verify fuel pressure is maintained at 40-50 psi (2.8-3.5 kg/cm²) with ignition switch in the "ON" position and engine off. See FUEL SYSTEM DIAGNOSIS chart. Fuel pressure should bleed down gradually with ignition off.

3) Verify EGR valve is not stuck open. Inspect exhaust system for possible restriction or internal failure. Check MAP sensor vacuum circuit for restrictions or trapped vacuum. Remove cylinder carbon with Top Engine Cleaner (12345089), following manufacturer's instructions. Check alternator voltage output. Voltage should be 9-16 volts. Check valve timing and compression pressure. Check TPS for proper operation.

BACKFIRE

Check for loose wiring connector or air duct at MAF sensor. Check compression pressure. Check for proper valve timing. Check crank and cam sensors for proper operation. Check spark plugs and proper routing of plug wires. Check for faulty spark plugs and/or plug wires, boots. Check for flooded condition. Check output voltage of ignition coils.

ADJUSTMENTS

MINIMUM IDLE SPEED

NOTE
- *This adjustment should be necessary only when throttle body parts have been replaced or when required by TPS adjustment. Engine should be at normal operating temperature.*

1) Enter diagnostics to verify absence of trouble codes. If trouble codes are recorded, perform repairs as necessary.

2) Start engine and warm to operating temperature. Position the ISC motor fully retracted by selecting ECM override ES03. Press the "COOLER" button on the CCDIC and wait for ISC to retract.

3) When the ISC plunger is fully retracted, the plunger should not touch the throttle lever. If contact is noticed, turn plunger inward. Verify that throttle lever is unrestricted and is resting on the minimum idle screw.

4) Check minimum idle speed. Minimum engine speed is 450-550 RPM. If RPM is above specification, thoroughly inspect intake manifold, throttle body, and all vacuum lines for leaks. If adjustment is made with a vacuum leak in the system, calibration throughout the driving range will be offset.

5) If vacuum system is okay, and idle adjustment is necessary, pierce the idle stop screw plug with an awl. Remove and discard the minimum idle speed screw plug. Use Torx Driver (T-20) and adjust idle to approximately 500 RPM.

THROTTLE POSITION SENSOR (TPS)

Allanté

1) With ignition on, attach high impedance digital voltmeter positive lead to TPS test point, circuit No. 417 (Dark Blue wire). Connect negative lead to circuit No. 476 (Black/White wire).

2) Loosen TPS screws and adjust sensor with Torx Driver (T-25) until voltmeter reads .475-.525 volts. Tighten screws, then recheck reading to make sure it has not changed.

Reatta, Riviera, Toronado

1) Install 3 jumper wires between TPS and harness connector or use a "Scan Tester. With ignition on, attach high impedance digital voltmeter to terminals "B" and ground.

2) Loosen TPS screws and adjust sensor with Torx Driver (T-25) until voltmeter reads .36-.44 volts. Tighten screws, then recheck reading to make sure it has not changed.

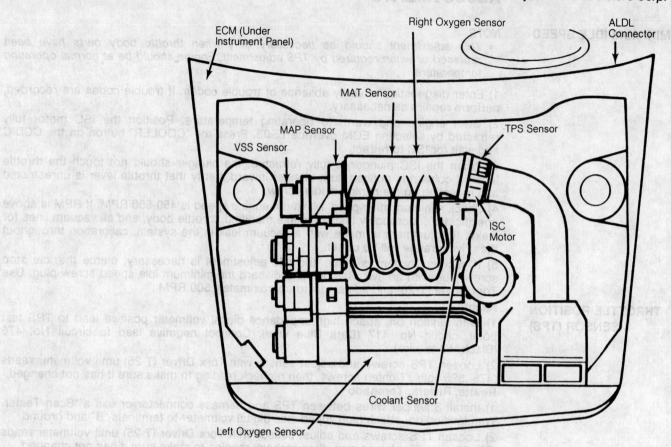

ECM (Under Instrument Panel)

Right Oxygen Sensor

ALDL Connector

MAT Sensor

MAP Sensor

TPS Sensor

VSS Sensor

ISC Motor

Coolant Sensor

Left Oxygen Sensor

Fig. 17: Allanté Component Location

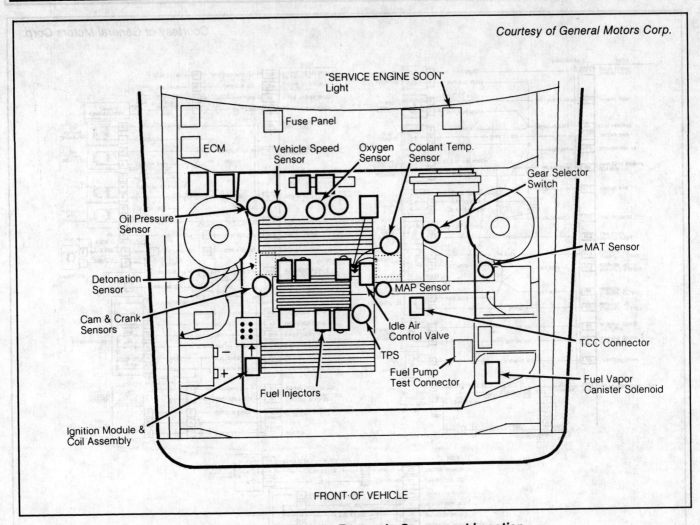

Fig. 18: Reatta, Riviera, Toronado Component Location

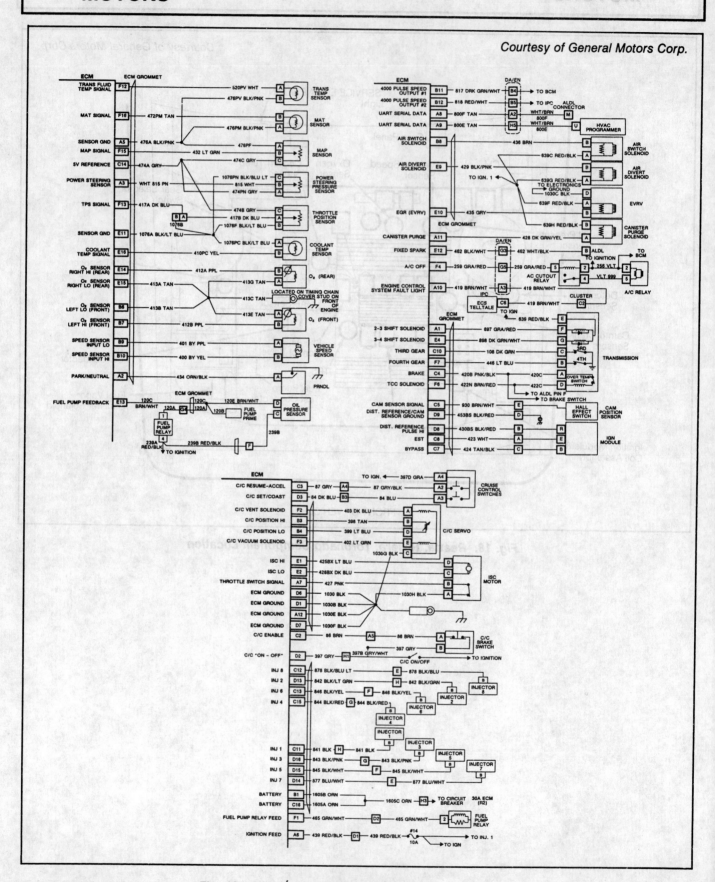

Fig. 19: Allanté ECM Circuit Schematics

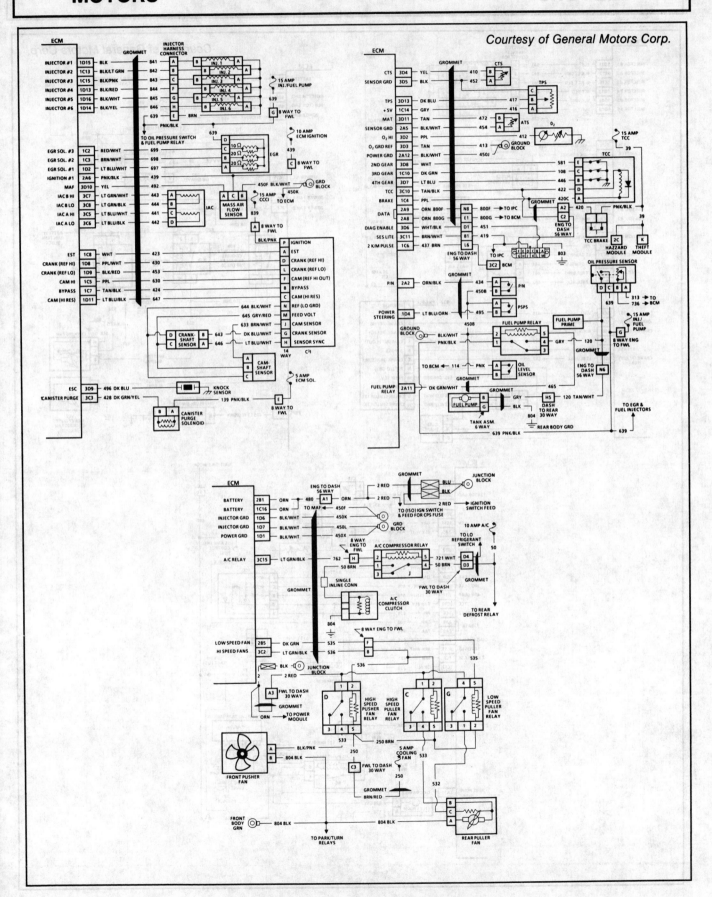

Courtesy of General Motors Corp.

Fig. 20: Reatta & Riviera ECM Circuit Schematics

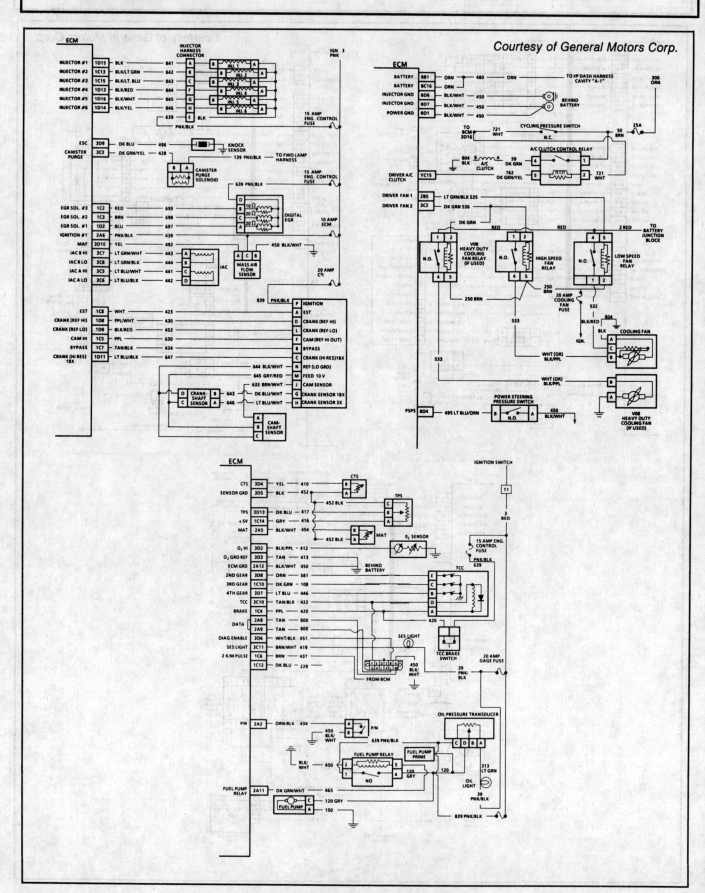

Fig. 21: Toronado ECM Circuit Schematics

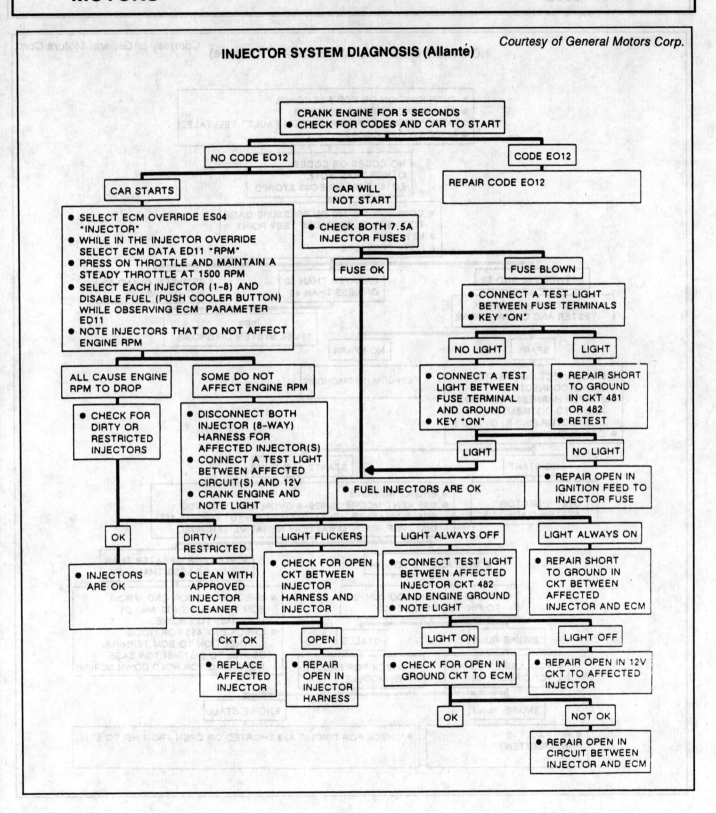

INJECTOR SYSTEM DIAGNOSIS (Allanté)

Courtesy of General Motors Corp.

NO START OR STALL AFTER START (Allanté) *Courtesy of General Motors Corp.*

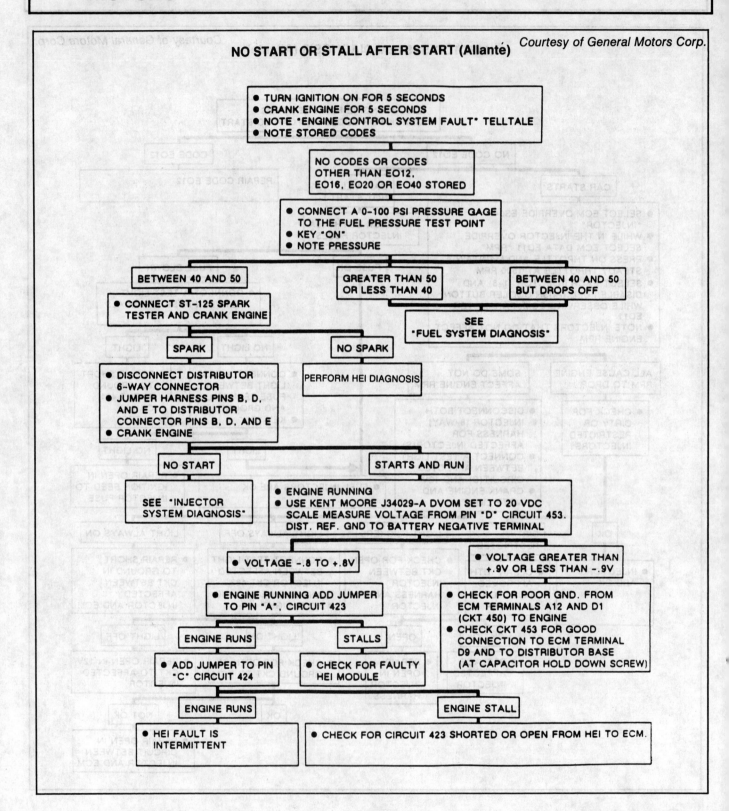

ECM REPLACEMENT CHECK (Allanté)

Courtesy of General Motors Corp.

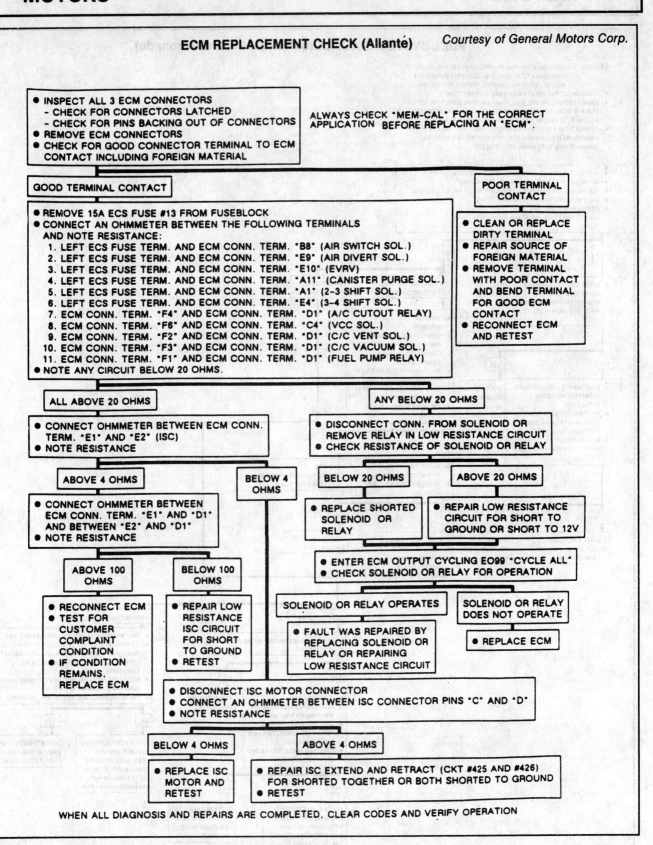

- INSPECT ALL 3 ECM CONNECTORS
  - CHECK FOR CONNECTORS LATCHED
  - CHECK FOR PINS BACKING OUT OF CONNECTORS
- REMOVE ECM CONNECTORS
- CHECK FOR GOOD CONNECTOR TERMINAL TO ECM CONTACT INCLUDING FOREIGN MATERIAL

ALWAYS CHECK "MEM-CAL" FOR THE CORRECT APPLICATION BEFORE REPLACING AN "ECM".

GOOD TERMINAL CONTACT

- REMOVE 15A ECS FUSE #13 FROM FUSEBLOCK
- CONNECT AN OHMMETER BETWEEN THE FOLLOWING TERMINALS AND NOTE RESISTANCE:
  1. LEFT ECS FUSE TERM. AND ECM CONN. TERM. "B8" (AIR SWITCH SOL.)
  2. LEFT ECS FUSE TERM. AND ECM CONN. TERM. "E9" (AIR DIVERT SOL.)
  3. LEFT ECS FUSE TERM. AND ECM CONN. TERM. "E10" (EVRV)
  4. LEFT ECS FUSE TERM. AND ECM CONN. TERM. "A11" (CANISTER PURGE SOL.)
  5. LEFT ECS FUSE TERM. AND ECM CONN. TERM. "A1" (2-3 SHIFT SOL.)
  6. LEFT ECS FUSE TERM. AND ECM CONN. TERM. "E4" (3-4 SHIFT SOL.)
  7. ECM CONN. TERM. "F4" AND ECM CONN. TERM. "D1" (A/C CUTOUT RELAY)
  8. ECM CONN. TERM. "F6" AND ECM CONN. TERM. "C4" (VCC SOL.)
  9. ECM CONN. TERM. "F2" AND ECM CONN. TERM. "D1" (C/C VENT SOL.)
  10. ECM CONN. TERM. "F3" AND ECM CONN. TERM. "D1" (C/C VACUUM SOL.)
  11. ECM CONN. TERM. "F1" AND ECM CONN. TERM. "D1" (FUEL PUMP RELAY)
- NOTE ANY CIRCUIT BELOW 20 OHMS.

POOR TERMINAL CONTACT

- CLEAN OR REPLACE DIRTY TERMINAL
- REPAIR SOURCE OF FOREIGN MATERIAL
- REMOVE TERMINAL WITH POOR CONTACT AND BEND TERMINAL FOR GOOD ECM CONTACT
- RECONNECT ECM AND RETEST

ALL ABOVE 20 OHMS

- CONNECT OHMMETER BETWEEN ECM CONN. TERM. "E1" AND "E2" (ISC)
- NOTE RESISTANCE

ANY BELOW 20 OHMS

- DISCONNECT CONN. FROM SOLENOID OR REMOVE RELAY IN LOW RESISTANCE CIRCUIT
- CHECK RESISTANCE OF SOLENOID OR RELAY

ABOVE 4 OHMS

- CONNECT OHMMETER BETWEEN ECM CONN. TERM. "E1" AND "D1" AND BETWEEN "E2" AND "D1"
- NOTE RESISTANCE

BELOW 4 OHMS

BELOW 20 OHMS

- REPLACE SHORTED SOLENOID OR RELAY

ABOVE 20 OHMS

- REPAIR LOW RESISTANCE CIRCUIT FOR SHORT TO GROUND OR SHORT TO 12V

ABOVE 100 OHMS

- RECONNECT ECM
- TEST FOR CUSTOMER COMPLAINT CONDITION
- IF CONDITION REMAINS, REPLACE ECM

BELOW 100 OHMS

- REPAIR LOW RESISTANCE ISC CIRCUIT FOR SHORT TO GROUND
- RETEST

- ENTER ECM OUTPUT CYCLING EO99 "CYCLE ALL"
- CHECK SOLENOID OR RELAY FOR OPERATION

SOLENOID OR RELAY OPERATES

- FAULT WAS REPAIRED BY REPLACING SOLENOID OR RELAY OR REPAIRING LOW RESISTANCE CIRCUIT

SOLENOID OR RELAY DOES NOT OPERATE

- REPLACE ECM

- DISCONNECT ISC MOTOR CONNECTOR
- CONNECT AN OHMMETER BETWEEN ISC CONNECTOR PINS "C" AND "D"
- NOTE RESISTANCE

BELOW 4 OHMS

- REPLACE ISC MOTOR AND RETEST

ABOVE 4 OHMS

- REPAIR ISC EXTEND AND RETRACT (CKT #425 AND #426) FOR SHORTED TOGETHER OR BOTH SHORTED TO GROUND
- RETEST

WHEN ALL DIAGNOSIS AND REPAIRS ARE COMPLETED, CLEAR CODES AND VERIFY OPERATION

FUEL SYSTEM DIAGNOSIS (Reatta, Riviera, Toronado)

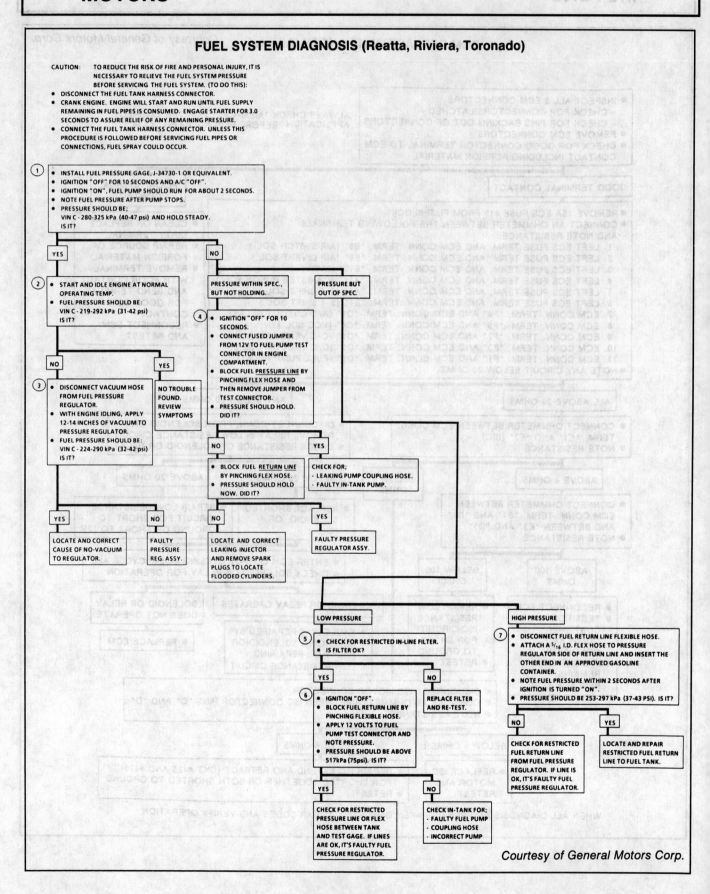

CAUTION: TO REDUCE THE RISK OF FIRE AND PERSONAL INJURY, IT IS
NECESSARY TO RELIEVE THE FUEL SYSTEM PRESSURE
BEFORE SERVICING THE FUEL SYSTEM. (TO DO THIS):
- DISCONNECT THE FUEL TANK HARNESS CONNECTOR.
- CRANK ENGINE. ENGINE WILL START AND RUN UNTIL FUEL SUPPLY
REMAINING IN FUEL PIPES IS CONSUMED. ENGAGE STARTER FOR 3.0
SECONDS TO ASSURE RELIEF OF ANY REMAINING PRESSURE.
- CONNECT THE FUEL TANK HARNESS CONNECTOR. UNLESS THIS
PROCEDURE IS FOLLOWED BEFORE SERVICING FUEL PIPES OR
CONNECTIONS, FUEL SPRAY COULD OCCUR.

①
- INSTALL FUEL PRESSURE GAGE, J-34730-1 OR EQUIVALENT.
- IGNITION "OFF" FOR 10 SECONDS AND A/C "OFF".
- IGNITION "ON", FUEL PUMP SHOULD RUN FOR ABOUT 2 SECONDS.
- NOTE FUEL PRESSURE AFTER PUMP STOPS.
- PRESSURE SHOULD BE;
VIN C - 280-325 kPa (40-47 psi) AND HOLD STEADY.
IS IT?

YES

NO

②
- START AND IDLE ENGINE AT NORMAL
OPERATING TEMP.
- FUEL PRESSURE SHOULD BE:
VIN C - 219-292 kPa (31-42 psi)
IS IT?

PRESSURE WITHIN SPEC.,
BUT NOT HOLDING.

PRESSURE BUT
OUT OF SPEC.

NO

YES

④
- IGNITION "OFF" FOR 10
SECONDS.
- CONNECT FUSED JUMPER
FROM 12V TO FUEL PUMP TEST
CONNECTOR IN ENGINE
COMPARTMENT.
- BLOCK FUEL PRESSURE LINE BY
PINCHING FLEX HOSE AND
THEN REMOVE JUMPER FROM
TEST CONNECTOR.
- PRESSURE SHOULD HOLD.
DID IT?

③
- DISCONNECT VACUUM HOSE
FROM FUEL PRESSURE
REGULATOR.
- WITH ENGINE IDLING, APPLY
12-14 INCHES OF VACUUM TO
PRESSURE REGULATOR.
- FUEL PRESSURE SHOULD BE:
VIN C - 224-290 kPa (32-42 psi)
IS IT?

NO TROUBLE
FOUND.
REVIEW
SYMPTOMS

NO

YES

- BLOCK FUEL RETURN LINE
BY PINCHING FLEX HOSE.
- PRESSURE SHOULD HOLD
NOW. DID IT?

CHECK FOR;
- LEAKING PUMP COUPLING HOSE.
- FAULTY IN-TANK PUMP.

YES

NO

LOCATE AND CORRECT
CAUSE OF NO-VACUUM
TO REGULATOR.

FAULTY
PRESSURE
REG. ASSY.

NO

YES

LOCATE AND CORRECT
LEAKING INJECTOR
AND REMOVE SPARK
PLUGS TO LOCATE
FLOODED CYLINDERS.

FAULTY PRESSURE
REGULATOR ASSY.

LOW PRESSURE

HIGH PRESSURE

⑤
- CHECK FOR RESTRICTED IN-LINE FILTER.
- IS FILTER OK?

⑦
- DISCONNECT FUEL RETURN LINE FLEXIBLE HOSE.
- ATTACH A 5/16 I.D. FLEX HOSE TO PRESSURE
REGULATOR SIDE OF RETURN LINE AND INSERT THE
OTHER END IN AN APPROVED GASOLINE
CONTAINER.
- NOTE FUEL PRESSURE WITHIN 2 SECONDS AFTER
IGNITION IS TURNED "ON".
- PRESSURE SHOULD BE 253-297 kPa (37-43 PSI). IS IT?

YES

NO

NO

YES

⑥
- IGNITION "OFF".
- BLOCK FUEL RETURN LINE BY
PINCHING FLEXIBLE HOSE.
- APPLY 12 VOLTS TO FUEL
PUMP TEST CONNECTOR AND
NOTE PRESSURE.
- PRESSURE SHOULD BE ABOVE
517kPa (75psi). IS IT?

REPLACE FILTER
AND RE-TEST.

CHECK FOR RESTRICTED
FUEL RETURN LINE
FROM FUEL PRESSURE
REGULATOR. IF LINE IS
OK, IT'S FAULTY FUEL
PRESSURE REGULATOR.

LOCATE AND REPAIR
RESTRICTED FUEL RETURN
LINE TO FUEL TANK.

YES

NO

CHECK FOR RESTRICTED
PRESSURE LINE OR FLEX
HOSE BETWEEN TANK
AND TEST GAGE. IF LINES
ARE OK, IT'S FAULTY FUEL
PRESSURE REGULATOR.

CHECK IN-TANK FOR;
- FAULTY FUEL PUMP
- COUPLING HOSE
- INCORRECT PUMP

Courtesy of General Motors Corp.

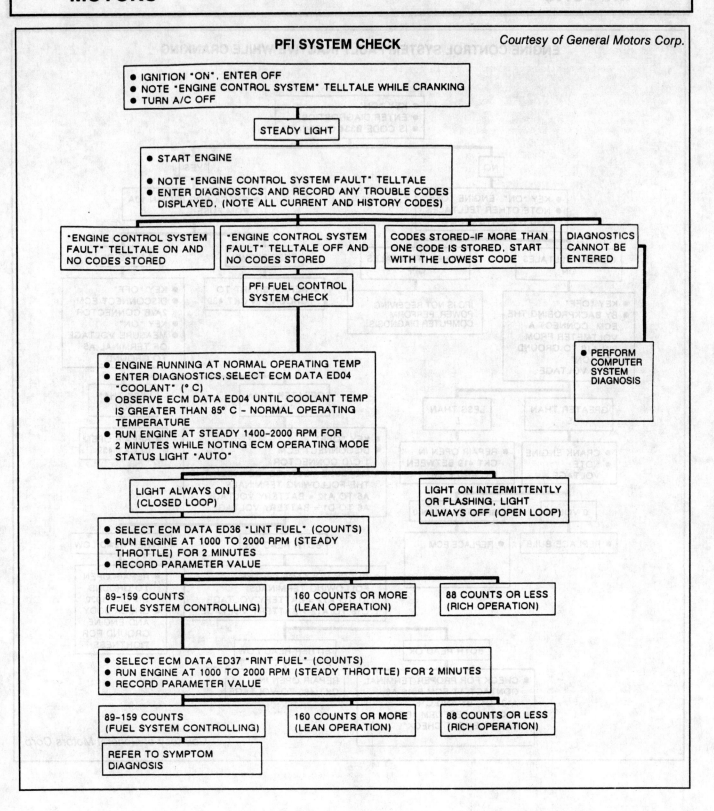

PFI SYSTEM CHECK

Courtesy of General Motors Corp.

- IGNITION "ON", ENTER OFF
- NOTE "ENGINE CONTROL SYSTEM" TELLTALE WHILE CRANKING
- TURN A/C OFF

STEADY LIGHT

- START ENGINE
- NOTE "ENGINE CONTROL SYSTEM FAULT" TELLTALE
- ENTER DIAGNOSTICS AND RECORD ANY TROUBLE CODES DISPLAYED. (NOTE ALL CURRENT AND HISTORY CODES)

| "ENGINE CONTROL SYSTEM FAULT" TELLTALE ON AND NO CODES STORED | "ENGINE CONTROL SYSTEM FAULT" TELLTALE OFF AND NO CODES STORED | CODES STORED-IF MORE THAN ONE CODE IS STORED, START WITH THE LOWEST CODE | DIAGNOSTICS CANNOT BE ENTERED |

PFI FUEL CONTROL SYSTEM CHECK

- PERFORM COMPUTER SYSTEM DIAGNOSIS

- ENGINE RUNNING AT NORMAL OPERATING TEMP
- ENTER DIAGNOSTICS, SELECT ECM DATA ED04 "COOLANT" (° C)
- OBSERVE ECM DATA ED04 UNTIL COOLANT TEMP IS GREATER THAN 85° C - NORMAL OPERATING TEMPERATURE
- RUN ENGINE AT STEADY 1400-2000 RPM FOR 2 MINUTES WHILE NOTING ECM OPERATING MODE STATUS LIGHT "AUTO"

LIGHT ALWAYS ON (CLOSED LOOP)

LIGHT ON INTERMITTENTLY OR FLASHING, LIGHT ALWAYS OFF (OPEN LOOP)

- SELECT ECM DATA ED36 "LINT FUEL" (COUNTS)
- RUN ENGINE AT 1000 TO 2000 RPM (STEADY THROTTLE) FOR 2 MINUTES
- RECORD PARAMETER VALUE

| 89-159 COUNTS (FUEL SYSTEM CONTROLLING) | 160 COUNTS OR MORE (LEAN OPERATION) | 88 COUNTS OR LESS (RICH OPERATION) |

- SELECT ECM DATA ED37 "RINT FUEL" (COUNTS)
- RUN ENGINE AT 1000 TO 2000 RPM (STEADY THROTTLE) FOR 2 MINUTES
- RECORD PARAMETER VALUE

| 89-159 COUNTS (FUEL SYSTEM CONTROLLING) | 160 COUNTS OR MORE (LEAN OPERATION) | 88 COUNTS OR LESS (RICH OPERATION) |

REFER TO SYMPTOM DIAGNOSIS

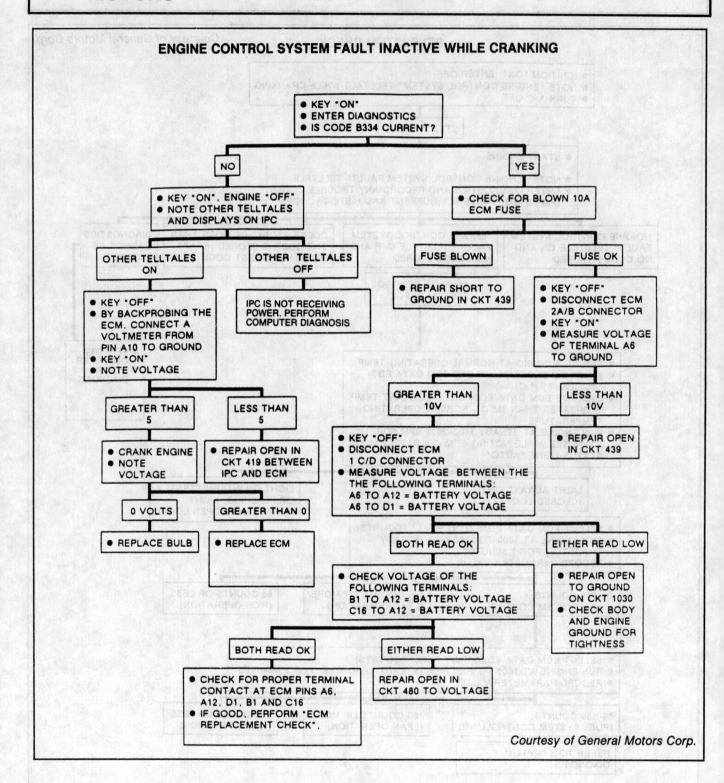

ENGINE CONTROL SYSTEM FAULT INACTIVE WHILE CRANKING

Courtesy of General Motors Corp.

NO "SERVICE ENGINE SOON" LIGHT (Reatta, Riviera, Toronado)

IF "SES" LIGHT IS FLASHING RAPIDLY WITH KEY "ON" AND ENGINE "OFF", CHECK THAT ALL PROM PINS ARE FULLY INSERTED IN THE SOCKET. IF OK, REPLACE PROM, CLEAR MEMORY AND CHECK. IF "SERVICE ENGINE SOON" LIGHT STILL FLASHES RAPIDLY, REPLACE ECM.

IF "SES" LIGHT IS "ON" STEADY WITH KEY "OFF" CKT 419 IS GROUNDED BETWEEN IPC AND ECM.

1
- IGNITION "ON", ENGINE STOPPED.
- JUMPER ALDL TERMINAL "D" TO GROUND.
- "SERVICE ENGINE SOON" LIGHT SHOULD BE "ON". IS IT?

NO

YES

2
- BACKPROBE ECM TERMINAL "2A6" (CKT 439) WITH A TEST LIGHT TO GROUND.
- "TEST LIGHT SHOULD BE "ON". IS IT?

4
- BACKPROBE ECM TERMINAL "3C11" WITH A JUMPER TO GROUND.
- "SERVICE ENGINE SOON" LIGHT SHOULD BE "ON". IS IT?

YES

NO

NO

YES

3
- IGNITION "OFF", REMOVE IPC.
- IGN. "ON", CONNECT A TEST LIGHT BETWEEN IPC HARNESS TERMINALS "D2" AND "C9".
- TEST LIGHT SHOULD BE "ON". IS IT?

CKT 439 OPEN OR SHORTED TO GROUND.

OPEN CKT 419 FROM ECM TO ALDL.

POOR CONTACT AT ECM TERMINAL "3C11" OR FAULTY ECM.

YES

NO

CHECK IPC BULB. IF BULB IS OK, IT'S POOR CONTACT AT IPC TERM. "C9" OR FAULTY IPC.

OPEN CKT 419 FROM IPC TERM. "C9" TO ALDL TERM. "D".

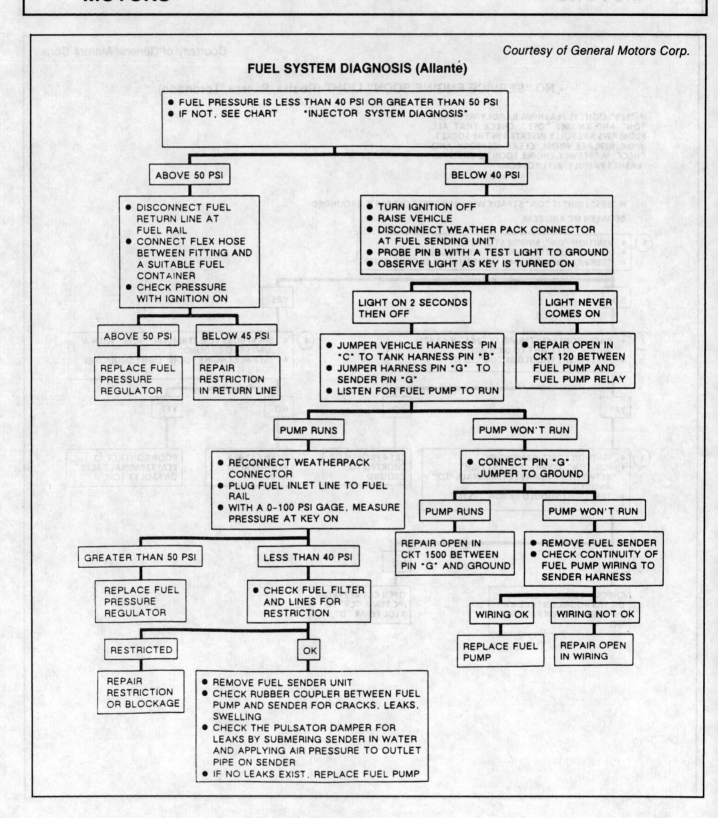

Courtesy of General Motors Corp.

FUEL SYSTEM DIAGNOSIS (Allanté)

- FUEL PRESSURE IS LESS THAN 40 PSI OR GREATER THAN 50 PSI
- IF NOT, SEE CHART   "INJECTOR SYSTEM DIAGNOSIS"

ABOVE 50 PSI

- DISCONNECT FUEL RETURN LINE AT FUEL RAIL
- CONNECT FLEX HOSE BETWEEN FITTING AND A SUITABLE FUEL CONTAINER
- CHECK PRESSURE WITH IGNITION ON

ABOVE 50 PSI

REPLACE FUEL PRESSURE REGULATOR

BELOW 45 PSI

REPAIR RESTRICTION IN RETURN LINE

BELOW 40 PSI

- TURN IGNITION OFF
- RAISE VEHICLE
- DISCONNECT WEATHER PACK CONNECTOR AT FUEL SENDING UNIT
- PROBE PIN B WITH A TEST LIGHT TO GROUND
- OBSERVE LIGHT AS KEY IS TURNED ON

LIGHT ON 2 SECONDS THEN OFF

- JUMPER VEHICLE HARNESS PIN "C" TO TANK HARNESS PIN "B"
- JUMPER HARNESS PIN "G" TO SENDER PIN "G"
- LISTEN FOR FUEL PUMP TO RUN

LIGHT NEVER COMES ON

- REPAIR OPEN IN CKT 120 BETWEEN FUEL PUMP AND FUEL PUMP RELAY

PUMP RUNS

- RECONNECT WEATHERPACK CONNECTOR
- PLUG FUEL INLET LINE TO FUEL RAIL
- WITH A 0-100 PSI GAGE, MEASURE PRESSURE AT KEY ON

GREATER THAN 50 PSI

REPLACE FUEL PRESSURE REGULATOR

LESS THAN 40 PSI

- CHECK FUEL FILTER AND LINES FOR RESTRICTION

RESTRICTED

REPAIR RESTRICTION OR BLOCKAGE

OK

- REMOVE FUEL SENDER UNIT
- CHECK RUBBER COUPLER BETWEEN FUEL PUMP AND SENDER FOR CRACKS, LEAKS, SWELLING
- CHECK THE PULSATOR DAMPER FOR LEAKS BY SUBMERING SENDER IN WATER AND APPLYING AIR PRESSURE TO OUTLET PIPE ON SENDER
- IF NO LEAKS EXIST, REPLACE FUEL PUMP

PUMP WON'T RUN

- CONNECT PIN "G" JUMPER TO GROUND

PUMP RUNS

REPAIR OPEN IN CKT 1500 BETWEEN PIN "G" AND GROUND

PUMP WON'T RUN

- REMOVE FUEL SENDER
- CHECK CONTINUITY OF FUEL PUMP WIRING TO SENDER HARNESS

WIRING OK

REPLACE FUEL PUMP

WIRING NOT OK

REPAIR OPEN IN WIRING

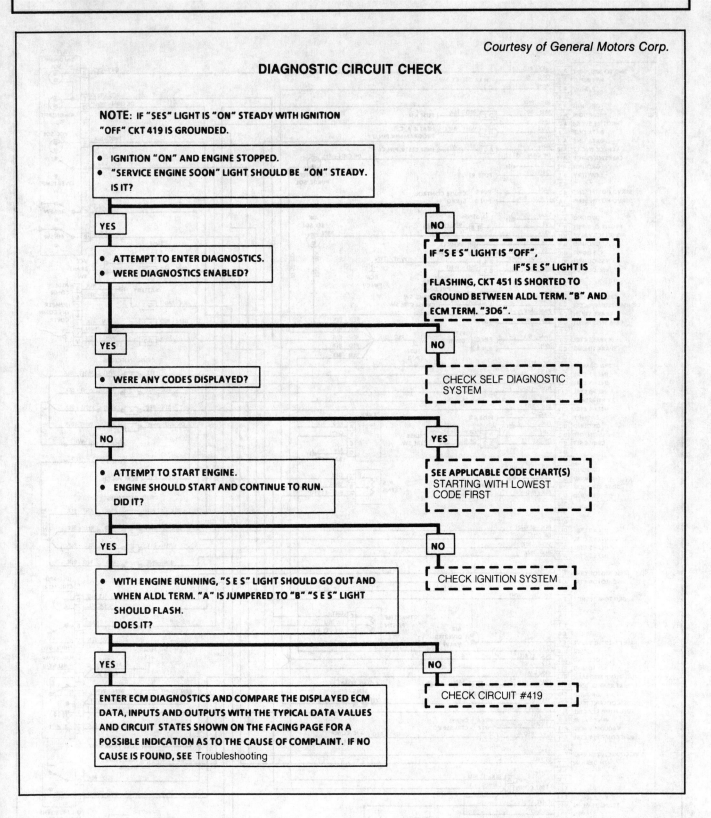

Courtesy of General Motors Corp.

DIAGNOSTIC CIRCUIT CHECK

NOTE: IF "SES" LIGHT IS "ON" STEADY WITH IGNITION "OFF" CKT 419 IS GROUNDED.

- IGNITION "ON" AND ENGINE STOPPED.
- "SERVICE ENGINE SOON" LIGHT SHOULD BE "ON" STEADY. IS IT?

YES

- ATTEMPT TO ENTER DIAGNOSTICS.
- WERE DIAGNOSTICS ENABLED?

YES

- WERE ANY CODES DISPLAYED?

NO

- ATTEMPT TO START ENGINE.
- ENGINE SHOULD START AND CONTINUE TO RUN. DID IT?

YES

- WITH ENGINE RUNNING, "S E S" LIGHT SHOULD GO OUT AND WHEN ALDL TERM. "A" IS JUMPERED TO "B" "S E S" LIGHT SHOULD FLASH. DOES IT?

YES

ENTER ECM DIAGNOSTICS AND COMPARE THE DISPLAYED ECM DATA, INPUTS AND OUTPUTS WITH THE TYPICAL DATA VALUES AND CIRCUIT STATES SHOWN ON THE FACING PAGE FOR A POSSIBLE INDICATION AS TO THE CAUSE OF COMPLAINT. IF NO CAUSE IS FOUND, SEE Troubleshooting

NO

IF "S E S" LIGHT IS "OFF", IF "S E S" LIGHT IS FLASHING, CKT 451 IS SHORTED TO GROUND BETWEEN ALDL TERM. "B" AND ECM TERM. "3D6".

NO

CHECK SELF DIAGNOSTIC SYSTEM

YES

SEE APPLICABLE CODE CHART(S) STARTING WITH LOWEST CODE FIRST

NO

CHECK IGNITION SYSTEM

NO

CHECK CIRCUIT #419

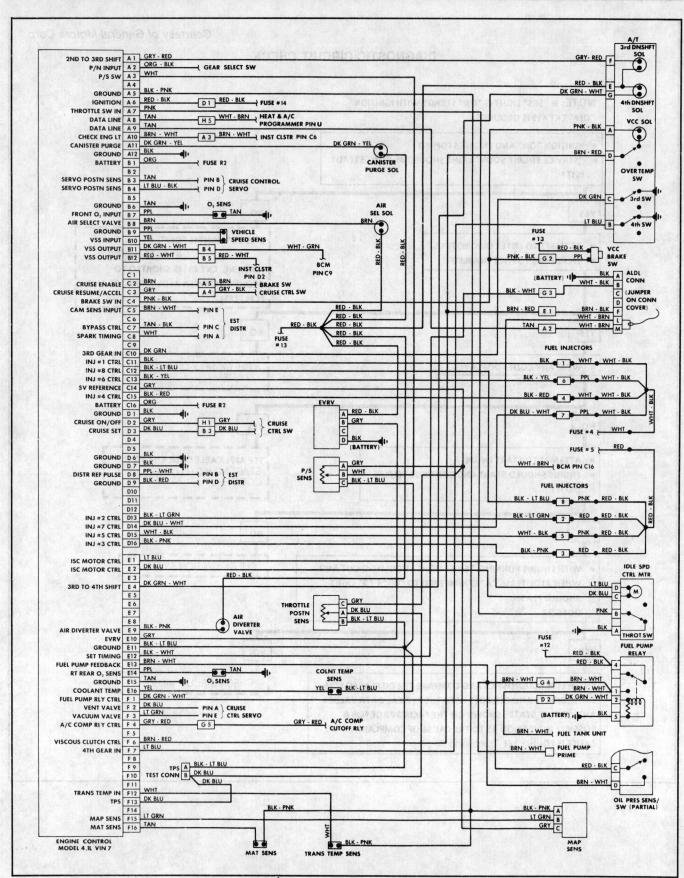

Fig. 22: Allanté Fuel Injection Wiring Diagram

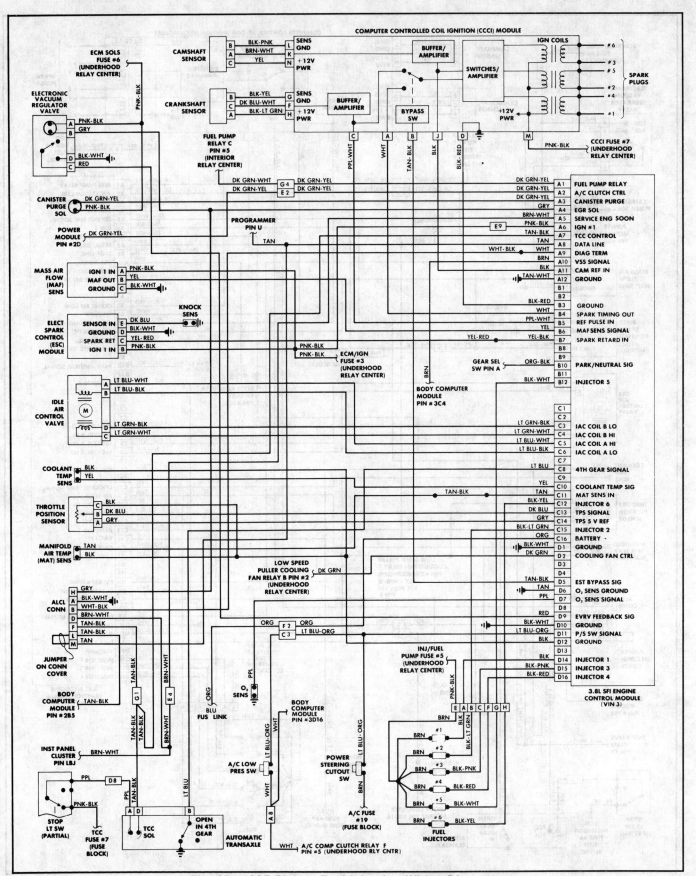

Fig. 23: 1987 Riviera Fuel Injection Wiring Diagram

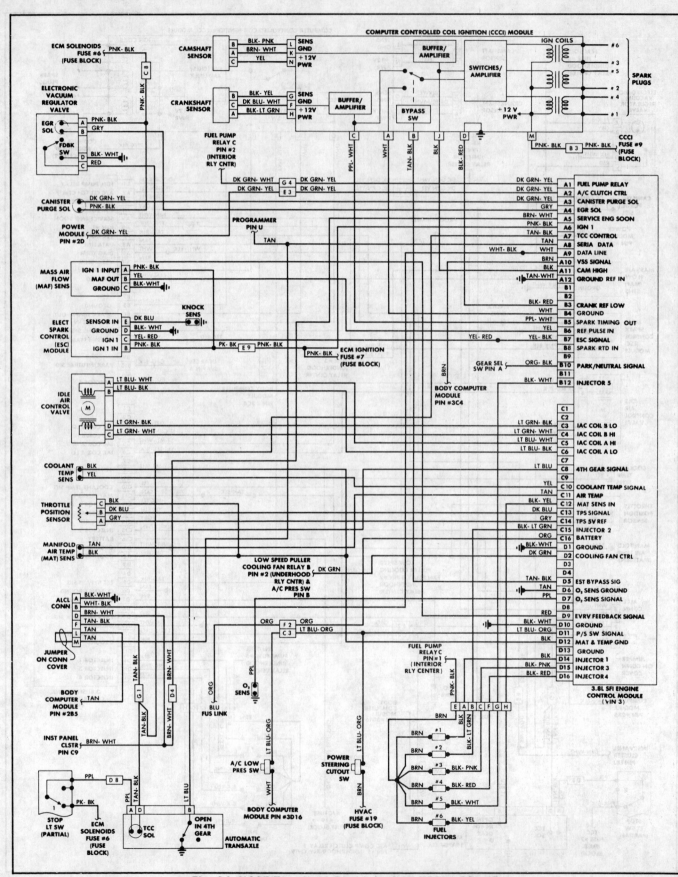

Fig. 24: 1987 Toronado Fuel Injection Wiring Diagram

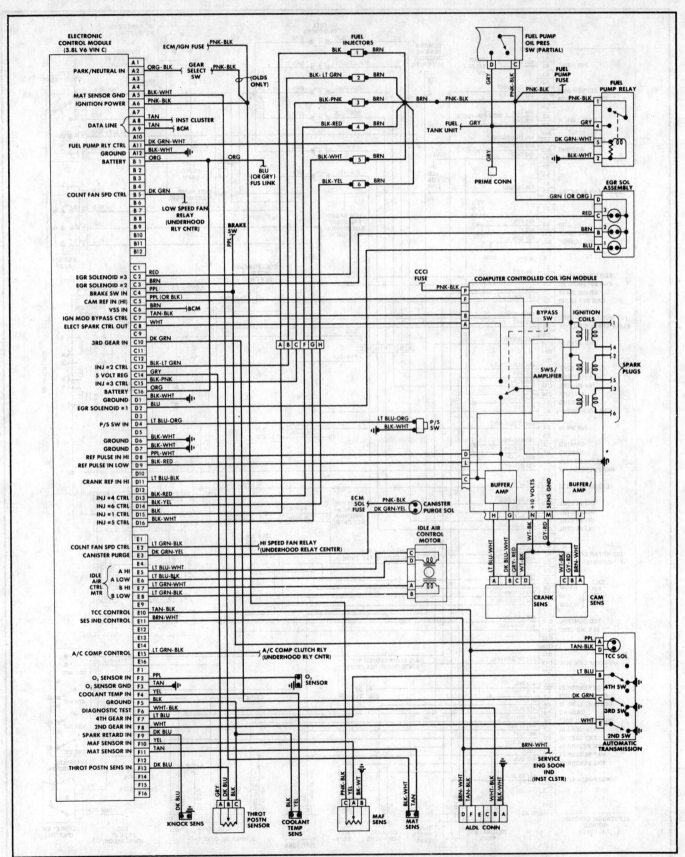

Fig. 25: *1988 Riviera & Toronado Fuel Injection Wiring Diagram*
(1988 Reatta Wiring Diagram Unavailable)

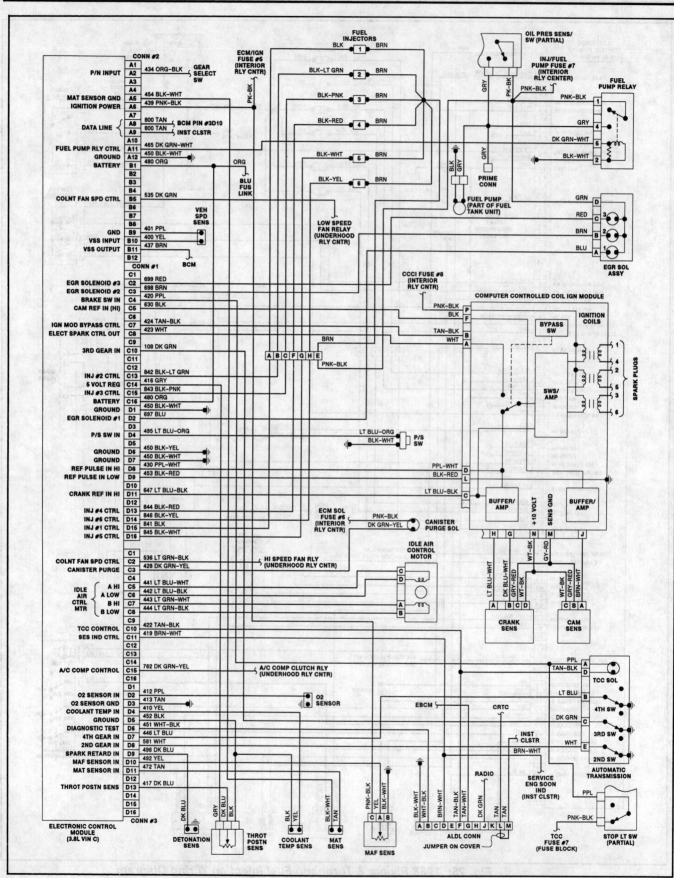

Fig. 26: 1989 Reatta & Riviera Fuel Injection Wiring Diagram

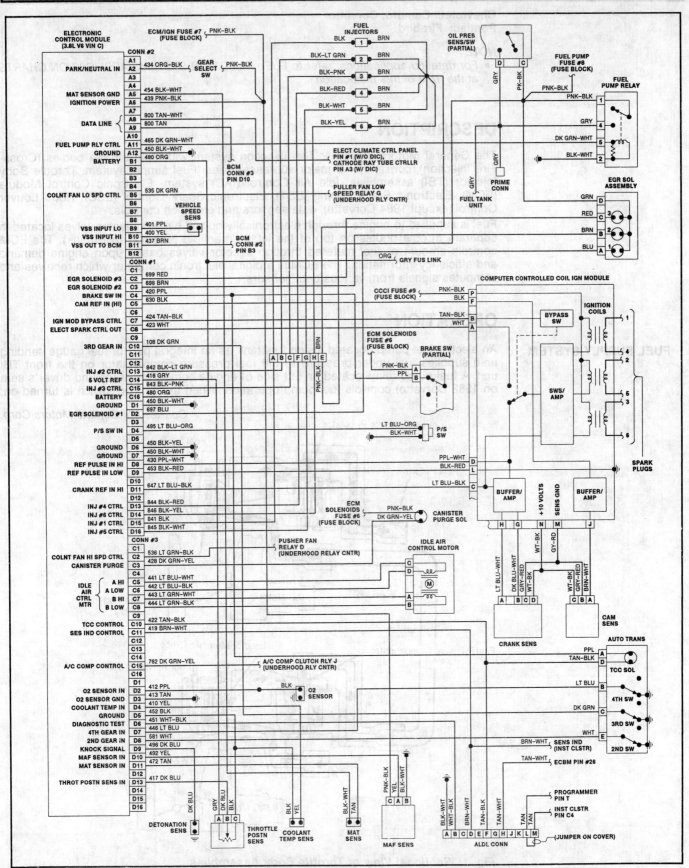

Fig. 27: 1989 Toronado Fuel Injection Wiring Diagram

Chevrolet: Camaro, Corvette
Pontiac: Firebird

NOTE
- *For detailed applications, refer to FUEL INJECTION SYSTEMS APPLICATION CHARTS at the front of this publication.*

DESCRIPTION

The General Motors electronic fuel injection system with dual throttle bodies (Cross-Fire Injection), consists of 9 major sub-assemblies: Fuel supply system, Throttle Body Injector (TBI) assemblies, Idle Air Control (IAC) system, Electronic Control Module (ECM), Electronic Spark Timing (EST), Electronic Spark Control (ESC), Hood Louver Control (except 1984 Corvette), data sensors and emission controls.

Fuel is supplied to engine through electronically pulsed (timed) injector valves located in separate throttle bodies on top of the intake manifold (one for each bank). The ECM controls amount of fuel metered through injector valves based upon engine demand and efficiency information. The ECM is a digital electronic computer which receives and computes signals from various data sensors.

OPERATION

FUEL SUPPLY SYSTEM

An electric fuel pump (located inside fuel tank as an integral part of fuel gauge sending unit) supplies fuel under pressure to the fuel pressure compensator on the front TBI unit. A fuel pump relay located on left side of engine compartment (behind driver's seat on 1982 Corvette) controls fuel pump operation. When the ignition switch is turned on,

Courtesy of General Motors Corp.

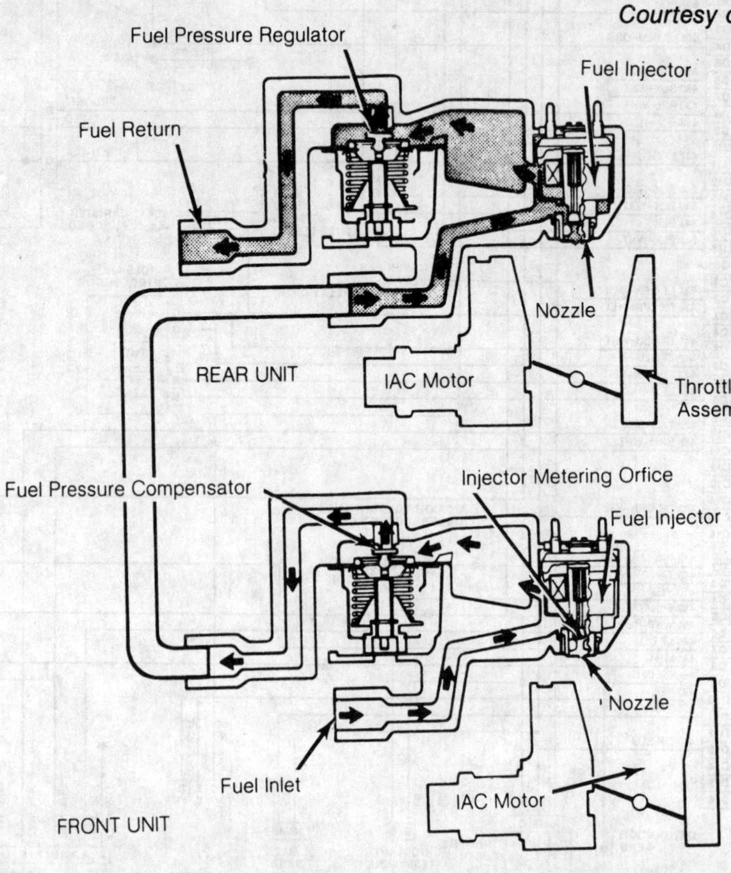

Fig. 1: Sectional View of Throttle Body Assemblies

the fuel pump relay activates the fuel pump for 1 1/2-2 seconds to prime the injectors. If the ECM does not receive reference pulses from the distributor after this period, the ECM deactivates the fuel pump circuit. The fuel pump circuit will be activated again through the relay when the ECM receives distributor reference pulses.

NOTE
- *Power to relay from battery is provided when oil pressure is above 4 psi (.3 kg/cm²). If pump relay malfunctions, engine can still be started after oil pressure has reached 4 psi (.3 kg/cm²).*

THROTTLE BODY

Each Throttle Body Injector (TBI) unit is composed of 2 casting assemblies: a throttle body with an Idle Air Control valve to control airflow and Throttle Position Sensor (TPS), and a fuel body meter cover with an built in pressure regulator (rear unit) or pressure compensator (front unit) and fuel injector. The throttle body casting may contain vacuum ports for EGR valve, MAP sensor and canister purge system.

The pressure regulator is a diaphragm-operated relief valve with injector pressure on one side and air cleaner pressure on the other side. The pressure regulator maintains a constant pressure drop of about 10 psi (.7 kg/cm²) across both injectors throughout all engine operating conditions by controlling the return of excess fuel to tank. *See Fig. 1.* The pressure compensator is of similar design to the pressure regulator. The compensator makes up for the momentary fuel pressure drop between the front and rear TBI units to maintain consistent operating pressures.

Each fuel injector is a solenoid-operated device controlled by the ECM. Fuel is supplied at the lower end of the injector by the fuel supply system. The ECM activates the solenoid which lifts a normally closed ball valve off its seat. Fuel under pressure is injected in a conical spray pattern at the walls of the throttle bore above the throttle valve. Excess fuel passes through the pressure compensator of the front unit, to the pressure regulator of the rear unit and is returned to tank.

IDLE AIR CONTROL (IAC) SYSTEM

The IAC system consists of an electrically controlled motor which positions the IAC valve in the air by-pass channel around the throttle plate of each TBI unit. The ECM calculates the desired position of each IAC valve based upon battery voltage, coolant temperature, engine load and engine speed. It controls idle speed while preventing stalls due to engine load changes.

If engine speed is lower than desired, the ECM activates the motor to retract the IAC valve. When the IAC valve is retracted, more air is diverted around the throttle plate, increasing engine speed. If engine speed is higher than desired, the ECM activates the motor to extend the IAC valve. When the IAC valve is extended, less air is diverted around the throttle plate, decreasing engine speed. If engine speed falls below a preset value and the throttle plate is closed, the ECM senses a near-stall condition. To prevent stalling, the ECM calculates an IAC valve position based upon barometric pressure.

Three different valve designs are used: single taper, dual taper and blunt. Whenever a valve is replaced, a new valve of the same design must be used. *See Fig. 2.*

Courtesy of General Motors Corp.

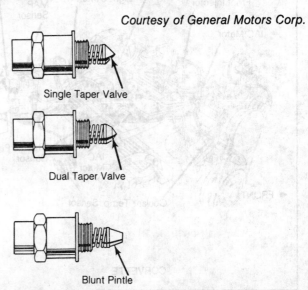

Single Taper Valve

Dual Taper Valve

Blunt Pintle

Fig. 2: IAC Valve Designs

ELECTRONIC CONTROL MODULE (ECM)

The electronic control module (ECM) is the "brain" of the EFI and Computer Command Control systems. It is located behind the instrument panel in Camaro and Firebird models; behind the driver's seat on Corvette models. Information from all data sensors is received and processed by the ECM to produce proper pulse duration for each injector, correct idle speed and proper ignition timing. The ECM performs calculations to control the following EFI operating conditions: Engine start, engine running, fuel enrichment during acceleration, lean fuel mixture during deceleration, fuel cut-off and battery voltage correction.

During engine starting, the first signal sent to the injectors is a "prime" pulse. This pulse charges the intake manifold with fuel during or just prior to engine starting. The pulse width (injector "on" time) is not synchronized with HEI distributor reference pulses. Prime pulses are delivered for a length of time determined by coolant temperature. After delivering prime pulses, the ECM will deliver one pulse for every distributor reference pulse received. The air/fuel ratio is determined by the ECM based upon throttle position and coolant temperature. At low coolant temperatures, injector pulse width is relatively long (richer air/fuel ratio). As coolant temperature increases, injector pulse width becomes gradually shorter (leaner air/fuel ratio).

If engine is flooded, the accelerator pedal must be depressed enough to set wide open throttle position. At wide open throttle, the ECM calculates injector pulse width equal to an air/fuel ratio of 20:1. This air/fuel ratio is maintained as long as the throttle remains wide open and engine speed is below 600 RPM. If throttle position is reduced to less than 80 percent, injector pulse width is increased to that used during engine starting.

When the engine is running above 600 RPM, the ECM operates in the open loop mode. In open loop, the ECM calculates injector pulse width based upon coolant temperature and manifold absolute pressure. The engine will remain in open loop operation until the oxygen sensor reaches operating temperature, coolant temperature reaches a preset level and a specific period of time elapses after engine start. When all conditions have been met, the ECM operates in the closed loop mode. In closed loop, the ECM controls the injector pulse width according to oxygen sensor signals, maintaining the air/fuel ratio at 14.7:1. In either mode, the injectors are pulsed alternately for each distributor reference pulse.

The ECM also provides for fuel enrichment during acceleration. Sudden opening of the throttle plates causes a rapid increase in manifold pressure. Pulse width is directly related to manifold pressure, throttle position and coolant temperature. As throttle angle is increased, manifold pressure climbs and pulse width is increased (richer mixture). During enrichment, the injector pulses are not in proportion with distributor reference signals. Any reduction in throttle angle will cancel fuel enrichment.

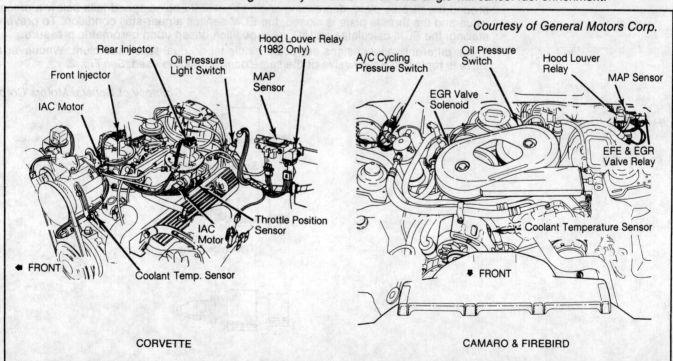

Courtesy of General Motors Corp.

CORVETTE

CAMARO & FIREBIRD

Fig. 3: EFI System Component Locations

During deceleration, the air/fuel mixture must be leaner. The ECM calculates injector pulse width similar to that during fuel enrichment. Fuel output is reduced due to fuel remaining in the intake manifold. During sudden deceleration, when manifold pressure, throttle position and engine speed are at preset specifications, injection stops, to remove fuel from the engine. This deceleration fuel cut-off overrides the normal deceleration mode. During either deceleration mode, the injector pulses are not in proportion to distributor reference signals.

Battery voltage corrections by the ECM are performed during all operating modes of the EFI system. As battery voltage decreases, the ECM increases the injector pulse width with a correction factor stored in the ECM's memory.

The Hood Louver Control (HLC) system is controlled by the ECM. When engine coolant temperature and throttle position meet preset specifications, the ECM activates the HLC relay which opens the hood louver to allow additional air into the engine.

DATA SENSORS

Each sensor furnishes an electrical signal to the ECM. The ECM modifies injector pulse width in response to sensor signals. Engine sensor functions are as follows:

Coolant Temperature Sensor (CTS)

The CTS is located in the thermostat housing. This sensor is a variable resistance type which transmits an electrical signal (proportionate to engine temperature) to the ECM. Sensor resistance decreases as coolant temperature increases. Voltage signal to ECM is interpreted as coolant temperature. This information is used to control fuel management, idle air control, spark timing, EGR, canister purge and other engine operating conditions.

Oxygen Sensor

The oxygen sensor is mounted in the exhaust manifold, directly behind the cross-over pipe. This sensor is similar to a small battery in that it supplies the ECM with a small voltage signal. The strength of the signal varies with the oxygen content of the exhaust gases. The ECM corrects the air/fuel ratio (according to signals received from the oxygen sensor) only when the system is operating in closed loop.

CAUTION

- *No attempt should be made to measure oxygen sensor voltage output. Current drain of conventional voltmeter could permanently damage sensor. Do not connect jumper wire, test leads or other electrical connectors to sensor.*

Manifold Absolute Pressure (MAP) Sensor

The MAP sensor is mounted on the left side of the engine compartment. This sensor is a variable resistance type which has a vacuum hose connected to the throttle body. The sensor monitors changes in intake manifold pressure which result from engine load and speed changes. As manifold pressure changes, the electrical resistance of the sensor changes. The ECM uses the resistance value of the sensor to control injector pulse width and to automatically adjust for changes in altitude.

Vehicle Speed Sensor (VSS)

The VSS is mounted behind the speedometer in the instrument cluster. This sensor provides the ECM with pulses to determine vehicle speed. This information is used by the ECM to control the idle air control motor, canister purge, and Torque Converter Clutch (TCC).

NOTE

- *Vehicle should not be driven without the VSS installed or speedometer cable disconnected as idle quality may be affected.*

Throttle Position Sensor (TPS)

The TPS is mounted on side of rear throttle body and is connected to throttle shaft. A throttle rod connects both the front and rear throttle body units to ensure that throttle valve of each unit is positioned the same. This sensor converts throttle angle to an electrical signal for use by the ECM to determine engine fuel requirements.

Engine RPM Reference

The "R" terminal of the HEI module is used to send engine RPM signals to ECM.

TROUBLE SHOOTING

PRELIMINARY CHECKS

The following systems and components must be in good condition and operating properly before beginning diagnosis of the fuel injection system.
- All support systems and wiring.
- Battery connections and specific gravity.
- Ignition system.
- Compression pressure.
- Fuel supply system pressure and flow.
- All electrical connections and terminals.
- Vacuum line, fuel hose and pipe connections.

TROUBLE SHOOTING

Runs Rough, Hesitates or Poor Mileage

1) Visually check MAP hose for leaks or restriction and TPS for sticking or binding. Ensure fuel pressure is steady at all operating ranges. Ensure base timing is correct.

2) With injector connectors disconnected, check for fuel leakage from injectors while cranking. Check for HEI ground circuit for ground. Check fuel injector fuel filters for blockage. Check TBI balance adjustment.

Cuts Out or Stalls

1) Check for intermittent open or short to ground in the following circuits: 5 volt reference (416), HEI reference, fuel pump signal, injector drive circuits, IAC drive circuits (4).

2) Check for clogged fuel filter. Ensure fuel pressure is steady at all operating ranges. Inspect fuel injector "O" rings for damage. Ensure steel back-up washer is located beneath large "O" ring of each injector.

Engine Surges

Check for intermittent open or short to ground in the following circuits: transmission converter clutch, and HEI by-pass and EST.

Hard Starting (Hot or Cold)

1) Check for high resistance in coolant sensor circuit. Visually check TPS for sticking or binding. Ensure fuel pressure is 9-13 psi (.6-.9 kg/cm²) in all operating ranges.

2) Check fuel pump relay. Disconnect oil pressure switch. If engine cranks but will not start, perform fuel system diagnosis (at point where fuel pump fuse proves okay).

3) Check injectors. With injector harness connectors disconnected, check for fuel leakage while cranking.

4) Check cranking circuit.

TESTING & DIAGNOSIS

FUEL SYSTEM PRESSURE

1) Before beginning any testing of the fuel system, pressure in the fuel lines must be released. Remove fuel pump fuse from fuse block in passenger compartment. Crank engine. Engine will start and run until fuel supply remaining in fuel lines is used. Engage starter again for about 3 seconds to ensure all fuel is out of lines. Turn ignition off and replace fuse.

2) Remove air cleaner and plug thermal vacuum port on throttle body. Remove steel fuel line between front and rear throttle body units. When removing fuel line, always use 2 wrenches. Install Fuel Pressure Gauge (J-29658) between throttle body units.

3) Start car and observe fuel pressure reading. Use FUEL SYSTEM DIAGNOSIS chart if pressure is not correct. Use INJECTOR SYSTEM DIAGNOSIS chart if pressure is within indicated range.

4) Depressurize fuel system as described in step **1)**. Remove fuel pressure gauge and reinstall steel line between throttle bodies. Start car and watch for leaks. Remove plug from throttle body thermal vacuum port and reinstall air cleaner.

ADJUSTMENTS

TBI BALANCE

The throttle position of each throttle body must be balanced so that the throttle plates are synchronized and open simultaneously. Adjustment should be performed only when a throttle body has been replaced or if there is some indication of tampering with the minimum air adjustment screw on TBI unit or idle balance screw on throttle linkage.

Perform adjustment procedures as follows:

1) Remove air cleaner and air cleaner-to-TBI gaskets. Plug vacuum port on rear TBI unit to air cleaner. *See Fig. 4.* Remove air adjustment screw plug. Block drive wheels and apply parking brake.

2) Connect tachometer. Disconnect IAC electrical connector. Plug idle air passages of each throttle body with Plugs (J-33047). Be certain that plugs are seated fully in passage and no air leaks exist.

3) Start engine and allow engine RPM to stabilize at normal operating temperature. Place transmission in Drive. Engine RPM should decrease below curb idle speed. If engine RPM does not decrease, check for vacuum leak.

4) Remove cap from ported tube on rear TBI unit and connect Water Manometer (J-23951). Adjust throttle stop screw to obtain approximately 6" water on manometer. Remove manometer from rear TBI unit and install cap on ported tube. Remove cap from ported tube on front TBI unit and connect manometer. Reading should also be approximately 6" water.

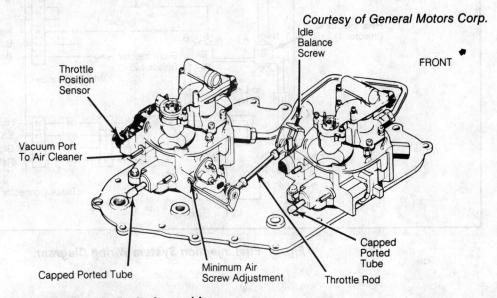

Courtesy of General Motors Corp.

Fig. 4: EFI Throttle Body Assembly

MINIMUM IDLE SPEED

1) If adjustment is required, set-up procedure used for TBI balance test must be used before adjustment can be made. Locate idle balance screw on throttle linkage. If screw is welded for tamper resistance, break weld and install new screw with thread sealing compound applied.

2) Adjust screw to obtain approximately 6" water on manometer. Remove manometer from front TBI unit and install cap on ported tube.

3) Adjust minimum air adjustment screw on rear unit to obtain 475 RPM. Turn ignition switch off and place transmission in Neutral. Remove idle air passage plugs and reconnect IAC assemblies.

4) Start engine. Engine may be running at high RPM but RPM will decrease when IAC assemblies close air passages. Stop engine when idle RPM has decreased. Check TPS voltage and adjust as needed.

5) Install air cleaner gasket, connect vacuum line to TBI unit and install air cleaner. Reset IAC Motors.

THROTTLE POSITION SENSOR (TPS)

1) Remove air cleaner, EGR valve and heat shield from engine. Using three 6" jumper wires, connect TPS harness to TPS. With ignition on and engine stopped, use a voltmeter to measure voltage between Dark Blue and Black wires.

2) Loosen 2 TPS attaching screws and rotate throttle position sensor to obtain a voltage reading of .450-.600 volts. Tighten screws. Turn ignition off, remove jumpers and reconnect TPS harness to TPS. Install EGR valve and heat shield, using new gasket if necessary.

3) Install air cleaner gasket, connect vacuum line to throttle body and install air cleaner. Reset IAC motors by driving vehicle to 30 MPH or, if equipped with cruise control, disconnect speedometer cable at transducer, turn ignition switch to "ON" position and rotate cable to obtain 30 MPH.

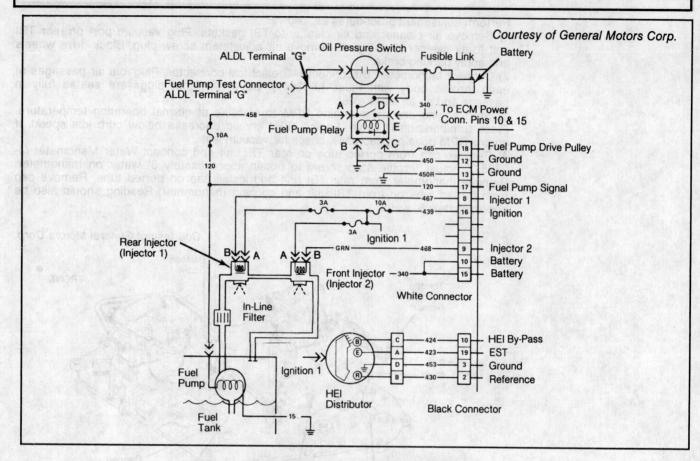

Courtesy of General Motors Corp.

Fig. 5: Fuel Injection System Wiring Diagram

1982-83 FUEL SYSTEM DIAGNOSIS
Courtesy of General Motors Corp.

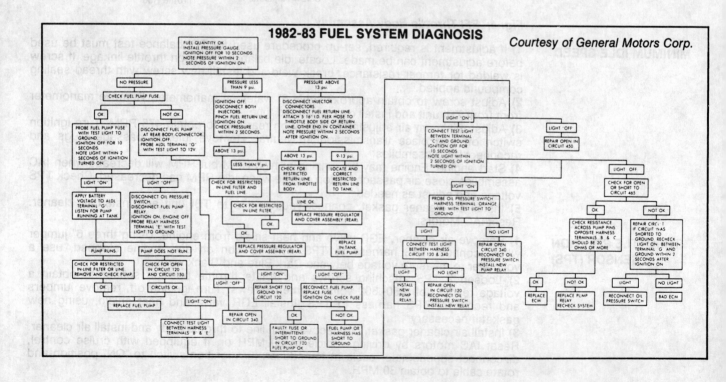

Courtesy of General Motors Corp.

1982-83 INJECTOR SYSTEM DIAGNOSIS

NOTE: Diagnostic chart assumes proper operation of "Check Engine" light.

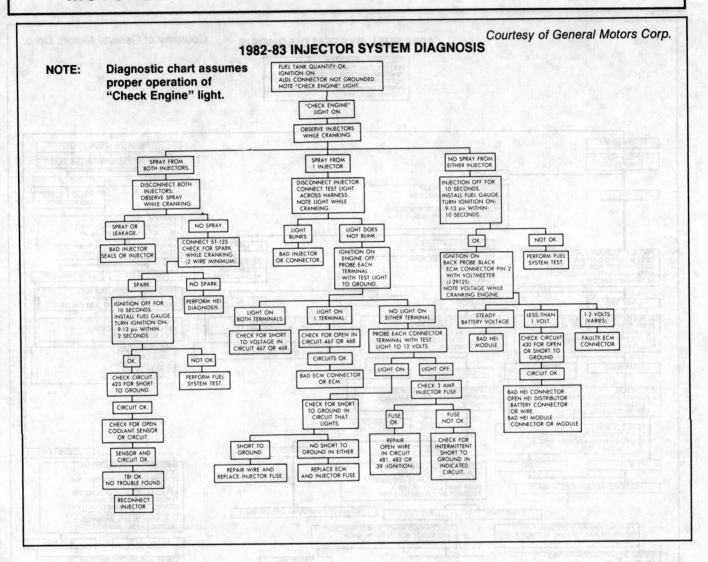

1984 FUEL SYSTEM DIAGNOSIS

Courtesy of General Motors Corp.

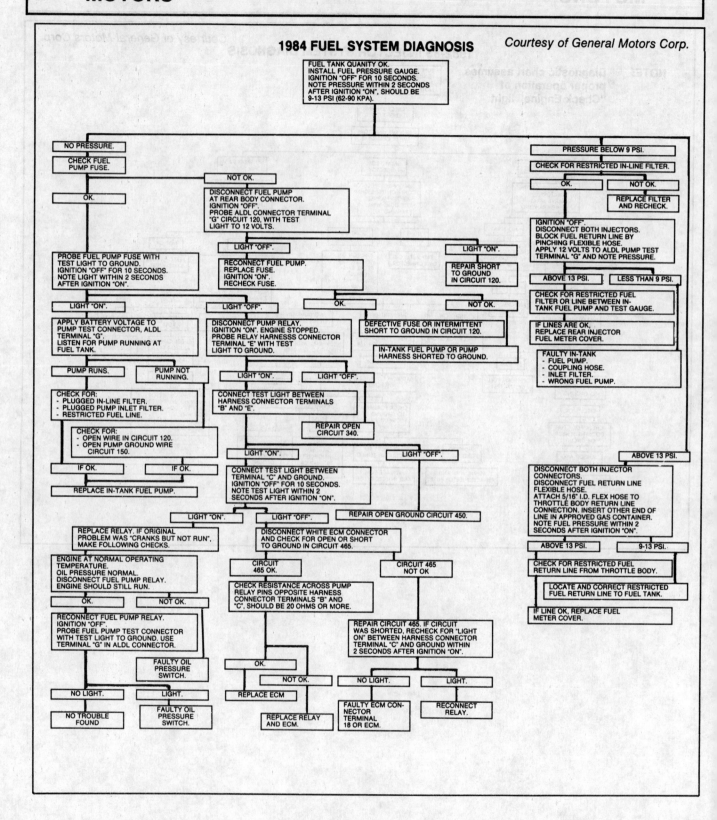

FUEL TANK QUANITY OK.
INSTALL FUEL PRESSURE GAUGE.
IGNITION "OFF" FOR 10 SECONDS.
NOTE PRESSURE WITHIN 2 SECONDS
AFTER IGNITION "ON", SHOULD BE
9-13 PSI (62-90 KPA).

NO PRESSURE.

CHECK FUEL PUMP FUSE.

OK.

NOT OK.

DISCONNECT FUEL PUMP
AT REAR BODY CONNECTOR.
IGNITION "OFF".
PROBE ALDL CONNECTOR TERMINAL
"G" CIRCUIT 120, WITH TEST
LIGHT TO 12 VOLTS.

PROBE FUEL PUMP FUSE WITH
TEST LIGHT TO GROUND.
IGNITION "OFF" FOR 10 SECONDS.
NOTE LIGHT WITHIN 2 SECONDS
AFTER IGNITION "ON".

LIGHT "OFF".

LIGHT "ON".

LIGHT "ON".

RECONNECT FUEL PUMP.
REPLACE FUSE.
IGNITION "ON".
RECHECK FUSE.

REPAIR SHORT
TO GROUND
IN CIRCUIT 120.

APPLY BATTERY VOLTAGE TO
PUMP TEST CONNECTOR, ALDL
TERMINAL "G".
LISTEN FOR PUMP RUNNING AT
FUEL TANK.

LIGHT "OFF".

OK.

NOT OK.

DISCONNECT PUMP RELAY.
IGNITION "ON". ENGINE STOPPED.
PROBE RELAY HARNESSS CONNECTOR
TERMINAL "E" WITH TEST
LIGHT TO GROUND.

DEFECTIVE FUSE OR INTERMITTENT
SHORT TO GROUND IN CIRCUIT 120.

IN-TANK FUEL PUMP OR PUMP
HARNESS SHORTED TO GROUND.

PUMP RUNS.

PUMP NOT
RUNNING.

LIGHT "ON".

LIGHT "OFF".

CHECK FOR:
- PLUGGED IN-LINE FILTER.
- PLUGGED PUMP INLET FILTER.
- RESTRICTED FUEL LINE.

CONNECT TEST LIGHT BETWEEN
HARNESS CONNECTOR TERMINALS
"B" AND "E".

REPAIR OPEN
CIRCUIT 340.

CHECK FOR:
- OPEN WIRE IN CIRCUIT 120.
- OPEN PUMP GROUND WIRE
  CIRCUIT 150.

IF OK.

IF OK.

LIGHT "ON".

LIGHT "OFF".

REPLACE IN-TANK FUEL PUMP.

CONNECT TEST LIGHT BETWEEN
TERMINAL "C" AND GROUND.
IGNITION "OFF" FOR 10 SECONDS.
NOTE TEST LIGHT WITHIN 2
SECONDS AFTER IGNITION "ON".

LIGHT "ON".

LIGHT "OFF".

REPAIR OPEN GROUND CIRCUIT 450.

REPLACE RELAY. IF ORIGINAL
PROBLEM WAS "CRANKS BUT NOT RUN",
MAKE FOLLOWING CHECKS.

DISCONNECT WHITE ECM CONNECTOR
AND CHECK FOR OPEN OR SHORT
TO GROUND IN CIRCUIT 465.

ENGINE AT NORMAL OPERATING
TEMPERATURE.
OIL PRESSURE NORMAL.
DISCONNECT FUEL PUMP RELAY.
ENGINE SHOULD STILL RUN.

CIRCUIT
465 OK.

CIRCUIT 465
NOT OK

OK.

NOT OK.

CHECK RESISTANCE ACROSS PUMP
RELAY PINS OPPOSITE HARNESS
CONNECTOR TERMINALS "B" AND
"C", SHOULD BE 20 OHMS OR MORE.

RECONNECT FUEL PUMP RELAY.
IGNITION "OFF".
PROBE FUEL PUMP TEST CONNECTOR
WITH TEST LIGHT TO GROUND. USE
TERMINAL "G" IN ALDL CONNECTOR.

REPAIR CIRCUIT 465. IF CIRCUIT
WAS SHORTED, RECHECK FOR "LIGHT
ON" BETWEEN HARNESS CONNECTOR
TERMINAL "C" AND GROUND WITHIN
2 SECONDS AFTER IGNITION "ON".

FAULTY OIL
PRESSURE
SWITCH.

OK.

NOT OK.

NO LIGHT.

LIGHT.

NO LIGHT.

LIGHT.

NO TROUBLE
FOUND

FAULTY OIL
PRESSURE
SWITCH.

REPLACE ECM

REPLACE RELAY
AND ECM.

FAULTY ECM CON-
NECTOR
TERMINAL
18 OR ECM.

RECONNECT
RELAY.

PRESSURE BELOW 9 PSI.

CHECK FOR RESTRICTED IN-LINE FILTER.

OK.

NOT OK.

REPLACE FILTER
AND RECHECK.

IGNITION "OFF".
DISCONNECT BOTH INJECTORS.
BLOCK FUEL RETURN LINE BY
PINCHING FLEXIBLE HOSE.
APPLY 12 VOLTS TO ALDL PUMP TEST
TERMINAL "G" AND NOTE PRESSURE.

ABOVE 13 PSI.

LESS THAN 9 PSI.

CHECK FOR RESTRICTED FUEL
FILTER OR LINE BETWEEN IN-
TANK FUEL PUMP AND TEST GAUGE.

IF LINES ARE OK,
REPLACE REAR INJECTOR
FUEL METER COVER.

FAULTY IN-TANK
- FUEL PUMP.
- COUPLING HOSE.
- INLET FILTER.
- WRONG FUEL PUMP.

ABOVE 13 PSI.

DISCONNECT BOTH INJECTOR
CONNECTORS.
DISCONNECT FUEL RETURN LINE
FLEXIBLE HOSE.
ATTACH 5/16" I.D. FLEX HOSE TO
THROTTLE BODY RETURN LINE
CONNECTION. INSERT OTHER END OF
LINE IN APPROVED GAS CONTAINER.
NOTE FUEL PRESSURE WITHIN 2
SECONDS AFTER IGNITION "ON".

ABOVE 13 PSI.

9-13 PSI.

CHECK FOR RESTRICTED FUEL
RETURN LINE FROM THROTTLE BODY.

LOCATE AND CORRECT RESTRICTED
FUEL RETURN LINE TO FUEL TANK.

IF LINE OK, REPLACE FUEL
METER COVER.

1984 INJECTOR SYSTEM DIAGNOSIS

Courtesy of General Motors Corp.

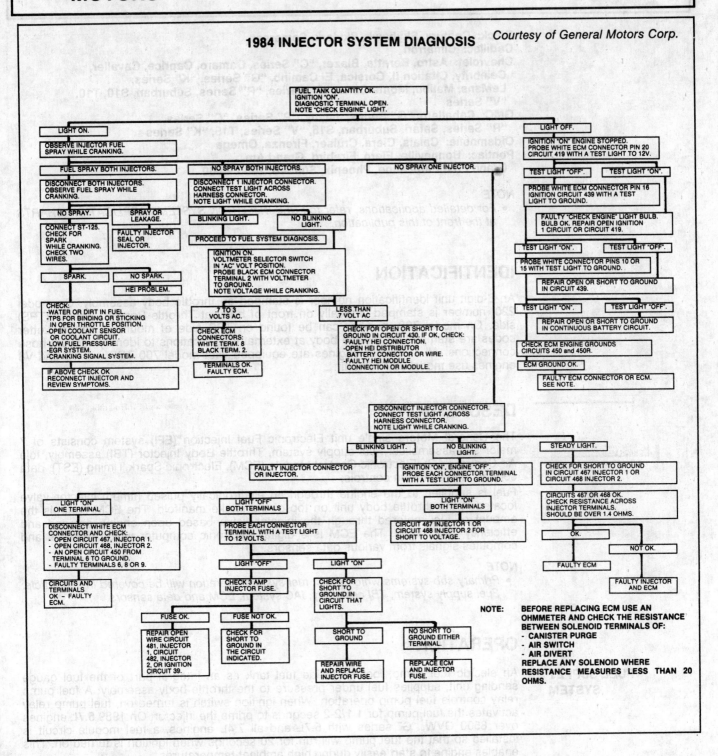

FUEL TANK QUANTITY OK.
IGNITION "ON".
DIAGNOSTIC TERMINAL OPEN.
NOTE "CHECK ENGINE" LIGHT.

LIGHT ON.

OBSERVE INJECTOR FUEL SPRAY WHILE CRANKING.

FUEL SPRAY BOTH INJECTORS.

NO SPRAY BOTH INJECTORS.

NO SPRAY ONE INJECTOR.

LIGHT OFF.

IGNITION "ON" ENGINE STOPPED. PROBE WHITE ECM CONNECTOR PIN 20 CIRCUIT 419 WITH A TEST LIGHT TO 12V.

DISCONNECT BOTH INJECTORS. OBSERVE FUEL SPRAY WHILE CRANKING.

DISCONNECT 1 INJECTOR CONNECTOR. CONNECT TEST LIGHT ACROSS HARNESS CONNECTOR. NOTE LIGHT WHILE CRANKING.

TEST LIGHT "OFF".

TEST LIGHT "ON".

NO SPRAY.

SPRAY OR LEAKAGE.

BLINKING LIGHT.

NO BLINKING LIGHT.

PROBE WHITE ECM CONNECTOR PIN 16 IGNITION CIRCUIT 439 WITH A TEST LIGHT TO GROUND.

CONNECT ST-125. CHECK FOR SPARK WHILE CRANKING. CHECK TWO WIRES.

FAULTY INJECTOR SEAL OR INJECTOR.

PROCEED TO FUEL SYSTEM DIAGNOSIS.

FAULTY "CHECK ENGINE" LIGHT BULB. BULB OK. REPAIR OPEN IGNITION 1 CIRCUIT OR CIRCUIT 419.

SPARK.

NO SPARK.

HEI PROBLEM.

IGNITION ON. VOLTMETER SELECTOR SWITCH IN "AC VOLT POSITION. PROBE BLACK ECM CONNECTOR TERMINAL 2 WITH VOLTMETER TO GROUND. NOTE VOLTAGE WHILE CRANKING.

TEST LIGHT "ON".

TEST LIGHT "OFF".

PROBE WHITE CONNECTOR PINS 10 OR 15 WITH TEST LIGHT TO GROUND.

REPAIR OPEN OR SHORT TO GROUND IN CIRCUIT 439.

CHECK:
-WATER OR DIRT IN FUEL.
-TPS FOR BINDING OR STICKING IN OPEN THROTTLE POSITION.
-OPEN COOLANT SENSOR OR COOLANT CIRCUIT.
-LOW FUEL PRESSURE.
-EGR SYSTEM.
-CRANKING SIGNAL SYSTEM.

.7 TO 3 VOLTS AC.

LESS THAN .7 VOLT AC

TEST LIGHT "ON".

TEST LIGHT "OFF".

CHECK ECM CONNECTORS: WHITE TERM. 8 BLACK TERM. 2.

CHECK FOR OPEN OR SHORT TO GROUND IN CIRCUIT 430. IF OK, CHECK:
-FAULTY HEI CONNECTION.
-OPEN HEI DISTRIBUTOR BATTERY CONECTOR OR WIRE.
-FAULTY HEI MODULE CONNECTION OR MODULE.

REPAIR OPEN OR SHORT TO GROUND IN CONTINUOUS BATTERY CIRCUIT.

IF ABOVE CHECK OK RECONNECT INJECTOR AND REVIEW SYMPTOMS.

TERMINALS OK. FAULTY ECM.

CHECK ECM ENGINE GROUNDS CIRCUITS 450 AND 450R.

ECM GROUND OK.

FAULTY ECM CONNECTOR OR ECM. SEE NOTE.

DISCONNECT INJECTOR CONNECTOR. CONNECT TEST LIGHT ACROSS HARNESS CONNECTOR. NOTE LIGHT WHILE CRANKING.

BLINKING LIGHT.

NO BLINKING LIGHT.

STEADY LIGHT.

FAULTY INJECTOR CONNECTOR OR INJECTOR.

IGNITION "ON", ENGINE "OFF". PROBE EACH CONNECTOR TERMINAL WITH A TEST LIGHT TO GROUND.

CHECK FOR SHORT TO GROUND IN CIRCUIT 467 INJECTOR 1 OR CIRCUIT 468 INJECTOR 2.

LIGHT "ON" ONE TERMINAL

LIGHT "OFF" BOTH TERMINALS

LIGHT "ON" BOTH TERMINALS

CIRCUITS 467 OR 468 OK. CHECK RESISTANCE ACROSS INJECTOR TERMINALS. SHOULD BE OVER 1.4 OHMS.

DISCONNECT WHITE ECM CONNECTOR AND CHECK:
- OPEN CIRCUIT 467, INJECTOR 1.
- OPEN CIRCUIT 468, INJECTOR 2.
- AN OPEN CIRCUIT 450 FROM TERMINAL 6 TO GROUND.
- FAULTY TERMINALS 6, 8 OR 9.

PROBE EACH CONNECTOR TERMINAL WITH A TEST LIGHT TO 12 VOLTS.

CIRCUIT 467 INJECTOR 1 OR CIRCUIT 468 INJECTOR 2 FOR SHORT TO VOLTAGE.

OK.

NOT OK.

CIRCUITS AND TERMINALS OK – FAULTY ECM.

LIGHT "OFF"

LIGHT "ON"

FAULTY ECM

FAULTY INJECTOR AND ECM

CHECK 3 AMP INJECTOR FUSE.

CHECK FOR SHORT TO GROUND IN CIRCUIT THAT LIGHTS.

FUSE OK.

FUSE NOT OK.

SHORT TO GROUND

NO SHORT TO GROUND EITHER TERMINAL.

REPAIR OPEN WIRE CIRCUIT 481, INJECTOR 1, CIRCUIT 482, INJECTOR 2, OR IGNITION CIRCUIT 39.

CHECK FOR SHORT TO GROUND IN THE CIRCUIT INDICATED.

REPAIR WIRE AND REPLACE INJECTOR FUSE.

REPLACE ECM AND INJECTOR FUSE.

NOTE: BEFORE REPLACING ECM USE AN OHMMETER AND CHECK THE RESISTANCE BETWEEN SOLENOID TERMINALS OF:
- CANISTER PURGE
- AIR SWITCH
- AIR DIVERT
REPLACE ANY SOLENOID WHERE RESISTANCE MEASURES LESS THAN 20 OHMS.

Buick: Century, Skyhawk, Skylark, Somerset
Cadillac: Cimarron
Chevrolet: Astro, Beretta, Blazer, "C" Series, Camaro, Caprice, Cavalier,
  Celebrity, Citation II, Corsica, El Camino, "G" Series, "K" Series,
  LeMans, Malibu, Monte Carlo, "P" Series, "R" Series, Suburban, S10, T10,
  "V" Series
GMC: Caballero, "G" Series, Jimmy, "P" Series, "C" Series,
  "R" Series, Safari, Suburban, S15, "V" Series, T15, "K" Series
Oldsmobile: Calais, Ciera, Cruiser, Firenza, Omega
Pontiac: Bonneville, Fiero, Firebird, Grand Am,
  Grand Prix, Parisienne, Phoenix, Sunbird, 2000, 6000

NOTE
 * *For detailed applications, refer to FUEL INJECTION SYSTEMS APPLICATION CHARTS at the front of this publication.*

IDENTIFICATION

An 8-digit unit identification number is stamped on throttle body assembly. On model 220, number is stamped vertically on front of body at Throttle Position Sensor (TPS) side. On model 700, number can be found on TPS side of mounting flange. Letter codes are stamped on throttle body at external tube locations to identify vacuum hose connections. All 4-cylinder engines are equipped with model 700 TBI unit. V6 and V8 engines use model 220 TBI unit.

DESCRIPTION

The General Motors single unit Electronic Fuel Injection (EFI) system consists of 7 major sub-assemblies: Fuel supply system, Throttle Body Injector (TBI) assembly, Idle Air Control (IAC), Electronic Control Module (ECM), Electronic Spark Timing (EST), data sensors and emission controls.

Fuel is supplied to the engine through an electronically pulsed (timed) injector valve located in the throttle body unit on top of the intake manifold. The ECM controls the amount of fuel metered through the injector valve based upon engine demand and efficiency information. The ECM is a digital electronic computer which receives and computes signals from various data sensors.

NOTE
 * *Primary sub-systems which affect fuel system operation will be covered in this article: Fuel supply system, TBI assembly, IAC system, ECM and data sensors.*

OPERATION

FUEL SUPPLY SYSTEM

An electric fuel pump, located inside fuel tank as an integral part of the fuel gauge sending unit, supplies fuel under pressure to the throttle body assembly. A fuel pump relay controls fuel pump operation. When ignition switch is turned on, fuel pump relay activates the fuel pump for 1 1/2-2 seconds to prime the injector. On 1989 5.7L engines over 8500 GVW, "G" series with 5.7L and all 7.4L engines, a fuel module circuit is installed so that the fuel pump will run for 20 seconds when ignition is turned on. This enables engine to start easily during high ambient temperatures.

If the ECM does not receive reference pulses (engine cranking) from the crankshaft sensor or distributor after this period, the ECM deactivates the fuel pump relay. The fuel pump relay will be activated again when the ECM receives engine speed reference pulses.

As a back-up system to the fuel pump relay for 1984 and later models, the fuel pump can also be turned on by the oil pressure sender. The sender has 2 circuits internally. One circuit operates the oil pressure indicator in the instrument panel. The second circuit is normally an open switch which closes when the oil pressure reaches approximately 4 psi (.3 kg/cm²). If the fuel pump relay fails, the oil pressure sender will close and run the fuel pump through this second circuit.

THROTTLE BODY INJECTOR ASSEMBLY

The Throttle Body Injector (TBI) assembly is composed of 2 castings: a throttle body with a throttle valve to control air flow and a fuel body with an integral pressure regulator and fuel injector. Throttle body has throttle valve, Idle Air Control (IAC), and Throttle Position Sensor (TPS). The throttle body casting may contain ports to generate vacuum signals for EGR valve, MAP sensor and canister purge system.

The pressure regulator is a diaphragm-operated relief valve with fuel pump pressure acting on one side of the valve and atmospheric pressure acting on the other side of the valve. The pressure regulator maintains a constant pressure drop of about 10.0 psi (.7 kg/cm²) across the injector, throughout all engine operating conditions. *See Fig. 1.*

The pressure regulator on a TBI 220 unit is serviced as part of the fuel meter cover and should not be disassembled. The model 700 fuel pressure regulator may be disassembled and repaired.

Courtesy of General Motors Corp.

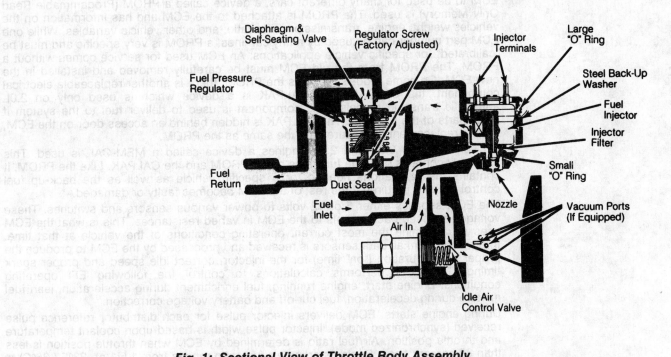

Fig. 1: Sectional View of Throttle Body Assembly

The fuel injector is a solenoid-operated device controlled by the ECM. Fuel is supplied at the lower end of the injector by the fuel supply system. Pressure to injector is maintained at 9-13 psi (.6-.9 kg/cm²) by the pressure regulator. Battery voltage is supplied to the injector when the ignition is on. The ECM activates the solenoid which lifts a normally closed ball valve off its seat. Fuel under pressure is injected in a conical spray pattern at the walls of the throttle bore, above the throttle valve. Excess fuel passes through the pressure regulator and is returned to the fuel tank.

During engine cranking, the fuel injector is pulsed (sprayed) once for each crankshaft or distributor reference pulse received by the ECM. This is referred to as the synchronized mode. In the non-synchronized mode, the injector is sprayed once every 6.25-12.5 milliseconds depending upon engine calibration and operating conditions. In this mode, the spray is totally independent of reference signal pulses.

On the 1989 5.0L and 5.7L model 220 TBI units, 2 injectors are used in the throttle body. Unlike the model 700 TBI unit, injectors are alternately sprayed, using ignition module reference pulses.

IDLE AIR CONTROL (IAC) SYSTEM

The IAC system consists of an electrically controlled motor (actuator) which positions the IAC valve in the air by-pass channel around the throttle valve. The IAC valve is a part of the throttle body casting. In 1989, IAC valve is threaded into throttle body casting on model 220 TBI. IAC is held in by screws on the 700 TBI (1989 models). The ECM calculates the desired position of the IAC valve based upon battery voltage, coolant temperature, engine load and engine speed to control idle speed while preventing stalls due to engine load changes.

If engine speed is lower than desired, the ECM activates the IAC motor to retract the IAC valve. If engine speed falls below a preset level and the throttle valve is closed, the ECM senses a near stall condition. To prevent stalling, the ECM will calculate an IAC valve position based upon barometric pressure. When the IAC valve is retracted, more air is diverted around the throttle valve to increase engine speed. If engine speed is higher than desired, the ECM activates the IAC motor to extend the IAC valve. When the IAC valve is extended, less air is diverted around the throttle valve, decreasing engine speed.

There are several different conical IAC pintle designs used. Ensure that correct design and type is used if IAC is replaced.

**ELECTRONIC
CONTROL
MODULE (ECM)**

The Electronic Control Module (ECM) is located in the passenger compartment and is the control center of the EFI system and Computer Command Control system. Locations vary, but the ECM is generally located under the instrument panel behind the glove compartment or behind the passenger footwell kick panel. To allow one model of ECM to be used for many different cars, a device called a PROM (Progammable Read Only Memory) is used. The PROM is attached to the ECM and has information on the vehicles weight, engine, transmission, axle ratio, and other vehicle variables. While one ECM part number can be used by many car lines, a PROM is very specific and must be "calibrated" for specific vehicle applications. An ECM used for service comes without a PROM. The PROM from the old ECM must be carefully removed and installed in the new ECM. On some models, besides the PROM, there is another replaceable electrical component, the CALPAK. The CALPAK is a device which is used only on 2.0L automatic transaxle models. This component is used to deliver fuel to the system if other parts of the ECM fail. The CALPAK is hidden behind an access door on the ECM, yet the replacement procedures are the same as the PROM.

On vehicles with 2.0L and 2.5L engines a device called a MEM-CAL is used. This assembly contains both the functions of the PROM and the CALPAK. Like the PROM, it contains the calibrations needed for a specific vehicle as well as the back-up fuel control circuitry required if the rest of the ECM becomes faulty or damaged.

The ECM supplies either 5 or 12 volts to power various sensors and switches. These voltages are feed back signals to the ECM in varied resistances. This is what the ECM uses to determine the most current operating conditions of the vehicle at that time. Information from all data sensors is received and processed by the ECM to produce the proper pulse duration ("on" time) for the injector, correct idle speed and proper spark timing. The ECM performs calculations to control the following EFI operating conditions: Engine start, engine running, fuel enrichment during acceleration, lean fuel mixture during deceleration, fuel cut-off and battery voltage correction.

During engine starts, ECM delivers injector pulse for each distributor reference pulse received (synchronized mode). Injector pulse width is based upon coolant temperature and throttle position. Air/fuel ratio is determined by ECM when throttle position is less than 80 percent open. Engine starting air/fuel ratio ranges from 1.5:1 at -33°F (-36°C) to 14.7:1 at 201°F (94°C). At lower coolant temperatures, injector pulse width is longer (richer air/fuel mixture ratio). When coolant temperature is high, injector pulse width becomes shorter (leaner air/fuel ratio).

If the engine is flooded, the driver must depress the accelerator pedal enough to set the wide open throttle position. At this position, the ECM calculates injector pulse width equal to an air/fuel ratio of 20:1. This air/fuel ratio will be maintained as long as the throttle remains in the wide open position and engine speed is below 600 RPM. If throttle position becomes less than 80 percent open and/or the engine speed exceeds 600 RPM, the ECM changes the injector pulse width to that used during engine starting (based upon coolant temperature and manifold vacuum).

When the engine is cold and running above 400 RPM (1989) and 600 RPM (1988 & earlier), the ECM operates in the open loop mode. In open loop, the ECM calculates injector pulse width based upon coolant temperature and Manifold Absolute Pressure (MAP) sensors input. The engine will remain in open loop operation until the oxygen sensor reaches operating temperature, the coolant temperature reaches a preset temperature, and a specific period of time has elapsed after the engine starts. When all these conditions are met, the ECM operates in the closed loop mode. In closed loop, the ECM controls injector pulse width based upon oxygen sensor signals, to maintain the air/fuel mixture ratio close to 14.7:1.

Fuel enrichment during acceleration is provided by the ECM. Sudden opening of the throttle valve causes a rapid increase in manifold absolute pressure. Pulse width is directly related to manifold absolute pressure, throttle position and coolant temperature. The higher the manifold absolute pressure and the wider the throttle angle, the wider the pulse width (richer mixture). During enrichment, the injector pulses are not in proportion to distributor reference signals (non-synchronized). Any reduction in throttle angle cancels fuel enrichment.

During normal deceleration, the air/fuel mixture must be leaner. The ECM calculates the injector pulse width in a manner similar to that used for fuel enrichment, and fuel output is reduced. This reduction in available fuel serves to remove residual fuel from intake manifold. During sudden deceleration, when manifold absolute pressure, throttle position and engine speed are reduced to preset levels, fuel flow is cut off completely to remove fuel from the engine. This deceleration fuel cut-off overrides the normal deceleration mode. During either deceleration mode, injector pulses are not in proportion to engine speed reference signals.

Battery voltage corrections by the ECM are performed during all operating modes of the EFI system. As battery voltage decreases, the ECM increases the injector pulse width with a correction factor stored in the ECM's memory.

The ECM used on EFI vehicles has a "learning" capacity. If the battery is disconnected, the "learning" process must begin all over again. During this period, a change may be noted in vehicle performance. To "teach" the vehicle, ensure the vehicle is at normal operating temperature and then drive the vehicle at part throttle, moderate acceleration and idle conditions until performance returns.

No fuel is delivered when ignition is turned off so that dieseling is prevented. No fuel is delivered if no reference pulses are sent by distributor even with ignition on. This prevents flooding before starting. Fuel cut-off will also occur at high engine RPM to prevent internal damage to engine.

DATA SENSORS

Each sensor furnishes an electrical signal to the ECM which modifies injector pulse width to conform to engine operating conditions. These sensors are as follows:

Coolant Temperature Sensor (CTS)

NOTE
- *This sensor looks much like the air temperature switch on some models and can be identified by having two grooves near the top.*

The CTS, located in the thermostat housing, is a variable resistor (thermistor) type sensor which transmits an electrical signal to the ECM proportionate to engine temperature. Low coolant temperature produces high resistance while high coolant temperature produces low resistance.

The ECM supplies a 5-volt signal to the CTS and measures the voltage that returns. By measuring the voltage resistance between the 2 readings, the ECM is informed of engine coolant temperature. Coolant temperature is used for fuel management, idle air control, spark timing, EGR operation, canister purge operation and other engine operating functions.

Oxygen Sensor

The oxygen sensor is mounted in exhaust pipe were it monitors oxygen content of exhaust gas stream. Oxygen in exhaust reacts with oxygen sensor to produce a voltage output to ECM. Voltage ranges from 100 millivolts (high oxygen, lean mixture) to 900 millivolts (low oxygen, rich mixture). The ECM monitors voltage output of oxygen sensor and determines what fuel mixture command to signal to fuel injector. The ECM interprets the electrical signal and adjusts the injector pulse width to maintain the air/fuel ratio close to 14.7:1.

The O_2 sensor will not function properly (produce voltage) until its temperature reaches 600°F (316°C). At temperatures less than the normal operating range of the sensor, vehicle will function in "open loop" mode and EMC will not make air/fuel adjustments based upon O_2 sensor signals but will use TPS and MAP values to determine air/fuel ratio from a table built into memory. When ECM reads a voltage signal of more than .45 volts from the O_2 sensor, ECM will begin to alter commands to injector or M/C solenoid to produce either a leaner or richer mixture. Once vehicle has entered "closed loop", a fault in the O_2 circuit (cooled-down, open or shorted O_2 sensor) is the only thing which can return it to "open loop".

NOTE
- *No attempt should be made to measure oxygen sensor voltage output. Current drain of conventional voltmeter could permanently damage sensor, shift sensor calibration and/or render sensor unusable. Do not connect jumper wire, test leads or other electrical connectors to sensor.*

Manifold Absolute Pressure (MAP) Sensor

The MAP sensor is mounted on the right side of the engine compartment. This sensor is a variable resistance type which measures the changes in the intake manifold pressure which result from engine load and speed changes.

The pressure measured by the MAP sensor is the difference between barometric pressure (atmospheric air) and manifold pressure (vacuum). A closed throttle condition (engine coast down) would produce a low manifold absolute pressure reading and a wide open throttle condition (engine acceleration) would produce a high manifold absolute pressure reading. The high value is produced because the pressure inside the intake manifold (vacuum) is the same as the pressure outside the manifold (atmospheric air).

The MAP sensor allows the ECM to adjust automatically at different altitudes. The ECM supplies a 5-volt reference signal to the MAP sensor. As manifold absolute pressure changes, the electrical resistance of the sensor also changes. By monitoring the sensor output voltage (similar to the CTS), the ECM is informed of intake manifold pressure. A higher pressure (high voltage) requires more fuel, while a lower pressure (low voltage) requires less fuel.

Vehicle Speed Sensor (VSS)

The speed sensor, mounted behind the speedometer in the instrument cluster, provides the ECM with pulses to determine vehicle speed. *See Fig. 2.* This information is used by the ECM to control the IAC motor, canister purge, and transmission clutch converter (TCC).

NOTE
- *The vehicle should not be driven without the vehicle speed sensor installed.*

Courtesy of General Motors Corp.

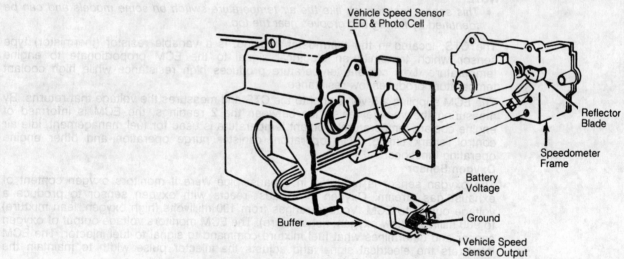

Fig. 2: Vehicle Speed Sensor

Throttle Position Sensor (TPS)

The TPS is mounted on the side of the throttle body and is connected to the throttle shaft. As the throttle valve angle changes (accelerator pedal moved), the resistance of the sensor also changes. The ECM supplies a 5-volt reference signal to the TPS. A closed throttle condition produces high resistance at the sensor and the output signal to the ECM will be low (about .5 volts). A wide open throttle condition produces low resistance at the sensor and the output signal to the ECM will be high (about 5 volts).

By monitoring the output voltage of the TPS and comparing that value to the reference signal, the ECM can calculate fuel requirements based upon throttle valve angle (driver demand).

Engine Speed Sensor

On models not using Direct Ignition System (DIS), the engine speed reference signal comes from the Hall Effect Unit mounted above the pickup coil in distributor (2.5L engine) or from terminal "R" of the conventional HEI module in the distributor. Pulses from the distributor are sent to the ECM where the time between these pulses is used to calculate the engine speed. The ECM adds spark advance modifications to the signal and sends the signal back to the distributor.

DIS ignitions do not have a distributor and, therefore, the ECM cannot obtain engine speed information from this source. Engine speed and crankshaft location are determined by the use of an internal crankshaft reluctor ring and a magnetic pick-up inserted through the side of the engine block.

Park/Neutral Switch

The Park/Neutral switch indicates to the ECM when the transmission is in Park or Neutral. This information is used by the idle air control valve and torque converter clutch system.

NOTE

- *Vehicle should not be driven with Park/Neutral switch disconnected as idle quality may be affected in Park or Neutral.*

Cranking Signal

The ECM is connected to the starter solenoid to determine engine cranking. This tells the ECM the engine is in the start mode. If this signal is not available, the engine may be hard to start in extremely cold weather.

Air Temperature Sensor

NOTE

- *This sensor looks much the same as the coolant temperature sensor on some models, it can be identified by one groove near the top.*

The air temperature sensor is mounted in the upper throttle body and sends information on incoming air temperature to the ECM.

TROUBLE SHOOTING

PRELIMINARY CHECKS

The following systems and components must be in good condition and operating properly before beginning diagnosis of the fuel injection system:

- All support systems and wiring
- Battery connections and specific gravity
- Engine state of tune
- Emission control devices
- Ignition system
- Compression pressure
- Fuel supply system pressure and flow
- All electrical connections and terminals
- Vacuum line, fuel hose and pipe connections
- Air leaks at throttle body mounting and intake manifold

TROUBLE SHOOTING

Hesitates, Sluggish, Sags or Poor Mileage

1) Visually check the MAP sensor hose for leaks or restrictions (water in hose). Replace hose if required. Check the TPS for sticking or binding and repair or replace as required. Ensure fuel pressure is a steady 9-13 psi (.6-.9 kg/cm²) at all operating ranges. If pressure is incorrect, go to FUEL SYSTEM DIAGNOSIS chart in this article. Ensure base engine timing is correct. Check vacuum hoses for splits, kinks and proper connections, as shown on Vehicle Emission Control Information label. Check ignition wires for cracking, hardness and proper connections at both distributor cap and spark plugs. Check for fouled spark plugs.

2) With fuel injector electrical connector disconnected, check for fuel leakage from injector when cranking. If leakage occurs, replace injector. Check fuel injector fuel filter screen for blockage and replace as necessary. Check for an open in HEI ground circuit and repair as required. Check wires for pinches, cuts and proper connections. Also check for poor quality or water contaminated fuel.

3) Check alternator output voltage. Alternator output voltage should be 9-16 volts. Check canister purge system and EGR valve for proper operation. Check for correct ECM PROM. Check the operation of the A/C compressor control and torque converter clutch (TCC) system. Repair as required. Make sure that ECM controlled idle speed is correct. On the 2.5L engine, check the operation of the cooling fan control circuit and repair as required.

Cuts Out or Misses

1) Check for an intermittent open or short to ground in the following circuits: 5 volt reference, HEI reference, fuel pump circuit, injector drive circuits and IAC drive circuits. Repair as required.

2) Check fuel system for restricted fuel filter and replace if required. Ensure fuel pressure is 9-13 psi (.6-.9 kg/cm²) at all operating ranges. If pressure is incorrect, go to FUEL SYSTEM DIAGNOSIS chart in this article. Inspect fuel injector "O" rings for damage and replace as required.

3) Inspect fuel injector screen for restriction. Ensure steel back-up washer is located beneath large "O" ring of fuel injector assembly. Check spark plug wires with ohmmeter. If over 30,000 ohms, replace wire(s).

4) Check ignition coil and secondary voltage using Spark Tester (J-26792) or equivalent. Check for proper valve timing. Perform compression check on questionable cylinders. Remove rocker covers and check for bent push rods, worn rocker arms, broken or weak valve springs and worn camshaft lobes.

Rough, Unstable, or Incorrect Idle, Stalling

1) Check ignition timing and adjust as necessary. Perform compression check on questionable cylinders. Check injector for leaking. Remove injector from engine, disconnect power wires from injector and crank engine. No fuel should be emitted from the tip of the injector.

2) Check for leaking or kinked vacuum lines. Check Park/Neutral switch operation. Check for sticking or binding of TPS. Check EGR operation while idling. There should be no EGR at idle. Check Idle Actuator Control (IAC) operation. Erratic voltage can cause a change in idle. Check engine idle speed, both base idle and ECM idle.

3) Check MAP sensor voltage. Compare suspect TPS voltage with known good unit. Check PCV valve for proper operation by placing finger over inlet hole and removing. Valve should snap back.

4) Inspect oxygen sensor for silicone contamination from incorrect fuel usage. The sensor will have a White powdery coating, resulting in a high but false signal voltage (rich exhaust indication) to the ECM. Inspect power steering pressure switch circuit. On 2.5L engines, ECM should compensate for power steering loads. Loss of this signal would be more noticeable under parking conditions where power steering loads are greatest. Check exhaust system for restrictions, such as a damaged or collapsed pipe, muffler or catalytic converter.

5) Inspect A/C compressor or relay operation and refrigerant pressure (too high). Inspect cooling fan operation. Ensure battery cables and ground straps are clean and secure.

Surges and/or Chuggle

1) Check for intermittent EGR operation at idle. EGR valve should be closed at idle at all times. Check ignition timing and adjust if necessary. Check for proper operation of canister purge system. Check for adequate spark output using Spark Tester (ST-125).

2) Check in-line fuel filter for dirt or restriction. Replace as required. Check fuel pressure. Operating pressure should be between 9-13 psi (.6-.9 kg/cm²) at all times. Check fuel for water contamination.

3) Check alternator output voltage. Repair if less than 9 or more than 16 volts. Inspect oxygen sensor for silicone contamination from improper fuel usage.

4) The oxygen sensor may have a White powdery coating, resulting in a high but false signal voltage (rich exhaust indication) to the ECM. Check spark plugs for cracks, wear, improper gap or burnt electrodes and inspect complete ignition system. Check A/C for excessive charge. Check for restricted exhaust system.

5) Ensure VSS reading matches actual vehicle speed on speedometer. Check air cleaner damper door (THERMAC). Check Park/Neutral switch adjustment. Check for intermittent open, or short to ground in Transmission Clutch Converter (TCC), HEI By-Pass, Electronic Spark Timing (EST) circuits. Check Electronic Spark Control (ESC) operation.

Hard Start

1) Check for high resistance in the coolant sensor circuit. Repair circuit or replace sensor as required. Visually check TPS for sticking or binding and repair as necessary. Ensure fuel pressure is 9-13 psi (.6-.9 kg/cm²) under all operating conditions.

2) On the 2.5L engine, fuel pressure leak down should be gradual after ignition is turned off . An instant drop in pressure indicates a leaking in-tank fuel pump coupling hose or check valve.

3) To check this valve with ignition off, disconnect fuel line at filter and remove the tank filler cap. Connect a radiator test pump to the line and apply 15.0 psi (1.1kg/cm²) pressure. If the pressure will hold for 60 seconds, the valve is good.

4) With injector harness connector disconnected, check for fuel leakage while cranking. Replace injector if required. Check cranking circuit for short or open and repair as required. Check for poor quality or water contaminated fuel.

5) Check that PROM in vehicle is correct. Check EGR operation. Inspect fuel pump relay. Connect a test light to test terminal and ground. With ignition on, test light should glow for 2 seconds. Check complete ignition system. Check for restricted exhaust system.

ON-BOARD DIAGNOSTICS

The ECM uses sensors to monitor many engine operating conditions. It contains a memory and is programmed to know what certain sensor readings should be. If the sensor reading is not what the ECM thinks it should be, the ECM will turn on the "Check Engine" light on the instrument panel, and also will store a trouble code in memory.

ECM TROUBLE CODES

To call up trouble codes stored in the ECM, the Assembly Line Communication Link (ALCL) must be used. The ALCL is located under the left or right side of instrument panel depending upon vehicle. *See Fig. 3.*

Courtesy of General Motors Corp.

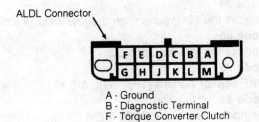

ALDL Connector

A - Ground
B - Diagnostic Terminal
F - Torque Converter Clutch
G - Fuel Pump
M - Serial Data

Fig. 3: ALCL Connector Location In Passenger Cars

To display all trouble codes stored in the ECM, do the following:

1) Turn ignition on but do not start engine. "Check Engine" light should glow. Locate ALCL connector attached to ECM wiring harness under dash. Insert spade lug terminal from terminal "B" to terminal "A", test terminal or ground.

2) "Check Engine" light should flash code "12". Code "12" consists of "Flash", pause "Flash", "Flash" followed by a longer pause. Trouble code "12" indicates the system is working and will repeat 2 more times. If any trouble codes are stored in the ECM memory they will be displayed in the same manner. Trouble codes are read by counting the flashes of the "Check Engine" light. For example, "Flash", "Flash", pause, "Flash", longer pause, identifies code 21.

3) Trouble codes will be displayed from the lowest to the highest numbered codes (3 times each) and be repeated as long as the test lead to the ALCL connector is grounded to terminal "A".

The following list represents these codes:

Code 12

No RPM reference. This may occur if attempting to retrieve trouble code when engine is not running.

Code 13

Checks the oxygen sensor circuit.

Code 14

Checks the coolant sensor on the low voltage side.

Code 15

Checks the coolant sensor on the high voltage side.

Code 21

Checks the throttle position sensor on the high voltage side.

Code 22

Checks the throttle position sensor on the low voltage side.

Code 23

Checks the MAT sensor on low temperature.

Code 24

Checks vehicle speed sensor.

Code 25

Checks the MAT sensor on high temperature.

Code 26

Quad-Driver error (1989 models).

Code 32

Checks for EGR system failure.

Code 33

Checks the MAP sensor on the high voltage side.

Code 34

Checks the MAP sensor on the low voltage side.

Code 35

Checks the Idle Air Control.

Code 36

MAF sensor burnoff (1989 5.0L & 5.7L).

Code 41

Cylinder select error (1989 MEM-CAL).

Code 42

Checks the Electronic Spark Timing (EST) circuit.

Code 43

Checks for Electronic Spark Control (ESC) retard circuit.

Code 44

Checks for lean exhaust indication.

Code 45

Checks for a rich exhaust indication.

Code 46

Anti-theft fault (1989 5.7).

Code 51

Checks PROM/MEMCAL and ECM condition.

Code 52

Checks CALPAK condition.

Code 53

System over voltage or EGR operation.

Code 54

Checks the fuel pump circuit.

Code 55

Checks on ECM condition.

CLEARING ECM TROUBLE CODES

Turn ignition on and ground test terminal at ALCL connector. Turn ignition off and remove ECM fuse from fuse block for 10 seconds. Replace fuse and remove test lead ground from ALCL connector. Removing battery voltage for 30 seconds will clear all stored trouble codes.

TESTING & DIAGNOSIS

FUEL SYSTEM PRESSURE TEST

NOTE
- *Diagnosis of the fuel system should begin with determining fuel system pressure. Before performing any test on fuel system, pressure must be released from the system.*

1) Remove fuel pump fuse from the fuse block in passenger compartment. Crank engine. Engine will start and run until fuel supply remaining in fuel lines is used. Engage starter again for about 3 seconds to ensure that all fuel is out of lines. Turn ignition off and replace fuse.

2) Remove air cleaner and plug Thermostatic Air Cleaner (THERMAC) vacuum port on throttle body unit. Remove the steel fuel line between the throttle body and the fuel filter. When removing the fuel line, always use 2 wrenches to prevent damage. Install Fuel Pressure Guage (J-29658) between throttle body and fuel filter.

3) Start vehicle and observe fuel pressure reading. Fuel pressure should be 9-13 psi (.6-.9 kg.cm²). If not, use FUEL SYSTEM DIAGNOSIS chart.

4) Depressurize fuel system as described in step **1)**. Remove fuel pressure gauge and reinstall steel line between filter and throttle body. Start car and check for leaks. Remove plug from throttle body THERMAC vacuum port and reinstall air cleaner.

ADJUSTMENTS

IDLE SPEED

Monitoring and control of engine idle is handled by the Electronic Control Module (ECM) and is not adjustable.

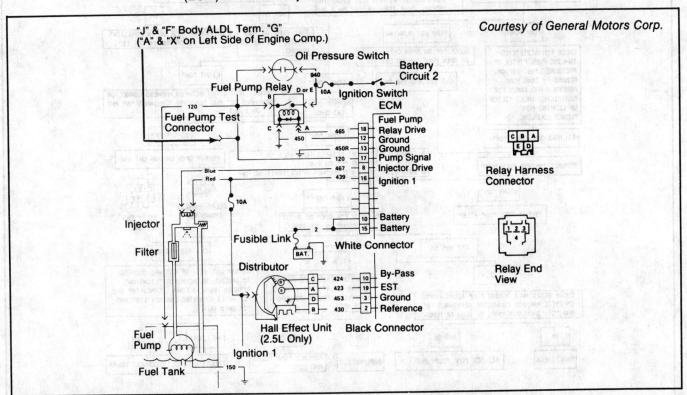

Fig. 4: 1982 Fuel Injection System Wiring Diagram

Courtesy of General Motors Corp.

FUEL SYSTEM DIAGNOSIS (1982-83 MODELS)

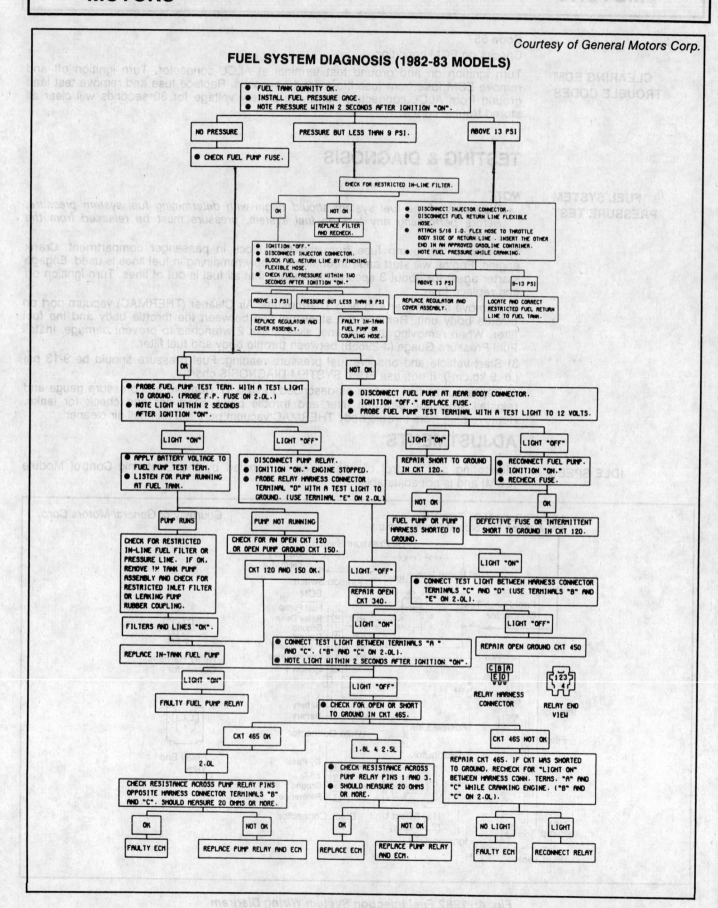

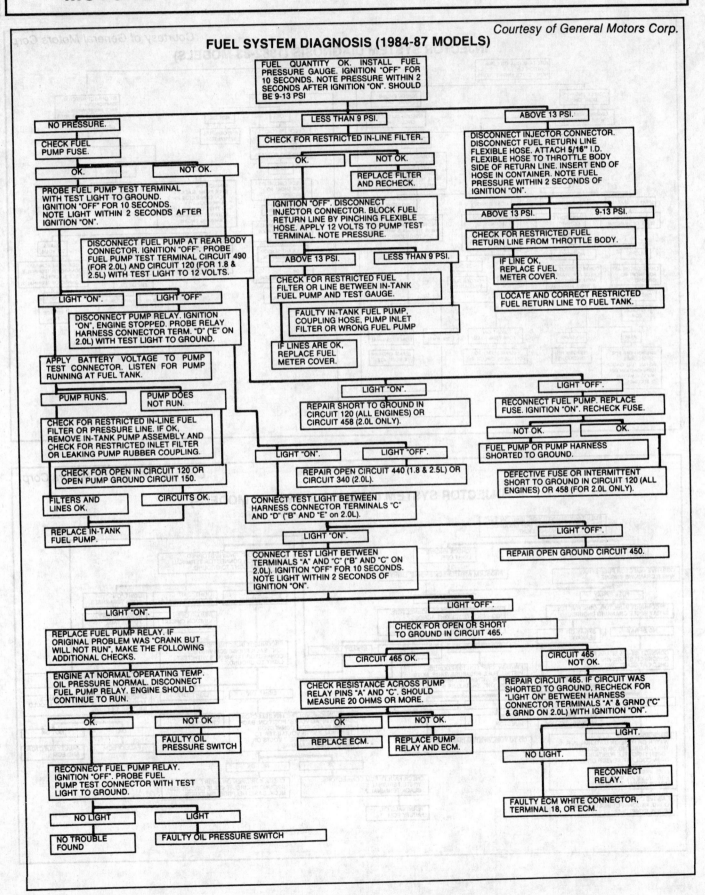

Courtesy of General Motors Corp.

FUEL SYSTEM DIAGNOSIS (1984-87 MODELS)

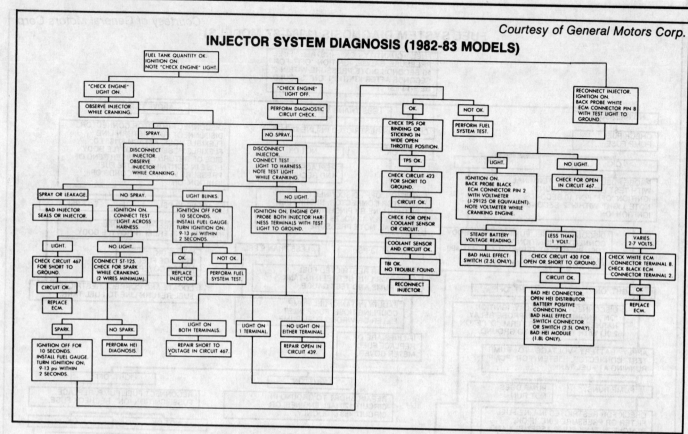

Courtesy of General Motors Corp.

INJECTOR SYSTEM DIAGNOSIS (1982-83 MODELS)

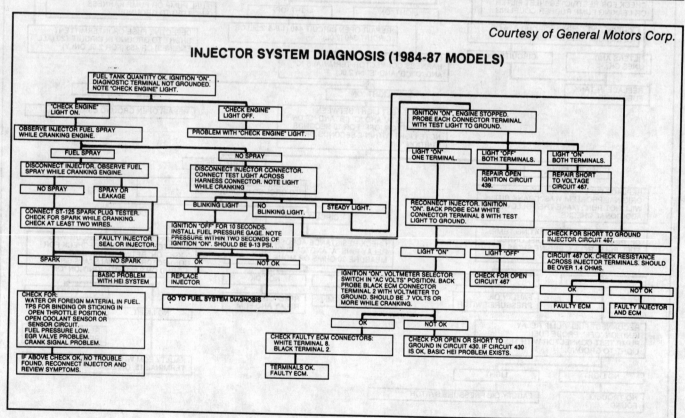

Courtesy of General Motors Corp.

INJECTOR SYSTEM DIAGNOSIS (1984-87 MODELS)

Courtesy of General Motors Corp.

INJECTOR DIAGNOSIS (LEMANS)

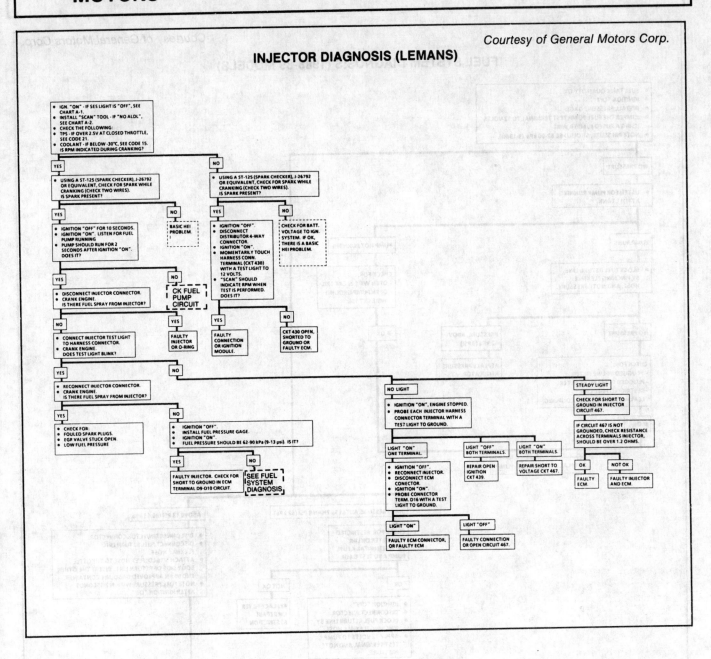

FUEL SYSTEM DIAGNOSIS (1988-89 MODELS)

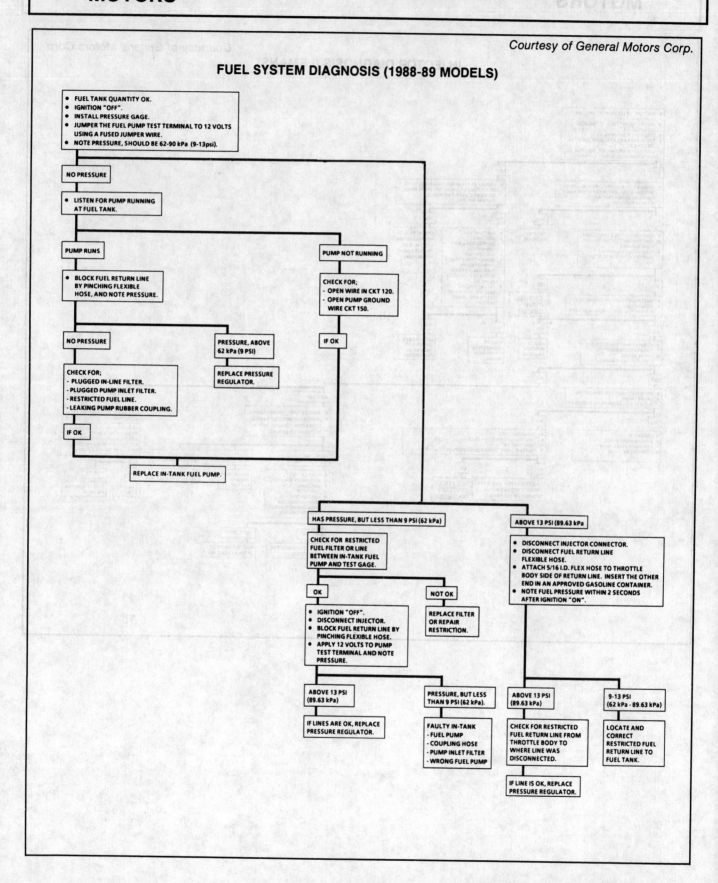

INJECTOR SYSTEM DIAGNOSIS (1988-89 MODELS (EXCEPT 1989 2.0L & 2.5L))

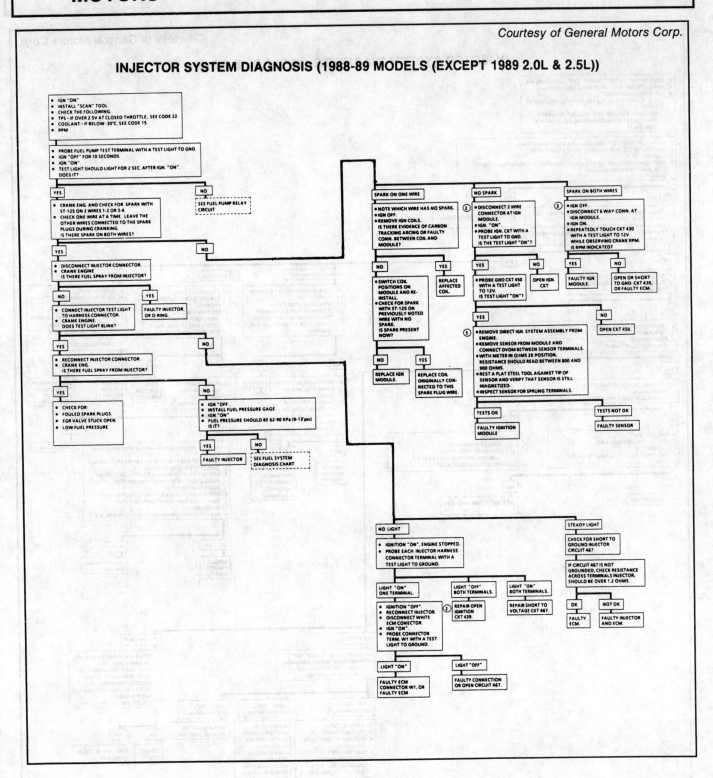

Courtesy of General Motors Corp.

INJECTOR SYSTEM DIAGNOSIS (1989 2.0L & 2.5L)

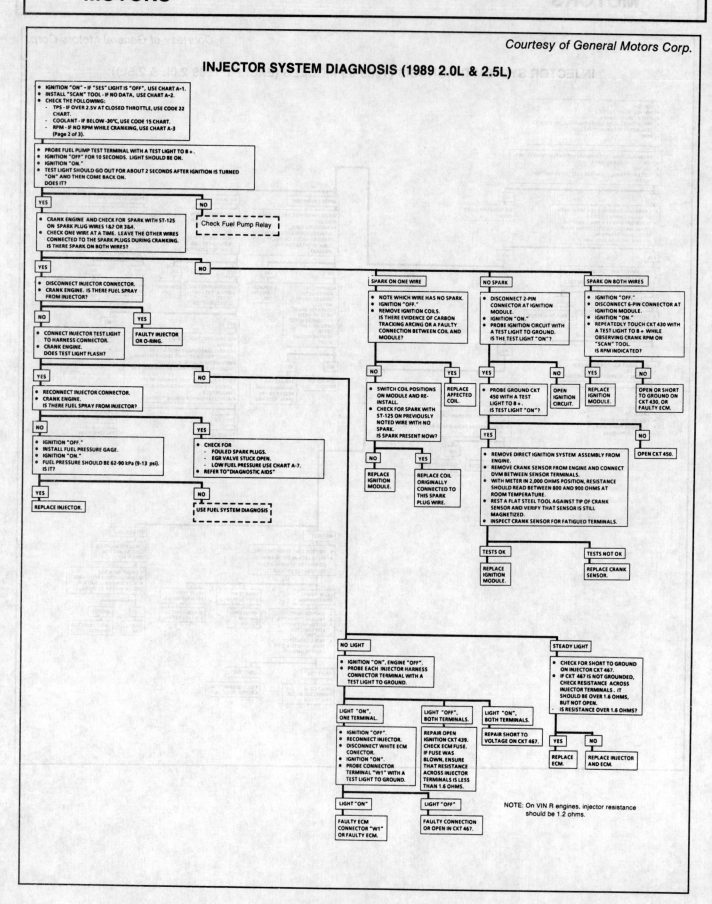

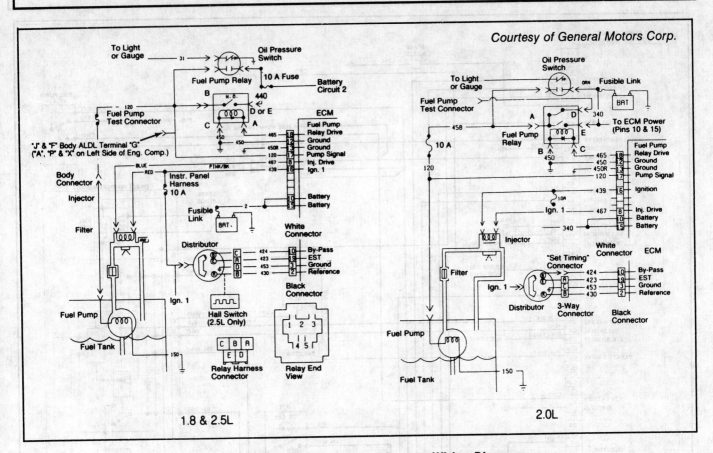

Courtesy of General Motors Corp.

1.8 & 2.5L

2.0L

Fig. 5: 1983-86 Fuel Injection System Wiring Diagram

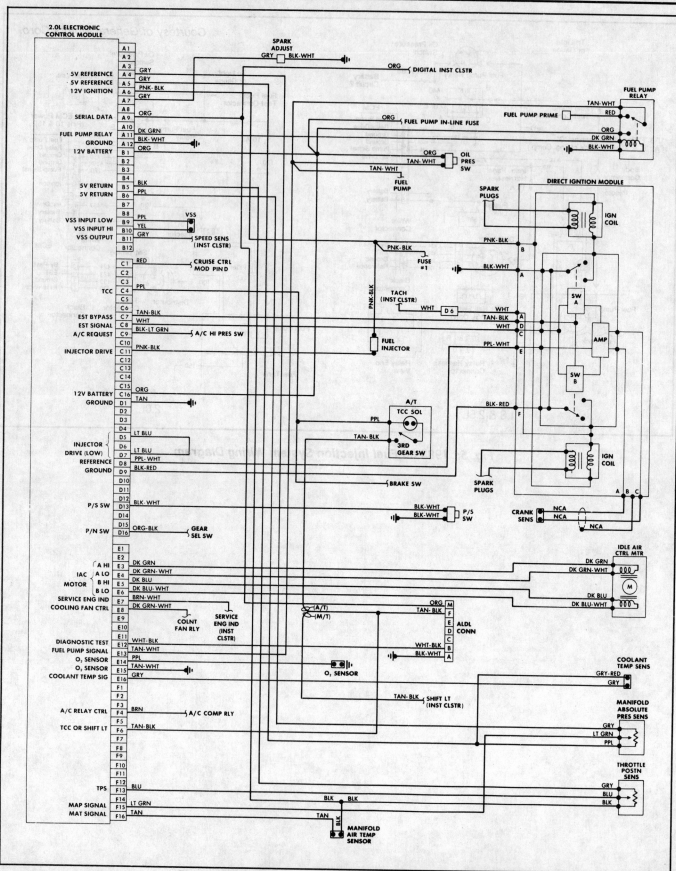

Fig. 6: 1987 Beretta & Corsica 2.0L (VIN 1) Fuel Injection System Wiring Diagram

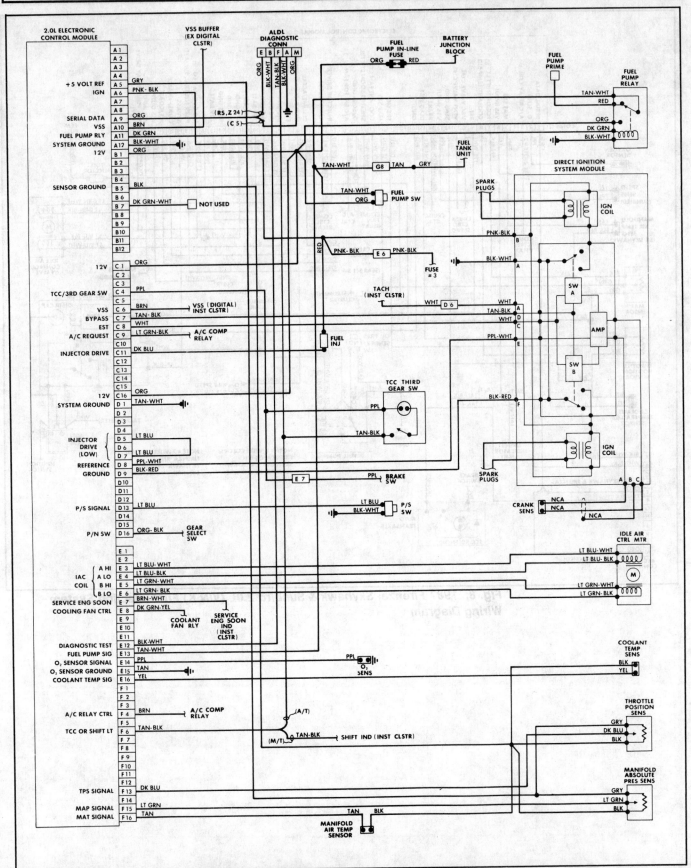

Fig. 7: *1987 Cavalier, Firenza & Skyhawk 2.0L (VIN 1) Fuel Injection System Wiring Diagram*

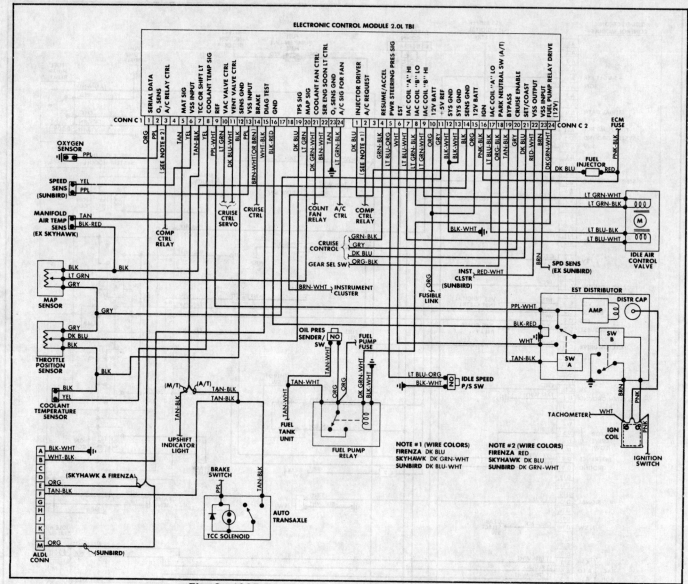

Fig. 8: 1987 Firenza, Skyhawk & Sunbird 2.0L (VIN K) Fuel Injection System Wiring Diagram

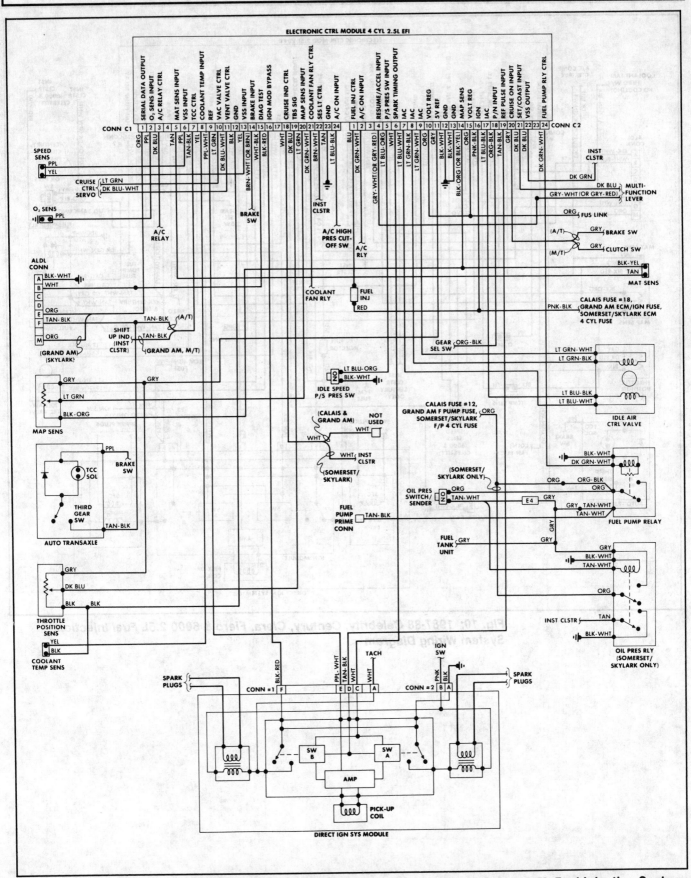

Fig. 9: *1987-88 Calais, Grand Am, Skylark & Somerset 2.5L Fuel Injection System Wiring Diagram*

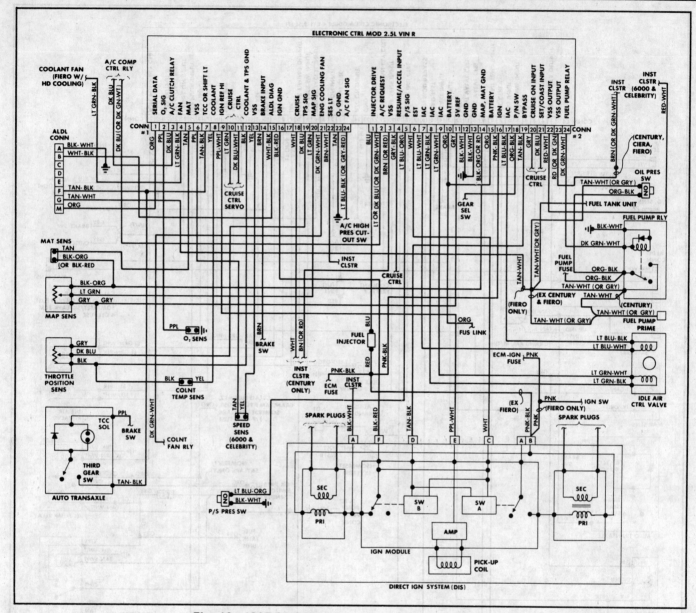

Fig. 10: 1987-88 Celebrity, Century, Ciera, Fiero & 6000 2.5L Fuel Injection System Wiring Diagram

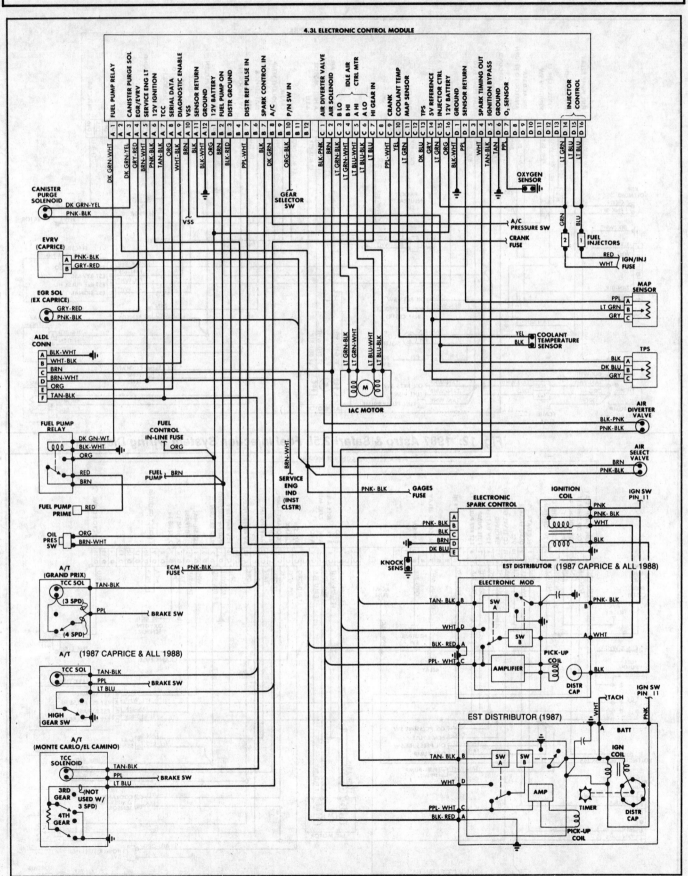

Fig. 11: 1987-88 Caballero, Caprice, El Camino, Grand Prix & Monte Carlo 4.3L Fuel Injection System Wiring Diagram

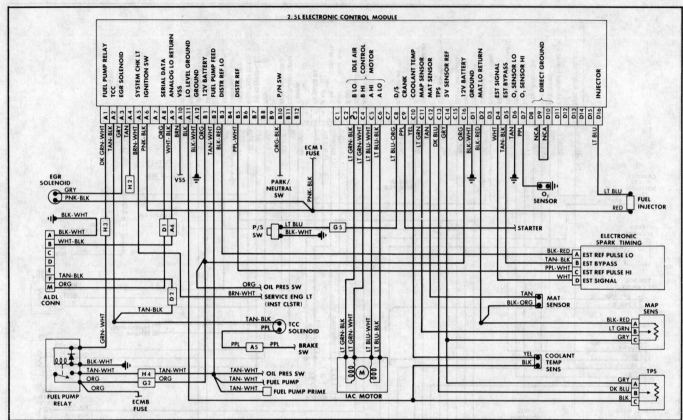

Fig. 12: 1987 Astro & Safari 2.5L Fuel Injection System Wiring Diagram

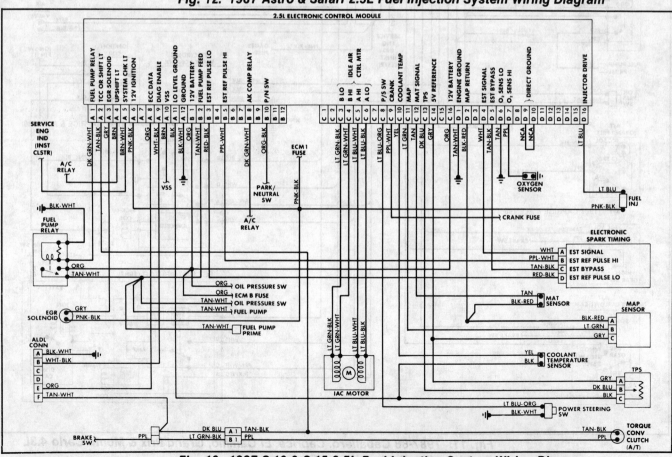

Fig. 13: 1987 S-10 & S-15 2.5L Fuel Injection System Wiring Diagram

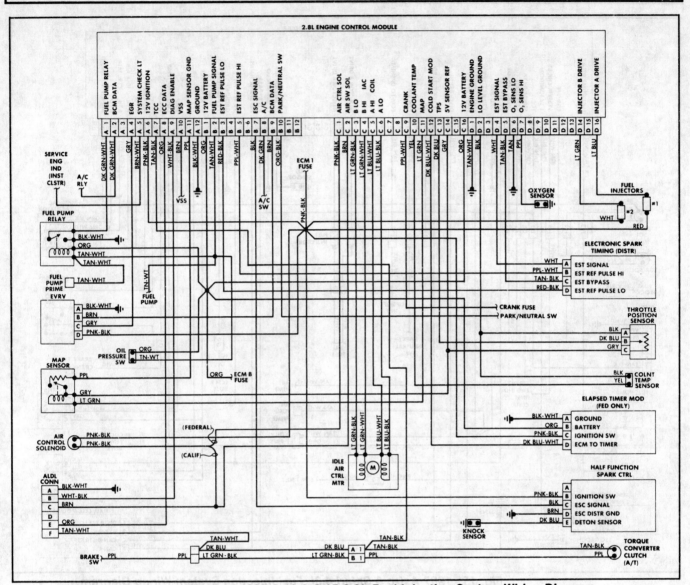

Fig. 14: 1987 S-10 & S-15 2.8L Fuel Injection System Wiring Diagram

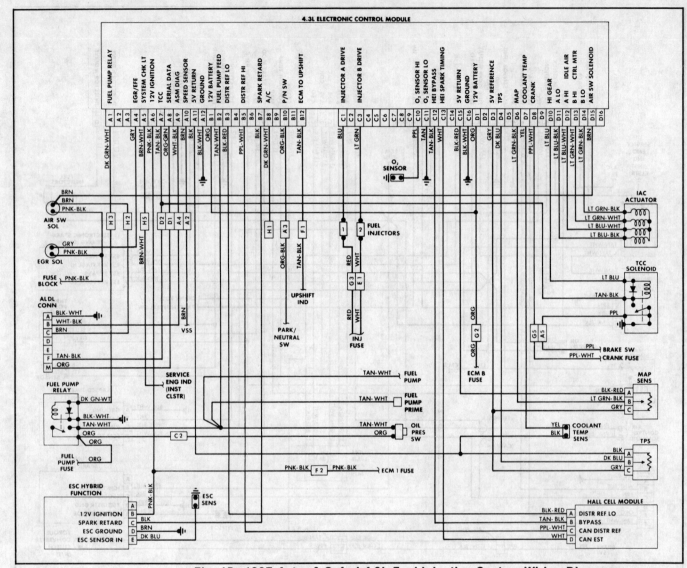

Fig. 15: 1987 Astro & Safari 4.3L Fuel Injection System Wiring Diagram

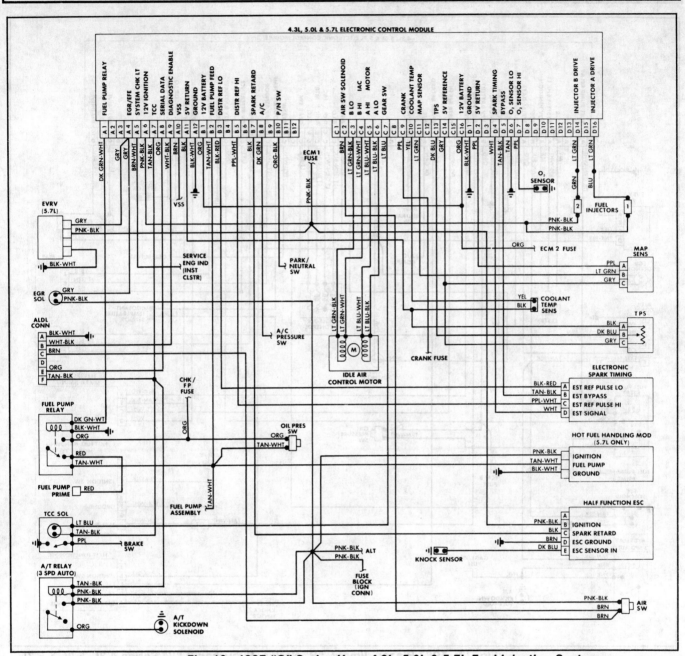

Fig. 16: 1987 "G" Series Vans 4.3L, 5.0L & 5.7L Fuel Injection System Wiring Diagram

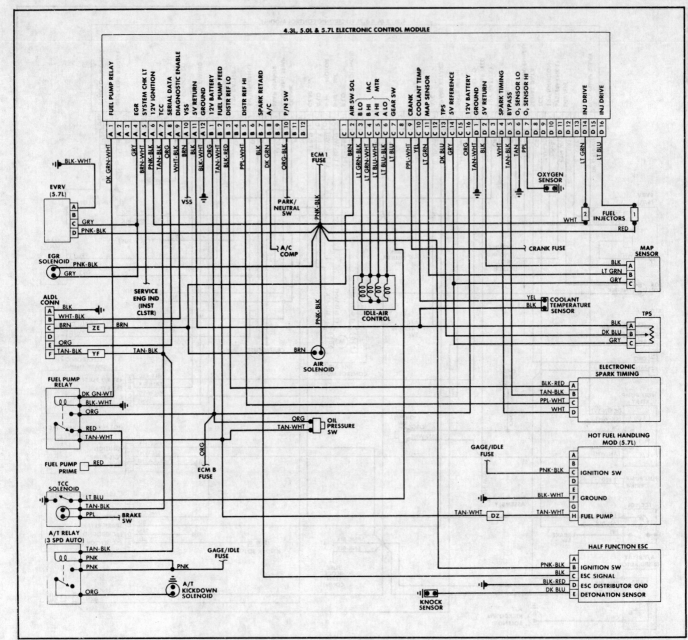

Fig. 17: 1987 Blazer, Jimmy, "R" Series, Suburban & "V" Series 4.3L, 5.0L & 5.7L Fuel Injection System Wiring Diagram

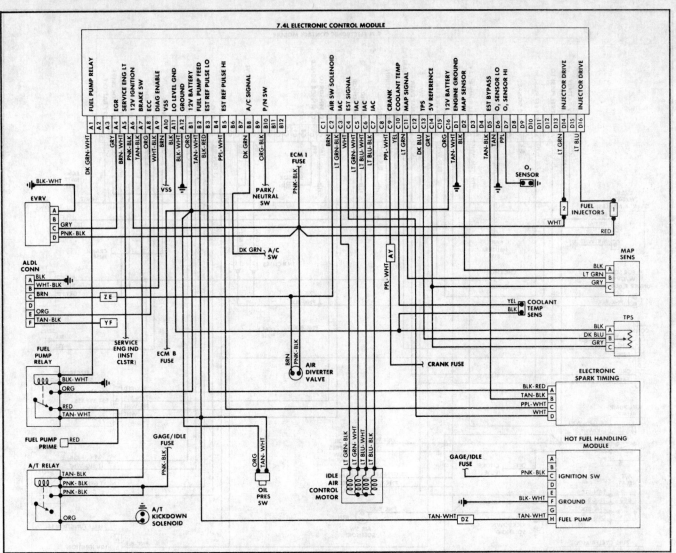

Fig. 18: 1987 "R" Series & "V" Series 7.4L Fuel Injection System Wiring Diagram

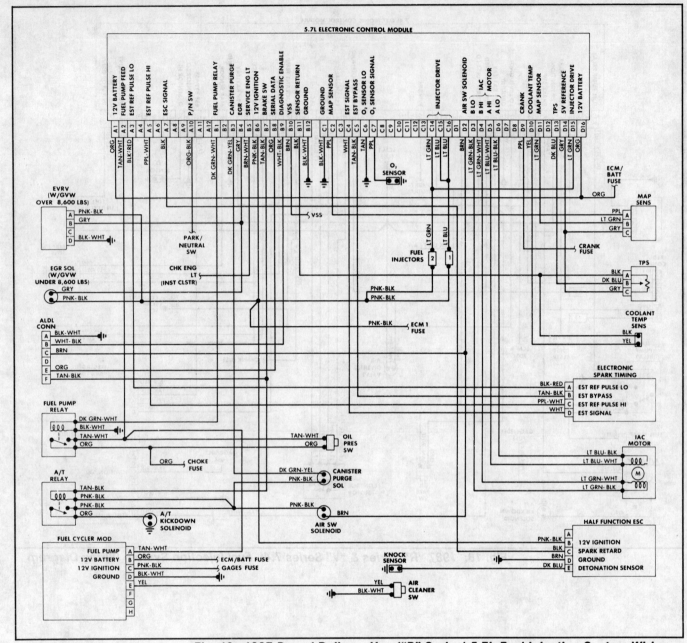

Fig. 19: 1987 Parcel Delivery Van ("P" Series) 5.7L Fuel Injection System Wiring Diagram

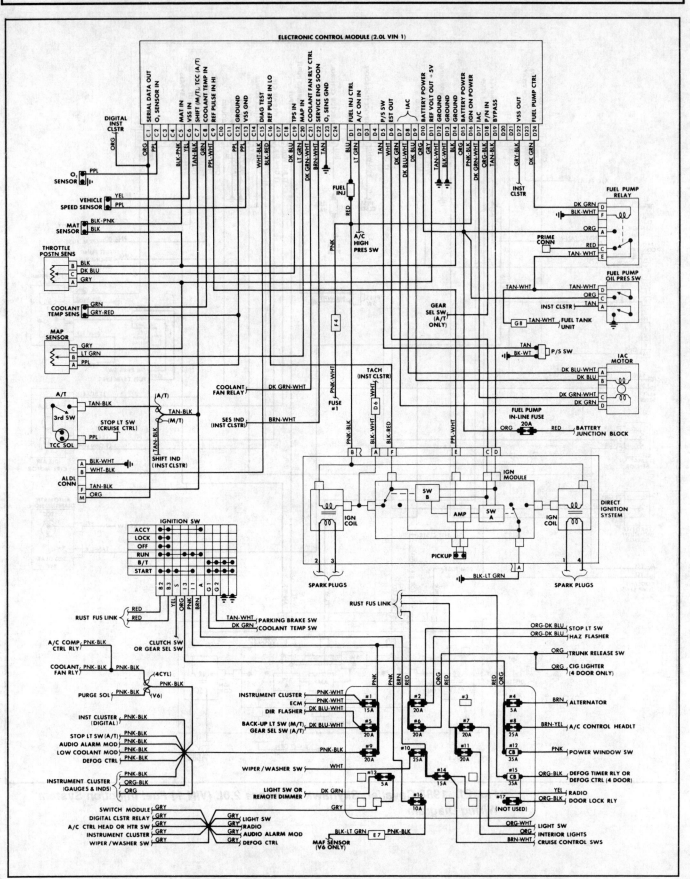

Fig. 20: 1988 Beretta & Corsica 2.0L (VIN 1) Fuel Injection System Wiring Diagram

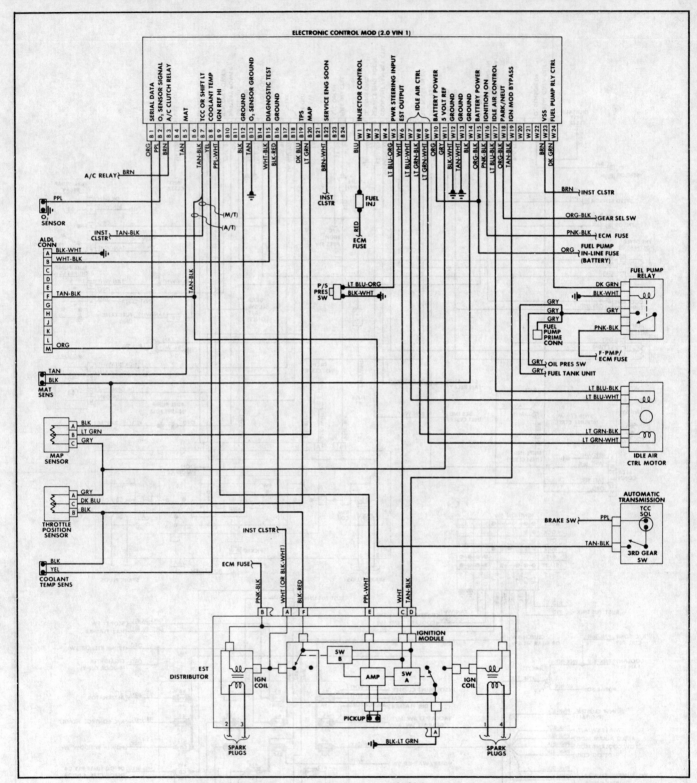

Fig. 21: 1988 Cavalier, Skyhawk & Firenza 2.0L (VIN 1) Fuel Injection System Wiring Diagram

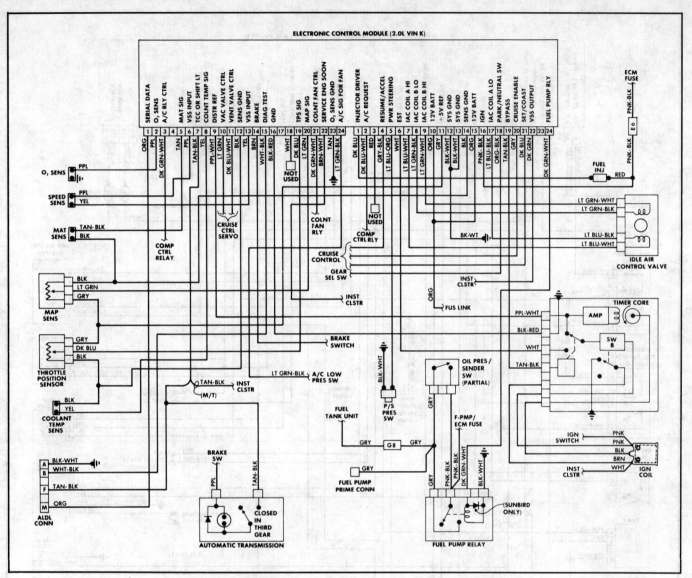

Fig. 22: 1988 Firenza, Skyhawk & Sunbird 2.0L (VIN K) Fuel Injection System Wiring Diagram

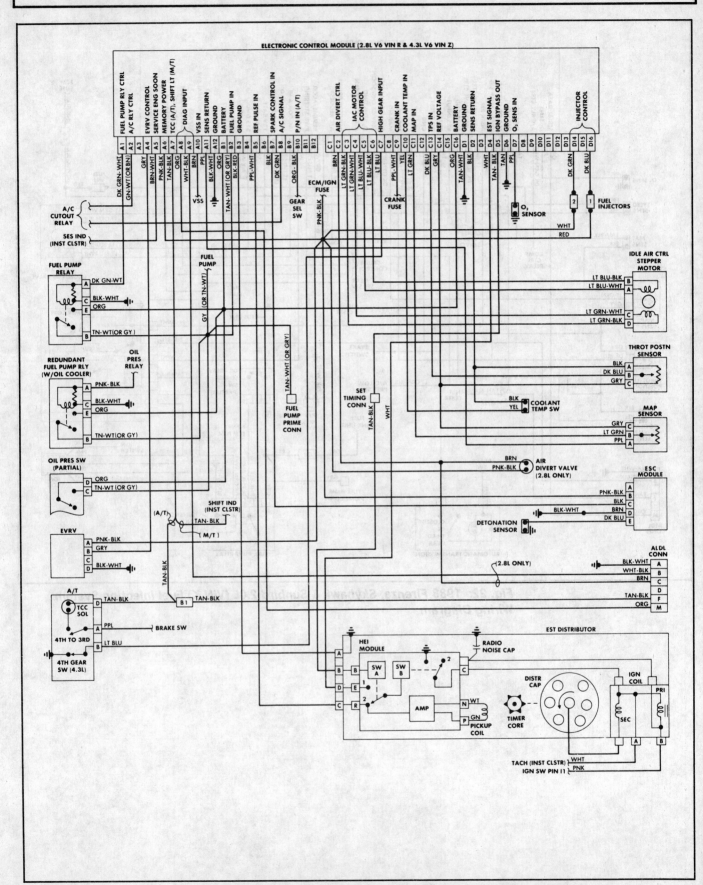

Fig. 23: *1988 S-10, S-15, T-10 & T-15 2.8L Fuel Injection System*
Wiring Diagram

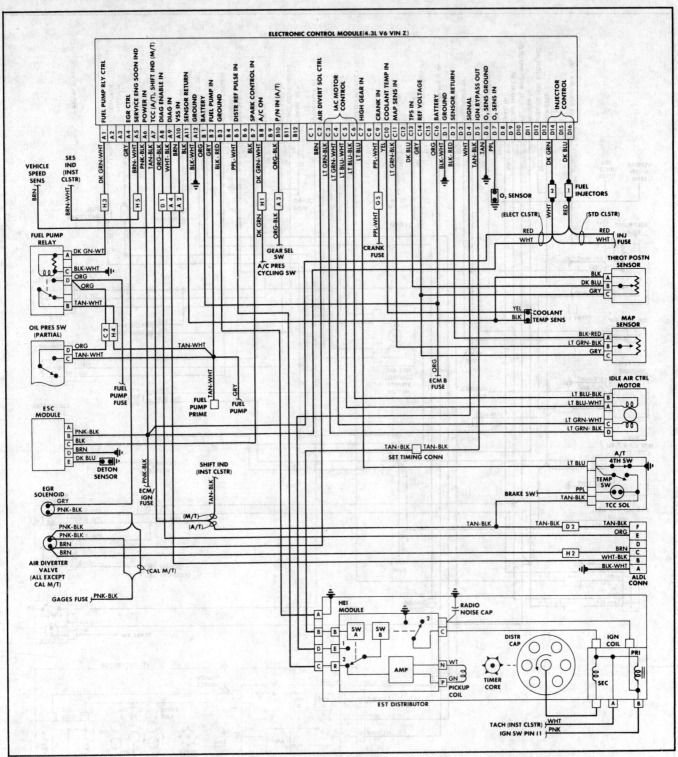

Fig. 24: 1988 Astro & Safari 4.3L Fuel Injection System Wiring Diagram

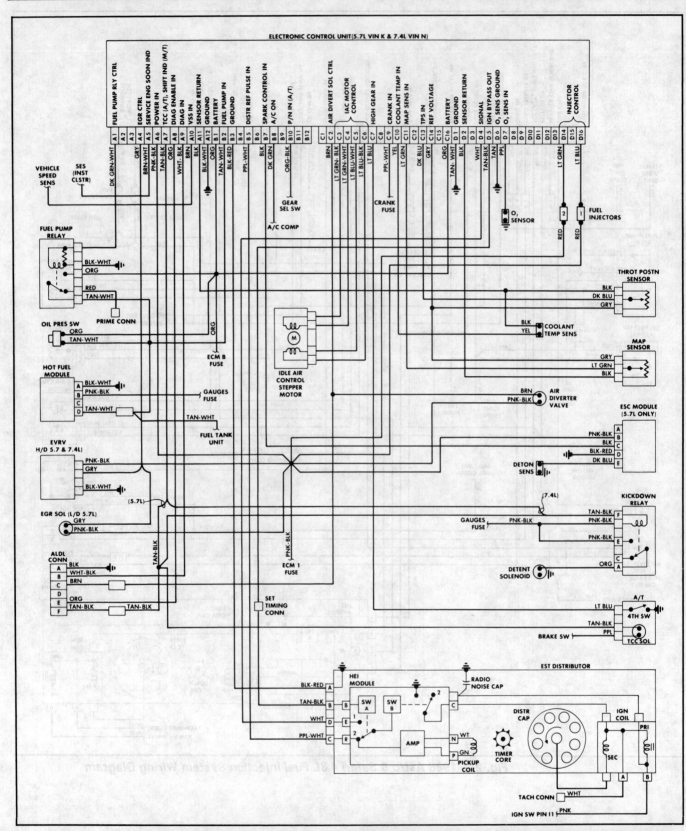

Fig. 25: 1988 Blazer, "G" Series Vans, Jimmy, "R" Series, Suburban & "V" Series 4.3L, 5.0L, 5.7L & 7.4L Fuel Injection System Wiring Diagram

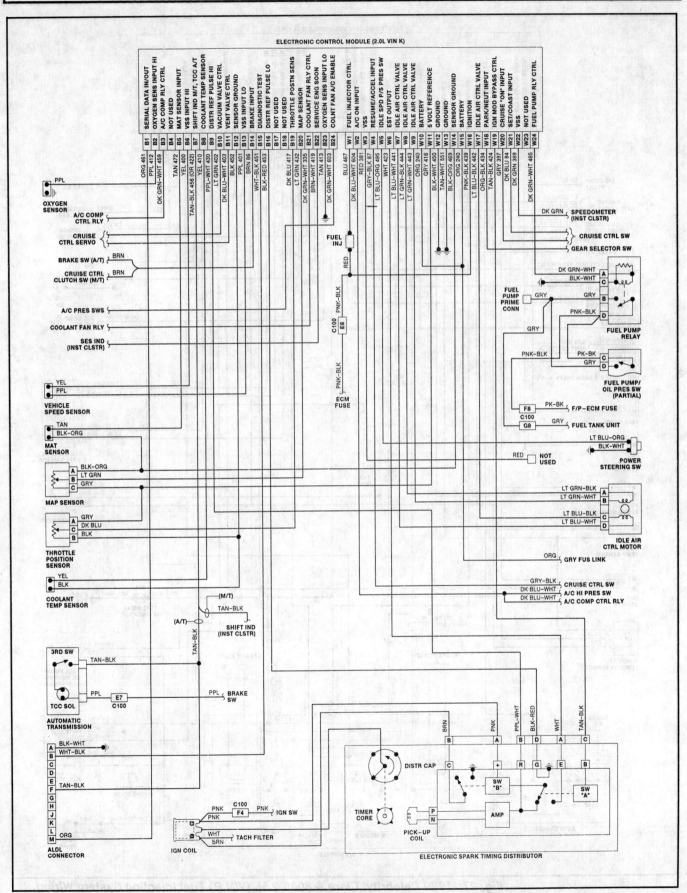

Fig. 26: 1989 "J" Body 2.0L (VIN K) Fuel Injection System Wiring Diagram

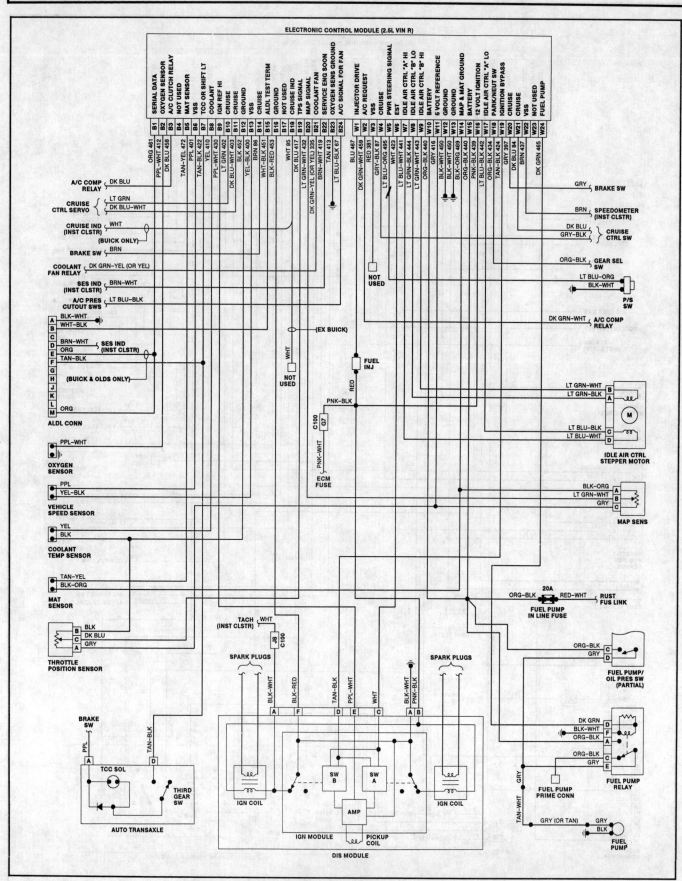

Fig. 27: *1989 Celebrity, Ciera & 6000 2.5L (VIN R) Fuel Injection System Wiring Diagram*

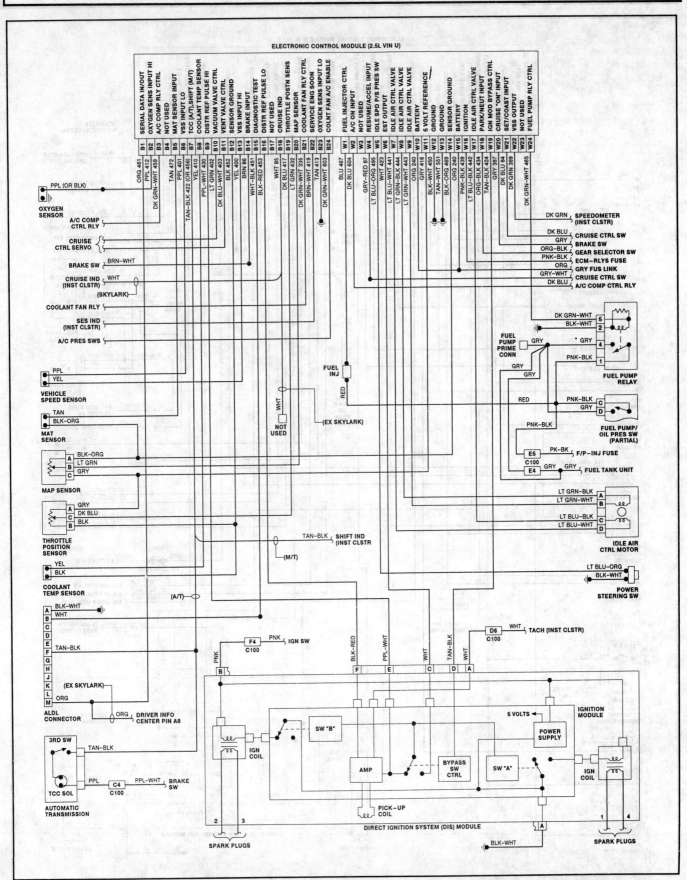

Fig. 28: *1989 Calais, Grand Am & Skylark 2.5L (VIN U) Fuel Injection System Wiring Diagram*

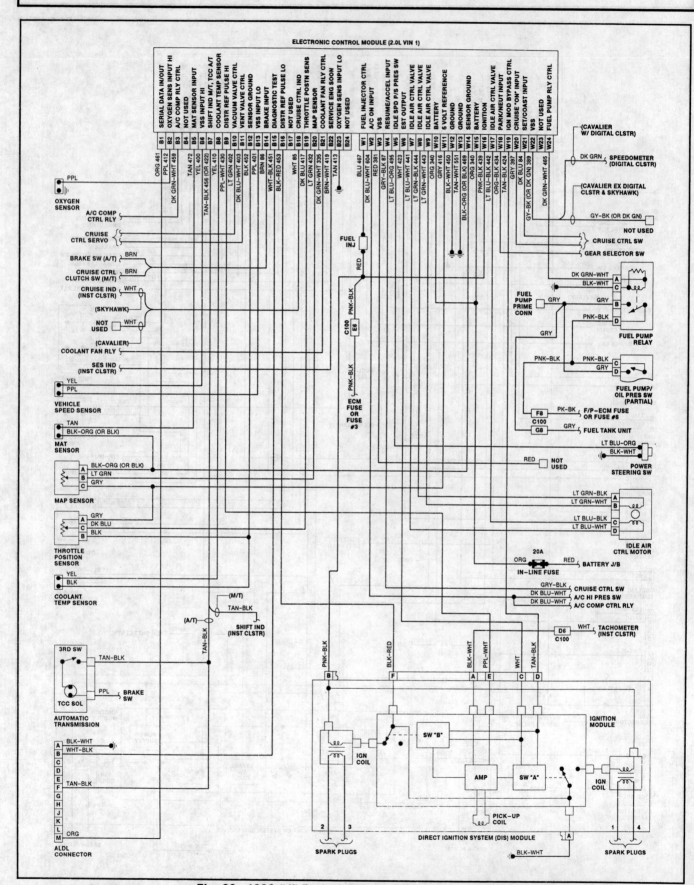

Fig. 29: 1989 "J" Body 2.0L (VIN 1) Fuel Injection System Wiring Diagram

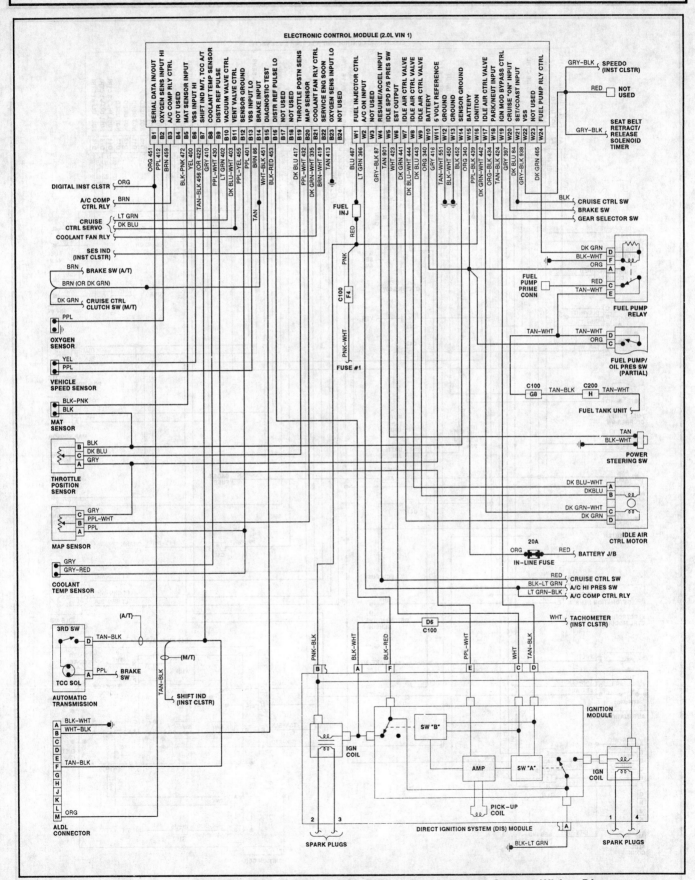

Fig. 30: 1989 "L" Body 2.0L (VIN 1) Fuel Injection System Wiring Diagram

437

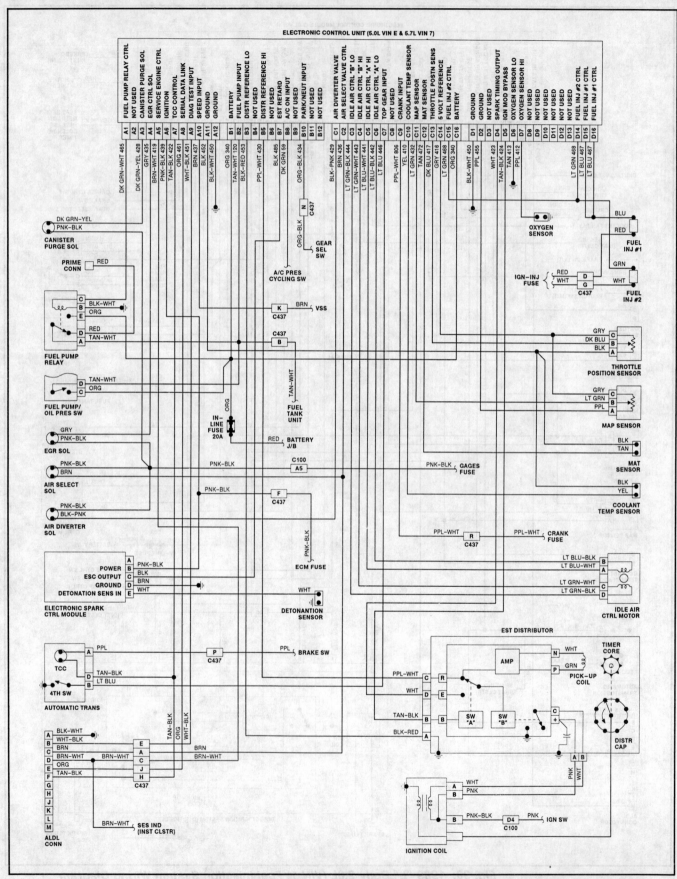

Fig. 31: 1989 "B" Body 5.0L (VIN E) & 5.7L (VIN 7) Fuel Injection System Wiring Diagram

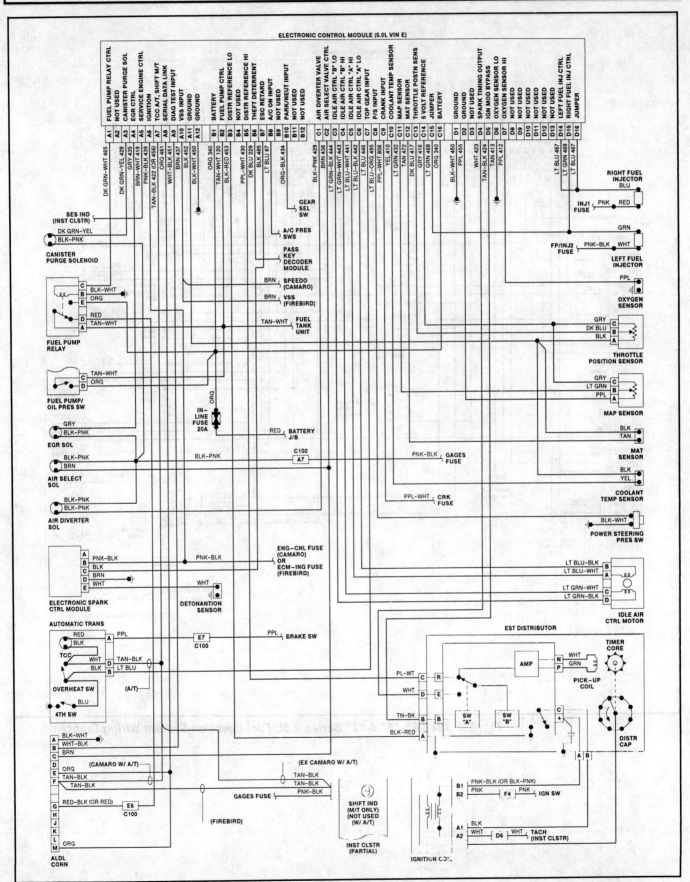

Fig. 32: 1988-89 "F" Body 5.0L (VIN E) Fuel Injection System Wiring Diagram

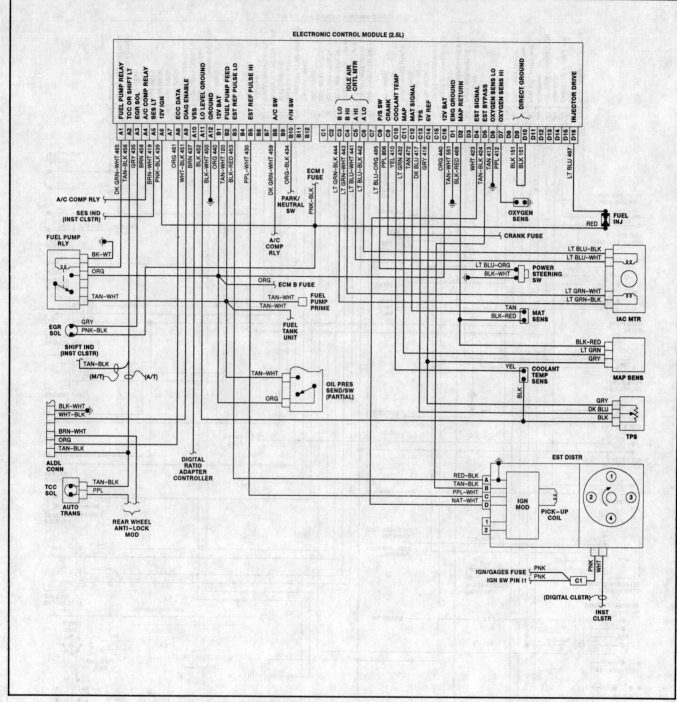

Fig. 33: 1988-89 "S" & "T" Series 2.5L Fuel Injection System Wiring Diagram

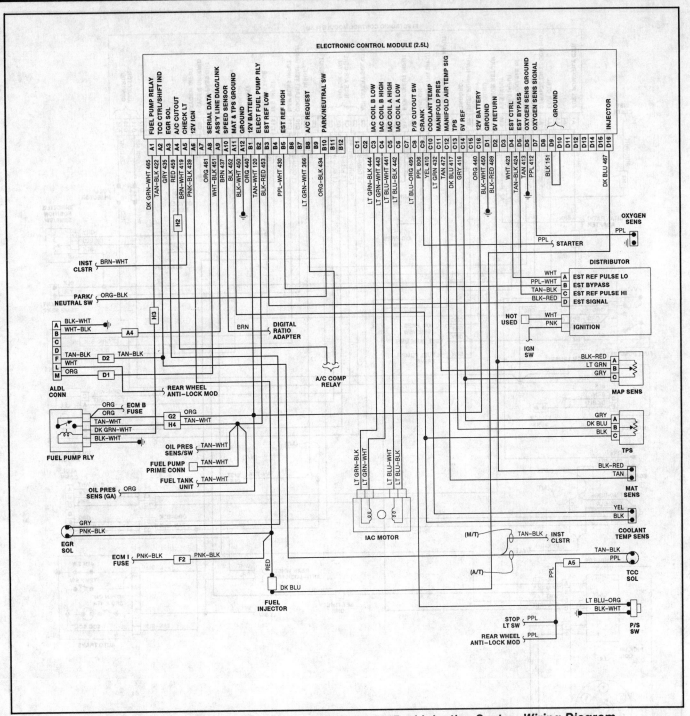

Fig. 34: 1989 Astro & Safari 2.5L Fuel Injection System Wiring Diagram

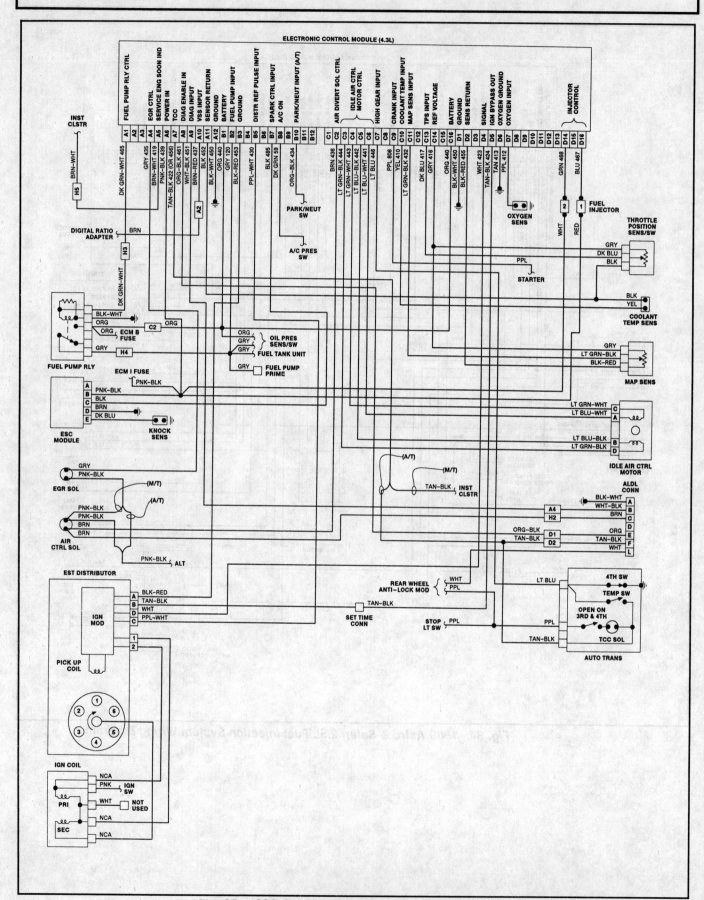

Fig. 35: 1989 Astro & Safari 4.3L Fuel Injection System Wiring Diagram

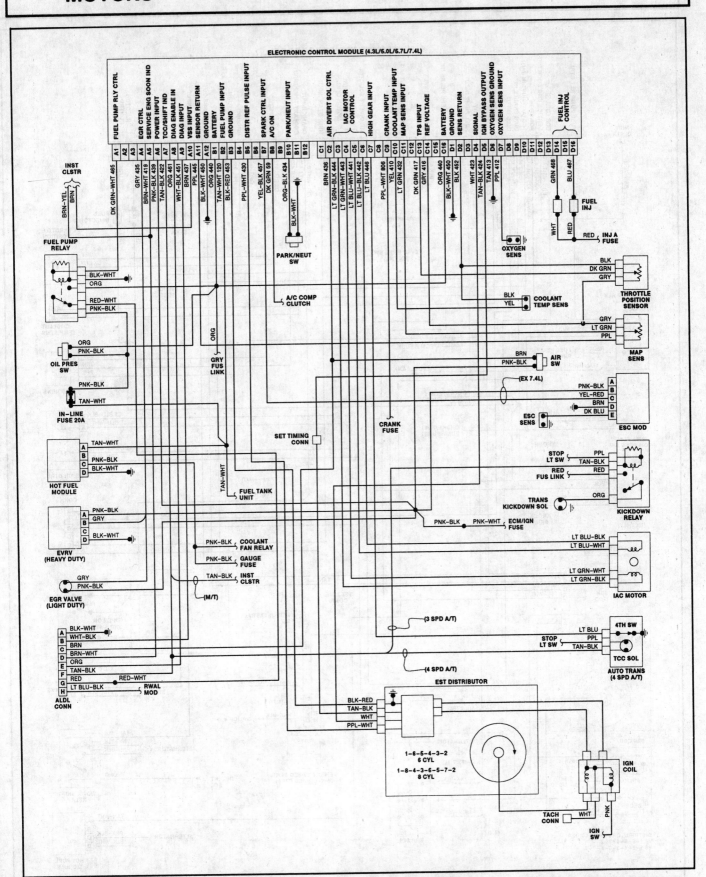

Fig. 36: 1988-89 "C" & "K" Series 4.3L, 5.0L, 5.7L & 7.4L Fuel Injection System Wiring Diagram

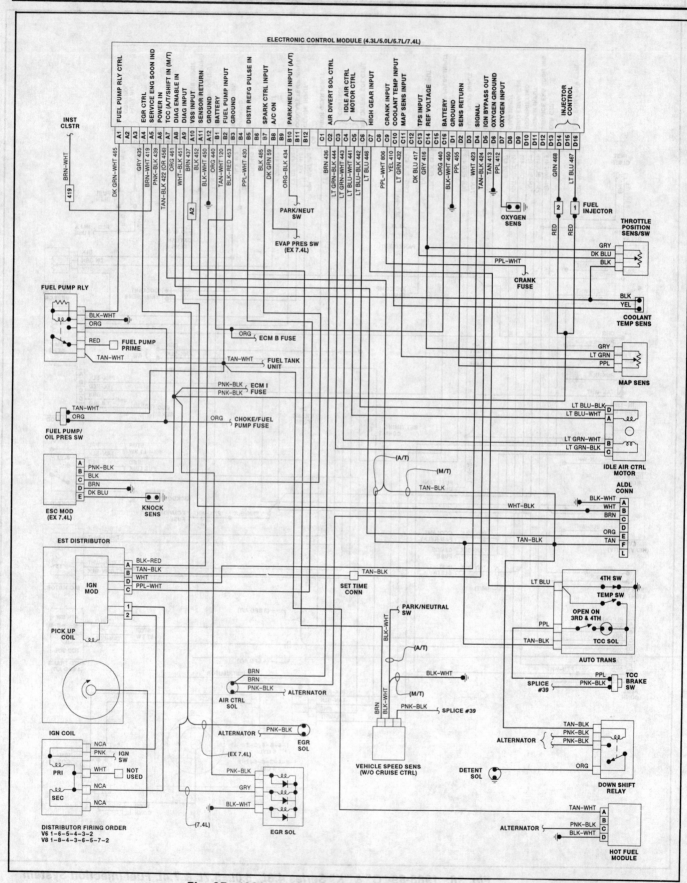

Fig. 37: 1989 "G" Series 4.3L, 5.0L, 5.7L & 7.4L Fuel Injection System Wiring Diagram

444

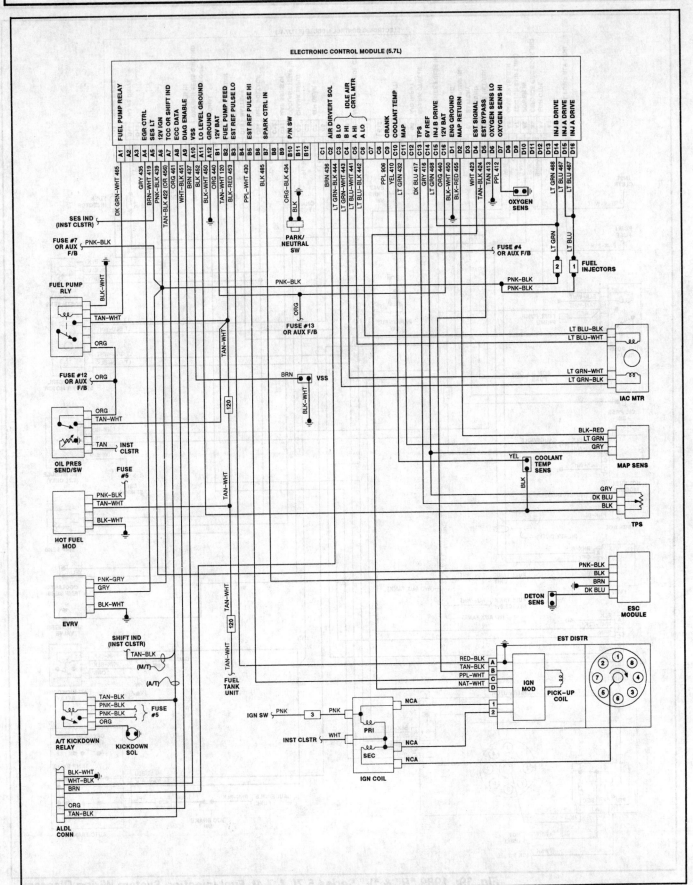

Fig. 38: 1988-89 "P" Series 5.7L Engine Fuel Injection System Wiring Diagram

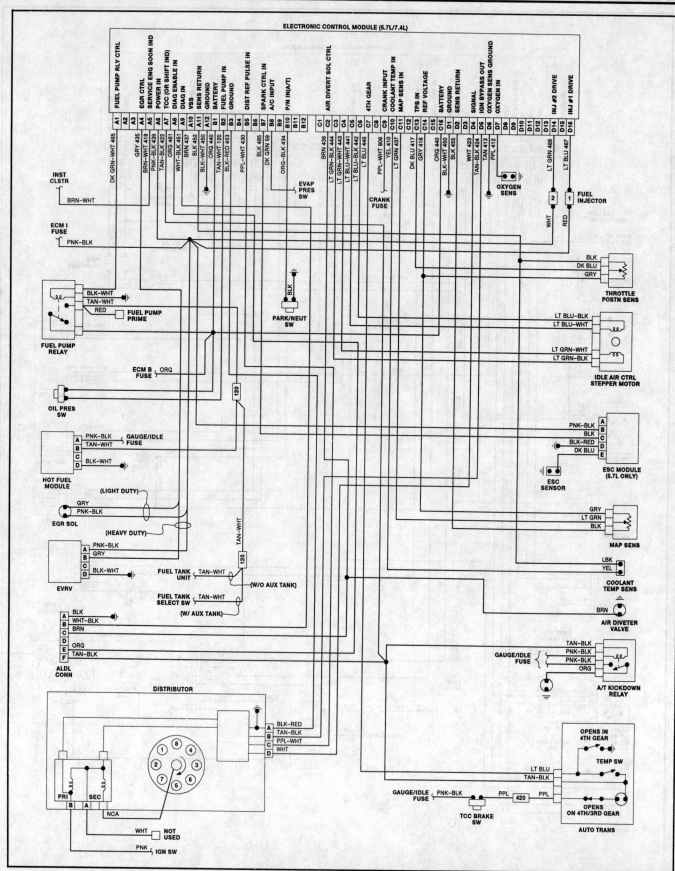

Fig. 39: 1989 "R" & "V" Series 5.7L & 7.4L Fuel Injection System Wiring Diagram

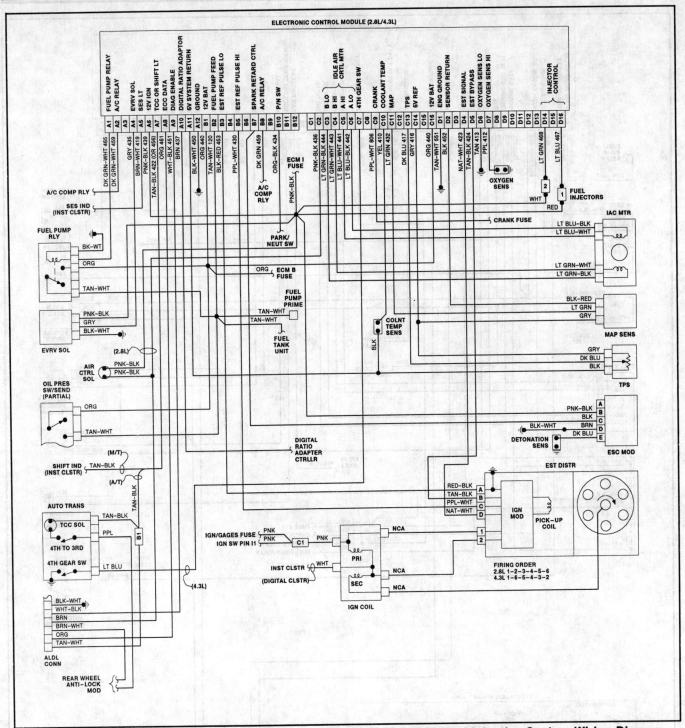

Fig. 40: 1989 "S" & "T" Series 2.8L & 4.3L Fuel Injection System Wiring Diagram

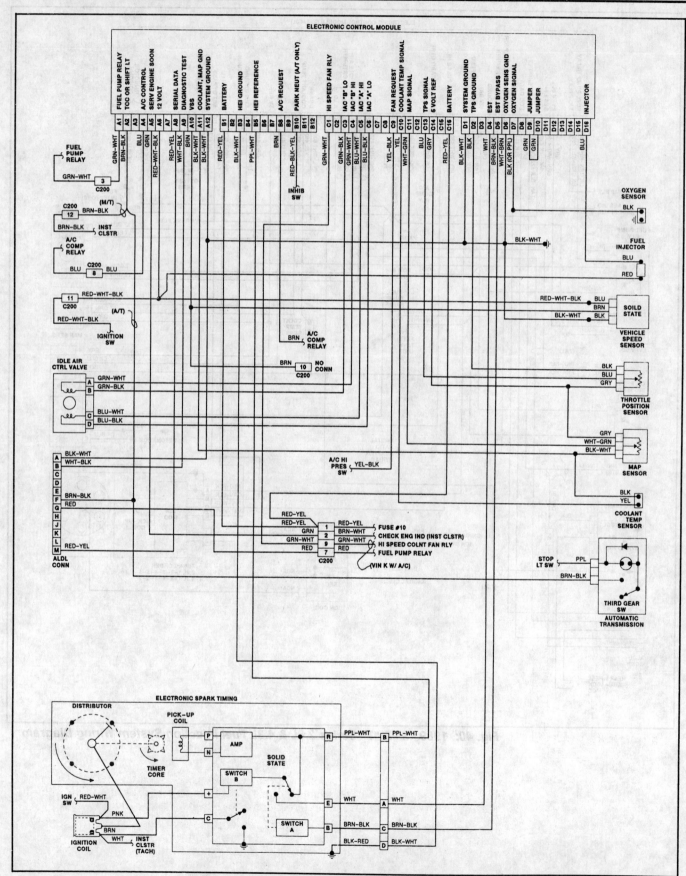

Fig. 41: LeMans Body Injection Wiring Diagram

Chevrolet: Nova

NOTE

- *For detailed applications, refer to FUEL INJECTION SYSTEMS APPLICATION CHARTS at the front of this publication.*

DESCRIPTION

The Bosch Airflow Controlled (AFC) fuel injection system is an electronically controlled system operated by incoming airflow. The AFC fuel injection system also contains a feedback system which measures oxygen content of exhaust gases and maintains the air/fuel ratio at about 14.7:1.

The AFC fuel injection system consists of 3 major sub-systems. These are the air intake system, the fuel system, and the Electronic Control Module (ECM). The fuel system and the control system are only connected at the fuel pump and the injectors.

Courtesy of General Motors Corp.

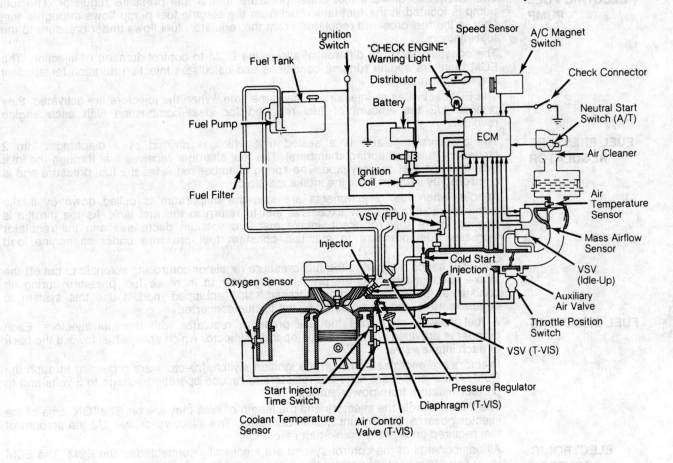

Fig. 1: Nova Fuel Injection System Diagram

AIR INTAKE SYSTEM

The air intake system cleans the incoming air, measures the amount and temperature and regulates the flow of air to the engine. The system consists of the air filter, airflow meter, throttle plate(s), and intake manifold.

FUEL SYSTEM

The fuel system provides a constant, pressurized supply of clean fuel to the cylinder intake port injectors. The fuel system consists of the fuel tank, fuel pump, fuel filter, fuel pressure regulator, fuel rail, cold start injector, and cylinder port injectors.

ELECTRONIC CONTROL SYSTEM

The electronic control system monitors several engine operating conditions. The computer processes the information and controls the engine for optimum performance and minimum emissions. Input signals are generated by the ignition switch, throttle position sensor, mass airflow sensor, air temperature sensor, thermo time switch, oxygen sensor, engine speed and engine temperature.

The ECM is the central component of the electronic control system. The ECM is a small computer, programmed to regulate fuel injection based on several sensor inputs.

The ECM generates control signals for the fuel pump relay, auxiliary air valve, cold start injector coil, and the cylinder port injector coils. These devices control cold idle, curb idle, air/fuel ratio and fuel supply.

AUXILIARY SYSTEMS

The AFC system has a cold start system to aid in cold engine starts. The cold start system consists of an auxiliary air valve, cold start injector, thermo time switch and a dashpot located on the throttle chamber to prevent stalling when the throttle is closed abruptly. The ECM can also be combined with computerized engine control systems for full control of fuel, ignition, emission control devices and vehicle accessories.

OPERATION

ELECTRIC FUEL PUMP

The fuel pump delivers fuel under pressure to the fuel pressure regulator. The fuel pump is located in the fuel tank. Fuel from the electric fuel pump flows through a fuel filter to the fuel pressure regulator. From the regulator, fuel flows under pressure to the fuel injectors.

The constant pressure differential allows the ECM to control duration of injection. The ECM monitors engine running conditions and calculates injection duration for efficient engine operation.

The ECM activates all injectors at the same time. When the injectors are activated, they provide 1/2 the amount of fuel required for ideal combustion with each engine revolution.

FUEL PRESSURE REGULATOR

The pressure regulator is a sealed unit which is divided by a diaphragm into 2 chambers (fuel and spring chambers). The fuel chamber receives fuel through the inlet side from the fuel delivery pipe. The spring chamber regulates the fuel pressure and is controlled by vacuum from the intake manifold.

At idle, when fuel requirements are low, the diaphragm is pulled down by intake manifold vacuum to allow excessive fuel to return to the fuel tank. As the throttle is opened (more fuel required), intake manifold vacuum decreases and the regulator increases fuel pressure to maintain constant fuel pressure under all engine load conditions.

The ECU sends a signal to the fuel pressure regulator controlling solenoid to cut off the intake vacuum signal to the pressure regulator to increase fuel pressure during air conditioner operation. Automatic transmission equipped models use this system to compensate for the load created by the torque converter.

FUEL INJECTORS

A fuel delivery pipe links the fuel pressure regulator with the fuel injectors. Each cylinder is provided with a solenoid-operated injector which sprays fuel toward the back of each intake valve.

Injectors are energized through the ignition switch. Injectors are grounded through the ECM. The injectors are linked to resistors to reduce operating voltage to 3 volts and to protect injectors from power surges.

The ECM controls the injectors and the length of time they are on. The "ON" time of the injector governs the amount of fuel delivered. The injectors deliver 1/2 the amount of fuel required once every crankshaft revolution.

ELECTRONIC CONTROL MODULE (ECM)

All components of the control system are electrically connected to the ECM. The ECM is a pre-programmed computer which receives and interprets data from various sensors. It then calculates the amount of fuel required by the engine to maintain efficiency with minimum exhaust emissions.

The oxygen sensor informs the ECM of oxygen content of exhaust gases and the ECM constantly adjusts the air/fuel ratio by controlling the injector "ON" time.

An automatic function of the ECM is to provide fuel enrichment whenever engine is cranked, regardless of engine temperature. The ECM is a sealed unit, and no service is required.

MASS AIRFLOW SENSOR

The mass airflow sensor is located between the air filter and the throttle plate(s). All engine air is drawn through the mass airflow sensor. The sensor consists of a tunnel containing a measuring and a dampening flap, offset 90 degrees on the same casting. The measuring flap swings in the air stream against the pressure of a spiral spring and is connected to a potentiometer. The potentiometer transmits an electrical signal, determined by the measuring flap angle, to inform the ECM of engine load. *See Fig. 2.*

In addition to monitoring airflow, the sensor also controls the fuel pump operation and engine idle. At idle, the measuring flap is almost closed due to spiral spring pressure.

Therefore an idle air by-pass receives air from the main airflow passage through a small hole, the size of which is controlled by the idle mixture screw. This adjustable air by-pass influences CO levels at low engine speeds.

Courtesy of General Motors Corp.

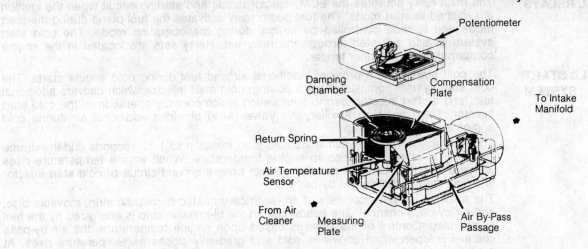

Fig. 2: Bosch AFC Airflow Meter

AIR TEMPERATURE SENSOR

The air temperature sensor is an integral component of the mass airflow sensor which converts temperature of incoming air into electrical signals. These signals are received by the ECM and processed to adjust the amount of fuel delivered by the injectors. The air temperature sensor is not serviceable.

THROTTLE BODY

The throttle body controls the intake airflow in response to accelerator pedal movement. Located between the airflow meter and the intake manifold, the throttle body shaft is connected to throttle position switch.

THROTTLE POSITION SWITCH

A contact-type throttle position switch is installed on the throttle chamber. It converts throttle position into electrical signals to inform ECM of throttle position, or angle. Signals are sent to ECM when throttle is fully open or at idle. *See Fig. 3.* The potentiometer prevents loss of power during sudden acceleration/deceleration by signaling the ECU of necessary fuel enrichment requirements. A dashpot mounted on the throttle body is used to prevent abrupt closing of throttle valve during sudden deceleration.

Courtesy of General Motors Corp.

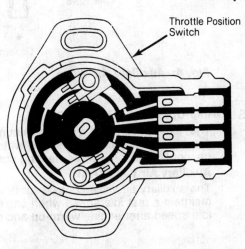

Fig. 3: Contact-Type Throttle Switch

COOLANT TEMPERATURE SENSOR

This sensor provides ECM with engine temperature information relating to warm-up enrichment operation. During warm-up period after a cold engine start, additional fuel is required to maintain engine performance. As engine temperature increases, the ECM decreases fuel enrichment until engine reaches normal operating temperature.

ELECTRICAL RELAYS

The main relay activates the ECM, injector circuit and starting circuit when the ignition is switched to start mode. The fuel pump relay activates the fuel pump during the start mode and is then controlled by airflow during the operating mode. The cold start system is also activated through the relay set. Relay sets are located in the engine compartment on left inner fender.

COLD START SYSTEM

The cold start system provides additional air and fuel during cold engine starts. The cold start system consists of the following: cold start injector which delivers additional fuel, and a start injector thermo time switch which controls operation of the cold start injector. The use of an Auxiliary Air Valve (AAV) provides additional air during cold engine starts and warm-up.

The start injector time switch limits cold start injection to 1-12 seconds under extreme cold engine starts in relation to engine temperature. When engine temperature rises above a specified point, bi-metallic contact breaks ground circuit of cold start injector and cold start enrichment is by-passed.

The auxiliary air valve consists of an electrically heated bi-metallic strip, movable disc, and air by-pass channel. The heater coil on the bi-metallic strip is energized by the fuel pump relay. Control of the valve is based upon engine temperature; the air by-pass channel is open when engine is cold and gradually closes as temperature rises. At predetermined temperatures, air by-pass channel is blocked and additional airflow stops. *See Fig. 4.*

Courtesy of General Motors Corp.

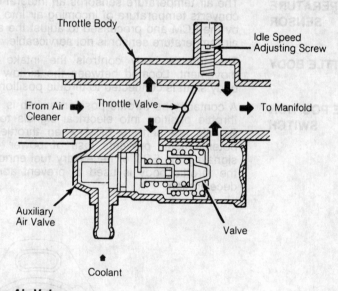

Fig. 4: Auxiliary Air Valve

SPECIAL FEATURES

Idle-Up Solenoid Valve (VSV)

The idle-up solenoid valve is mounted to the intake collector. The solenoid valve actutates to stabilize idle speed when engine is under heavy load such as headlights on, cooling fan on, power steering pump load or when battery voltage is below 12 volts.

Auxiliary Air Valve (AAV)

The auxiliary air valve controls air flow through the throttle body air by-pass passage to maintain a fast idle speed when engine is cold. It also provides by-pass air to maintain idle speed after engine warm-up and during air conditioner operation.

TROUBLE SHOOTING

PRELIMINARY CHECKS

Prior to diagnosing the fuel injection system, the following systems and components must be in good condition and operating properly:

- Fusible links, fuses and wiring.
- Battery, battery connections and body ground.
- All components of ignition system.
- Compression pressure.
- Fuel supply pressure and flow.
- Air induction system.
- Vacuum lines, fuel hoses and pipe connections.
- Ignition timing, idle speed.

TROUBLE SHOOTING

Engine Cranks, But Will Not Run

If engine turns over, but will not fire when cold after repeated cranking, test the following items: ECM diagnosis system for codes, vacuum leaks in air induction system, ignition spark and timing, fuel flow to injectors, fuel pump switch in Mass Airflow Sensor, spark plug gap, compression pressure, valve clearance (if necessary), auxiliary air valve, ECM circuit wiring and sensors.

Engine Stalls

If engine stalls repeatedly, test the following items: ECM diagnosis system for codes, vacuum leaks in air intake line, fuel flow to injectors, air cleaner element, idle speed, ignition timing, spark plug gap, compression pressure, valve clearance (if necessary), cold start injector, auxiliary air valve, fuel pressure and fuel injectors, ECM circuit wiring and sensors.

Engine Stalls Intermittently

If engine stalls often, test the following items: ECM diagnosis system for codes, Mass Airflow Sensor, wiring connectors and EFI main and circuit opening relays.

Rough Running Or Missing

If engine runs rough or misses, test the following items: ECM diagnosis system for codes, vacuum leaks in air intake line, air filter element, idle speed, ignition timing, spark plug gap, compression pressure, valve clearance (if necessary), intake air control valve, cold start injector, injectors, fuel pressure, ECM circuit wiring and sensors.

High Engine Idle Speed

If engine runs at a high idle speed, test the following items: accelerator linkage, auxiliary air valve, air conditioning and power steering idle-up circuit, ECM diagnosis system for codes, throttle position sensor, fuel pressure, cold start injector, fuel injectors, ECM circuit wiring and sensors.

Engine Backfires (Lean Mixture)

If engine backfires and has a lean mixture, test the following items: ECM diagnosis system for codes, vacuum leaks in air intake line, ignition timing, idle speed, cold start injector, fuel pressure, fuel injectors, ECM circuit wiring and sensors.

Engine Backfires & Misses (Rich Mixture)

If the engine backfires and misses, test the following items: ECM diagnosis system for codes, ignition timing, idle speed, cold start injector, fuel pressure, fuel injectors, spark plug gap, compression pressure, valve clearance (if necessary), ECM circuit wiring and sensors.

Engine Hesitates & Poor Acceleration

If engine hesitates and has poor acceleration at all speeds, test the following items: clutch and brake operation, vacuum leaks in air intake line, air filter element, ECM diagnosis system for codes, ignition spark, ignition timing, fuel pressure, fuel injectors, spark plug gap, compression pressure, valve clearance (if necessary), intake air control valve, ECM circuit wiring and sensors.

DIAGNOSIS & TESTING

FUEL PUMP

1) Turn ignition on, DO NOT start engine. Using a jumper wire, jump both terminals of fuel pump test connector. Test connector is located near wiper motor.

2) Check fuel pressure to delivery pipe. Fuel pressure noise should be heard. Remove jumper wire from test connector. Turn ignition off. If there is no fuel pressure to delivery pipe, check fusible link, EFI fuse, EFI main and circuit opening relays, fuel pump and wiring connections.

FUEL PRESSURE

1) Ensure battery voltage is above 12 volts. Disconnect negative battery cable and cold start injector wiring connector. Place container or towel under cold start injector pipe.

2) Slowly loosen cold start injector union bolt and remove gaskets and bolt from delivery pipe. Drain fuel from delivery pipe. Install Fuel Pressure Gauge (J-34730-1) and Adapter (J-37144) to delivery pipe.

3) Wipe off any splattered fuel. Reinstall negative battery cable. Using a jumper wire, jump both terminals of fuel pump test connector. Test connector is located near wiper motor. Turn ignition on and measure fuel pressure.

4) Fuel pressure should be 38-44 psi (2.7-3.1 kg/cm²). If fuel pressure is higher than specification, replace fuel pressure regulator. If fuel pressure is lower than specification, check fuel hose and connections, fuel pump, fuel filter and fuel pressure regulator.

5) Remove jumper wire from fuel pump test connnector. Start engine and disconnect vacuum sensing hose from fuel pressure regulator and pinch hose closed. Measure fuel pressure at idle speed. Fuel pressure should be 38-44 (2.7-3.1 kg/cm²).

6) Reinstall vacuum sensing hose to fuel pressure regulator. Measure fuel pressure at idle speed after 90 seconds. Fuel pressure should be 30-33 psi (2.1-2.3 kg/cm²).

7) If there is no fuel pressure, check vacuum sensing hose and fuel pressure regulator. Turn ignition off, check that fuel pressure remains above 21 psi (1.5 kg/cm²) for 5 minutes.

8) If fuel pressure is not within specifications, check fuel pump, pressure regulator and fuel injectors. Disconnect negative battery cable and carefully remove fuel pressure gauge and adapter to prevent fuel from splashing.

9) Using new gaskets, reconnect cold start injector hose to delivery pipe. Connect cold start injector wiring connector. Start engine and check for fuel leaks.

COLD START INJECTOR

1) Remove intake air hose from throttle body. Install Cold Start Injector Test Harness (J-36674) to injector. Rotate throttle lever so that cold start injector nozzle can be seen in throttle body.

2) Connect test lead to battery for 2 seconds. Fuel should be seen spraying from injector nozzle. If no fuel is sprayed from nozzle, replace cold start injector.

FUEL INJECTOR BALANCE TEST

NOTE
- *If fuel injectors are suspected of being dirty, injectors should be cleaned and fuel pressure should be checked before this test is performed.*

1) If engine is at normal operating temperature, allow 10 minutes to cool down. Turn ignition off. Connect Fuel Pressure Gauge (J-34730-1), Adapter (J-37144) and Injector Tester (J-34730-B). Turn ignition on.

2) Bleed off air in gauge and repeat bleed off procedure until all air is bled from gauge. Turn ignition off for 10 seconds. Turn ignition on and note gauge pressure. Fuel pressure must hold steady.

3) If fuel pressure does not hold steady, check all electrical connections. Depress injector tester button and note fuel pressure the instant the gauge needle stops.

4) Repeat injector pressure test and note fuel pressure drop on each injector. Retest injectors that have more or less than 147 psi (10.3 kg/cm²) in pressure as an average. *See Fig. 5.*

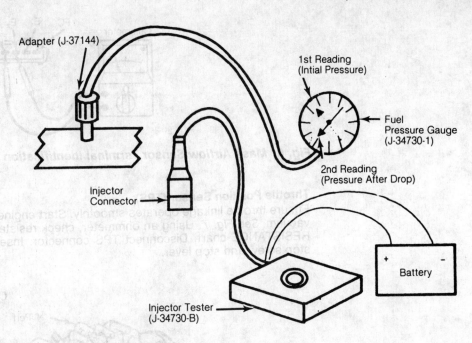

Fig. 5: Injector Balance Test

Mass Airflow Sensor (MAF)

Disconnect MAF sensor connector. Using an ohmmeter, measure resistance between each connector terminal. See MAF SENSOR RESISTANCE chart. *See Fig. 6*. If resistance is not within specification, replace MAF sensor. Using an ohmmeter, also measure resistance between terminals by moving measuring plate. See MAF SENSOR PLATE RESISTANCE chart.

MAF SENSOR RESISTANCE

Between Terminals	Resistance (Ohms)	Temperature °F (°C)
VS-E²	20-3000	[1]
VC-E²	1100-300	[1]
VB-E²	200-400	[1]
THA-E²	10,000-20,000	-4 (20)
THA-E²	4000-7000	32 (0)
THA-E²	2000-3000	68 (20)
THA-E²	900-1300	104 (40)
THA-E²	400-700	204 (60)
FC-E¹	Infinity	[1]

[1] – At Any Temperature

MAF SENSOR PLATE RESISTANCE

Between Terminals	Resistance (Ohms)	Measuring Plate Opening
FC-E¹	Infinity	Fully Closed [1]
FC-E¹	Zero	
VS-E²	20-400	Fully Closed
VS-E²	20-3000	[2]

[1] – Except Closed Position.
[2] – Fully Closed To Fully Open Position.

Courtesy of General Motors Corp.

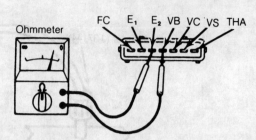

Fig. 6: Mass Airflow Sensor Terminal Identification

Throttle Position Sensor (TPS)

Ensure throttle linkage operates smoothly. Start engine, check that vacuum port "N" has vacuum. *See Fig. 7.* Using an ohmmeter, check resistance between terminals. See TPS RESISTANCE chart. Disconnect TPS connector. Insert feeler gauge between throttle stop screw and stop lever.

Courtesy of General Motors Corp.

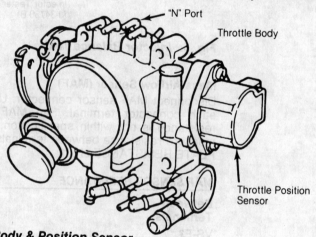

Fig. 7: Throttle Body & Position Sensor

TPS RESISTANCE

Clearance	Terminals	Ohms
0"	VTA-E^2	200-800
.0138"	IDL-E^2	[1]
.0232"	IDL-E^2	Infinity
[2]	VTA-E^2	3300-10,000
[3]	VCC-E^2	3000-7000

[1] – Less than 2,300 Ohms.
[2] – Throttle valve in fully opened position.
[3] – Throttle valve at any position.

Auxiliary Air Valve

Check engine RPM by fully screwing in idle speed adjusting screw. With engine below 176°F (80°C) and idle speed adjusting screw in, engine RPM should drop. When engine is at normal operating temperature and idle speed adjusting screw in, engine RPM should drop below idle speed stop.

EFI Main Relay

Turn ignition on, noise should be heard from relay. If no noise is heard from relay, check for continuity between terminals "A" and "C". There should be no continuity between terminals "B" and "D" and "C" and "D". If continuity check is not as specified, replace relay.

To check operation of relay, apply 12 volts across terminals "A" and "C". Check for continuity between terminals "B" and "D". If continuity check is not as specified, replace relay. *See Fig. 8.*

Courtesy of General Motors Corp.

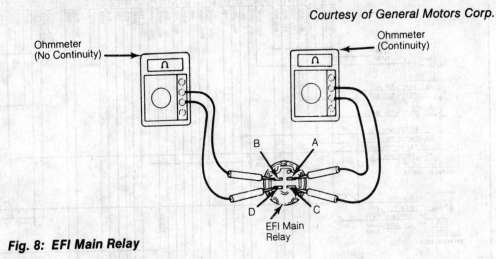

Fig. 8: EFI Main Relay

Circuit Opening Relay

Using an ohmmeter, check that there is continuity between terminals "STA" and "E". Check for continuity between terminals "B" and "FC" and no continuity between terminals "B" and "FC" and no continuity between terminals "B" and "FP". *See Fig. 9.*

Courtesy of General Motors Corp.

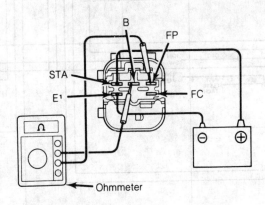

Fig. 9: Circuit Opening Relay

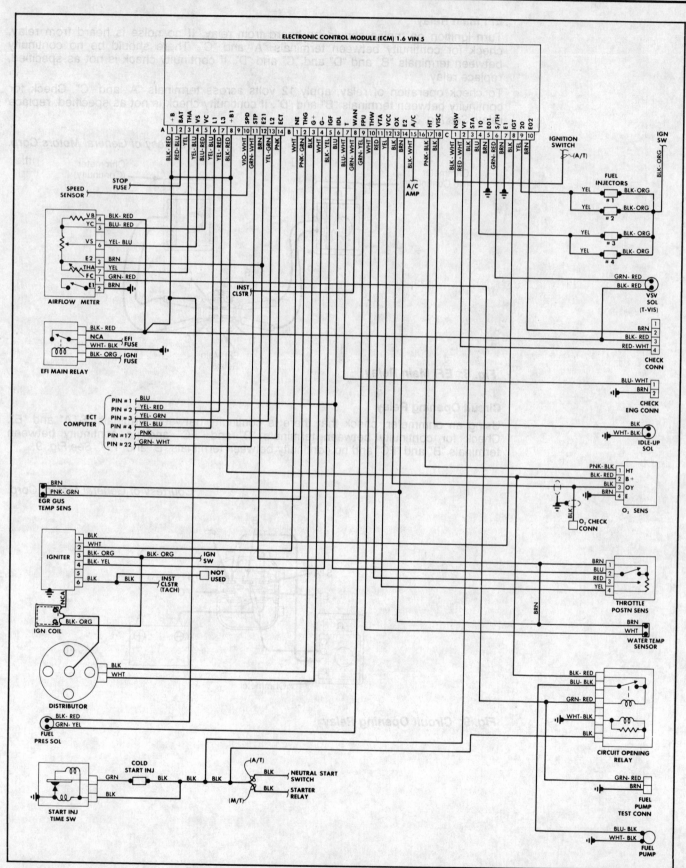

Fig. 10: Nova Fuel Injection Wiring Diagram

General Motors: Spectrum Turbo (EFI)

DESCRIPTION

The Electronic Control Module (ECM) controls the fuel injection system, ignition system and turbocharger control system. This lowers emissions while maintaining fuel economy and driveability.

The ECM governs fuel injection operation by reading data from the following sensors: manifold absolute pressure sensor, oxygen sensor, detonation sensor, vehicle speed sensor, coolant temperature sensor, manifold air temperature sensor, and throttle position sensor. The fuel delivery system consists of fuel tank, electric fuel pump, pressure regulator, and 4 fuel injectors.

OPERATION

ELECTRONIC CONTROL MODULE (ECM)

The ECM is located under the glove box. The ECM has self-diagnosis capability that allows for troubleshooting the system through a check engine light. The check engine light on the dash will illuminate to inform driver of a system malfunction. The ECM stores a maximum of 3 system malfunctions, and will display problem(s) via a 2-digit code displayed by the check engine light.

IDLE AIR CONTROL VALVE

The air control valve is an electrically driven motor controlled by the ECM to maintain idle speed. The ECM will command the air control valve to move its pintle in or out to by-pass air around the throttle valve.

FUEL DELIVERY SYSTEM

Dropping Resistors

Dropping resistors reduce current flow to fuel injectors. They are in-line with fuel injector wiring.

Fuel Injectors

Fuel is metered into cylinders by electrically controlled solenoid valves in injectors. Injectors are energized through the computer.

Fuel Pressure Regulator

Pressure regulator governs flow of fuel to injectors. Pressure varies depending on different vehicle speed and load conditions.

Fuel Pump

Fuel pump is located inside fuel tank. Pump is integral in design and must be replaced as assembly.

INPUT SENSORS

Coolant Temperature Sensor (CTS)

The CTS is threaded into the intake manifold coolant passage and informs ECM of engine coolant temperature.

Detonation Sensor

The detonation sensor is attached to the side of engine block. If this sensor hears any detonation, it retards ignition timing through a controller incorporated in the ignition control module.

Engine Speed Sensor

This signal is taken from the ignition distributor and informs ECM of engine speed.

Manifold Absolute Pressure Sensor (MAP)

The MAP sensor monitors intake manifold vacuum and informs ECM of engine load.

Manifold Air Temperature Sensor (MAT)

The MAT sensor is located in the lower part of the intake manifold and informs ECM of intake air temperature.

Vehicle Speed Sensor (VSS)

The VSS is located in the speedometer and informs ECM of vehicle road speed.

Oxygen (O_2) Sensor

The O_2 sensor is threaded into the exhaust manifold and informs ECM of amount of oxygen in exhaust gases.

Throttle Position Sensor (TPS)

The TPS is mounted on the throttle body throttle shaft and informs ECM of changes in throttle position.

VACUUM SWITCHING VALVE (VSV) — Controlled by ECM, the VSV's are used to regulate fuel pressure, canister purge, turbo operation and air induction control.

TROUBLE SHOOTING

PRELIMINARY CHECKS — Ensure all basic systems, engine mechanical, charging system, ignition system, etc. are functioning properly before attempting to diagnosis computer control system.

HARD OR NO START (ENGINE CRANKS) — Low or no fuel pressure. No electrical power feed to ECM. Defective fuel injector or harness.

POOR ENGINE IDLE — Defective TPS. Improperly adjusted idle speed screw. Defective IAC valve. Leaking or defective fuel injector. Vacuum leaks.

IDLE SPEED TOO HIGH (ENGINE WARM) — Fuel pressure too high. Leaking injector(s). IAC valve stuck open. Defective ECM.

POOR PERFORMANCE (NO POWER) — Fuel pressure too low. Defective injector(s). Defective turbocharger. Defective detonation sensor. Defective ECM.

ENGINE KNOCKING — Defective detonation sensor, detonation sensor wiring or ECM.

ENGINE RUNNING TOO RICH OR TOO LEAN — Defective injector(s). Fuel pressure too high or too low. Defective O_2 sensor signal. Defective ECM.

DIAGNOSIS & TESTING

SELF-DIAGNOSTIC SYSTEM CHECK — 1) Locate ALDL diagnostic connector near ECM. Using a jumper wire, jump between terminals "A" and "C". See Fig. 1. Turn ignition on, but do not start engine. CHECK ENGINE light will begin to flash diagnostic code 12 indicating the system is working.

2) Code 12 will flash 3 times, then any additional code will flash 3 times and continue to cycle in this sequence up to a maximum of 3 codes. A code 12 consists of one flash, a short pause and 2 flashes. See Fig. 2. To clear codes, remove the 10A ECM fuse in the fuse box for 10 seconds.

Courtesy of General Motors Corp.

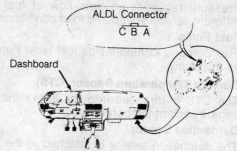

Fig. 1: Location of Diagnostic Leads

Courtesy of General Motors Corp.

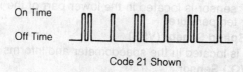

Fig. 2: Reading Diagnostic Codes

DIAGNOSTIC CODES

Code 12 – No distributor reference pulse to ECM.
Code 13 – O₂ sensor circuit is open (lean fuel condition).
Code 14 – CTS voltage signal low.
Code 15 – CTS voltage signal high.
Code 21 – TPS voltage signal high.
Code 22 – TPS voltage signal low.
Code 23 – MAT sensor voltage signal high.
Code 24 – VSS malfunction.
Code 25 – MAT sensor voltage signal low.
Code 31 – Turbocharger wastegate control malfunction.
Code 32 – EGR system control malfunction.
Code 33 – MAP sensor voltage signal high.
Code 34 – MAP sensor voltage signal low.
Code 42 – Electronic Spark Timing (EST) malfunction.
Code 43 – Detonation sensor malfunction.
Code 44 – Lean exhaust indication.
Code 45 – Rich exhaust indication.
Code 51 – MEM-CAL failure.

COMPONENT TESTING BY TROUBLE CODE

Code 13, 44 & 45 (O₂) – 1) Jumper ALDL. With engine at normal operating temperature, above 158°F (70°C), run engine at 2000 RPM for 2 minutes. If check engine light is flashing closed loop (light flashes once per second), go to step **4)**.

2) If check engine light is flashing open loop (light flashes 25 times per second), turn engine off. Disconnect O₂ sensor harness and ground Yellow wire. Start engine and note check engine light. Light should flash open loop for about 2 minutes then go out. If light goes out in 2 minutes, replace O₂ sensor.

3) If light stays on, check continuity in O₂ sensor harness. Check that continuity exists between ECM harness connector B23 and Yellow connector wire at O₂ sensor, and between B22 and ground. If harness checks bad, repair or replace. If harness checks good, replace ECM.

4) If check engine light is flashing closed loop, remove ALDL jumper wire and clear code. Run engine above 2000 RPM for one minute. Note check engine light.

5) If check engine light is on, turn engine off. Jumper ALDL and turn ignition on. Note code and refer to applicable code diagnosis. If check engine light is off, problem may be intermittent.

Code 14 & 15 (CTS) – 1) Remove CTS connector and check voltage across harness terminals. Turn ignition on. If reading is below 4 volts, go to step **2)**. If reading is above 4 volts, connect an ohmmeter across CTS terminals. Read resistance on ohmmeter. See COOLANT TEMPERATURE SENSOR RESISTANCE table. If resistance is not equal to table reading, replace CTS.

COOLANT TEMPERATURE SENSOR RESISTANCE

Temperature	Ohms
-22°F (-30°C)	26,000
-4°F (-20°C)	15,000
32°F (0°C)	5600
68°F (20°C)	2500
104°F (40°C)	1200
140°F (60°C)	600
176°F (80°C)	320
212°F (100°C)	180

2) Disconnect B connector at ECM. Using an ohmmeter, check that continuity exists in CTS harness B10 and Green/Black wire at CTS, and B18 and Blue/Green wire at CTS.

3) Check that no continuity exists between B10 and B18, and between B10 and ground or B18 and ground.

4) Using a voltmeter, check voltage in CTS harness Green/Black wire and ground. Reading should be 4-6 volts. If harness checks bad, repair or replace. If harness checks good, replace ECM.

Code 21 (TPS) – 1) Clear code. Start engine and idle for one minute with A/C off. If check engine light is off, problem is intermittent.

2) If check engine light is on, jumper ALDL and note codes. If code 22 is present, go to appropriate test. If code 21 is present, turn engine off. Remove jumper from ALDL and clear codes.

3) Disconnect TPS sensor. Start engine and idle with A/C off for one minute or until check engine light comes on. Stop engine. Jumper ALDL and turn ignition on. Note code. If code 21 is present, check that continuity exists between TPS harness connector B13 to Red wire connector at TPS.

4) With ECM connector installed and ignition on, check that no voltage is present between Red wire at TPS connector and ground. If harness tests bad, repair or replace.

5) With ECM connector installed and ignition on, probe TPS Green wire connector with a test light to 12 volts. If light is on, TPS or harness connection is faulty. If test light is off, harness from Green wire connector to ECM connector A11 is faulty or bad ECM.

Code 22 (TPS) – 1) Clear codes. Start engine and idle for one minute or until check engine light comes on. If light will not come on, problem is intermittent.

2) If check engine light comes on, turn off engine. Jumper ALDL and turn ignition on. Note codes. If code 22 is present, turn off engine. Clear code and remove ALDL jumper.

3) Disconnect TPS connector and jumper between Red and Light Green/Red wire connectors. Start engine and idle for one minute or until light comes on. Turn engine off. Jumper ALDL and turn ignition on. Note code.

4) If code 21 is present, remove jumper wire and connect harness. Adjust TPS to proper specification. If TPS cannot be adjusted, replace TPS.

5) If code 22 is present, remove jumper wire. Using a digital voltmeter, check voltage between Green and Light Green/Red wire terminals at TPS connector. With ignition on, reading should be 4-6 volts.

6) If reading is above 6 volts, check TPS harness Red wire connector to ECM connector B13 for short or ground, faulty connection or bad ECM.

7) If reading is below 4 volts, repeat step **6)** for Light Green/Red harness connector and B14.

Code 23 & 25 (MAT) – 1) Clear codes. Start engine and run for one minute or until light comes on. Turn engine off and turn ignition on. Note codes.

2) If code 23 is present, disconnect MAT sensor harness. Turn ignition on and check voltage between MAT sensor harness terminals. If voltage is below 4 volts, check voltage between MAT Black/Yellow harness connector and ground.

3) If voltage check is below 4 volts, check for open in harness, faulty connection or ECM. If voltage in Black/Yellow wire is over 4 volts, check for faulty circuit in Red/Black wire to ECM harness connector A11.

4) If reading across harness terminals in step **2)** was over 4 volts, check resistance across MAT sensor terminals. See MAT SENSOR RESISTANCE table. If resistance values are not approximately as indicated, replace MAT sensor.

MAT SENSOR RESISTANCE

Temperature	Ohms
-4°F (-20°C)	28,677
32°F (0°C)	9423
68°F (20°C)	3515
104°F (40°C)	1459
140°F (60°C)	667
176°F (80°C)	332
212°F (100°C)	177

Code 24 (VSS) – 1) Verify speedometer operation. Disconnect speedometer cable at transaxle. Turn ignition on and check voltage at ECM harness connector A10 and ground while turning speedometer cable slowly.

2) If voltage varies from 1-6 volts, problem is intermittent. If voltage is 8-12 volts and steady, check VSS harness for open or short. If harness is good, replace VSS.

3) If voltage is below one volt, disconnect harness at VSS and check voltage. With ignition on, if voltage is above 10 volts, replace VSS.

4) If voltage is below 10 volts, check VSS harness to ECM connector A10 for shorting to ground. If harness tests good, repair harness connector or replace ECM.

Code 31 (Turbocharger Wastegate Control) – 1) Clear codes. Disconnect hoses at turbocharger Vacuum Solenoid Valve (VSV) and air duct. Turn ignition on and blow air through VSV. If air passes through valve, check that VSV harness White wire and ECM harness connector B2 is not shorted to ground. If harness tests good, replace VSV.

2) If air will not pass through VSV, jumper ALDL and turn ignition on. Blow air through VSV. If air will not pass through VSV, check circuit B2 for open or faulty ECM.

3) If air will pass through VSV in step **2)**, check turbocharger wastegate valve diaphragm. If diaphragm is defective, replace turbocharger. If diaphragm tests good, check for clogged or defective hoses.

Code 32 (EGR System Control) – 1) With engine off and ignition on, jumper ALDL and check for additional codes. Repair other codes first. If only code 32 is present, continue with test.

2) With engine at normal operating temperature, raise RPM to 3000 and check for movement of EGR valve diaphragm. If EGR is operating, go to step **5)**.

3) If EGR is inoperative, shut off engine and disconnect vacuum hose from Back Pressure Transducer (BPT) to EGR valve. Connect vacuum gauge to BPT. Start engine and run at 3000 RPM. If vacuum is present, replace EGR valve.

4) If vacuum is not present in step **3)**, check VSV by applying battery voltage to Blue/Yellow wire terminal while grounding White/Green wire terminal of VSV. If VSV operates when voltage is applied, VSV is good. Start engine and check vacuum signal to VSV. Correct if improper. Check for faulty BPT or bad hoses.

5) If EGR operation was proper in step **2)**, stop engine and remove EGR Gas Temperature Sensor (GTS) from EGR valve. Heat the sensing portion of the GTS and check resistance. See EGR GTS SENSOR RESISTANCE table. If GTS values are improper, replace GTS.

6) If GTS check good, check for clogged exhaust passage. If exhaust passage is clear, replace ECM.

EGR GTS SENSOR RESISTANCE

Temperature	Ohms
32°F (0°C)	8,200,000
122°F (50°C)	635,000
212°F (100°C)	85,300
392°F (200°C)	5000

Code 33 (MAP) – 1) Clear codes. Start engine and run for one minute or until check engine light comes on. Turn engine off. Jumper ALDL and turn ignition on.

2) If code 33 is present, clear code. Disconnect MAP sensor connector. Remove ALDL jumper. Start engine and run engine for one minute. If check engine light comes on, jumper ALDL and note code.

3) If code 33 is present, check for short circuit in Green wire MAP sensor connector and B11. Repair harness or replace ECM.

4) If code 34 is present in step **2)**, check for plugged or leaking vacuum hoses to MAP sensor. If vacuum hoses are good, check for open in circuit B18 to MAP sensor Red wire harness terminal. If harness good, replace ECM.

Code 34 (MAP) – 1) Clear codes. Disconnect MAP sensor connector and jumper connector terminal Green to Light Green/Yellow. With ALDL jumper removed, start engine and run for one minute. If check engine light comes on, stop engine. Jumper ALDL and turn ignition on.

2) If code 33 is present, replace MAT sensor. If code 34 is present, remove jumper wire from connector. With ignition on, check voltage between Red wire and Light Green/Yellow wire at connector.

3) If reading is 4-6 volts, check for open in Green wire at connector to B11 at ECM harness connector. If harness tests good, repair harness connector or replace ECM.

4) If reading in step **2)** was below 4 volts, check for open or short in Light Green/Yellow wire at connector and B14 at ECM harness connector. If harness tests good, repair harness connector or replace ECM.

Code 42 (Electronic Spark Timing) – 1) Clear code. Turn ignition off. Disconnect ECM harness connectors. Turn ignition on. Using an ohmmeter, probe harness connector terminal B20 to ground. Reading should be less than 500 ohms.

2) If resistance test is not proper, check circuit B20 for open. If circuit B20 has continuity, check connectors for poor connections. If connections are good, replace ECM.

3) If resistance reading in step **1)** was proper, probe ECM harness connector terminal B21 with a test light connected to 12 volts. If test light comes on, go to step **6)**

4) If test light does not light, connect an ohmmeter between ECM harness connector B20 and ground. Probe harness connector B21 with test light connected to 12 volts.

Ohmmeter reading should change from less than 500 ohms to over 5000 ohms when B21 is probed.

5) If resistance does not change, disconnect distributor 4-way connector. Ohmmeter should indicate an open circuit (high reading). If open circuit is not indicated, repair circuit B20. If open circuit is indicated, check circuit B21 for open, poor connection at distributor or bad ECM.

6) If test light came on in step 3), disconnect distributor 4-way connector. If test light is still on, circuit B21 is shorted to ground.

7) If test light goes off in step 6), check for short to ground between 4-way connector and Green/White wire ignition module terminal. If no short is present, replace ignition module.

Code 43 (EST Detonation Control) – 1) Check connection between detonation sensor and harness. If connection good, disconnect wire at sensor and turn ignition on. Check voltage between sensor harness connector and ground.

2) If reading is less than 4 volts or over 6 volts, check circuit A23 for shorting to ground or shorting to voltage.

3) If reading in step 1) is 4-6 volts, check resistance across detonation sensor terminal to ground. Reading should be 2550-3450 ohms. If reading is bad, replace detonation sensor. If reading is good, check for weak connections at terminals.

Code 51 (MEM-CAL Failure) – Check that all pins are fully inserted into socket. If good, replace MEM-CAL. Clear code. If code 51 reappears, replace ECM.

PRELIMINARY INSPECTION

Handling Precautions

1) EFI system components and connectors are very sensitive to shock, distortion and water. Turn ignition switch to "OFF" position before unplugging any electrical connectors.

2) All EFI cables must be placed at least 4" (100 mm) away from high tension cables. Do not operate fuel pump without fuel. The fuel is used to lubricate pump.

Fuel System

1) Ensure all fuel hoses are tight and in good order. Check fuel lines and hoses for corrosion, cracks, or other signs of damage. Ensure that hoses are not collapsed, and that fuel filter is not clogged.

2) Make sure that there is no water in fuel tank or in any other part of system.

FUEL PRESSURE

CAUTION
- *Fuel line is under pressure. When installing fuel pressure gauge, caution must be used when disconnecting fuel lines.*

1) Install Fuel Pressure Gauge (J-33945) between pressure regulator and fuel distributor pipe. Unplug VSV at connector. Start engine and measure fuel pressure under the following conditions:

2) Disconnect pressure regulator vacuum hose. Plug intake manifold side of hose. Fuel pump pressure should be about 35.6 psi (2.5 kg/cm²). Reconnect pressure regulator vacuum hose.

3) Run engine at 900 RPM. Fuel pump pressure should be about 28.4 psi (2.0 kg/cm²). Remove fuel pressure gauge. Reconnect fuel line.

Low Fuel Pressure

1) Apply battery voltage directly to fuel pump connector with jumper wires. If pressure is normal, check pump circuit and relay. If fuel pressure remains unchanged, check for restrictions on intake side of fuel pump.

2) Restrictions in intake hose may cause abnormal vibrations in hose. Remove restrictions as necessary. If no restrictions are found, replace defective fuel pump and/or pressure regulator.

Loss of Pressure Immediately After Pump Stops

Replace defective fuel pump and/or pressure regulator. Check fuel injectors for internal leaks and replace as necessary.

High Fuel Pressure

Check for restrictions in fuel return lines. Clean or replace as necessary. Replace defective fuel pump and/or pressure regulator.

FUEL INJECTORS

1) Unplug fuel injector harness at injector and measure resistance across terminals. All injectors must have 2-3 ohms of resistance. If resistance is not to specification, replace injector.

2) With engine running, check fuel injector operating noise with stethoscope. Normal operation of injectors is indicated when a regular "click" is heard which varies with engine speed. If "click" is not heard, replace injector.

3) Remove common chamber assembly. Remove all injectors with fuel hoses connected. Check for fuel leakage by operating fuel pump with battery voltage applied directly to fuel pump relay terminal (Black/Green wire). Leakage must be less than 2 drops per minute. If leakage exceeds limit, replace injector(s).

DROPPING RESISTOR

1) Disconnect dropping resistor wiring at connector. Measure resistance between center terminal of resistor connector and other terminals. *See Fig. 3.*

2) Standard resistance at all terminals should be 5-7 ohms. If any of measurements deviates from standard value, dropping resistor assembly must be replaced.

Courtesy of General Motors Corp.

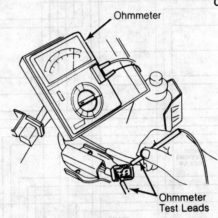

Ohmmeter

Ohmmeter Test Leads

Fig. 3: Checking Dropping Resistor

IDLE AIR CONTROL VALVE

1) Warm engine to normal operating temperature. Record idle RPM in Neutral. Turn ignition off. Disconnect IAC valve. Start engine. Record idle RPM in Neutral. If RPM increases, go to next step. If RPM does NOT increase, go to step **3)**.

2) Turn ignition off. Reconnect IAC valve. Start engine. Record engine RPM. If RPM returned to specification recorded in step **1)**, IAC valve is functioning properly. If RPM did not return to specification recorded in step **1)**, go to next step.

3) Turn ignition on with engine off. Ground diagnostic test terminal. Connect a test light between each IAC valve harness terminal and ground. If test light illuminates on all terminals, repair faulty IAC valve terminal connectors or replace faulty IAC valve. If test light does not illuminate on one or more terminals, check circuit that did not light for an open or short.

4) If circuits that did not light are okay, check resistance across IAC terminals. Connect an ohmmeter between IAC terminals "A" and "B", then "C" and "D". Ohmmeter should read more than 20 ohms.

5) If ohmmeter reads more than 20 ohms, repair faulty ECM connector or replace faulty ECM. If ohmmeter reads less than 20 ohms, replace faulty IAC valve and replace ECM.

ADJUSTMENTS

THROTTLE POSITION SENSOR (TPS)

1) Ensure that throttle cable moves freely. Check that valve returns and contacts throttle valve stopper screw. Throttle valve stopper screw is factory set and should NOT be disturbed.

2) Ensure throttle valve is completely closed. Loosen TPS screws and cover on TPS connector. DO NOT unplug connector. Insert digital voltmeter positive probe in Red wire. Connect negative lead to ground.

3) Turn ignition on and measure TPS voltage. Reading should be .31-.39 volts. If incorrect, loosen screw fixing TPS to throttle valve and adjust TPS until reading is correct.

IDLE SPEED

1) Start and warm engine to normal operating temperature. Ensure front wheels are in straight-ahead position, transaxle is in Neutral or "P" (auto. trans.), and all accessories and cooling fan are off.

2) Connect tachometer and timing light to engine. Ensure "CHECK ENGINE" light is not on and ground test terminal. Increase engine speed to 2000 RPM to reset Idle Air Control (IAC) valve. Set idle to 950 RPM. Unground test terminal, clear trouble codes, and remove test equipment.

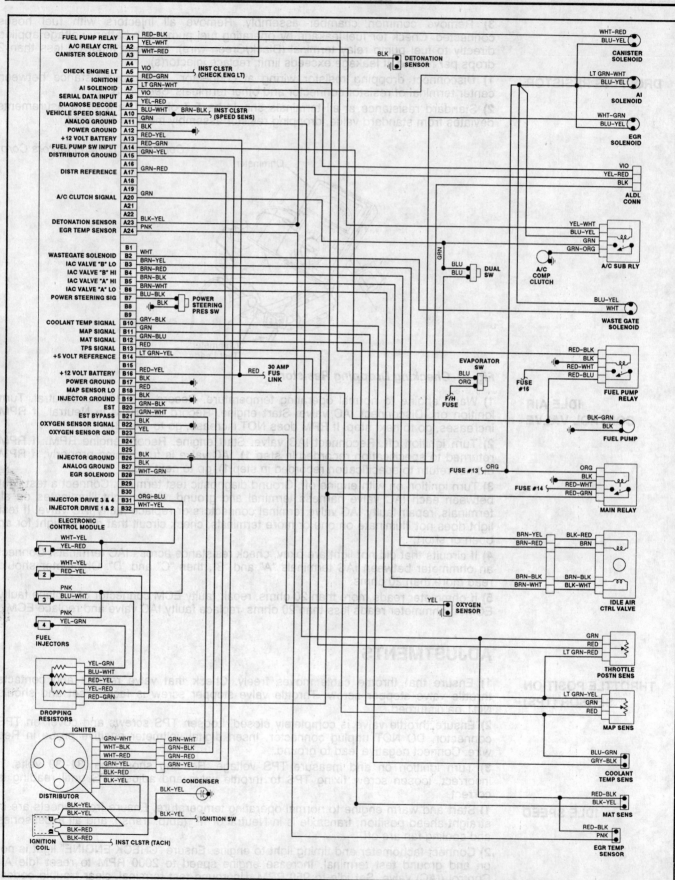

Fig. 4: Spectrum Turbo Electronic Fuel Injection Wiring Diagram

Buick: Century, Electra, LeSabre, Reatta,
 Regal, Riviera, Skylark, Skyhawk, Somerset
Cadillac: Cimarron
Chevrolet: Beretta, Corsica, Camaro,
 Cavalier, Celebrity, Citation II, Corvette
Oldsmobile: Calais, Ciera, Cutlass Cruiser,
 Cutlass Supreme, Delta 88, Firenza,
  Ninety-Eight, Toronado
Pontiac: Bonneville, Fiero, Firebird,
 Grand Am, Grand Prix, Sunbird, 2000, 6000

DESCRIPTION

The Port Fuel Injection (PFI) system is controlled by an on-board computer known as the Electronic Control Module (ECM). The ECM monitors engine operations and environmental conditions. ECM output signals provide correct air/fuel mixture, ignition timing and engine idle speed.

Port fuel injection features simultaneous double-fire injection. Fuel rail provides fuel for the injectors and is mounted on the intake manifold. Fuel is constantly circulated through the rail during engine operation. Injectors pulse once for each engine revolution providing fuel directly to the combustion chamber with precise timing. On Sequential Fuel Injection (SFI) models, injectors pulse sequentially in spark plug firing order. Differences between these systems are the injectors, wiring and the ECM.

In all systems, constant fuel pressure is maintained to the injectors. Air/fuel ratio is adjusted by the time that injector stays open (pulse width). Various sensors provide information to the ECM to control the pulse width. There are 2 major sub-systems; fuel system and electronic control system.

Mechanical portion of the fuel injection system consists of fuel injectors, cold start valve, throttle body, fuel rail, fuel pressure regulator, Idle Air Control (IAC) valve, fuel pump and fuel pump relay.

FUEL SYSTEM
Fuel system provides constant pressurized fuel supply to the injectors. Fuel system consists of fuel tank, in-tank electric fuel pump, fuel pump relay, fuel pressure regulator, in-line fuel filter, fuel rail, fuel injectors, and Idle Air Control (IAC) valve.

CAUTION
- *Fuel system is under pressure. Pressure must be carefully released prior to fuel system repairs to prevent personal injury.*

ELECTRONIC CONTROL SYSTEM
Electronic control system monitors engine operating conditions, processes the information, and controls engine for optimum performance and minimum emissions.

Input signals are generated by the Coolant Temperature Sensor (CTS), Mass Airflow (MAF) sensor, exhaust Oxygen (O_2) sensor, Throttle Position Sensor (TPS), Power Steering Pressure Switch (PSPS), Manifold Air Temperature (MAT), Manifold Absolute Pressure (MAP), Vehicle Speed Sensor (VSS) and Park/Neutral (P/N) switch on A/T applications.

Some engines may use a combination of MAT and MAP sensors in place of MAF sensor. The MAT and MAP sensors are used to determine airflow. The ECM may also receive signals from starter solenoid during cranking mode, air conditioning selector switch and distributor.

OPERATION

FUEL CONTROL SYSTEM
Fuel control system contains an electric fuel pump, located in the fuel tank with the gauge sending unit. Fuel is pumped to the fuel rail through an in-line fuel filter. Fuel pump provides fuel at a pressure and volume which exceeds the engine's requirements. Pressure regulator, mounted on the fuel rail, maintains a proper constant fuel pressure to the injectors. Pressure regulator increases fuel pressure when engine vacuum decreases. Unused fuel is returned to the fuel tank by a fuel return line. For correct fuel injector delivery at the ECM command, the fuel supply system maintains a constant pressure of approximately 36.0 psi (2.5 kg/cm²) drop across the injectors.

The fuel pressure accumulator, used on some models, isolates fuel line noises. The fuel rail provides the upper mount for the fuel injectors. Spring-loaded pressure tap may be used for testing the fuel system.

Courtesy of General Motors Corp.

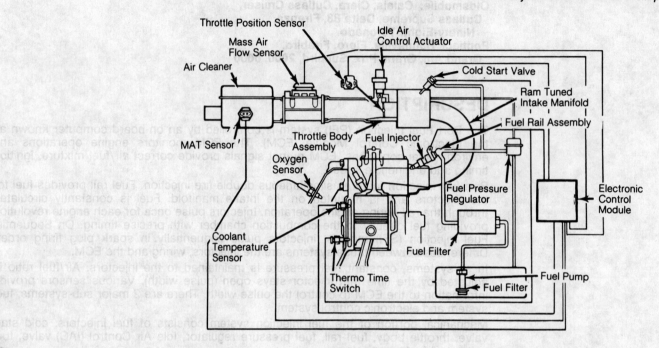

Fig. 1: Typical Port Fuel Injection System

FUEL PUMP

Fuel is supplied by an in-tank positive displacement roller vane pump. The pump supplies fuel through an in-line fuel filter to the fuel rail assembly. The pump is removed for service along with the fuel gauge sending unit. Once removed from the tank, pump and sending unit are serviced separately.

The pressure relief valve in the fuel pump, controls fuel pump maximum pressure to 60-90 psi (4.2-6.3 kg/cm²). Excess fuel flows through the pressure regulator and returns to the tank.

The constant flow of cool fuel prevents fuel vapor bubbles. When ignition is turned on, fuel pump relay activates the fuel pump for 2 seconds to prime the injectors. If engine does not start within this time, the ECM deactivates the fuel pump until the engine starts.

After engine starts, the ECM will close the fuel pump relay to activate the fuel pump. As a back-up system to the fuel pump relay, the fuel pump can also be activated by an oil pressure switch. The oil pressure switch is normally open until oil pressure reaches approximately 4 psi (.28 kg/cm²). If fuel pump relay fails, the oil pressure switch will close when oil pressure is obtained. This operates the fuel pump. Inoperative fuel pump relay may result in extended cranking time. *See Fig. 2.*

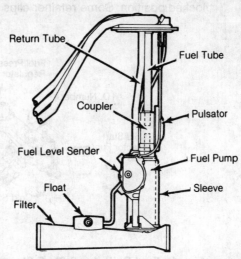

Fig. 2: In-Tank Fuel Pump Assembly

FUEL PRESSURE REGULATOR

Fuel pressure regulator contains a pressure chamber separated by a diaphragm relief valve assembly with a calibrated spring in the vacuum chamber side. Fuel pressure is regulated when pump pressure acting on the bottom side of the diaphragm overcomes the spring force action on the top side.

The diaphragm relief valve moves, opening or closing fuel chamber orifice, controlling the amount of fuel returned to the tank. Fuel pressure regulator controls system pressure to approximately 34-46 psi (2.4-3.2 kg/cm²) across the fuel injectors at all intake manifold vacuum readings.

Vacuum and spring pressure acting on the top side of the diaphragm controls the fuel pressure. Increased manifold vacuum decreases the fuel pressure. Under heavy load conditions more fuel is required and manifold vacuum decreases. Decrease in vacuum allows an increase in pressure to the top side of the pressure relief valve to increase fuel pressure. *See Fig. 3.*

NOTE
- *The SFI turbo engine pulses only one injector at a time, so fuel pressure requirements are reduced. Therefore, the SFI system does not use a fuel pressure accumulator.*

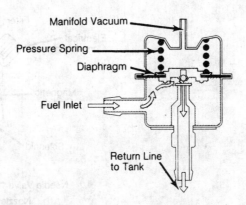

Fig. 3: Sectional View of Fuel Pressure Regulator

FUEL RAILS

Fuel rail assembly includes a fuel pressure regulator, individual high pressure fuel injectors and on some models, a cold start injector. Fuel rail assembly seats in the "V" between the upper plenum and the crossover runner section. The injectors fit into individual sockets in the base plate. Fuel rails used on 1.8L, 2.0L, 3.0L and 3.8L engines are manufactured, assembled and flow tested with the injectors as an assembly.

The injectors are force fit into the opening on the rail. Retainer clip is pushed into its locked position. Some retainer clips must be rotated in the proper direction. *See Fig. 4.*

Courtesy of General Motors Corp.

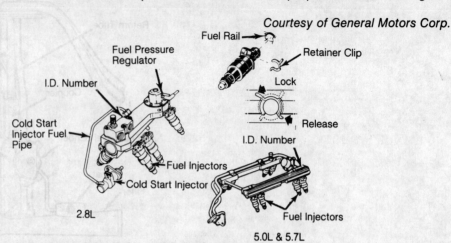

Fig. 4: Fuel Rails for 2.8L, 5.0L and 5.7L Engines

FUEL INJECTORS

Fuel injector is installed in intake manifold at each cylinder. Injector sealing is provided with "O" ring seals. The lower "O" ring seals the injector at the intake manifold. Upper "O" ring seals injector at fuel rail. The "O" rings must be replaced when injector is removed from intake manifold. *See Fig. 5.*

Injectors are identified by an ID number cast near the top side of injector. Injectors manufactured by Rochester Products have an "RP" positioned near the top side in addition to the ID number.

The solenoid-operated injector consists of a valve body with a specially ground nozzle valve. The moveable armature is attached to the nozzle valve. Nozzle valve is held against the nozzle body sealing seat by spring pressure. Solenoid winding is located in the front section of injector body.

Each injector contains a wire connector. On some 3.0L and 3.8L engines, wire connectors have a spring clip that must be released prior to removal of connector. One wire supplies voltage from the fuse panel. Second wire connects to the ECM ground and controls injector pulse width.

The electric pulses generate a magnetic field in injector solenoid winding. This draws the armature backward against spring pressure and lifts the nozzle valve from its seat. Spring pressure forces the nozzle valve closed.

Courtesy of General Motors Corp.

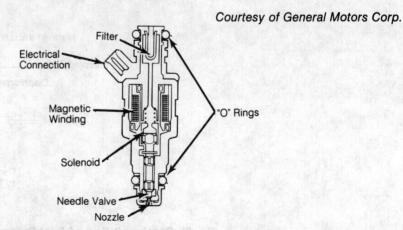

Fig. 5: Typical Fuel Injector

NOTE
- *On Sequential Fuel Injection (SFI) systems, wiring harness MUST be connected to the correct injector. One wire connects battery voltage to injectors. The ECM ground circuit for each injector is color coded differently for identification.*

COLD START VALVE

The cold start valve provides additional vaporized fuel during the crank mode to improve cold start-up. This circuit is required with low coolant temperature due to short injector pulse width not providing enough fuel to start the engine.

The starter solenoid supplied circuit is activated only during the crank mode. A thermal switch provides a valve ground path when engine coolant temperature is below 95°F (35°C).

Thermal switch consists of a bi-metallic material which opens at a specified coolant temperature. This bi-metallic material is also heated by the thermal switch windings. This allows valve to remain on for 8 seconds at -5°F (-20°C) coolant temperature.

The thermal switch closed time varies with the coolant temperature. As the coolant temperature rises, cold start valve operation time decreases.

AIR INDUCTION SYSTEM

Air is drawn into the induction system through an air pick-up mounted in front of the radiator support. This ensures intake air is lower temperature than engine compartment temperatures, resulting in denser air supply to combustion chambers.

Air cleaners contain replaceable paper elements. Air intake duct provides turbulant free airflow for accurate measurement at the MAF sensor. Air entering behind the MAF sensor would not be measured, allowing improper reading to the computer and a lean air/fuel ratio. Rubber boot between MAF sensor and throttle body prevents this air leakage. Duct between MAF sensor and throttle body must be air tight. Air leakage behind MAF sensor will not be measured, causing a lean air/fuel mixture.

THROTTLE BODY

Throttle body controls the amount of air entering the intake manifold. Mounted on the throttle body are thye TPS and IAC valves. The TPS enables the ECM to determine throttle position under all operating conditions. The throttle body contains vacuum ports used to supply signals to various components. On some models, engine coolant is routed through bottom of throttle body to warm housing and prevent icing. *See Fig. 6.*

Courtesy of General Motors Corp.

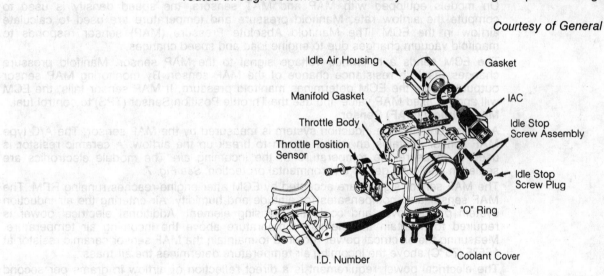

Fig. 6: Exploded View of Typical Throttle Body

IDLE AIR CONTROL

The Idle Air Control (IAC) valve controls engine idle speed during changes in engine load to prevent stalling. Mounted in the throttle body, the IAC controls by-pass air around the throttle valve. Movement of the IAC conical valve inward toward the seat decreases airflow while movement away from the seat increases the airflow around the throttle plate.

More air is by-passed around the throttle valve to increase RPM. Less air is by-passed to decrease RPM. The IAC valve moves in small steps called "COUNTS", and can be measured with test equipment plugged into the Assembly Line Diagnostic Link (ALDL). Increase of the "COUNTS" indicates more air passing the IAC valve.

Proper positioning of IAC valve at idle is determined by the ECM based on battery voltage, coolant temperature, engine load, and engine RPM. Drop in engine speed below a specified RPM, with throttle plate closed, the ECM senses a near stall condition. The ECM then repositions IAC valve to prevent stalls.

Disconnecting and reconnecting of IAC valve with engine running will result in incorrect idle RPM. The IAC must be reset. On some models the IAC resets when ignition switch is turned from the "ON" to "OFF" position. Some models require that engine be operated at 2500 RPM or over 35 MPH. The IAC should only be disconnected or connected with ignition off.

The IAC valve affects only the idle system. If valve is stuck fully open, excessive airflow into the manifold creates an high idle speed. Valves stuck closed allows insufficient air flow resulting in low idle speed.

Partially stuck open valve creates a rough idle and will not respond to engine load changes. Different design IAC valves are used. Ensure proper design valve is used during replacement.

POWER STEERING PRESSURE SWITCH (PSPS) PARK/NEUTRAL (P/N) SWITCH THROTTLE POSITION SENSOR (TPS)

The PSPS indicates to the ECM that power steering unit is under load. The ECM uses this information to relocate the IAC valve to compensate for additional engine load. The ECM will also turn off the A/C clutch when pressure is noted.

The P/N switch indicates gear position to the ECM. This information is used for the Torque Converter Clutch (TCC), EGR valve and the IAC valve operation.

The TPS is mounted on the throttle body shaft. The TPS receives 5 volts supplied by the ECM with one terminal connected to the ECM ground. Remaining terminal is connected to the ECM to determine voltage reading from the TPS.

Movement of throttle shaft alters TPS output voltage to ECM. Closed throttle provides approximately .5 volts while wide open throttle provides 5 volts. Voltage monitoring by the ECM controls the fuel delivery due to engine demand. Broken or damaged TPS may create bursts of fuel delivery causing unstable idle.

AIRFLOW SENSING

Speed Density

On models equipped with MAP and MAT sensors, the speed density is used to compute the airflow rate. Manifold pressure and temperature are used to calculate airflow in the ECM. The Manifold Absolute Pressure (MAP) sensor responds to manifold vacuum changes due to engine load and speed changes.

The ECM sends a reference voltage signal to the MAP sensor. Manifold pressure changes result in resistance change of the MAP sensor. By monitoring MAP sensor output voltage, the ECM determines manifold pressure. If MAP sensor fails, the ECM will apply a fixed MAP value and use the Throttle Position Sensor (TPS) to control fuel.

Mass Airflow (MAF) Sensor

Air mass entering the induction system is measured by the MAF sensor. The A/C type MAF sensor contains an intake screen to break up the airflow. A ceramic resistor is used to measure the temperature of the incoming air. The module electronics are sealed in a silicone gel for environmental protection. *See Fig. 7.*

The MAF sensor signals are accepted by ECM after engine reaches running RPM. The MAF sensor also compensates for altitude and humidity. Air entering the air induction system passes over and cools the sensing element. Additional electrical power is required to maintain the element temperature above the incoming air temperature. Measuring the electrical power required to maintain the MAF sensor ceramic resistor at 165°F (75°C) above the incoming air temperature determines the air mass.

The electrical power requirement is a direct reflection of airflow in grams per second and is converted to a digital signal (32-150 Hz) which is sent to the ECM. This is then used to calculate engine load.

Increasing the air mass results in an increase in the 32-150 Hz frequency. Using mass airflow calculations, engine temperature and RPM, the ECM calculates the amount of fuel required to provide a proper air/fuel ratio of 14.7:1.

Some MAF units are the Bosch hot wire type. This type contains a sensing wire. Current is supplied to maintain a calibrated temperature. Current varies with amount of airflow. Current changes are processed by the ECM to provide proper air/fuel ratio.

Sensing wire contamination is prevented by heating sensing wire to 1000°F (538°C) after engine shutdown. This is know as the burn off cycle. Burn off cycle is controlled by the ECM. The ECM controls the grounds circuit for the burn off relay, providing engine was operated for a certain time period. If burn off cycle did not occur, the Service Engine Soon light will activate and stay on after next engine start.

Courtesy of General Motors Corp.

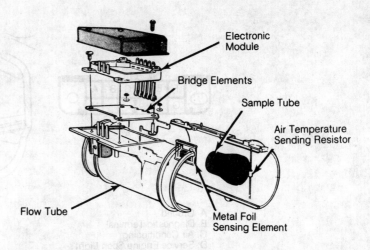

Electronic Module

Bridge Elements

Sample Tube

Air Temperature Sending Resistor

Flow Tube

Metal Foil Sensing Element

Fig. 7: *Exploded View of Typical Mass Airflow Sensor*

Manifold Absolute Pressure Sensor (MAP)

The MAP sensor responds to manifold vacuum changes and also measures barometric pressure. These readings allow the ECM to adjust for different altitudes. Signal voltage from MAP sensor will vary from approximately .5-1.0 volts at idle to 4-4.5 volts at wide open throttle.

OXYGEN SENSOR (O₂)

The O$_2$ sensor monitors exhaust gas oxygen content to produce voltage output. The ECM monitors the voltage output to determine the fuel mixture. Low voltage indicates lean fuel mixture. High voltage indicates rich fuel mixture.

TROUBLE SHOOTING

ECM TROUBLE CODES

Entering Diagnostic Mode

The ECM uses various sensors to monitor engine operating conditions. The ECM memory relates certain sensor readings to certain conditions. If sensor reading varies from ECM memory, the ECM activates the "CHECK ENGINE" or "SERVICE ENGINE SOON" light on the instrument panel. A trouble code is then stored in the memory.

Trouble codes may be either intermittent or hard. A hard code is stored, activating the "CHECK ENGINE" light and exists during vehicle repair. An intermittent code does not always store a code or activate the "CHECK ENGINE" light and does not exist during vehicle repair. Intermittent codes are often caused by loose connections. Connecting to the Assembly Line Diagnostic Link (ALDL) allows trouble codes to be obtained from the ECM.

NOTE

• *Usage of the term "CHECK ENGINE" light in this article also includes the "SERVICE ENGINE SOON" light used on later model vehicles.*

View "A"

A. Ground
B. Diagnostic Terminal
C. Air Conditioning
D. Service Engine Soon Light

E. Serial Data
F. Torque Converter Clutch (TCC)
G. Fuel Pump
M. Serial Data (Late 1987 & All 1988 Models)

Fig. 8: ALDL Connector

The following procedure is designed to obtain trouble codes stored in the ECM memory.

1) Turn ignition on. DO NOT start engine. The "CHECK ENGINE" light should glow. Locate ALDL connector attached to ECM wiring harness under instrument panel. Install jumper wire between terminals "A" and "B". *See Fig. 8.*

2) The "CHECK ENGINE" light should flash a code "12" which is indicated by a "flash", pause, "flash", "flash" followed by a longer pause. Trouble code "12" indicates system is working correctly and will repeat 2 more times. If any trouble codes are stored in ECM memory, they will be displayed in the same manner.

3) Trouble codes are read by counting the flashes of the "CHECK ENGINE" light. A "flash", "flash", pause, "flash", longer pause, indicates "21". Trouble codes are displayed from lowest to the highest numbered codes 3 times each. Codes are repeated as long as the jumper wire is connected to the ALDL.

The following is a list and description of these codes:

Code 12
No RPM reference.
Code 13
Open oxygen sensor circuit.
Code 14
Coolant sensor reading too low.
Code 15
Coolant sensor reading too high.
Code 16
High system voltage.
Code 21
TPS reading too high.
Code 22
TPS reading too low.
Code 23
MAT sensor voltage high.
Code 24
VSS circuit.
Code 25
MAT sensor voltage low.
Code 26
Quad-Driver error.
Code 27, 28 & 29

Gear switch problem.

Code 31

Wastegate Solenoid (Turbo).

Park/Neutral switch.

Code 32

EGR vacuum control signal.

Code 33

MAF or MAP sensor reading high.

Code 34

MAF or MAP sensor reading low.

Code 35

IAC error.

Code 36

MAF sensor burn off (5.0L & 5.7L).

Closed throttle airflow high (2.3L VIN D).

Code 38

Brake switch.

Code 39

TCC operation.

Code 41

Cylinder select error (MEM-CAL).

C^3I cam sensor circuit.

Code 42

EST circuit open or grounded.

C^3I ignition cam sensor loss.

Code 43

Electronic Spark Control (ESC) failure.

Code 44

Oxygen sensor lean too long.

Code 45

Oxygen sensor rich too long.

Code 46

Anti-theft fault (5.7L).

Power steering pressure switch (1989 3.3L & 3.8L).

Code 48

Misfire diagnosis.

Code 51

Faulty PROM, MEM-CAL or ECM.

Code 52

Faulty/missing CALPAK or MEM-CAL.

Code 53

Faulty alternator voltage.

EGR system malfunction.

Anti-theft circuit fault.

Code 54

Low fuel pump voltage.

Code 55

Internal ECM error.

Code 61

Degraded O_2 sensor (2.8L & 3.1L).

Code 62

Gear switch error.

Code 63

MAP sensor voltage high (2.8L).

EGR flow error.

Code 64

MAP sensor voltage low (2.8L).

EGR flow error (3800 VIN C).

Code 65

Fuel injector current low.

EGR flow error (3800 VIN C).

Code 66

A/C pressure sensor voltage out of spec.

MAP sensor high (2.8L).

EGR flow error (3800 VIN C).

Entering Field Service Mode

Installation of jumper wire between termials "A" and "B" of ALDL with engine running will enter Field Service Mode. This test should be performed after every repair to ensure closed loop operation. Clear all codes.

Apply parking brake and block wheels. Locate ALDL connector under instrument panel. Install jumper wire between terminals "A" and "B" with engine running. *See Fig. 8.* During this mode, the "CHECK ENGINE" light will indicate if system is in open or closed loop operation.

Open loop operation does not use O_2 sensor input to the ECM. Closed loop relies on O_2 sensor input for ECM fuel control. During open loop "CHECK ENGINE LIGHT" will flash 2 1/2 times per second. On closed loop "CHECK ENGINE LIGHT" will flash once per second. During closed loop "CHECK ENGINE LIGHT" may stay off most of the time if mixture is lean or remain on if mixture is rich.

Clearing ECM Trouble Codes

Trouble codes must be removed from ECM memory after diagnosis and repairs are made. Battery voltage must be removed from ECM. Turn ignition on and install jumper wire between termials "A" and "B" of ALDL. *See Fig. 8.* Turn ignition off and remove ECM fuse from fuse block for at least 30 seconds. Remove test lead from ALDL connector.

CAUTION

- *Ignition must be off when disconnecting or reconnecting ECM battery voltage to prevent ECM damage.*

DIAGNOSIS & TESTING

PRELIMINARY CHECKS

Prior to diagnosing the fuel injection system, following systems and components must be in good condition and operating properly:

- ECM "CHECK ENGINE" light must be operational.
- No trouble codes stored in memory.
- ECM grounds should be tight and clean.
- All support systems and wiring operational.
- Battery connections and specific gravity to specification.
- Compression pressure should be even.
- Fuel supply system pressure and flow unrestricted.
- All electrical connections clean and tight.
- Air filter must be clean and correct application.
- Vacuum lines, fuel hoses and pipe connections.
- Check for vacuum leaks at throttle body and intake manifold.

CAUTION

- *Fuel system pressure is under high pressure. Pressure must be released prior to disconnecting fuel system components.*

FUEL PRESSURE RELIEF

Install Fuel Pressure Gauge (J-34730-1) to fuel system. Cover fuel pressure tap with shop towel to prevent fuel leakage. Install bleed hose in container. Open bleed valve to release pressure.

FUEL SYSTEM DIAGNOSIS

NOTE

- *For diagnosis on all models, see appropriate FUEL SYSTEM DIAGNOSIS chart in this article.*

Restricted Flow

Normally, a vehicle with fuel pressure below 24 psi (1.7 kg/cm²) at idle will not be driveable. If pressure drop occurs only while driving, engine surge will occur and then quit as pressure drops rapidly. Restricting of fuel return line allows fuel pump to develop maximum pressure (dead head pressure). With battery voltage applied the pump test terminal, pressure should exceed 75 psi (5.3 kg/cm²). Test system to determine if high fuel pressure is due to a restricted fuel return line or a pressure regulator problem.

NOTE

- *Following symptoms should be checked when no hard codes are stored. DO NOT use hard codes for system repair.*

HARD START

1) If engine starts and then dies immediately, see ENGINE CRANKS BUT WILL NOT RUN diagnostic chart. Check fuel pump relay by attaching a test light between fuel pump test terminal and ground.

2) Turn ignition off for 10 seconds, then turn the ignition on. Test light should light for approximately 2 seconds. Fuel pump relay is defective if test light did not operate.

3) Check for sticking or binding TPS. Check for high TPS voltage with throttle closed. Voltage should be less than .7 volts. Check for high resistance in coolant sensor circuit or coolant sensor.

4) Check fuel pressure. Check for contaminated fuel. Check in-tank fuel pump check valve operation. If problem exists in cold weather only, check cold start valve operation. Check fuel evaporation system.

5) Check EGR operation. On 1988-89 2.8L VIN S, 5.0L VIN F and 5.7L VIN 8, if engine starts but then immediately stalls, open distributor by-pass line. If engine starts and runs, replace pickup coil. On all models if engine starts but then immediately stalls, disconnect MAF sensor. If engine then runs and sensor connections are OK, replace the sensor. Remove spark plugs and check for wet plugs, cracks, improper gap, burned electrodes or heavy carbon deposits. Also check plug wires for shorts.

STALL AFTER START

All Exc. 3.0L & 3.8L V6

Check IAC system operation. See IDLE AIR CONTROL TESTING chart. Check for plugged or restricted fuel lines. Check PCV valve operation. Check MAF sensor (if equipped). Unplug MAF sensor. If problem is eliminated, replace MAF sensor. Check EGR system.

3.0L & 3.8L

On all 3.0L and 3.8L V6 engines, check for air leaks at air duct between MAF sensor and throttle body. Inspect EGR system for correct operation.

HESITATION, SAG, STUMBLE

1) Check fuel pressure. See FUEL SYSTEM DIAGNOSIS chart. Check TPS for sticking or binding. Check vacuum hose to MAP sensor for damage or restrictions.

2) Verify alternator output voltage is between 9 and 16 volts. Repair or replace as necessary. Check canister purge system for proper operation.

3) Check for proper seal of oil filler cap and tube. Check for fuel contamination. Check for air leaks at air duct between MAF sensor and throttle body.

4) Check High Energy Ignition ground circuit. Check injector balance. See INJECTOR BALANCE TEST chart. Check for fouled spark plugs. Check that PROM in vehicle is correct. Make sure that initial ignition timing is properly set. Check that ECM controlled idle speed is correct.

ROUGH, UNSTABLE IDLE

1) Check for sticking or binding TPS or throttle linkage. Check idle speed and IAC system. See IDLE AIR CONTROL TESTING chart. Verify alternator output voltage is between 9 and 16 volts. Repair or replace as necessary. Make sure that ignition timing is properly set.

2) Check P/N switch circuit (if equipped). Check ECM ground circuits. Check injector balance. See INJECTOR BALANCE TEST chart. Check EGR system, PCV valve and fuel evaporation system. Check power steering pressure switch circuit.

3) Check A/C signal to ECM terminal. Check for loose or damaged MAF sensor duct between sensor and throttle body. Check AIR system. No air should exist at ports during closed loop operation. Check exhaust system for restrictions, such as a damaged or collapsed pipe, muffler or catalytic converter.

4) On all models, inspect O_2 sensor for silicon contamination from fuel or improper use of RTV sealant. Sensor is defective if coated with a white, powdery coating.

5) Check for fuel in pressure regulator hose. Replace pressure regulator if fuel exists. Disconnect MAF sensor. Replace sensor if condition is corrected.

CUTS OUT, MISSES

Check for restricted fuel filter or fuel contamination. Check for low fuel pressure. See FUEL SYSTEM DIAGNOSIS chart. Check injector balance. See INJECTOR BALANCE TEST chart. Check ignition system operation.

DETONATION

1) Check ignition timing. Check for improper fuel octane rating. Check fuel pressure. See FUEL SYSTEM DIAGNOSIS chart. Check cooling system for obvious overheating problems. Check TPS adjustment and operation.

2) Check EGR valve operation. Check for P/N switch (if equipped). Check Torque Converter Clutch (TCC) and Electronic Spark Control (ESC) operation. Make sure that correct PROM is installed in ECM.

3) Remove carbon from engine with top engine cleaner. If excessive carbon exists in combustion chamber, check for excessive oil burning due to leaking valve guide seals. Check for incorrect basic engine parts such as camshaft, cylinder heads and pistons.

SURGES AND/OR CHUGGLE

1) Check for loose or leaking air duct between MAF sensor and throttle body. Verify alternator output voltage is between 9 and 16 volts. Repair or replace as necessary. Check that P/N switch is properly adjusted. Check for proper operation of ESC system.

2) Check EGR filter condition and EGR valve operation. Check vacuum hoses for kinks or leaks. Check for dirty or plugged in-line fuel filter. Check fuel pressure. See FUEL SYSTEM DIAGNOSIS chart. Check ignition timing. Check for intermittent open or short to ground in Torque Converter Clutch (TCC) or HEI by-pass circuits. Check for adequate spark output using Spark Tester (ST-125).

3) Inspect oxygen sensor for silicon contamination from improper fuel usage. Sensors having a white powdery coating will result in a false high voltage signal to the ECM, creating a lean condition. Check for proper operation of canister purge system.

4) Check fuel for water contamination. Remove spark plugs and check for wet plugs, cracks, improper gap, burned electrodes or heavy carbon deposits. Check A/C for excessive charge. Check for restricted exhaust system.

LACK OF POWER, SLUGGISH, OR SPONGY

1) Check air filter. Check for loose or leaking air duct between MAF sensor and throttle body. Check ignition timing. Check for fuel contamination, or restricted fuel filter. Check fuel pressure. See FUEL SYSTEM DIAGNOSIS chart. Check for proper operation of TCC system. Check EST system for proper operation.

2) Check ECM ground circuits. Check EGR operation. Check for restricted exhaust system. Verify alternator output voltage is between 9 and 16 volts. Repair or replace as necessary. Check ignition system. Check ESC system for excessive retard.

3) Check MAP or pressure sensor output. Using Spark Tester (ST-125), check for available secondary voltage. Check engine valve timing and compression. Check engine for a worn camshaft.

BACKFIRE

1) Check for loose wiring connector or air duct at MAF sensor. Check EGR valve mounting gasket or faulty valve operation. Check ignition timing and ignition system. Check for proper valve timing.

2) Check fuel system. See FUEL SYSTEM DIAGNOSIS chart. Check injector balance. See INJECTOR BALANCE TEST chart. Check ignition system operation. Check engine for sticking or leaking valves.

3) Check for fuel or water in vacuum hose to MAP sensor. Using Spark Tester (ST-125), check available output voltage of ignition coil. Check for crossfire between spark plugs, distributor cap and spark plug wires.

POOR FUEL ECONOMY

Check air filter. Check fuel pressure. See FUEL SYSTEM DIAGNOSIS chart. Check ignition system and ignition timing. Verify engine reaches normal operating temperature. Check ignition timing. Check TCC operation. Check exhaust system for restrictions.

DIESELING

Check injectors for leakage. See FUEL SYSTEM DIAGNOSIS chart.

EXCESSIVE EXHAUST EMISSIONS OR ODORS

1) Check fuel pressure. See FUEL SYSTEM DIAGNOSIS chart. Check ignition timing. Check fuel evaporation and PCV system. Check injector balance. See INJECTOR BALANCE TEST chart. Check all systems and components that could cause engine to run rich.

2) Check lead contamination of catalytic converter. If system is running lean, check EGR operation. Check for vacuum leaks. Verify engine operating temperature.

ADJUSTMENTS

MINIMUM IDLE SPEED

CAUTION

- *Minimum idle speed adjustment should only be required when throttle body parts have been replaced or when required by TPS adjustment.*

1.8L (1984 & 1985)

1) Operate engine to normal operating temperature. Disconnect PCV hose to create a severe vacuum leak and substantial increase in idle speed. Wait 2 minutes and disconnect IAC electrical connector.

2) Reconnect PCV hose. Set minimum idle rate to 675-725 RPM with A/T or 700-750 RPM with M/T. Reconnect IAC connector. IAC will reset when vehicle is driven over 45 MPH.

2.3L (1988-89)

1) Install Scan Tool. Operate engine to normal operating temperature and be in closed loop. On A/T models, shift into Drive then back to Neutral. Check that no input is received on Scan Tool from power steering circuit.

2) Allow idle to stabilize and note IAC valve counts. If reading is between 15-45 counts, adjustment is correct. If counts are low, check for vacuum leak at hoses, throttle body and intake manifold. If no leaks are found, throttle plate stop screw requires adjustment.

3) If count reading was high, clean residue from throttle body bore and edges of throttle plate. Recheck IAC counts. If IAC counts are found, idle stop screw requires adjustment.

4) Pry idle stop screw plug from throttle body. Back stop screw out until an air gap exists between idle stop screw and throttle lever. Turn idle stop screw inward until it contacts the throttle lever.

5) Turn idle stop screw an additional 1 1/2 turns. Recheck IAC counts once idle speed stabilizes. Adjust to obtain correct IAC counts. Adjust TPS to obtain reading of .49-.59 volts at closed throttle.

All Others (1984-88)

1) Pry idle stop screw plug from throttle body. With IAC connected, ground diagnostic lead. Connect tachometer to engine.

2) Turn ignition on, but DO NOT start engine. Wait at least 45 seconds. With ignition on, disconnect IAC connector. Install Scan Tool.

3) On models equipped with EST by-pass, unplug the Set Timing connector. This connector contains a Tan/Black wire located near the distributor, blower motor or right valve cover. DO NOT disconnect 4-wire connector.

4) Start engine and allow to go closed loop. Remove ground from diagnostic lead. Adjust idle set screws to obtain 500-600 RPM with gear selector in Drive for 2.8L engines equipped with A/T or 600-700 RPM in Neutral for M/T. On 5.0L and 5.7L, adjust idle speed to 550 RPM in Drive on A/T or 650 RPM with M/T. Adjust all others to 450-550 RPM in Neutral.

5) Turn ignition off and reconnect connector at IAC and Set Timing connector. Adjust TPS to specification.

2.0L (1989)

1) Ensure engine is at normal operating temperature. Ground ALDL diagnostic "test" terminal. With ignition on and engine off, wait 30 seconds for IAC valve to seat in throttle body. Disconnect IAC valve connector. Connect a tachometer or Scan Tool to monitor engine RPM.

2) Disconnect ground from ALDL diagnostic "test" terminal. With transaxle in Park or Neutral, start engine and allow engine RPM to stabilize. Ensure all accessories are off. Check idle speed. See 1989 MINIMUM IDLE SPEED SPECIFICATIONS table. If idle speed is incorrect. remove air cleaner (if necessary). Plug all disconnected hoses.

3) Disconnect throttle and TV cables on automatic transaxle models. Remove minimum idle stop screw cover. Adjust minimum idle speed to specification. If screw has to be turned more than a half a turn to obtain specified idle speed, check for vacuum leaks or dirty throttle plate.. Clear any trouble codes.

V6 Models (1989)

Install Scan Tool. Start engine and run until normal operating temperature is obtained. Check IAC counts. See 1989 MINIMUM IDLE SPEED SPECIFICATIONS table. If setting is incorrect, remove minimum idle stop screw cover. Turn idle stop screw out until it clears the linkage. Turn screw in until it touches the linkage. Turn screw an additional 1 1/2 turns. turn idle screw in to decrease the IAC count. Turn screw out to increase IAC count. If screw must be turned more than 1/2 turn to obtain specified setting, check for vacuum leak or dirty throttle plate.

V8 Models (1989)

1) Pry idle stop screw plug from throttle body. With IAC connected, ground diagnostic lead. Connect tachometer to engine. Turn ignition on, but DO NOT start engine. Wait at least 45 seconds. With ignition on, disconnect IAC connector. Install Scan Tool.

2) On models equipped with EST by-pass, unplug the Set Timing connector. DO NOT disconnect 4-wire connector. Start engine and allow to go closed loop. Remove ground from diagnostic lead. Adjust idle set screw to obtain specification. See 1989 MINIMUM IDLE SPEED SPECIFICATIONS table. Turn ignition off and reconnect connector at IAC and Set Timing connector. Adjust TPS to specification.

1989 MINIMUM IDLE SPEED SPECIFICATIONS

Application	RPM or IAC Counts
2.0L	450-650 RPM
2.3L	15-45 Counts
2.8L	10-15 Counts
3.1L, 3.3L & 3.8L	10-20 Counts
5.0L & 5.7L (B Body)	450-500
5.0L & 5.7L (F & Y Bodies)	400-450

THROTTLE POSITION SENSOR

NOTE

- *Throttle Position Sensor (TPS) is non-adjustable on many models. If TPS reading is not as specified, replace TPS and recheck.*

1) Install 3 jumper wires between TPS and harness connector. With ignition on, attach digital voltmeter between terminals "B" and "C" on 1.8L Turbo and 3.8L non-turbo models and "A" and "B" on all other models.

2) Adjust TPS to obtain specified voltage, with throttle plates closed. See TPS ADJUSTMENT VOLTAGE table. Start engine and check for proper idle operation. Coat screw threads with Loctite and tighten. Recheck reading. With ignition off, remove jumper wires and reconnect TPS connector.

TPS ADJUSTMENT VOLTAGE

Application	[1] Voltage
1.8L	.5
2.0L (VIN M)	1.25
2.8L	
(VIN P)	.35-.45
(VIN 9)	Less Than 1.25
All Others (1984-88)	.5-.6
1989	Zero
3.0L	.5-.6
3.1L	Zero
3.3L	.38-.42
3.8L	.36-44
5.0L & 5.7L	
1984-88	.5
1989	Less Than 1.25

[1] – With throttle plates closed.

IDLE AIR CONTROL TESTING (1984-89 MODELS)

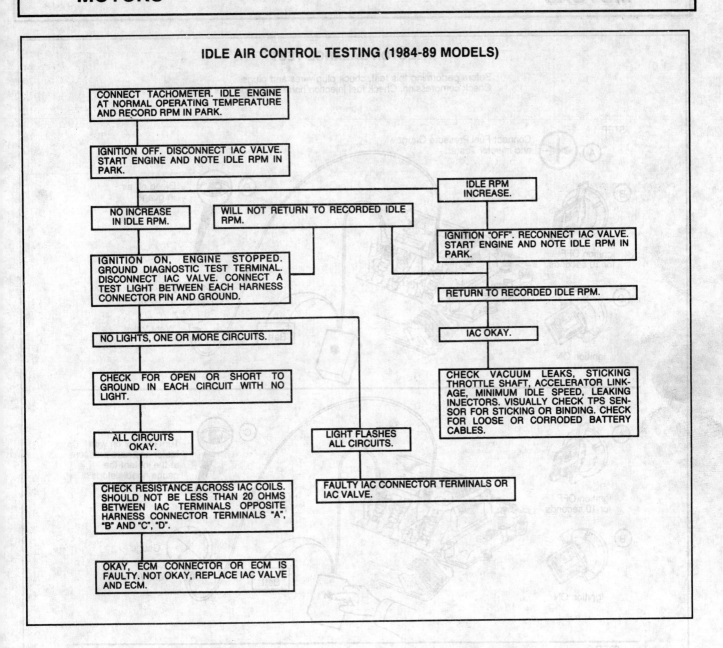

```
CONNECT TACHOMETER. IDLE ENGINE
AT NORMAL OPERATING TEMPERATURE
AND RECORD RPM IN PARK.
            │
            ▼
IGNITION OFF. DISCONNECT IAC VALVE.
START ENGINE AND NOTE IDLE RPM IN
PARK.
     │
     ├──────────────────────────────────────────────┐
     │                                                │
     ▼                                         IDLE RPM
                                               INCREASE.
                                                    │
NO INCREASE      WILL NOT RETURN TO RECORDED IDLE    ▼
IN IDLE RPM.     RPM.                          IGNITION "OFF". RECONNECT IAC VALVE.
     │                        │                START ENGINE AND NOTE IDLE RPM IN
     ▼                        │                PARK.
IGNITION ON, ENGINE STOPPED.  │                     │
GROUND DIAGNOSTIC TEST TERMINAL.                     ▼
DISCONNECT IAC VALVE. CONNECT A                RETURN TO RECORDED IDLE RPM.
TEST LIGHT BETWEEN EACH HARNESS                      │
CONNECTOR PIN AND GROUND.                            ▼
     │                                         IAC OKAY.
     ▼                                               │
NO LIGHTS, ONE OR MORE CIRCUITS.                     ▼
     │                                         CHECK VACUUM LEAKS, STICKING
     ▼                                         THROTTLE SHAFT, ACCELERATOR LINK-
CHECK FOR OPEN OR SHORT TO                     AGE, MINIMUM IDLE SPEED, LEAKING
GROUND IN EACH CIRCUIT WITH NO                 INJECTORS. VISUALLY CHECK TPS SEN-
LIGHT.                                         SOR FOR STICKING OR BINDING. CHECK
     │                                         FOR LOOSE OR CORRODED BATTERY
     ▼                                         CABLES.
ALL CIRCUITS        LIGHT FLASHES
OKAY.               ALL CIRCUITS.
     │                   │
     ▼                   ▼
CHECK RESISTANCE ACROSS IAC COILS.   FAULTY IAC CONNECTOR TERMINALS OR
SHOULD NOT BE LESS THAN 20 OHMS      IAC VALVE.
BETWEEN IAC TERMINALS OPPOSITE
HARNESS CONNECTOR TERMINALS "A",
"B" AND "C", "D".
     │
     ▼
OKAY, ECM CONNECTOR OR ECM IS
FAULTY. NOT OKAY, REPLACE IAC VALVE
AND ECM.
```

INJECTOR BALANCE TEST

Before performing this test, check plug wires and plugs.
Check compression. Check fuel injection harness for opens or shorts.

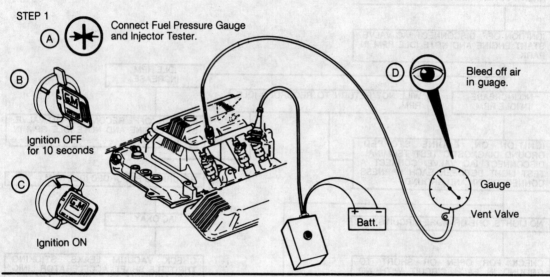

STEP 1

(A) Connect Fuel Pressure Gauge and Injector Tester.

(B) Ignition OFF for 10 seconds

(C) Ignition ON

(D) Bleed off air in guage.

Gauge

Vent Valve

Batt.

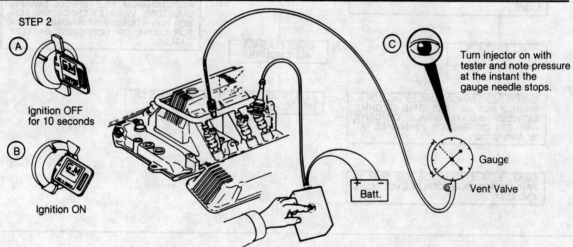

STEP 2

(A) Ignition OFF for 10 seconds

(B) Ignition ON

(C) Turn injector on with tester and note pressure at the instant the gauge needle stops.

Gauge

Vent Valve

Batt.

STEP 3

Repeat test as in Step 2 on all injectors and record pressure drop on each. Retest injectors that appear faulty. Replace injectors that have a 1.5 psi difference in pressure.

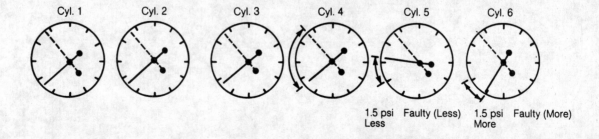

Cyl. 1   Cyl. 2   Cyl. 3   Cyl. 4   Cyl. 5   Cyl. 6

1.5 psi Less   Faulty (Less)   1.5 psi More   Faulty (More)

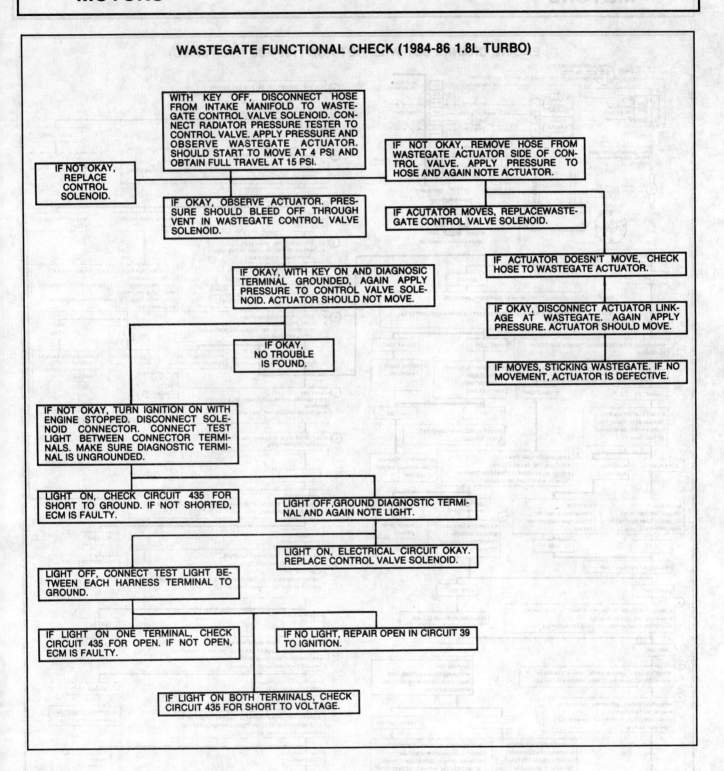

WASTEGATE FUNCTIONAL CHECK (1984-86 1.8L TURBO)

WITH KEY OFF, DISCONNECT HOSE FROM INTAKE MANIFOLD TO WASTE-GATE CONTROL VALVE SOLENOID. CONNECT RADIATOR PRESSURE TESTER TO CONTROL VALVE. APPLY PRESSURE AND OBSERVE WASTEGATE ACTUATOR. SHOULD START TO MOVE AT 4 PSI AND OBTAIN FULL TRAVEL AT 15 PSI.

IF NOT OKAY, REPLACE CONTROL SOLENOID.

IF NOT OKAY, REMOVE HOSE FROM WASTEGATE ACTUATOR SIDE OF CONTROL VALVE. APPLY PRESSURE TO HOSE AND AGAIN NOTE ACTUATOR.

IF OKAY, OBSERVE ACTUATOR. PRESSURE SHOULD BLEED OFF THROUGH VENT IN WASTEGATE CONTROL VALVE SOLENOID.

IF ACUTATOR MOVES, REPLACEWASTE-GATE CONTROL VALVE SOLENOID.

IF ACTUATOR DOESN'T MOVE, CHECK HOSE TO WASTEGATE ACTUATOR.

IF OKAY, WITH KEY ON AND DIAGNOSIC TERMINAL GROUNDED, AGAIN APPLY PRESSURE TO CONTROL VALVE SOLENOID. ACTUATOR SHOULD NOT MOVE.

IF OKAY, DISCONNECT ACTUATOR LINKAGE AT WASTEGATE. AGAIN APPLY PRESSURE. ACTUATOR SHOULD MOVE.

IF OKAY, NO TROUBLE IS FOUND.

IF MOVES, STICKING WASTEGATE. IF NO MOVEMENT, ACTUATOR IS DEFECTIVE.

IF NOT OKAY, TURN IGNITION ON WITH ENGINE STOPPED. DISCONNECT SOLENOID CONNECTOR. CONNECT TEST LIGHT BETWEEN CONNECTOR TERMINALS. MAKE SURE DIAGNOSTIC TERMINAL IS UNGROUNDED.

LIGHT ON, CHECK CIRCUIT 435 FOR SHORT TO GROUND. IF NOT SHORTED, ECM IS FAULTY.

LIGHT OFF, GROUND DIAGNOSTIC TERMINAL AND AGAIN NOTE LIGHT.

LIGHT ON, ELECTRICAL CIRCUIT OKAY. REPLACE CONTROL VALVE SOLENOID.

LIGHT OFF, CONNECT TEST LIGHT BETWEEN EACH HARNESS TERMINAL TO GROUND.

IF LIGHT ON ONE TERMINAL, CHECK CIRCUIT 435 FOR OPEN. IF NOT OPEN, ECM IS FAULTY.

IF NO LIGHT, REPAIR OPEN IN CIRCUIT 39 TO IGNITION.

IF LIGHT ON BOTH TERMINALS, CHECK CIRCUIT 435 FOR SHORT TO VOLTAGE.

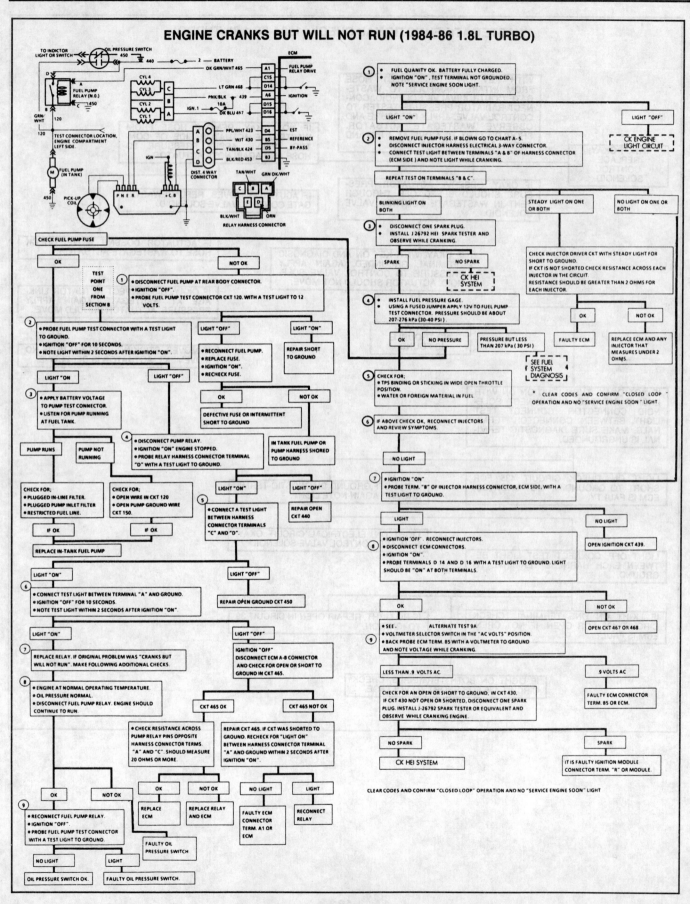

ENGINE CRANKS BUT WILL NOT RUN (1984-86 1.8L TURBO)

ENGINE CRANKS BUT WILL NOT RUN (1984-87 3.8L NON-TURBO)

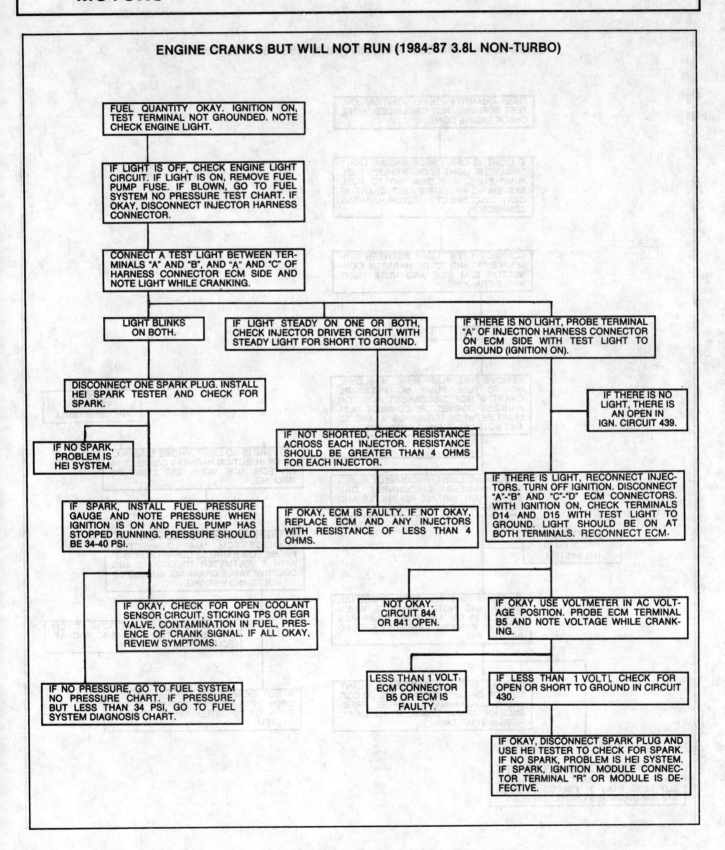

FUEL QUANTITY OKAY. IGNITION ON, TEST TERMINAL NOT GROUNDED. NOTE CHECK ENGINE LIGHT.

IF LIGHT IS OFF, CHECK ENGINE LIGHT CIRCUIT. IF LIGHT IS ON, REMOVE FUEL PUMP FUSE. IF BLOWN, GO TO FUEL SYSTEM NO PRESSURE TEST CHART. IF OKAY, DISCONNECT INJECTOR HARNESS CONNECTOR.

CONNECT A TEST LIGHT BETWEEN TERMINALS "A" AND "B", AND "A" AND "C" OF HARNESS CONNECTOR ECM SIDE AND NOTE LIGHT WHILE CRANKING.

LIGHT BLINKS ON BOTH.

IF LIGHT STEADY ON ONE OR BOTH, CHECK INJECTOR DRIVER CIRCUIT WITH STEADY LIGHT FOR SHORT TO GROUND.

IF THERE IS NO LIGHT, PROBE TERMINAL "A" OF INJECTION HARNESS CONNECTOR ON ECM SIDE WITH TEST LIGHT TO GROUND (IGNITION ON).

DISCONNECT ONE SPARK PLUG. INSTALL HEI SPARK TESTER AND CHECK FOR SPARK.

IF THERE IS NO LIGHT, THERE IS AN OPEN IN IGN. CIRCUIT 439.

IF NO SPARK, PROBLEM IS HEI SYSTEM.

IF NOT SHORTED, CHECK RESISTANCE ACROSS EACH INJECTOR. RESISTANCE SHOULD BE GREATER THAN 4 OHMS FOR EACH INJECTOR.

IF THERE IS LIGHT, RECONNECT INJECTORS. TURN OFF IGNITION. DISCONNECT "A"-"B" AND "C"-"D" ECM CONNECTORS. WITH IGNITION ON, CHECK TERMINALS D14 AND D15 WITH TEST LIGHT TO GROUND. LIGHT SHOULD BE ON AT BOTH TERMINALS. RECONNECT ECM.

IF SPARK, INSTALL FUEL PRESSURE GAUGE AND NOTE PRESSURE WHEN IGNITION IS ON AND FUEL PUMP HAS STOPPED RUNNING. PRESSURE SHOULD BE 34-40 PSI.

IF OKAY, ECM IS FAULTY. IF NOT OKAY, REPLACE ECM AND ANY INJECTORS WITH RESISTANCE OF LESS THAN 4 OHMS.

IF OKAY, CHECK FOR OPEN COOLANT SENSOR CIRCUIT, STICKING TPS OR EGR VALVE, CONTAMINATION IN FUEL, PRESENCE OF CRANK SIGNAL. IF ALL OKAY, REVIEW SYMPTOMS.

NOT OKAY, CIRCUIT 844 OR 841 OPEN.

IF OKAY, USE VOLTMETER IN AC VOLTAGE POSITION. PROBE ECM TERMINAL B5 AND NOTE VOLTAGE WHILE CRANKING.

IF NO PRESSURE, GO TO FUEL SYSTEM NO PRESSURE CHART. IF PRESSURE, BUT LESS THAN 34 PSI, GO TO FUEL SYSTEM DIAGNOSIS CHART.

LESS THAN 1 VOLT ECM CONNECTOR B5 OR ECM IS FAULTY.

IF LESS THAN 1 VOLT, CHECK FOR OPEN OR SHORT TO GROUND IN CIRCUIT 430.

IF OKAY, DISCONNECT SPARK PLUG AND USE HEI TESTER TO CHECK FOR SPARK. IF NO SPARK, PROBLEM IS HEI SYSTEM. IF SPARK, IGNITION MODULE CONNECTOR TERMINAL "R" OR MODULE IS DEFECTIVE.

ENGINE CRANKS BUT WILL NOT RUN (1984-87 3.8L TURBO)

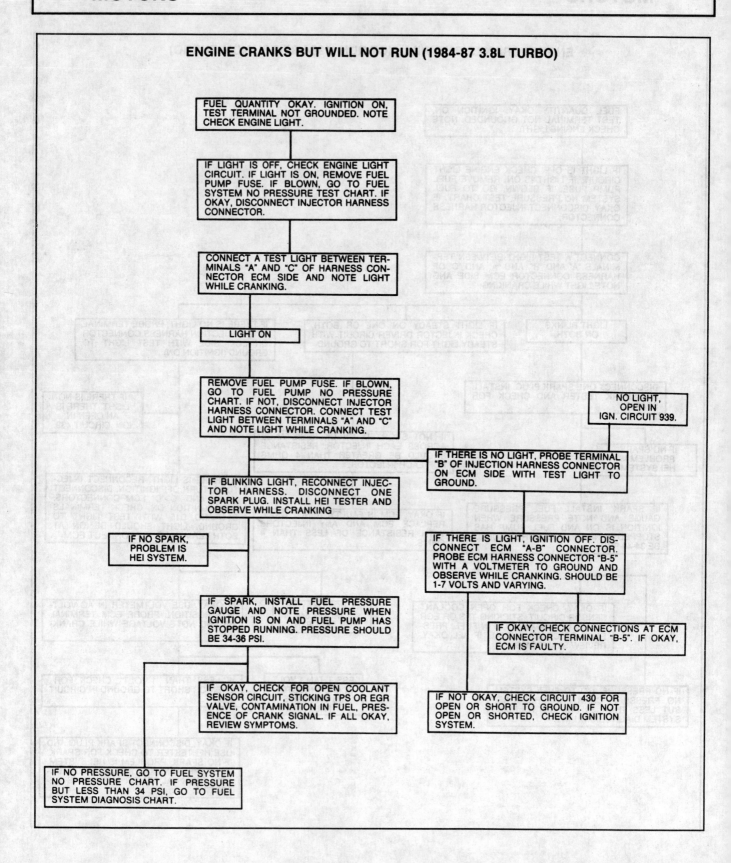

FUEL QUANTITY OKAY. IGNITION ON, TEST TERMINAL NOT GROUNDED. NOTE CHECK ENGINE LIGHT.

IF LIGHT IS OFF, CHECK ENGINE LIGHT CIRCUIT. IF LIGHT IS ON, REMOVE FUEL PUMP FUSE. IF BLOWN, GO TO FUEL SYSTEM NO PRESSURE TEST CHART. IF OKAY, DISCONNECT INJECTOR HARNESS CONNECTOR.

CONNECT A TEST LIGHT BETWEEN TERMINALS "A" AND "C" OF HARNESS CONNECTOR ECM SIDE AND NOTE LIGHT WHILE CRANKING.

LIGHT ON

REMOVE FUEL PUMP FUSE. IF BLOWN, GO TO FUEL PUMP NO PRESSURE CHART. IF NOT, DISCONNECT INJECTOR HARNESS CONNECTOR. CONNECT TEST LIGHT BETWEEN TERMINALS "A" AND "C" AND NOTE LIGHT WHILE CRANKING.

NO LIGHT, OPEN IN IGN. CIRCUIT 939.

IF THERE IS NO LIGHT, PROBE TERMINAL "B" OF INJECTION HARNESS CONNECTOR ON ECM SIDE WITH TEST LIGHT TO GROUND.

IF BLINKING LIGHT, RECONNECT INJECTOR HARNESS. DISCONNECT ONE SPARK PLUG. INSTALL HEI TESTER AND OBSERVE WHILE CRANKING

IF NO SPARK, PROBLEM IS HEI SYSTEM.

IF THERE IS LIGHT, IGNITION OFF. DISCONNECT ECM "A-B" CONNECTOR. PROBE ECM HARNESS CONNECTOR "B-5" WITH A VOLTMETER TO GROUND AND OBSERVE WHILE CRANKING. SHOULD BE 1-7 VOLTS AND VARYING.

IF SPARK, INSTALL FUEL PRESSURE GAUGE AND NOTE PRESSURE WHEN IGNITION IS ON AND FUEL PUMP HAS STOPPED RUNNING. PRESSURE SHOULD BE 34-36 PSI.

IF OKAY, CHECK CONNECTIONS AT ECM CONNECTOR TERMINAL "B-5". IF OKAY, ECM IS FAULTY.

IF OKAY, CHECK FOR OPEN COOLANT SENSOR CIRCUIT, STICKING TPS OR EGR VALVE, CONTAMINATION IN FUEL, PRESENCE OF CRANK SIGNAL. IF ALL OKAY, REVIEW SYMPTOMS.

IF NOT OKAY, CHECK CIRCUIT 430 FOR OPEN OR SHORT TO GROUND. IF NOT OPEN OR SHORTED, CHECK IGNITION SYSTEM.

IF NO PRESSURE, GO TO FUEL SYSTEM NO PRESSURE CHART. IF PRESSURE BUT LESS THAN 34 PSI, GO TO FUEL SYSTEM DIAGNOSIS CHART.

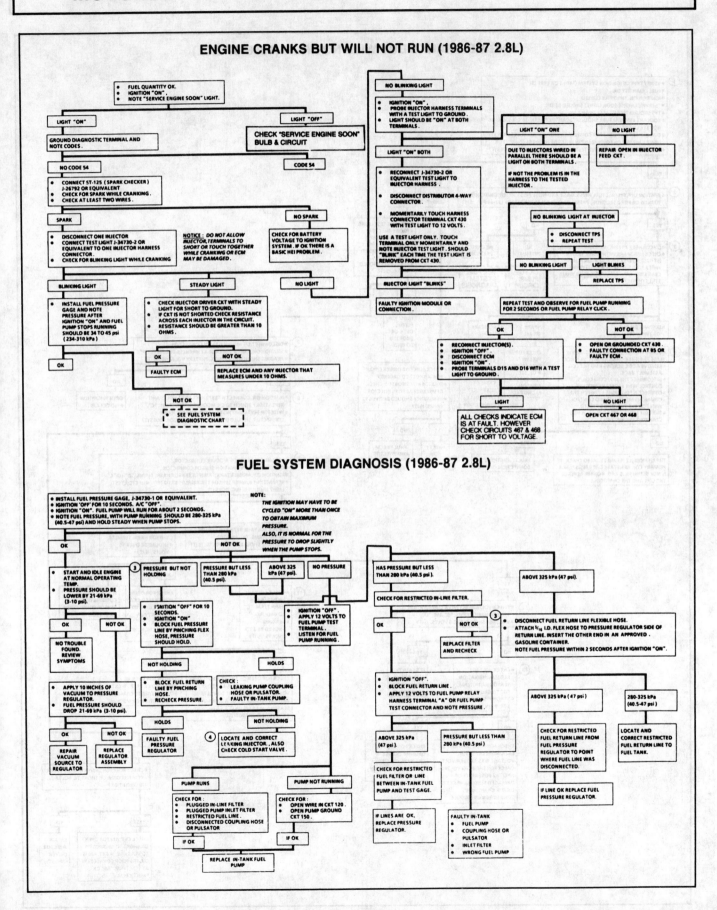

ENGINE CRANKS BUT WILL NOT RUN (1986-87 2.8L)

FUEL SYSTEM DIAGNOSIS (1986-87 2.8L)

ENGINE CRANKS BUT WILL NOT RUN (1986-87 3.0L)

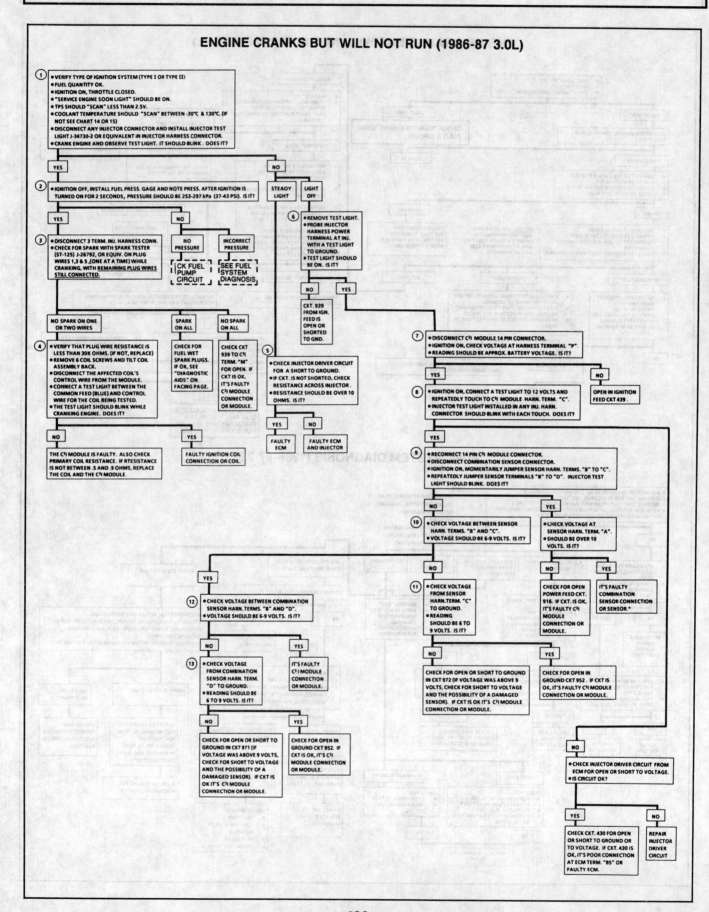

1 • VERIFY TYPE OF IGNITION SYSTEM (TYPE I OR TYPE II)
• FUEL QUANTITY OK.
• IGNITION ON, THROTTLE CLOSED.
• "SERVICE ENGINE SOON LIGHT" SHOULD BE ON.
• TPS SHOULD "SCAN" LESS THAN 2.5V.
• COOLANT TEMPERATURE SHOULD "SCAN" BETWEEN -30°C & 130°C. (IF NOT SEE CHART 14 OR 15)
• DISCONNECT ANY INJECTOR CONNECTOR AND INSTALL INJECTOR TEST LIGHT J-34730-2 OR EQUIVALENT IN INJECTOR HARNESS CONNECTOR.
• CRANK ENGINE AND OBSERVE TEST LIGHT. IT SHOULD BLINK. DOES IT?

YES | NO

STEADY LIGHT | LIGHT OFF

2 • IGNITION OFF, INSTALL FUEL PRESS. GAGE AND NOTE PRESS. AFTER IGNITION IS TURNED ON FOR 2 SECONDS, PRESSURE SHOULD BE 253-297 kPa (37-43 PSI). IS IT?

YES | NO

NO PRESSURE | INCORRECT PRESSURE

CK FUEL PUMP CIRCUIT | SEE FUEL SYSTEM DIAGNOSIS

6 • REMOVE TEST LIGHT.
• PROBE INJECTOR HARNESS POWER TERMINAL AT INJ. WITH A TEST LIGHT TO GROUND.
• TEST LIGHT SHOULD BE ON. IS IT?

NO | YES

CKT. 939 FROM IGN. FEED IS OPEN OR SHORTED TO GND.

3 • DISCONNECT 3 TERM. INJ. HARNESS CONN.
• CHECK FOR SPARK WITH SPARK TESTER (ST-125) J-26792, OR EQUIV. ON PLUG WIRES 1,3 & 5, (ONE AT A TIME) WHILE CRANKING, WITH REMAINING PLUG WIRES STILL CONNECTED.

NO SPARK ON ONE OR TWO WIRES | SPARK ON ALL | NO SPARK ON ALL

4 • VERIFY THAT PLUG WIRE RESISTANCE IS LESS THAN 30K OHMS. (IF NOT, REPLACE)
• REMOVE 6 COIL SCREWS AND TILT COIL ASSEMBLY BACK.
• DISCONNECT THE AFFECTED COIL'S CONTROL WIRE FROM THE MODULE.
• CONNECT A TEST LIGHT BETWEEN THE COMMON FEED (BLUE) AND CONTROL WIRE FOR THE COIL BEING TESTED.
• THE TEST LIGHT SHOULD BLINK WHILE CRANKING ENGINE. DOES IT?

CHECK FOR FUEL WET SPARK PLUGS. IF OK, SEE "DIAGNOSTIC AIDS" ON FACING PAGE.

CHECK CKT 939 TO C³ TERM. "M" FOR OPEN. IF CKT IS OK, IT'S FAULTY C³ MODULE CONNECTION OR MODULE.

5 • CHECK INJECTOR DRIVER CIRCUIT FOR A SHORT TO GROUND.
• IF CKT. IS NOT SHORTED, CHECK RESISTANCE ACROSS INJECTOR.
• RESISTANCE SHOULD BE OVER 10 OHMS. IS IT?

YES | NO

FAULTY ECM | FAULTY ECM AND INJECTOR

7 • DISCONNECT C³ MODULE 14 PIN CONNECTOR.
• IGNITION ON, CHECK VOLTAGE AT HARNESS TERMINAL "P".
• READING SHOULD BE APPROX. BATTERY VOLTAGE. IS IT?

YES | NO

OPEN IN IGNITION FEED CKT 439.

8 • IGNITION ON, CONNECT A TEST LIGHT TO 12 VOLTS AND REPEATEDLY TOUCH TO C³ MODULE HARN. TERM. "C".
• INJECTOR TEST LIGHT INSTALLED IN ANY INJ. HARN. CONNECTOR SHOULD BLINK WITH EACH TOUCH. DOES IT?

YES

NO | YES

THE C³ MODULE IS FAULTY. ALSO CHECK PRIMARY COIL RESISTANCE. IF RTESISTANCE IS NOT BETWEEN .5 AND .9 OHMS, REPLACE THE COIL AND THE C³ MODULE.

FAULTY IGNITION COIL CONNECTION OR COIL.

9 • RECONNECT 14 PIN C³ MODULE CONNECTOR.
• DISCONNECT COMBINATION SENSOR CONNECTOR.
• IGNITION ON, MOMENTARILY JUMPER SENSOR HARN. TERMS. "B" TO "C".
• REPEATEDLY JUMPER SENSOR TERMINALS "B" TO "D". INJECTOR TEST LIGHT SHOULD BLINK. DOES IT?

NO | YES

10 • CHECK VOLTAGE BETWEEN SENSOR HARN. TERMS. "B" AND "C".
• VOLTAGE SHOULD BE 6-9 VOLTS. IS IT?

• CHECK VOLTAGE AT SENSOR HARN. TERM. "A".
• SHOULD BE OVER 10 VOLTS. IS IT?

NO | NO | YES

11 • CHECK VOLTAGE FROM SENSOR HARN. TERM. "C" TO GROUND.
• READING SHOULD BE 6 TO 9 VOLTS. IS IT?

CHECK FOR OPEN POWER FEED CKT. 916. IF CKT. IS OK, IT'S FAULTY C³ MODULE CONNECTION OR MODULE.

IT'S FAULTY COMBINATION SENSOR CONNECTION OR SENSOR.*

YES

12 • CHECK VOLTAGE BETWEEN COMBINATION SENSOR HARN. TERMS. "B" AND "D".
• VOLTAGE SHOULD BE 6-9 VOLTS. IS IT?

NO | YES

13 • CHECK VOLTAGE FROM COMBINATION SENSOR HARN. TERM. "D" TO GROUND.
• READING SHOULD BE 6 TO 9 VOLTS. IS IT?

IT'S FAULTY C³ MODULE CONNECTION OR MODULE.

NO | YES

CHECK FOR OPEN OR SHORT TO GROUND IN CKT 971 (IF VOLTAGE WAS ABOVE 9 VOLTS, CHECK FOR SHORT TO VOLTAGE AND THE POSSIBILITY OF A DAMAGED SENSOR). IF CKT IS OK IT'S C³ MODULE CONNECTION OR MODULE.

CHECK FOR OPEN IN GROUND CKT 952. IF CKT IS OK, IT'S C³ MODULE CONNECTION OR MODULE.

NO | YES

CHECK FOR OPEN OR SHORT TO GROUND IN CKT 972 (IF VOLTAGE WAS ABOVE 9 VOLTS, CHECK FOR SHORT TO VOLTAGE AND THE POSSIBILITY OF A DAMAGED SENSOR). IF CKT IS OK IT'S C³ MODULE CONNECTION OR MODULE.

CHECK FOR OPEN IN GROUND CKT 952. IF CKT IS OK, IT'S FAULTY C³ MODULE CONNECTION OR MODULE.

NO

• CHECK INJECTOR DRIVER CIRCUIT FROM ECM FOR OPEN OR SHORT TO VOLTAGE.
• IS CIRCUIT OK?

YES | NO

CHECK CKT. 430 FOR OPEN OR SHORT TO GROUND OR TO VOLTAGE. IF CKT. 430 IS OK, IT'S POOR CONNECTION AT ECM TERM. "B5" OR FAULTY ECM.

REPAIR INJECTOR DRIVER CIRCUIT

ENGINE CRANKS BUT WILL NOT RUN (1986 5.0L & 5.7L)

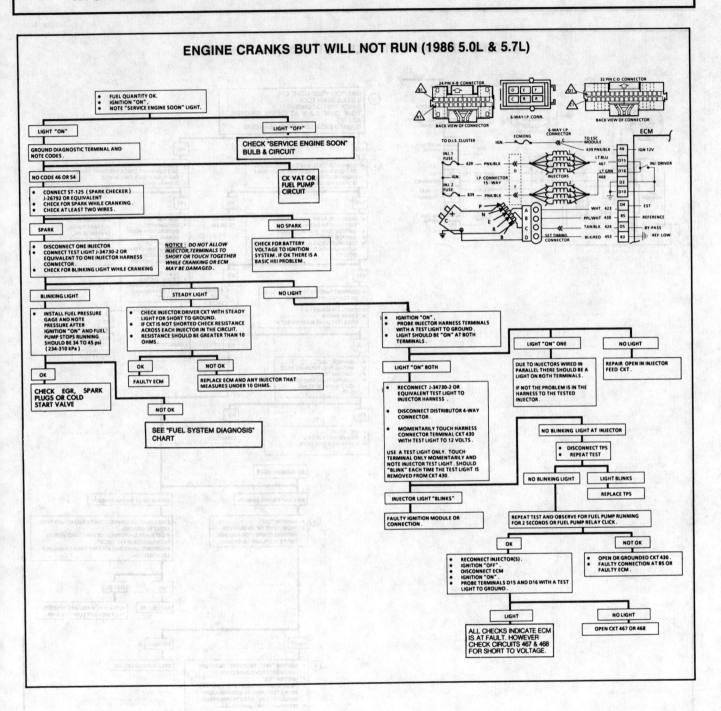

ENGINE CRANKS BUT WILL NOT RUN (1987-89 2.0L)

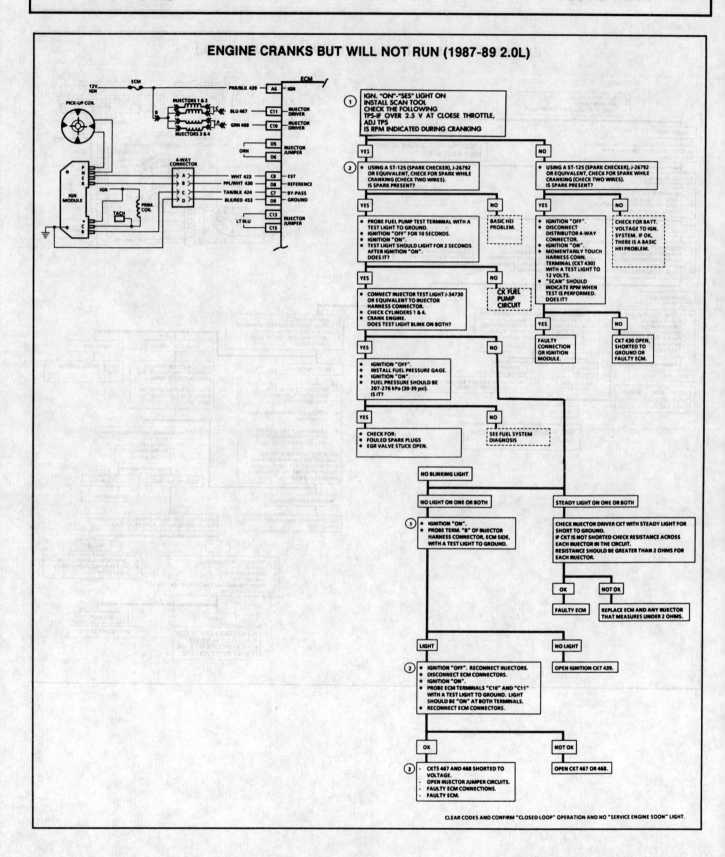

CLEAR CODES AND CONFIRM "CLOSED LOOP" OPERATION AND NO "SERVICE ENGINE SOON" LIGHT.

ENGINE CRANKS BUT WILL NOT RUN (1987-89 5.0L & 5.7L)

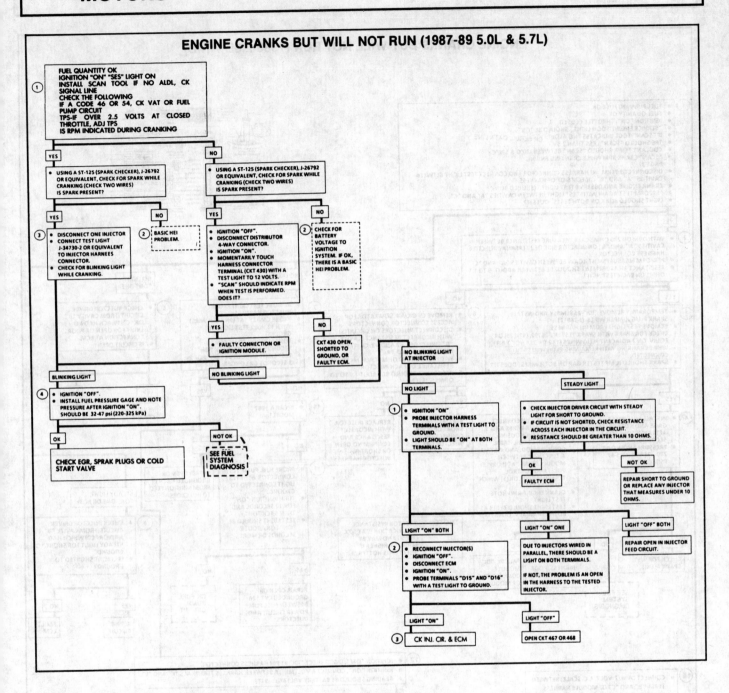

1
- FUEL QUANTITY OK
- IGNITION "ON" "SES" LIGHT ON
- INSTALL SCAN TOOL IF NO ALDL, CK SIGNAL LINE
- CHECK THE FOLLOWING
- IF A CODE 46 OR 54, CK VAT OR FUEL PUMP CIRCUIT
- TPS-IF OVER 2.5 VOLTS AT CLOSED THROTTLE, ADJ TPS
- IS RPM INDICATED DURING CRANKING

YES

- USING A ST-125 (SPARK CHECKER), J-26792 OR EQUIVALENT, CHECK FOR SPARK WHILE CRANKING (CHECK TWO WIRES) IS SPARK PRESENT?

YES

3
- DISCONNECT ONE INJECTOR
- CONNECT TEST LIGHT J-34730-2 OR EQUIVALENT TO INJECTOR HARNESS CONNECTOR.
- CHECK FOR BLINKING LIGHT WHILE CRANKING.

NO

2 BASIC HEI PROBLEM.

NO

- USING A ST-125 (SPARK CHECKER), J-26792 OR EQUIVALENT, CHECK FOR SPARK WHILE CRANKING (CHECK TWO WIRES) IS SPARK PRESENT?

YES

- IGNITION "OFF".
- DISCONNECT DISTRIBUTOR 4-WAY CONNECTOR.
- IGNITION "ON".
- MOMENTARILY TOUCH HARNESS CONNECTOR TERMINAL (CKT 430) WITH A TEST LIGHT TO 12 VOLTS.
- "SCAN" SHOULD INDICATE RPM WHEN TEST IS PERFORMED. DOES IT?

NO

2 CHECK FOR BATTERY VOLTAGE TO IGNITION SYSTEM. IF OK, THERE IS A BASIC HEI PROBLEM.

YES
- FAULTY CONNECTION OR IGNITION MODULE.

NO
- CKT 430 OPEN, SHORTED TO GROUND, OR FAULTY ECM.

BLINKING LIGHT

4
- IGNITION "OFF".
- INSTALL FUEL PRESSURE GAGE AND NOTE PRESSURE AFTER IGNITION "ON". SHOULD BE 32-47 psi (220-325 kPa)

OK
- CHECK EGR, SPRAK PLUGS OR COLD START VALVE

NOT OK
- SEE FUEL SYSTEM DIAGNOSIS

NO BLINKING LIGHT

NO BLINKING LIGHT AT INJECTOR

NO LIGHT

1
- IGNITION "ON"
- PROBE INJECTOR HARNESS TERMINALS WITH A TEST LIGHT TO GROUND.
- LIGHT SHOULD BE "ON" AT BOTH TERMINALS.

STEADY LIGHT

- CHECK INJECTOR DRIVER CIRCUIT WITH STEADY LIGHT FOR SHORT TO GROUND.
- IF CIRCUIT IS NOT SHORTED, CHECK RESISTANCE ACROSS EACH INJECTOR IN THE CIRCUIT.
- RESISTANCE SHOULD BE GREATER THAN 10 OHMS.

OK
- FAULTY ECM

NOT OK
- REPAIR SHORT TO GROUND OR REPLACE ANY INJECTOR THAT MEASURES UNDER 10 OHMS.

LIGHT "ON" BOTH

2
- RECONNECT INJECTOR(S)
- IGNITION "OFF".
- DISCONNECT ECM
- IGNITION "ON".
- PROBE TERMINALS "D15" AND "D16" WITH A TEST LIGHT TO GROUND.

LIGHT "ON" ONE

- DUE TO INJECTORS WIRED IN PARALLEL, THERE SHOULD BE A LIGHT ON BOTH TERMINALS.
- IF NOT, THE PROBLEM IS AN OPEN IN THE HARNESS TO THE TESTED INJECTOR.

LIGHT "OFF" BOTH

- REPAIR OPEN IN INJECTOR FEED CIRCUIT.

LIGHT "ON"

3 CK INJ. CIR. & ECM

LIGHT "OFF"

OPEN CKT 467 OR 468

ENGINE CRANKS BUT WILL NOT RUN (1988 2.3L VIN D)

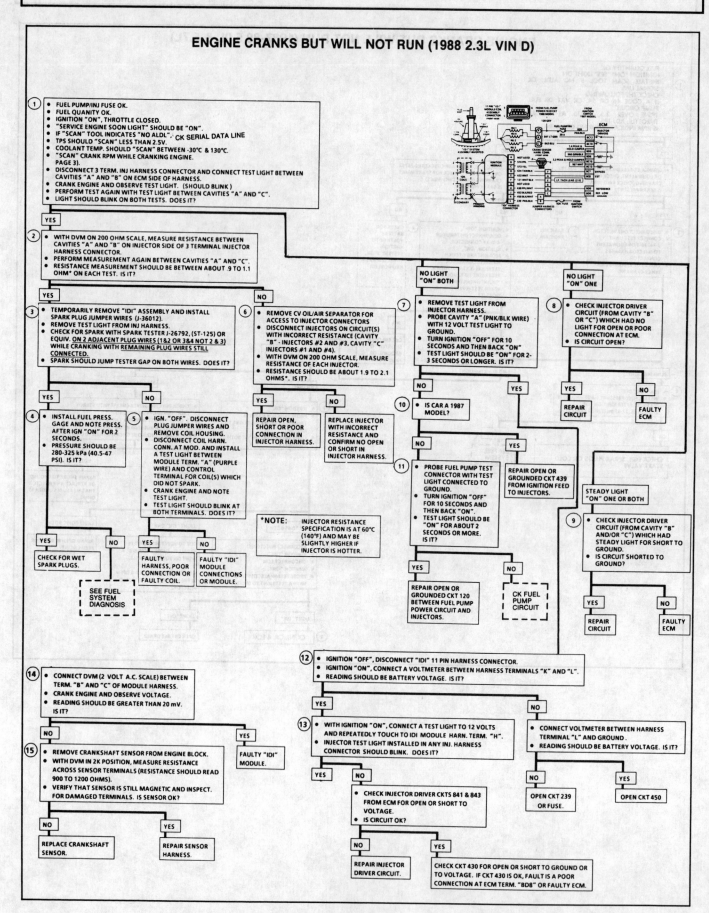

(1)
- FUEL PUMP/INJ FUSE OK.
- FUEL QUANITY OK.
- IGNITION "ON", THROTTLE CLOSED.
- "SERVICE ENGINE SOON LIGHT" SHOULD BE "ON".
- IF "SCAN" TOOL INDICATES "NO ALDL". CK SERIAL DATA LINE
- TPS SHOULD "SCAN" LESS THAN 2.5V.
- COOLANT TEMP. SHOULD "SCAN" BETWEEN -30°C & 130°C.
- "SCAN" CRANK RPM WHILE CRANKING ENGINE. PAGE 3).
- DISCONNECT 3 TERM. INJ HARNESS CONNECTOR AND CONNECT TEST LIGHT BETWEEN CAVITIES "A" AND "B" ON ECM SIDE OF HARNESS.
- CRANK ENGINE AND OBSERVE TEST LIGHT. (SHOULD BLINK)
- PERFORM TEST AGAIN WITH TEST LIGHT BETWEEN CAVITIES "A" AND "C".
- LIGHT SHOULD BLINK ON BOTH TESTS. DOES IT?

YES

(2)
- WITH DVM ON 200 OHM SCALE, MEASURE RESISTANCE BETWEEN CAVITIES "A" AND "B" ON INJECTOR SIDE OF 3 TERMINAL INJECTOR HARNESS CONNECTOR.
- PERFORM MEASUREMENT AGAIN BETWEEN CAVITIES "A" AND "C".
- RESISTANCE MEASUREMENT SHOULD BE BETWEEN ABOUT .9 TO 1.1 OHM* ON EACH TEST. IS IT?

YES / **NO**

(3)
- TEMPORARILY REMOVE "IDI" ASSEMBLY AND INSTALL SPARK PLUG JUMPER WIRES (J-36012).
- REMOVE TEST LIGHT FROM INJ HARNESS.
- CHECK FOR SPARK WITH SPARK TESTER J-26792, (ST-125) OR EQUIV. ON 2 ADJACENT PLUG WIRES (1&2 OR 3&4 NOT 2 & 3) WHILE CRANKING WITH REMAINING PLUG WIRES STILL CONNECTED.
- SPARK SHOULD JUMP TESTER GAP ON BOTH WIRES. DOES IT?

(6)
- REMOVE CV OIL/AIR SEPARATOR FOR ACCESS TO INJECTOR CONNECTORS
- DISCONNECT INJECTORS ON CIRCUIT(S) WITH INCORRECT RESISTANCE (CAVITY "B" - INJECTORS #2 AND #3, CAVITY "C" INJECTORS #1 AND #4).
- WITH DVM ON 200 OHM SCALE, MEASURE RESISTANCE OF EACH INJECTOR.
- RESISTANCE SHOULD BE ABOUT 1.9 TO 2.1 OHMS*. IS IT?

YES / **NO**

(7)
- REMOVE TEST LIGHT FROM INJECTOR HARNESS.
- PROBE CAVITY "A" (PNK/BLK WIRE) WITH 12 VOLT TEST LIGHT TO GROUND.
- TURN IGNITION "OFF" FOR 10 SECONDS AND THEN BACK "ON"
- TEST LIGHT SHOULD BE "ON" FOR 2-3 SECONDS OR LONGER. IS IT?

NO LIGHT "ON" BOTH

(8)
- CHECK INJECTOR DRIVER CIRCUIT (FROM CAVITY "B" OR "C") WHICH HAD NO LIGHT FOR OPEN OR POOR CONNECTION AT ECM.
- IS CIRCUIT OPEN?

NO LIGHT "ON" ONE

YES

(4)
- INSTALL FUEL PRESS. GAGE AND NOTE PRESS. AFTER IGN "ON" FOR 2 SECONDS.
- PRESSURE SHOULD BE 280-325 kPa (40.5-47 PSI). IS IT?

(5)
- IGN. "OFF". DISCONNECT PLUG JUMPER WIRES AND REMOVE COIL HOUSING.
- DISCONNECT COIL HARN. CONN. AT MOD. AND INSTALL A TEST LIGHT BETWEEN MODULE TERM. "A" (PURPLE WIRE) AND CONTROL TERMINAL FOR COIL(S) WHICH DID NOT SPARK.
- CRANK ENGINE AND NOTE TEST LIGHT.
- TEST LIGHT SHOULD BLINK AT BOTH TERMINALS. DOES IT?

REPAIR OPEN, SHORT OR POOR CONNECTION IN INJECTOR HARNESS.

REPLACE INJECTOR WITH INCORRECT RESISTANCE AND CONFIRM NO OPEN OR SHORT IN INJECTOR HARNESS.

(10) IS CAR A 1987 MODEL?

NO

(11)
- PROBE FUEL PUMP TEST CONNECTOR WITH TEST LIGHT CONNECTED TO GROUND.
- TURN IGNITION "OFF" FOR 10 SECONDS AND THEN BACK "ON"
- TEST LIGHT SHOULD BE "ON" FOR ABOUT 2 SECONDS OR MORE. IS IT?

YES

REPAIR OPEN OR GROUNDED CKT 439 FROM IGNITION FEED TO INJECTORS.

YES
REPAIR CIRCUIT

NO
FAULTY ECM

STEADY LIGHT "ON" ONE OR BOTH

(9)
- CHECK INJECTOR DRIVER CIRCUIT (FROM CAVITY "B" AND/OR "C") WHICH HAD STEADY LIGHT FOR SHORT TO GROUND.
- IS CIRCUIT SHORTED TO GROUND?

YES
CHECK FOR WET SPARK PLUGS.

NO

YES
FAULTY HARNESS, POOR CONNECTION OR FAULTY COIL.

NO
FAULTY "IDI" MODULE CONNECTIONS OR MODULE.

*NOTE: INJECTOR RESISTANCE SPECIFICATION IS AT 60°C (140°F) AND MAY BE SLIGHTLY HIGHER IF INJECTOR IS HOTTER.

YES
REPAIR OPEN OR GROUNDED CKT 120 BETWEEN FUEL PUMP POWER CIRCUIT AND INJECTORS.

NO
CK FUEL PUMP CIRCUIT

YES
REPAIR CIRCUIT

NO
FAULTY ECM

SEE FUEL SYSTEM DIAGNOSIS

(14)
- CONNECT DVM (2 VOLT A.C. SCALE) BETWEEN TERM. "B" AND "C" OF MODULE HARNESS.
- CRANK ENGINE AND OBSERVE VOLTAGE.
- READING SHOULD BE GREATER THAN 20 mV. IS IT?

(12)
- IGNITION "OFF", DISCONNECT "IDI" 11 PIN HARNESS CONNECTOR.
- IGNITION "ON", CONNECT A VOLTMETER BETWEEN HARNESS TERMINALS "K" AND "L".
- READING SHOULD BE BATTERY VOLTAGE. IS IT?

YES

NO

NO

YES
FAULTY "IDI" MODULE.

(13)
- WITH IGNITION "ON", CONNECT A TEST LIGHT TO 12 VOLTS AND REPEATEDLY TOUCH TO IDI MODULE HARN. TERM. "H".
- INJECTOR TEST LIGHT INSTALLED IN ANY INJ. HARNESS CONNECTOR SHOULD BLINK. DOES IT?

- CONNECT VOLTMETER BETWEEN HARNESS TERMINAL "L" AND GROUND .
- READING SHOULD BE BATTERY VOLTAGE. IS IT?

(15)
- REMOVE CRANKSHAFT SENSOR FROM ENGINE BLOCK.
- WITH DVM IN 2K POSITION, MEASURE RESISTANCE ACROSS SENSOR TERMINALS (RESISTANCE SHOULD READ 900 TO 1200 OHMS).
- VERIFY THAT SENSOR IS STILL MAGNETIC AND INSPECT. FOR DAMAGED TERMINALS. IS SENSOR OK?

YES / **NO**

- CHECK INJECTOR DRIVER CKTS 841 & 843 FROM ECM FOR OPEN OR SHORT TO VOLTAGE.
- IS CIRCUIT OK?

NO
OPEN CKT 239 OR FUSE.

YES
OPEN CKT 450

NO
REPLACE CRANKSHAFT SENSOR.

YES
REPAIR SENSOR HARNESS.

NO
REPAIR INJECTOR DRIVER CIRCUIT.

YES
CHECK CKT 430 FOR OPEN OR SHORT TO GROUND OR TO VOLTAGE. IF CKT 430 IS OK, FAULT IS A POOR CONNECTION AT ECM TERM. "BD8" OR FAULTY ECM.

ENGINE CRANKS BUT WILL NOT RUN (1988-89 2.8L VIN S, "F" BODY)

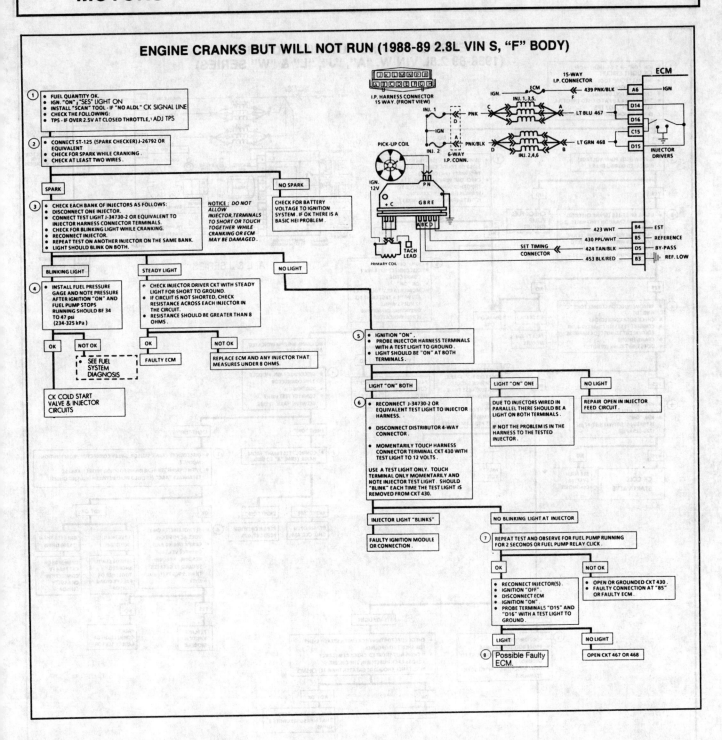

I.P. HARNESS CONNECTOR 15 WAY. (FRONT VIEW)
15-WAY I.P. CONNECTOR
ECM
IGN.
ECM
F
439 PNK/BLK
A6
IGN
INJ. 1, 3, 5.
INJ. 1
C
A
PNK
LT BLU 467
D14
D16
C15
D15
IGN
D
A
PNK/BLK
B
LT GRN 468
INJ. 2
6-WAY I.P. CONN.
INJ. 2,4,6
INJECTOR DRIVERS
PICK-UP COIL
IGN. 12V
P N
+ C
G B R E
A B C D
423 WHT
B4
EST
430 PPL/WHT
B5
REFERENCE
SET TIMING CONNECTOR
424 TAN/BLK
D5
BY-PASS
453 BLK/RED
B3
REF. LOW
TACH LEAD
PRIMARY COIL

1
- FUEL QUANTITY OK.
- IGN. "ON", "SES" LIGHT ON
- INSTALL "SCAN" TOOL - IF "NO ALDL" CK SIGNAL LINE
- CHECK THE FOLLOWING:
- TPS - IF OVER 2.5V AT CLOSED THROTTLE, ADJ TPS

2
- CONNECT ST-125 (SPARK CHECKER) J-26792 OR EQUIVALENT
- CHECK FOR SPARK WHILE CRANKING.
- CHECK AT LEAST TWO WIRES.

SPARK / **NO SPARK**

NO SPARK: CHECK FOR BATTERY VOLTAGE TO IGNITION SYSTEM. IF OK THERE IS A BASIC HEI PROBLEM.

SPARK:

3
- CHECK EACH BANK OF INJECTORS AS FOLLOWS:
- DISCONNECT ONE INJECTOR.
- CONNECT TEST LIGHT J-34730-2 OR EQUIVALENT TO INJECTOR HARNESS CONNECTOR TERMINALS.
- CHECK FOR BLINKING LIGHT WHILE CRANKING.
- RECONNECT INJECTOR.
- REPEAT TEST ON ANOTHER INJECTOR ON THE SAME BANK.
- LIGHT SHOULD BLINK ON BOTH.

NOTICE: DO NOT ALLOW INJECTOR, TERMINALS TO SHORT OR TOUCH TOGETHER WHILE CRANKING OR ECM MAY BE DAMAGED.

BLINKING LIGHT / **STEADY LIGHT** / **NO LIGHT**

BLINKING LIGHT:

4
- INSTALL FUEL PRESSURE GAGE AND NOTE PRESSURE AFTER IGNITION "ON" AND FUEL PUMP STOPS RUNNING SHOULD BE 34 TO 47 psi (234-325 kPa)

- **OK**: CK COLD START VALVE & INJECTOR CIRCUITS
- **NOT OK**: SEE FUEL SYSTEM DIAGNOSIS

STEADY LIGHT:
- CHECK INJECTOR DRIVER CKT WITH STEADY LIGHT FOR SHORT TO GROUND.
- IF CIRCUIT IS NOT SHORTED, CHECK RESISTANCE ACROSS EACH INJECTOR IN THE CIRCUIT.
- RESISTANCE SHOULD BE GREATER THAN 8 OHMS.

- **OK**: FAULTY ECM
- **NOT OK**: REPLACE ECM AND ANY INJECTOR THAT MEASURES UNDER 8 OHMS.

NO LIGHT:

5
- IGNITION "ON".
- PROBE INJECTOR HARNESS TERMINALS WITH A TEST LIGHT TO GROUND.
- LIGHT SHOULD BE "ON" AT BOTH TERMINALS.

LIGHT "ON" BOTH / **LIGHT "ON" ONE** / **NO LIGHT**

LIGHT "ON" ONE: DUE TO INJECTORS WIRED IN PARALLEL THERE SHOULD BE A LIGHT ON BOTH TERMINALS. IF NOT THE PROBLEM IS IN THE HARNESS TO THE TESTED INJECTOR.

NO LIGHT: REPAIR OPEN IN INJECTOR FEED CIRCUIT.

LIGHT "ON" BOTH:

6
- RECONNECT J-34730-2 OR EQUIVALENT TEST LIGHT TO INJECTOR HARNESS.
- DISCONNECT DISTRIBUTOR 4-WAY CONNECTOR.
- MOMENTARILY TOUCH HARNESS CONNECTOR TERMINAL CKT 430 WITH TEST LIGHT TO 12 VOLTS.

USE A TEST LIGHT ONLY. TOUCH TERMINAL ONLY MOMENTARILY AND NOTE INJECTOR TEST LIGHT. SHOULD "BLINK" EACH TIME THE TEST LIGHT IS REMOVED FROM CKT 430.

INJECTOR LIGHT "BLINKS": FAULTY IGNITION MODULE OR CONNECTION.

NO BLINKING LIGHT AT INJECTOR:

7 REPEAT TEST AND OBSERVE FOR FUEL PUMP RUNNING FOR 2 SECONDS OR FUEL PUMP RELAY CLICK.

OK:
- RECONNECT INJECTOR(S).
- IGNITION "OFF".
- DISCONNECT ECM
- IGNITION "ON".
- PROBE TERMINALS "D15" AND "D16" WITH A TEST LIGHT TO GROUND.

- **LIGHT**: **8** Possible Faulty ECM.
- **NO LIGHT**: OPEN CKT 467 OR 468

NOT OK:
- OPEN OR GROUNDED CKT 430.
- FAULTY CONNECTION AT "B5" OR FAULTY ECM.

ENGINE CRANKS BUT WILL NOT RUN
(1988-89 2.8L VIN W, "A", "J", "L" & "W" SERIES)

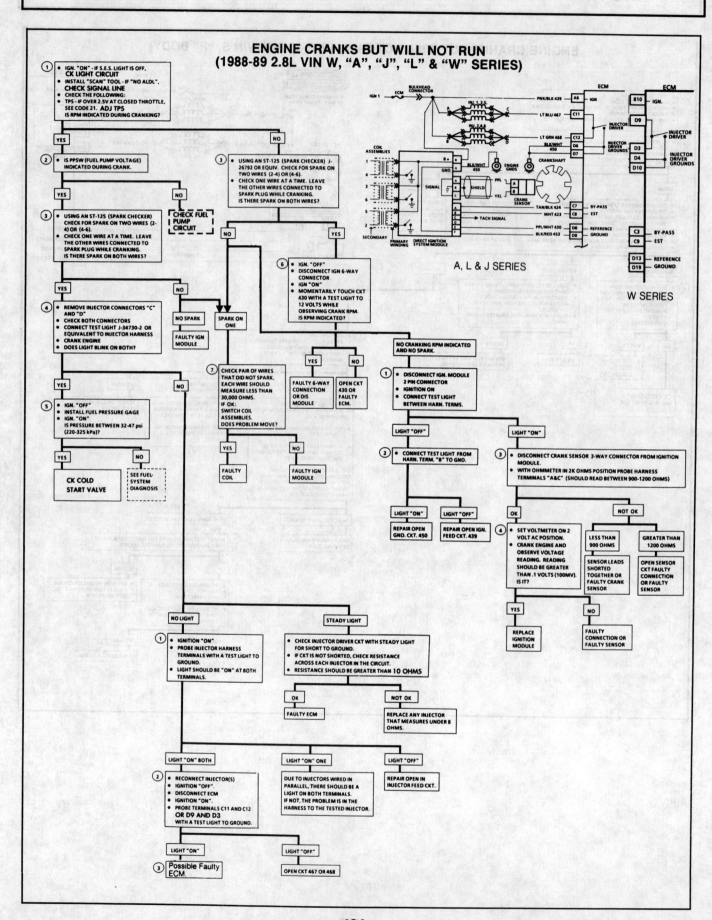

① IGN. "ON" - IF S.E.S. LIGHT IS OFF, CK LIGHT CIRCUIT
- INSTALL "SCAN" TOOL - IF "NO ALDL", CHECK SIGNAL LINE
- CHECK THE FOLLOWING:
- TPS - IF OVER 2.5V AT CLOSED THROTTLE, SEE CODE 21. ADJ TPS
IS RPM INDICATED DURING CRANKING?

② IS PPSW (FUEL PUMP VOLTAGE) INDICATED DURING CRANK.

③ USING AN ST-125 (SPARK CHECKER) CHECK FOR SPARK ON TWO WIRES (2-4) OR (4-6).
- CHECK ONE WIRE AT A TIME. LEAVE THE OTHER WIRES CONNECTED TO SPARK PLUG WHILE CRANKING. IS THERE SPARK ON BOTH WIRES?

④ REMOVE INJECTOR CONNECTORS "C" AND "D"
- CHECK BOTH CONNECTORS
- CONNECT TEST LIGHT J-34730-2 OR EQUIVALENT TO INJECTOR HARNESS
- CRANK ENGINE
- DOES LIGHT BLINK ON BOTH?

⑤ IGN. "OFF"
- INSTALL FUEL PRESSURE GAGE
- IGN. "ON"
IS PRESSURE BETWEEN 32-47 psi (220-325 kPa)?

CK COLD START VALVE

SEE FUEL-SYSTEM DIAGNOSIS

CHECK FUEL PUMP CIRCUIT

③ USING AN ST-125 (SPARK CHECKER) J-26792 OR EQUIV. CHECK FOR SPARK ON TWO WIRES (2-4) OR (4-6).
- CHECK ONE WIRE AT A TIME. LEAVE THE OTHER WIRES CONNECTED TO SPARK PLUG WHILE CRANKING. IS THERE SPARK ON BOTH WIRES?

NO SPARK — FAULTY IGN MODULE

SPARK ON ONE

⑦ CHECK PAIR OF WIRES THAT DID NOT SPARK. EACH WIRE SHOULD MEASURE LESS THAN 30,000 OHMS. IF OK: SWITCH COIL ASSEMBLIES. DOES PROBLEM MOVE?

FAULTY COIL

FAULTY IGN MODULE

⑥ IGN. "OFF"
- DISCONNECT IGN 6-WAY CONNECTOR
- IGN "ON"
- MOMENTARILY TOUCH CKT 430 WITH A TEST LIGHT TO 12 VOLTS WHILE OBSERVING CRANK RPM. IS RPM INDICATED?

FAULTY 6-WAY CONNECTION OR DIS MODULE

OPEN CKT 430 OR FAULTY ECM.

A, L & J SERIES

W SERIES

NO CRANKING RPM INDICATED AND NO SPARK.

① DISCONNECT IGN. MODULE 2 PIN CONNECTOR
- IGNITION ON
- CONNECT TEST LIGHT BETWEEN HARN. TERMS.

LIGHT "OFF"

LIGHT "ON"

② CONNECT TEST LIGHT FROM HARN. TERM. "B" TO GND.

③ DISCONNECT CRANK SENSOR 3-WAY CONNECTOR FROM IGNITION MODULE.
- WITH OHMMETER IN 2K OHMS POSITION PROBE HARNESS TERMINALS "A&C" (SHOULD READ BETWEEN 900-1200 OHMS)

LIGHT "ON" — REPAIR OPEN GND. CKT. 450

LIGHT "OFF" — REPAIR OPEN IGN. FEED CKT. 439

OK

NOT OK

④ SET VOLTMETER ON 2 VOLT AC POSITION.
- CRANK ENGINE AND OBSERVE VOLTAGE READING. READING SHOULD BE GREATER THAN .1 VOLTS (100MV). IS IT?

LESS THAN 900 OHMS — SENSOR LEADS SHORTED TOGETHER OR FAULTY CRANK SENSOR

GREATER THAN 1200 OHMS — OPEN SENSOR CKT FAULTY CONNECTION OR FAULTY SENSOR

YES — REPLACE IGNITION MODULE

NO — FAULTY CONNECTION OR FAULTY SENSOR

NO LIGHT

① IGNITION "ON"
- PROBE INJECTOR HARNESS TERMINALS WITH A TEST LIGHT TO GROUND.
- LIGHT SHOULD BE "ON" AT BOTH TERMINALS.

STEADY LIGHT
- CHECK INJECTOR DRIVER CKT WITH STEADY LIGHT FOR SHORT TO GROUND.
- IF CKT IS NOT SHORTED, CHECK RESISTANCE ACROSS EACH INJECTOR IN THE CIRCUIT.
- RESISTANCE SHOULD BE GREATER THAN 10 OHMS

OK — FAULTY ECM

NOT OK — REPLACE ANY INJECTOR THAT MEASURES UNDER 8 OHMS.

LIGHT "ON" BOTH

② RECONNECT INJECTOR(S)
- IGNITION "OFF".
- DISCONNECT ECM
- IGNITION "ON".
- PROBE TERMINALS C11 AND C12 OR D9 AND D3 WITH A TEST LIGHT TO GROUND.

LIGHT "ON" ONE — DUE TO INJECTORS WIRED IN PARALLEL, THERE SHOULD BE A LIGHT ON BOTH TERMINALS. IF NOT, THE PROBLEM IS IN THE HARNESS TO THE TESTED INJECTOR.

LIGHT "OFF" — REPAIR OPEN IN INJECTOR FEED CKT.

LIGHT "ON" — ③ Possible Faulty ECM.

LIGHT "OFF" — OPEN CKT 467 OR 468

ENGINE CRANKS BUT WILL NOT RUN (1988-89 2.8L VIN 9)

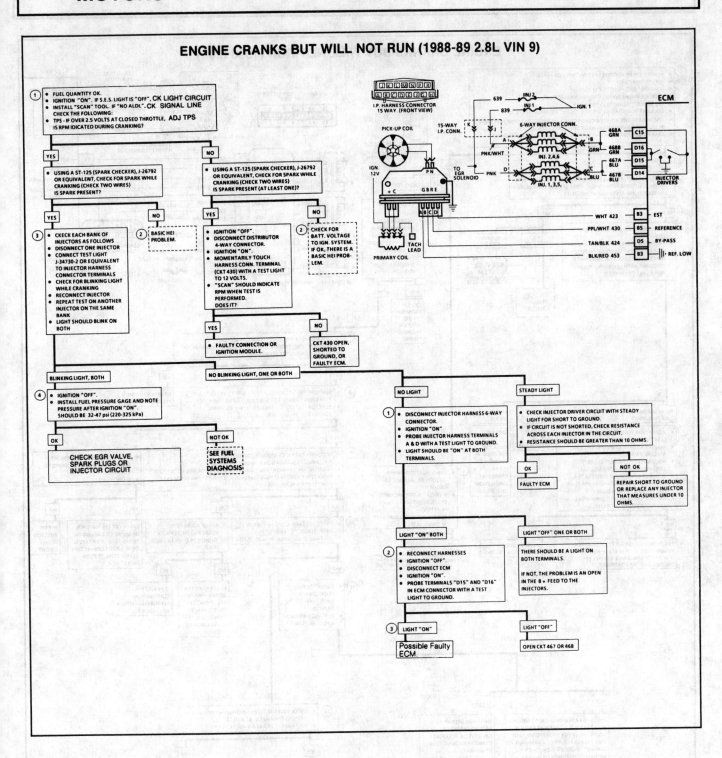

① · FUEL QUANTITY OK.
· IGNITION "ON". IF S.E.S. LIGHT IS "OFF", CK LIGHT CIRCUIT
· INSTALL "SCAN" TOOL. IF "NO ALDL", CK SIGNAL LINE
CHECK THE FOLLOWING:
· TPS - IF OVER 2.5 VOLTS AT CLOSED THROTTLE, ADJ TPS
IS RPM IDICATED DURING CRANKING?

YES / **NO**

· USING A ST-125 (SPARK CHECKER), J-26792
OR EQUIVALENT, CHECK FOR SPARK WHILE
CRANKING (CHECK TWO WIRES)
IS SPARK PRESENT?

· USING A ST-125 (SPARK CHECKER), J-26792
OR EQUIVALENT, CHECK FOR SPARK WHILE
CRANKING (CHECK TWO WIRES)
IS SPARK PRESENT (AT LEAST ONE)?

YES / **NO** / **YES** / **NO**

③ · CKECK EACH BANK OF
INJECTORS AS FOLLOWS
· DISCONNECT ONE INJECTOR
· CONNECT TEST LIGHT
J-34730-2 OR EQUIVALENT
TO INJECTOR HARNESS
CONNECTOR TERMINALS
· CHECK FOR BLINKING LIGHT
WHILE CRANKING
· RECONNECT INJECTOR
· REPEAT TEST ON ANOTHER
INJECTOR ON THE SAME
BANK
· LIGHT SHOULD BLINK ON
BOTH

② · BASIC HEI
PROBLEM.

· IGNITION "OFF".
· DISCONNECT DISTRIBUTOR
4-WAY CONNECTOR.
· IGNITION "ON".
· MOMENTARILY TOUCH
HARNESS CONN. TERMINAL
(CKT 430) WITH A TEST LIGHT
TO 12 VOLTS.
· "SCAN" SHOULD INDICATE
RPM WHEN TEST IS
PERFORMED.
DOES IT?

② · CHECK FOR
BATT. VOLTAGE
TO IGN. SYSTEM.
IF OK, THERE IS A
BASIC HEI PROB-
LEM.

YES / **NO**

· FAULTY CONNECTION OR
IGNITION MODULE.

· CKT 430 OPEN,
SHORTED TO
GROUND, OR
FAULTY ECM.

BLINKING LIGHT, BOTH / **NO BLINKING LIGHT, ONE OR BOTH**

④ · IGNITION "OFF".
· INSTALL FUEL PRESSURE GAGE AND NOTE
PRESSURE AFTER IGNITION "ON".
SHOULD BE 32-47 psi (220-325 kPa)

OK / **NOT OK**

CHECK EGR VALVE,
SPARK PLUGS OR
INJECTOR CIRCUIT

SEE FUEL
SYSTEMS
DIAGNOSIS

NO LIGHT / **STEADY LIGHT**

① · DISCONNECT INJECTOR HARNESS 6-WAY
CONNECTOR.
· IGNITION "ON"
· PROBE INJECTOR HARNESS TERMINALS
A & D WITH A TEST LIGHT TO GROUND.
· LIGHT SHOULD BE "ON" AT BOTH
TERMINALS.

· CHECK INJECTOR DRIVER CIRCUIT WITH STEADY
LIGHT FOR SHORT TO GROUND.
· IF CIRCUIT IS NOT SHORTED, CHECK RESISTANCE
ACROSS EACH INJECTOR IN THE CIRCUIT.
· RESISTANCE SHOULD BE GREATER THAN 10 OHMS.

OK / **NOT OK**

FAULTY ECM

REPAIR SHORT TO GROUND
OR REPLACE ANY INJECTOR
THAT MEASURES UNDER 10
OHMS.

LIGHT "ON" BOTH / **LIGHT "OFF" ONE OR BOTH**

② · RECONNECT HARNESSES
· IGNITION "OFF".
· DISCONNECT ECM
· IGNITION "ON".
· PROBE TERMINALS "D15" AND "D16"
IN ECM CONNECTOR WITH A TEST
LIGHT TO GROUND.

THERE SHOULD BE A LIGHT ON
BOTH TERMINALS.

IF NOT, THE PROBLEM IS AN OPEN
IN THE B + FEED TO THE
INJECTORS.

③ LIGHT "ON" / **LIGHT "OFF"**

Possible Faulty
ECM.

OPEN CKT 467 OR 468

**I.P. HARNESS CONNECTOR
15 WAY. (FRONT VIEW)**

PICK-UP COIL

IGN.
12V

PRIMARY COIL

TACH
LEAD

**15-WAY
I.P. CONN.**

TO
EGR
SOLENOID

PNK/WHT

PNK

639 — IINJ 2
839 — INJ 1 — IGN. 1

6-WAY INJECTOR CONN.

INJ. 2,4,6

INJ. 1,3,5,

468A GRN
468B GRN
467A BLU
467B BLU

ECM

C15
D16
D15
D14

INJECTOR
DRIVERS

WHT 423 — B3 — EST
PPL/WHT 430 — B5 — REFERENCE
TAN/BLK 424 — D5 — BY-PASS
BLK/RED 453 — B3 — REF. LOW

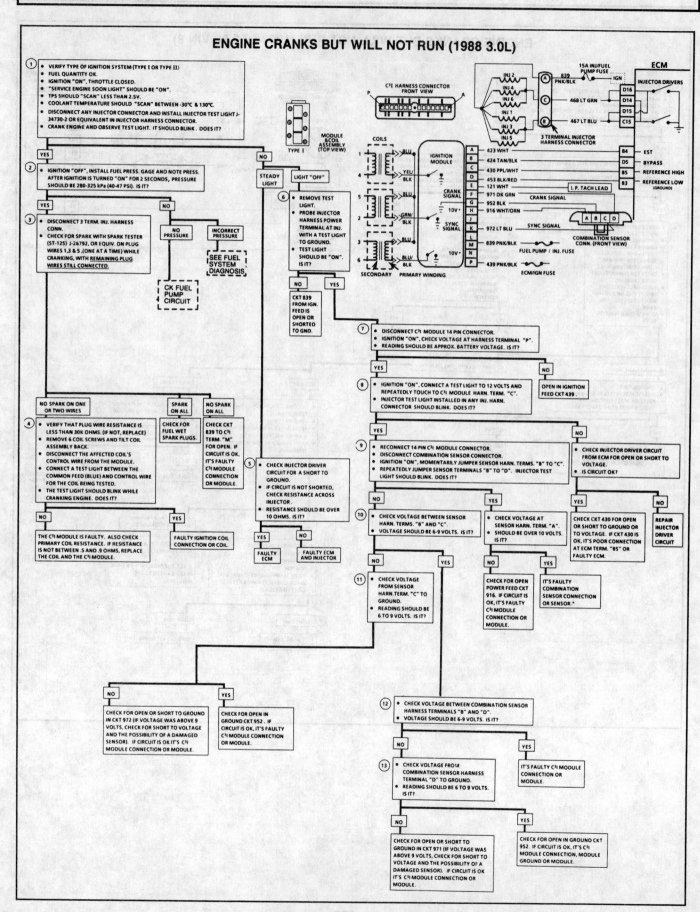

ENGINE CRANKS BUT WILL NOT RUN (1988 3.0L)

① • VERIFY TYPE OF IGNITION SYSTEM (TYPE I OR TYPE II)
• FUEL QUANTITY OK.
• IGNITION "ON", THROTTLE CLOSED.
• "SERVICE ENGINE SOON LIGHT" SHOULD BE "ON".
• TPS SHOULD "SCAN" LESS THAN 2.5V.
• COOLANT TEMPERATURE SHOULD "SCAN" BETWEEN -30°C & 130°C.
• DISCONNECT ANY INJECTOR CONNECTOR AND INSTALL INJECTOR TEST LIGHT J-34730-2 OR EQUIVALENT IN INJECTOR HARNESS CONNECTOR.
• CRANK ENGINE AND OBSERVE TEST LIGHT. IT SHOULD BLINK . DOES IT?

② • IGNITION "OFF", INSTALL FUEL PRESS. GAGE AND NOTE PRESS. AFTER IGNITION IS TURNED "ON" FOR 2 SECONDS, PRESSURE SHOULD BE 280-325 kPa (40-47 PSI). IS IT?

③ • DISCONNECT 3 TERM. INJ. HARNESS CONN.
• CHECK FOR SPARK WITH SPARK TESTER (ST-125) J-26792, OR EQUIV. ON PLUG WIRES 1,3 & 5 .(ONE AT A TIME) WHILE CRANKING, WITH REMAINING PLUG WIRES STILL CONNECTED.

NO PRESSURE

CK FUEL PUMP CIRCUIT

INCORRECT PRESSURE

SEE FUEL SYSTEM DIAGNOSIS

STEADY LIGHT

LIGHT "OFF"

⑥ • REMOVE TEST LIGHT.
• PROBE INJECTOR HARNESS POWER TERMINAL AT INJ. WITH A TEST LIGHT TO GROUND.
• TEST LIGHT SHOULD BE "ON". IS IT?

CKT 839 FROM IGN. FEED IS OPEN OR SHORTED TO GND.

MODULE & COIL ASSEMBLY (TOP VIEW)
TYPE I

C³I HARNESS CONNECTOR FRONT VIEW

COILS

IGNITION MODULE

A 423 WHT
B 424 TAN/BLK
C 430 PPL/WHT
D 453 BLK/RED
E 121 WHT
F 971 DK GRN   CRANK SIGNAL
G 952 BLK
H 916 WHT/ORN
J
K 972 LT BLU   SYNC SIGNAL
L
M 839 PNK/BLK   FUEL PUMP / INJ. FUSE
N
P 439 PNK/BLK   ECM/IGN FUSE

CRANK SIGNAL   10V+
SYNC SIGNAL   10V+

SECONDARY   PRIMARY WINDING

ECM
INJ 2
INJ 4
INJ 6
INJ 1
INJ 3
INJ 5

A 839 PNK/BLK   15A INJ/FUEL PUMP FUSE   IGN
C 468 LT GRN
B 467 LT BLU

3 TERMINAL INJECTOR HARNESS CONNECTOR

INJECTOR DRIVERS
D16
D14
D15
C15

B4 EST
D5 BYPASS
B5 REFERENCE HIGH
B3 REFERENCE LOW (GROUND)

I.P. TACH LEAD

COMBINATION SENSOR CONN. (FRONT VIEW)
A B C D

⑦ • DISCONNECT C³I MODULE 14 PIN CONNECTOR.
• IGNITION "ON", CHECK VOLTAGE AT HARNESS TERMINAL "P".
• READING SHOULD BE APPROX. BATTERY VOLTAGE. IS IT?

OPEN IN IGNITION FEED CKT 439 .

⑧ • IGNITION "ON", CONNECT A TEST LIGHT TO 12 VOLTS AND REPEATEDLY TOUCH TO C³I MODULE HARN. TERM. "C".
• INJECTOR TEST LIGHT INSTALLED IN ANY INJ. HARN. CONNECTOR SHOULD BLINK. DOES IT?

⑨ • RECONNECT 14 PIN C³I MODULE CONNECTOR.
• DISCONNECT COMBINATION SENSOR CONNECTOR.
• IGNITION "ON", MOMENTARILY JUMPER SENSOR HARN. TERMS. "B" TO "C".
• REPEATEDLY JUMPER SENSOR TERMINALS "B" TO "D". INJECTOR TEST LIGHT SHOULD BLINK. DOES IT?

• CHECK INJECTOR DRIVER CIRCUIT FROM ECM FOR OPEN OR SHORT TO VOLTAGE.
• IS CIRCUIT OK?

NO SPARK ON ONE OR TWO WIRES

SPARK ON ALL

NO SPARK ON ALL

④ • VERIFY THAT PLUG WIRE RESISTANCE IS LESS THAN 30K OHMS. (IF NOT, REPLACE)
• REMOVE 6 COIL SCREWS AND TILT COIL ASSEMBLY BACK.
• DISCONNECT THE AFFECTED COIL'S CONTROL WIRE FROM THE MODULE.
• CONNECT A TEST LIGHT BETWEEN THE COMMON FEED (BLUE) AND CONTROL WIRE FOR THE COIL BEING TESTED.
• THE TEST LIGHT SHOULD BLINK WHILE CRANKING ENGINE. DOES IT?

CHECK FOR FUEL WET SPARK PLUGS.

CHECK CKT 839 TO C³I TERM. "M" FOR OPEN. IF CIRCUIT IS OK, IT'S FAULTY C³I MODULE CONNECTION OR MODULE.

⑤ • CHECK INJECTOR DRIVER CIRCUIT FOR A SHORT TO GROUND.
• IF CIRCUIT IS NOT SHORTED, CHECK RESISTANCE ACROSS INJECTOR.
• RESISTANCE SHOULD BE OVER 10 OHMS. IS IT?

FAULTY ECM

FAULTY ECM AND INJECTOR

THE C³I MODULE IS FAULTY. ALSO CHECK PRIMARY COIL RESISTANCE. IF RESISTANCE IS NOT BETWEEN .5 AND .9 OHMS, REPLACE THE COIL AND THE C³I MODULE.

FAULTY IGNITION COIL CONNECTION OR COIL.

⑩ • CHECK VOLTAGE BETWEEN SENSOR HARN. TERMS. "B" AND "C".
• VOLTAGE SHOULD BE 6-9 VOLTS. IS IT?

• CHECK VOLTAGE AT SENSOR HARN. TERM. "A".
• SHOULD BE OVER 10 VOLTS. IS IT?

CHECK CKT 430 FOR OPEN OR SHORT TO GROUND OR TO VOLTAGE. IF CKT 430 IS OK, IT'S POOR CONNECTION AT ECM TERM. "B5" OR FAULTY ECM.

REPAIR INJECTOR DRIVER CIRCUIT

⑪ • CHECK VOLTAGE FROM SENSOR HARN. TERM. "C" TO GROUND.
• READING SHOULD BE 6 TO 9 VOLTS. IS IT?

CHECK FOR OPEN POWER FEED CKT 916. IF CIRCUIT IS OK, IT'S FAULTY C³I MODULE CONNECTION OR MODULE.

IT'S FAULTY COMBINATION SENSOR CONNECTION OR SENSOR.

CHECK FOR OPEN OR SHORT TO GROUND IN CKT 972 (IF VOLTAGE WAS ABOVE 9 VOLTS, CHECK FOR SHORT TO VOLTAGE AND THE POSSIBILITY OF A DAMAGED SENSOR). IF CIRCUIT IS OK IT'S C³I MODULE CONNECTION OR MODULE.

CHECK FOR OPEN IN GROUND CKT 952. IF CIRCUIT IS OK, IT'S FAULTY C³I MODULE CONNECTION OR MODULE.

⑫ • CHECK VOLTAGE BETWEEN COMBINATION SENSOR HARNESS TERMINALS "B" AND "D".
• VOLTAGE SHOULD BE 6-9 VOLTS. IS IT?

⑬ • CHECK VOLTAGE FROM COMBINATION SENSOR HARNESS TERMINAL "D" TO GROUND.
• READING SHOULD BE 6 TO 9 VOLTS. IS IT?

IT'S FAULTY C³I MODULE CONNECTION OR MODULE.

CHECK FOR OPEN OR SHORT TO GROUND IN CKT 971 (IF VOLTAGE WAS ABOVE 9 VOLTS, CHECK FOR SHORT TO VOLTAGE AND THE POSSIBILITY OF A DAMAGED SENSOR). IF CIRCUIT IS OK IT'S C³I MODULE CONNECTION OR MODULE.

CHECK FOR OPEN IN GROUND CKT 952. IF CIRCUIT IS OK, IT'S C³I MODULE CONNECTION, MODULE GROUND OR MODULE.

ENGINE CRANKS BUT WILL NOT RUN
(1988 3.8L "3800" VIN C)

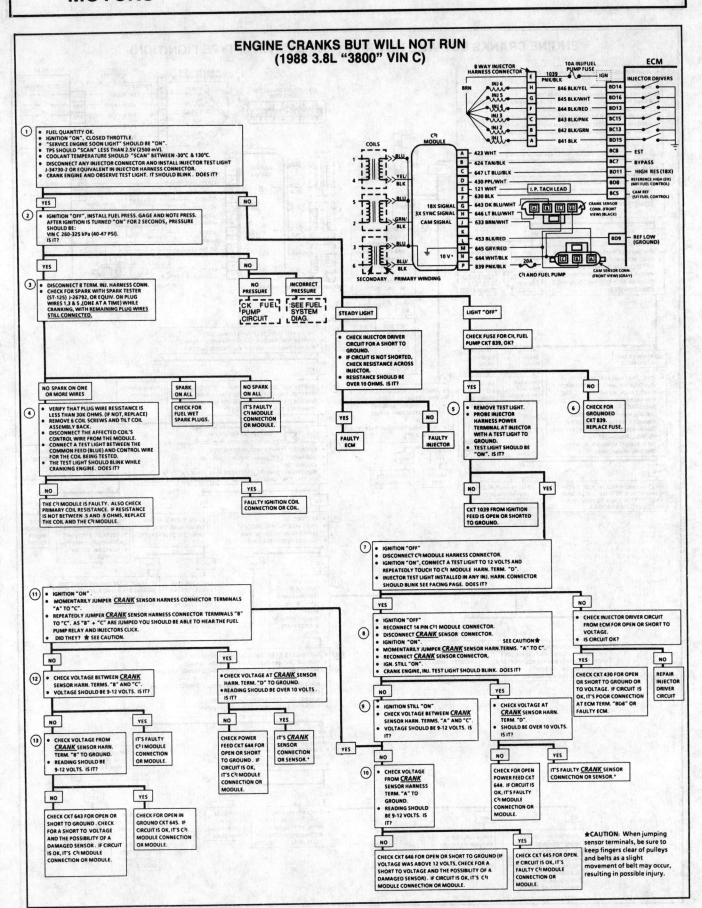

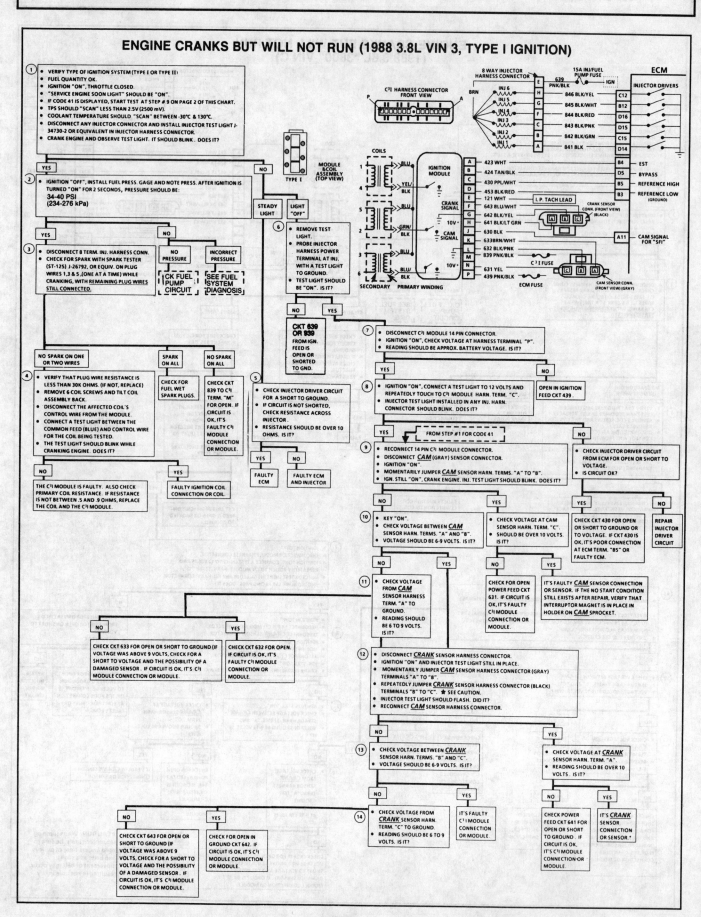

ENGINE CRANKS BUT WILL NOT RUN (1988 3.8L VIN 3, TYPE I IGNITION)

ENGINE CRANKS BUT WILL NOT RUN
(1988 3.8L VIN 3, TYPE II IGNITION)

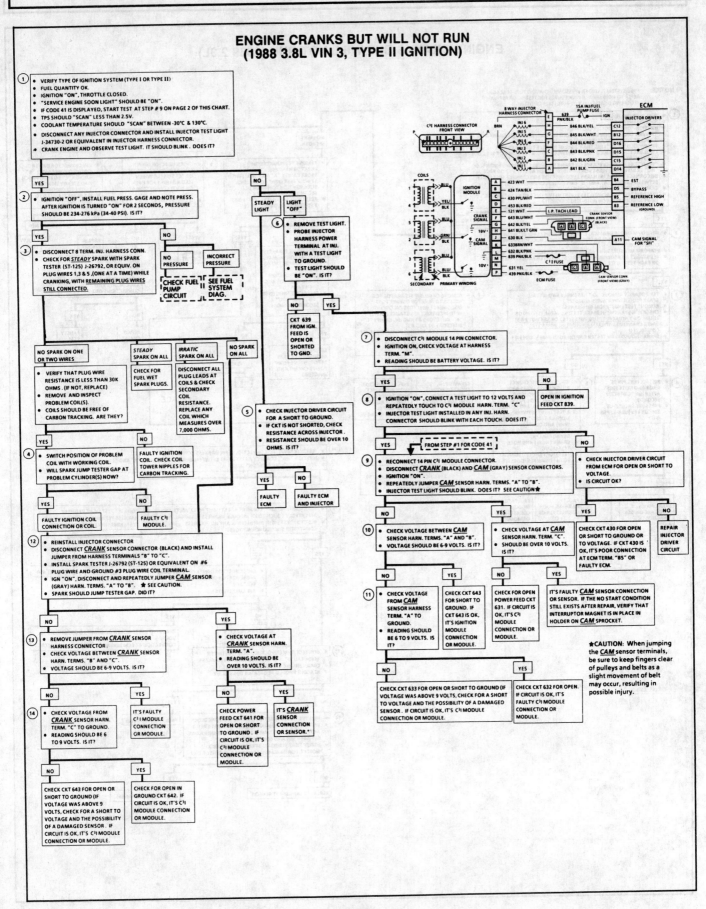

ENGINE CRANKS BUT WILL NOT RUN (1989 2.3L)

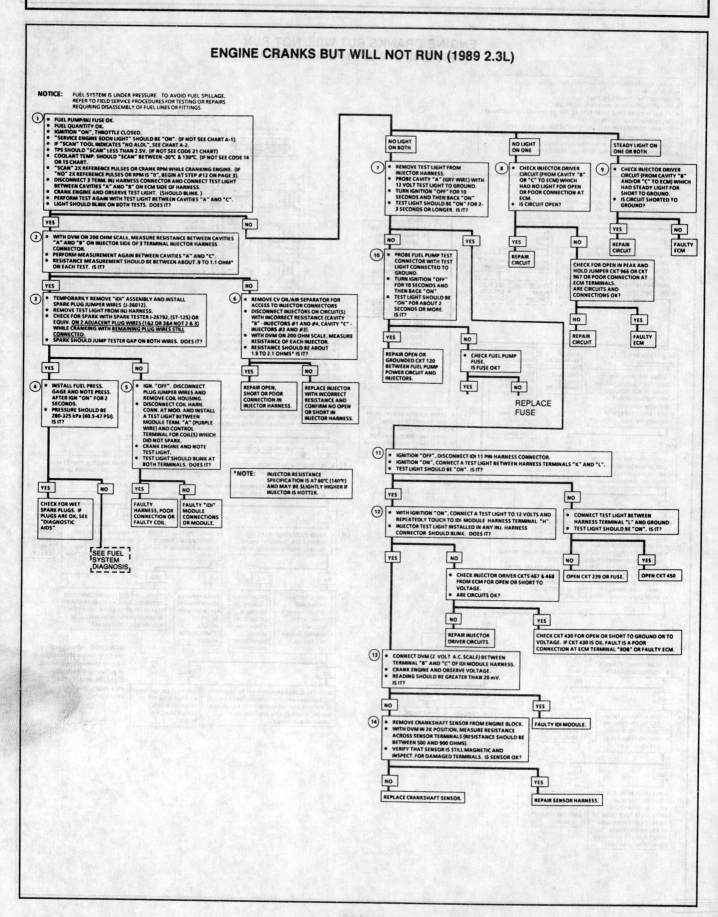

ENGINE CRANKS BUT WILL NOT RUN (1989 2.3L – CONT.)

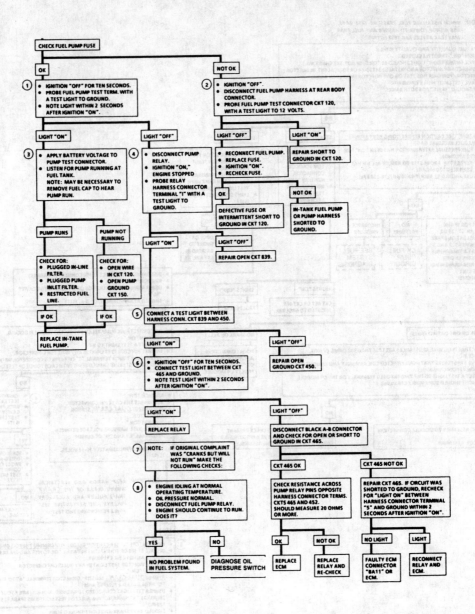

CHECK FUEL PUMP FUSE

OK

① • IGNITION "OFF" FOR TEN SECONDS.
• PROBE FUEL PUMP TEST TERM. WITH A TEST LIGHT TO GROUND.
• NOTE LIGHT WITHIN 2 SECONDS AFTER IGNITION "ON".

NOT OK

② • IGNITION "OFF".
• DISCONNECT FUEL PUMP HARNESS AT REAR BODY CONNECTOR.
• PROBE FUEL PUMP TEST CONNECTOR CKT 120, WITH A TEST LIGHT TO 12 VOLTS.

LIGHT "ON"

③ • APPLY BATTERY VOLTAGE TO PUMP TEST CONNECTOR.
• LISTEN FOR PUMP RUNNING AT FUEL TANK.
NOTE: MAY BE NECESSARY TO REMOVE FUEL CAP TO HEAR PUMP RUN.

LIGHT "OFF"

④ • DISCONNECT PUMP RELAY.
• IGNITION "ON," ENGINE STOPPED
• PROBE RELAY HARNESS CONNECTOR TERMINAL "I" WITH A TEST LIGHT TO GROUND.

LIGHT "OFF"

• RECONNECT FUEL PUMP.
• REPLACE FUSE.
• IGNITION "ON".
• RECHECK FUSE.

OK

DEFECTIVE FUSE OR INTERMITTENT SHORT TO GROUND IN CKT 120.

NOT OK

IN-TANK FUEL PUMP OR PUMP HARNESS SHORTED TO GROUND.

LIGHT "ON"

REPAIR SHORT TO GROUND IN CKT 120.

PUMP RUNS

CHECK FOR:
• PLUGGED IN-LINE FILTER.
• PLUGGED PUMP INLET FILTER.
• RESTRICTED FUEL LINE.

PUMP NOT RUNNING

CHECK FOR:
• OPEN WIRE IN CKT 120.
• OPEN PUMP GROUND CKT 150.

LIGHT "ON"

LIGHT "OFF"

REPAIR OPEN CKT 839.

IF OK

IF OK

REPLACE IN-TANK FUEL PUMP.

⑤ CONNECT A TEST LIGHT BETWEEN HARNESS CONN. CKT 839 AND 450.

LIGHT "ON"

⑥ • IGNITION "OFF" FOR TEN SECONDS.
• CONNECT TEST LIGHT BETWEEN CKT 465 AND GROUND.
• NOTE TEST LIGHT WITHIN 2 SECONDS AFTER IGNITION "ON".

LIGHT "OFF"

REPAIR OPEN GROUND CKT 450.

LIGHT "ON"

REPLACE RELAY

LIGHT "OFF"

DISCONNECT BLACK A-B CONNECTOR AND CHECK FOR OPEN OR SHORT TO GROUND IN CKT 465.

⑦ NOTE: IF ORIGINAL COMPLAINT WAS "CRANKS BUT WILL NOT RUN" MAKE THE FOLLOWING CHECKS:

⑧ • ENGINE IDLING AT NORMAL OPERATING TEMPERATURE.
• OIL PRESSURE NORMAL.
• DISCONNECT FUEL PUMP RELAY.
• ENGINE SHOULD CONTINUE TO RUN. DOES IT?

CKT 465 OK

CHECK RESISTANCE ACROSS PUMP RELAY PINS OPPOSITE HARNESS CONNECTOR TERMS. CKTS 465 AND 452. SHOULD MEASURE 20 OHMS OR MORE.

CKT 465 NOT OK

REPAIR CKT 465. IF CIRCUIT WAS SHORTED TO GROUND, RECHECK FOR "LIGHT ON" BETWEEN HARNESS CONNECTOR TERMINAL "S" AND GROUND WITHIN 2 SECONDS AFTER IGNITION "ON".

YES

NO PROBLEM FOUND IN FUEL SYSTEM.

NO

DIAGNOSE OIL PRESSURE SWITCH

OK

REPLACE ECM

NOT OK

REPLACE RELAY AND RE-CHECK

NO LIGHT

FAULTY ECM CONNECTOR "BA11" OR ECM.

LIGHT

RECONNECT RELAY AND ECM.

ENGINE CRANKS BUT WILL NOT RUN (1989 3.3L)

CAUTION: *WHEN INSTALLING FUEL PRESSURE TEST GAGE, USE A SHOP TOWEL TO ABSORB ANY FUEL THAT MAY LEAK AT FUEL RAIL TEST FITTING.*

1
- VERIFY FUEL QUANTITY AND QUALITY IS OK.
- IGNITION "ON," THROTTLE CLOSED.
- "SERVICE ENGINE SOON" LIGHT SHOULD BE "ON." (IF NOT, SEE CHART A-1.)
- DISCONNECT ANY INJECTOR AND INSTALL J 34730-2A OR EQUIVALENT IN INJECTOR CONNECTOR, (PERFORM THIS TEST ON ONE INJECTOR FROM EACH BANK).
- OBSERVE TEST LIGHT WHILE CRANKING.
- LIGHT SHOULD "BLINK" FOR BOTH BANKS. DOES IT?

YES / **NO**

2 IGNITION "OFF," SEE CAUTION BEFORE DOING NEXT STEP.
- INSTALL FUEL PRESSURE GAGE.
- NOTE FUEL PRESSURE AFTER IGNITION HAS BEEN "ON" FOR ABOUT 2 SECONDS.
- NOTE: REPEAT THE ABOVE STEP TO ASSURE ALL AIR HAS BEEN BLED FROM GAGE.
- PRESSURE SHOULD BE ABOUT 40-44 psi (276-303 kPa) IS IT?

YES / **NO**

STEADY LIGHT (S)
- IGNITION "OFF."
- DISCONNECT ECM BLACK C-D CONNECTOR.
- IGNITION "ON"
- NOTE INJECTOR TEST LIGHT ON EACH BANK.

LIGHT(S) "OFF"

5
- REMOVE INJECTOR TEST LIGHT. IGNITION "ON."
- PROBE INJECTOR CONNECTOR TERMINAL "A" OF EACH BANK WITH TEST LIGHT TO GROUND.
- LIGHT SHOULD BE "ON." IS IT?

YES / **NO**

3 CHECK FOR SPARK WITH AN ST 125 OR EQUIVALENT ON SPARK PLUG WIRES 1, 3 AND 5 (ONE AT A TIME) WHILE CRANKING ENGINE WITH REMAINING PLUG WIRES STILL CONNECTED.

NO PRESSURE → CK FUEL PUMP CIRCUIT

INCORRECT PRESSURE → SEE FUEL SYSTEM DIAGNOSIS

OPEN OR SHORT TO GROUND IN CKT 839.

LIGHTS "ON" → CKT 467 OR CKT 468 SHORTED TO GROUND.

LIGHTS "OFF" → FAULTY ECM.

- DISCONNECT C³ MODULE 14 PIN CONNECTOR HARNESS.
- IGNITION "ON."
- USING A DVM, CHECK VOLTAGE ON C³ CONNECTOR CAVITY "M."
- SHOULD BE APPROXIMATELY 8 +. IS IT?

YES / **NO** → OPEN IN CKT 839

NO SPARK ON ONE OR TWO WIRES

NO SPARK ON ALL.

4
- VERIFY THAT PLUG WIRE RESISTANCES ARE LESS THAN 30K OHMS. (IF NOT, REPLACE.)
- REMOVE 2 FASTENERS THAT HOLD EFFECTED COIL ASSEMBLY AND REMOVE COIL ASSEMBLY FROM C³ MODULE.
- CONNECT A TEST LIGHT TO THE TWO EXPOSED TERMINALS ON THE C³ MODULE.
- TEST LIGHT SHOULD BLINK WHILE CRANKING. DOES IT?

POOR CONNECTION AT C³ MODULE. OR FAULTY C³ MODULE.

1
- DISCONNECT FUEL PUMP RELAY TO PREVENT ENGINE FLOODING.
- IGNITION "ON."
- CONNECT A TEST LIGHT TO B +.
- INSTALL INJECTOR TEST LIGHT IN ANY INJECTOR CONNECTOR.
- REPEATEDLY TOUCH TERMINAL "C" OF 14 PIN CONNECTOR WITH TEST LIGHT.
- INJECTOR TEST LIGHT SHOULD BLINK WITH EACH TOUCH OF TEST LIGHT. DOES IT?

YES / **NO**

YES / **NO**

POOR COIL ASSEMBLY TO MODULE CONNECTION OR FAULTY COIL ASSEMBLY.

FAULTY C³ MODULE, ALSO CHECK PRIMARY COIL RESISTANCE, SHOULD BE APPROX. 0.5-0.9 OHMS, IF NOT, REPLACE COIL ASSEMBLY.

2
- IGNITION "OFF."
- RECONNECT THE C³ 14 PIN CONNECTOR.
- DISCONNECT *DUAL CRANK* SENSOR CONNECTOR.
- IGNITION "ON."
- USING A DVM, MEASURE VOLTAGE BETWEEN TERMINALS "C" AND "D" OF SENSOR CONNECTOR.
- SHOULD BE APPROXIMATELY 10-12 VOLTS. IS IT?

4
- IGNITION "ON."
- TEST LIGHT TO GROUND
- BACK PROBE BLACK ECM C-D CONNECTOR CAVITY "BC11" AND "BC12"
- TEST LIGHT SHOULD BE "ON" FOR BOTH. IS IT?

YES / **NO** → OPEN CKT 467 OR 468.

- IGNITION "ON."
- USING A TEST LIGHT TO GROUND
- PROBE CAVITY "H" OF THE INJECTOR HARNESS CONNECTOR. (CHECK ONE FROM EACH BANK)
- TEST LIGHT SHOULD BE "OFF." IS IT?

NO → OPEN CKT 467 OR CKT 468.

YES → SHORTED CKT 467 OR 468. OPEN OR SHORTED CKT 430 OR POOR CONNECTION OR FAULTY ECM.

CAUTION: *KEEP HANDS AND ELECTRICAL WIRING CLEAR OF THE CRANK SHAFT PULLEY AND ACCESSORY PULLEYS AND SERPENTINE BELT WHILE PERFORMING THIS TEST.*

POOR CONNECTION AT C³ MODULE OR OPEN IN CKT 644 OR CKT 645 OR SHORT TO GROUND IN CKTS 644 OR 645 OR FAULTY C³ MODULE.

3
- IGNITION "OFF." * SEE CAUTION BEFORE PERFORMING THE FOLLOWING TEST.
- INSTALL ST 125 OR EQUIVALENT BETWEEN COIL TOWER AND SPARK PLUG WIRE OF NUMBER SIX CYLINDER.
- INSTALL INJECTOR TEST LIGHT IN ANY INJECTOR CONNECTOR.
- IGNITION "ON."
- JUMPER *DUAL CRANK* SENSOR CONNECTOR TERMINAL "A" TO TERMINAL "B" WITH SUITABLE JUMPER WIRE.
- USING A TEST LIGHT CONNECTED TO GROUND, MOMENTARILY TOUCH TERMINAL "A" WHILE NOTING INJECTOR TEST LIGHT AND SPARK TESTER. SEE NOTE *
- SHOULD HAVE SPARK AND LIGHT SHOULD BLINK. DO THEY?

*NOTE: REPEATED TOUCHING OF TERMINAL "A" OR "C" TO GROUND MAY CAUSE A FLOODED ENGINE CONDITION.

YES / **NO**

POOR CONNECTION AT *DUAL CRANK* SENSOR OR FAULTY *DUAL CRANK* SENSOR. INSPECT *DUAL CRANK* SENSOR AND CRANKSHAFT PULLEY VANES FOR SIGNS OF HITTING.

POOR CONNECTION AT C³ MODULE OR FAULTY C³ MODULE.

CLEAR CODES AND CONFIRM "CLOSED LOOP" OPERATION AND NO "SERVICE ENGINE SOON" LIGHT.

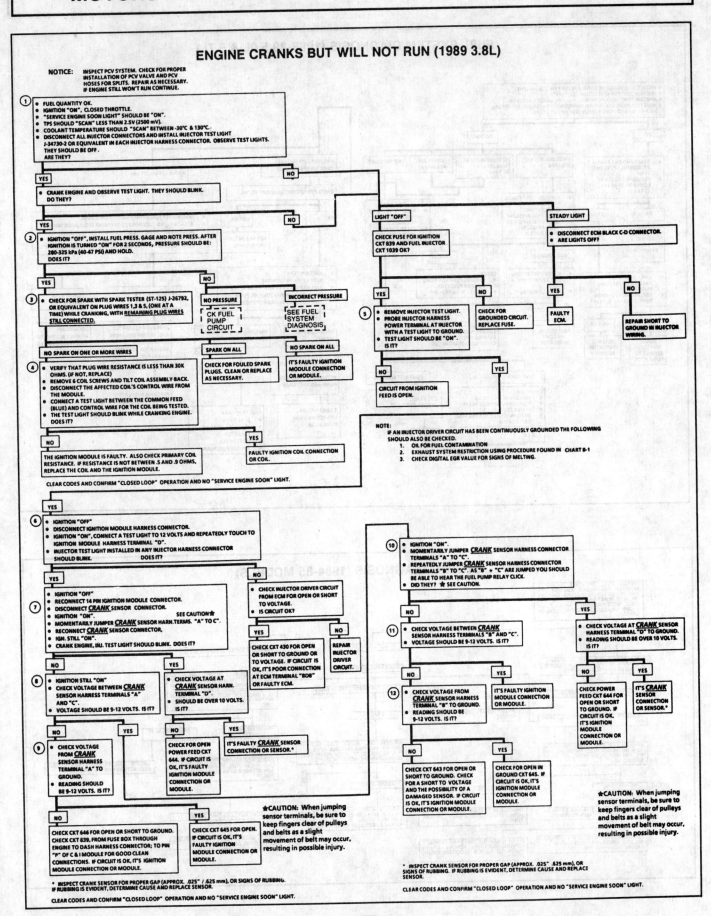

ENGINE CRANKS BUT WILL NOT RUN (1989 3.8L)

NOTICE: INSPECT PCV SYSTEM. CHECK FOR PROPER INSTALLATION OF PCV VALVE AND PCV HOSES FOR SPLITS. REPAIR AS NECESSARY. IF ENGINE STILL WON'T RUN CONTINUE.

1
- FUEL QUANTITY OK.
- IGNITION "ON", CLOSED THROTTLE.
- "SERVICE ENGINE SOON LIGHT" SHOULD BE "ON".
- TPS SHOULD "SCAN" LESS THAN 2.5V (2500 mV).
- COOLANT TEMPERATURE SHOULD "SCAN" BETWEEN -30°C & 130°C.
- DISCONNECT ALL INJECTOR CONNECTORS AND INSTALL INJECTOR TEST LIGHT J-34730-2 OR EQUIVALENT IN EACH INJECTOR HARNESS CONNECTOR. OBSERVE TEST LIGHTS. THEY SHOULD BE OFF. ARE THEY?

YES
- CRANK ENGINE AND OBSERVE TEST LIGHT. THEY SHOULD BLINK. DO THEY?

NO

LIGHT "OFF"
CHECK FUSE FOR IGNITION CKT 839 AND FUEL INJECTOR CKT 1039 OK?

STEADY LIGHT
- DISCONNECT ECM BLACK C-D CONNECTOR.
- ARE LIGHTS OFF?

YES
2
- IGNITION "OFF", INSTALL FUEL PRESS. GAGE AND NOTE PRESS. AFTER IGNITION IS TURNED "ON" FOR 2 SECONDS, PRESSURE SHOULD BE: 280-325 kPa (40-47 PSI) AND HOLD. DOES IT?

YES   NO

YES   NO

YES
FAULTY ECM.

NO
REPAIR SHORT TO GROUND IN INJECTOR WIRING.

YES
3
- CHECK FOR SPARK WITH SPARK TESTER (ST-125) J-26792, OR EQUIVALENT ON PLUG WIRES 1,3 & 5, (ONE AT A TIME) WHILE CRANKING, WITH REMAINING PLUG WIRES STILL CONNECTED.

NO PRESSURE
CK FUEL PUMP CIRCUIT

INCORRECT PRESSURE
SEE FUEL SYSTEM DIAGNOSIS

5
- REMOVE INJECTOR TEST LIGHT.
- PROBE INJECTOR HARNESS POWER TERMINAL AT INJECTOR WITH A TEST LIGHT TO GROUND.
- TEST LIGHT SHOULD BE "ON". IS IT?

CHECK FOR GROUNDED CIRCUIT. REPLACE FUSE.

NO SPARK ON ONE OR MORE WIRES

SPARK ON ALL

NO SPARK ON ALL

NO
CIRCUIT FROM IGNITION FEED IS OPEN.

YES

4
- VERIFY THAT PLUG WIRE RESISTANCE IS LESS THAN 30K OHMS. (IF HOT, REPLACE)
- REMOVE 6 COIL SCREWS AND TILT COIL ASSEMBLY BACK.
- DISCONNECT THE AFFECTED COIL'S CONTROL WIRE FROM THE MODULE.
- CONNECT A TEST LIGHT BETWEEN THE COMMON FEED (BLUE) AND CONTROL WIRE FOR THE COIL BEING TESTED.
- THE TEST LIGHT SHOULD BLINK WHILE CRANKING ENGINE. DOES IT?

CHECK FOR FOULED SPARK PLUGS. CLEAN OR REPLACE AS NECESSARY.

IT'S FAULTY IGNITION MODULE CONNECTION OR MODULE.

NOTE: IF AN INJECTOR DRIVER CIRCUIT HAS BEEN CONTINUOUSLY GROUNDED THE FOLLOWING SHOULD ALSO BE CHECKED.
1. OIL FOR FUEL CONTAMINATION
2. EXHAUST SYSTEM RESTRICTION USING PROCEDURE FOUND IN CHART B-1
3. CHECK DIGITAL EGR VALVE FOR SIGNS OF MELTING.

NO
THE IGNITION MODULE IS FAULTY. ALSO CHECK PRIMARY COIL RESISTANCE. IF RESISTANCE IS NOT BETWEEN .5 AND .9 OHMS, REPLACE THE COIL AND THE IGNITION MODULE.

YES
FAULTY IGNITION COIL CONNECTION OR COIL.

CLEAR CODES AND CONFIRM "CLOSED LOOP" OPERATION AND NO "SERVICE ENGINE SOON" LIGHT.

YES
6
- IGNITION "OFF"
- DISCONNECT IGNITION MODULE HARNESS CONNECTOR.
- IGNITION "ON", CONNECT A TEST LIGHT TO 12 VOLTS AND REPEATEDLY TOUCH TO IGNITION MODULE HARNESS TERMINAL "D".
- INJECTOR TEST LIGHT INSTALLED IN ANY INJECTOR HARNESS CONNECTOR SHOULD BLINK.   DOES IT?

10
- IGNITION "ON".
- MOMENTARILY JUMPER CRANK SENSOR HARNESS CONNECTOR TERMINALS "A" TO "C".
- REPEATEDLY JUMPER CRANK SENSOR HARNESS CONNECTOR TERMINALS "B" TO "C". AS "B" + "C" ARE JUMPED YOU SHOULD BE ABLE TO HEAR THE FUEL PUMP RELAY CLICK.
- DID THEY? ☆ SEE CAUTION.

YES   NO

NO   YES

7
- IGNITION "OFF".
- RECONNECT 14 PIN IGNITION MODULE CONNECTOR.
- DISCONNECT CRANK SENSOR CONNECTOR.
- IGNITION "ON".
- MOMENTARILY JUMPER CRANK SENSOR HARN.TERMS. "A" TO C".
- RECONNECT CRANK SENSOR CONNECTOR.
- IGN. STILL "ON".
- CRANK ENGINE, INJ. TEST LIGHT SHOULD BLINK. DOES IT?

SEE CAUTION ☆

- CHECK INJECTOR DRIVER CIRCUIT FROM ECM FOR OPEN OR SHORT TO VOLTAGE.
- IS CIRCUIT OK?

11
- CHECK VOLTAGE BETWEEN CRANK SENSOR HARNESS TERMINALS "B" AND "C".
- VOLTAGE SHOULD BE 9-12 VOLTS. IS IT?

- CHECK VOLTAGE AT CRANK SENSOR HARNESS TERMINAL "D" TO GROUND.
- READING SHOULD BE OVER 10 VOLTS. IS IT?

NO   YES

YES   NO

NO   YES

8
- IGNITION STILL "ON".
- CHECK VOLTAGE BETWEEN CRANK SENSOR HARNESS TERMINALS "A" AND "C".
- VOLTAGE SHOULD BE 9-12 VOLTS. IS IT?

- CHECK VOLTAGE AT CRANK SENSOR HARN. TERMINAL "D".
- SHOULD BE OVER 10 VOLTS. IS IT?

CHECK CKT 430 FOR OPEN OR SHORT TO GROUND OR TO VOLTAGE. IF CIRCUIT IS OK, IT'S POOR CONNECTION AT ECM TERMINAL "BD8" OR FAULTY ECM.

REPAIR INJECTOR DRIVER CIRCUIT.

12
- CHECK VOLTAGE FROM CRANK SENSOR HARNESS TERMINAL "B" TO GROUND.
- READING SHOULD BE 9-12 VOLTS. IS IT?

IT'S FAULTY IGNITION MODULE CONNECTION OR MODULE.

CHECK POWER FEED CKT 644 FOR OPEN OR SHORT TO GROUND. IF CIRCUIT IS OK, IT'S IGNITION MODULE CONNECTION OR MODULE.

IT'S CRANK SENSOR CONNECTION OR SENSOR.*

NO   YES

9
- CHECK VOLTAGE FROM CRANK SENSOR HARNESS TERMINAL "A" TO GROUND.
- READING SHOULD BE 9-12 VOLTS. IS IT?

CHECK FOR OPEN POWER FEED CKT 644. IF CIRCUIT IS OK, IT'S FAULTY IGNITION MODULE CONNECTION OR MODULE.

IT'S FAULTY CRANK SENSOR CONNECTION OR SENSOR.*

NO   YES

CHECK CKT 643 FOR OPEN OR SHORT TO GROUND. CHECK FOR A SHORT TO VOLTAGE AND THE POSSIBILITY OF A DAMAGED SENSOR. IF CIRCUIT IS OK, IT'S IGNITION MODULE CONNECTION OR MODULE.

CHECK FOR OPEN IN GROUND CKT 645. IF CIRCUIT IS OK, IT'S IGNITION MODULE CONNECTION OR MODULE.

☆CAUTION: When jumping sensor terminals, be sure to keep fingers clear of pulleys and belts as a slight movement of belt may occur, resulting in possible injury.

NO   YES

CHECK CKT 646 FOR OPEN OR SHORT TO GROUND. CHECK CKT 839, FROM FUSE BOX THROUGH ENGINE TO DASH HARNESS CONNECTOR; TO PIN "F" OF C & I MODULE FOR GOOD CLEAN CONNECTIONS. IF CIRCUIT IS OK, IT'S IGNITION MODULE CONNECTION OR MODULE.

CHECK CKT 645 FOR OPEN. IF CIRCUIT IS OK, IT'S FAULTY IGNITION MODULE CONNECTION OR MODULE.

☆CAUTION: When jumping sensor terminals, be sure to keep fingers clear of pulleys and belts as a slight movement of belt may occur, resulting in possible injury.

* INSPECT CRANK SENSOR FOR PROPER GAP (APPROX. .025" / .625 mm), OR SIGNS OF RUBBING. IF RUBBING IS EVIDENT, DETERMINE CAUSE AND REPLACE SENSOR.

* INSPECT CRANK SENSOR FOR PROPER GAP (APPROX. .025" / .625 mm), OR SIGNS OF RUBBING. IF RUBBING IS EVIDENT, DETERMINE CAUSE AND REPLACE SENSOR.

CLEAR CODES AND CONFIRM "CLOSED LOOP" OPERATION AND NO "SERVICE ENGINE SOON" LIGHT.

CLEAR CODES AND CONFIRM "CLOSED LOOP" OPERATION AND NO "SERVICE ENGINE SOON" LIGHT.

FUEL SYSTEM DIAGNOSIS (1986-89 5.0L & 5.7L)

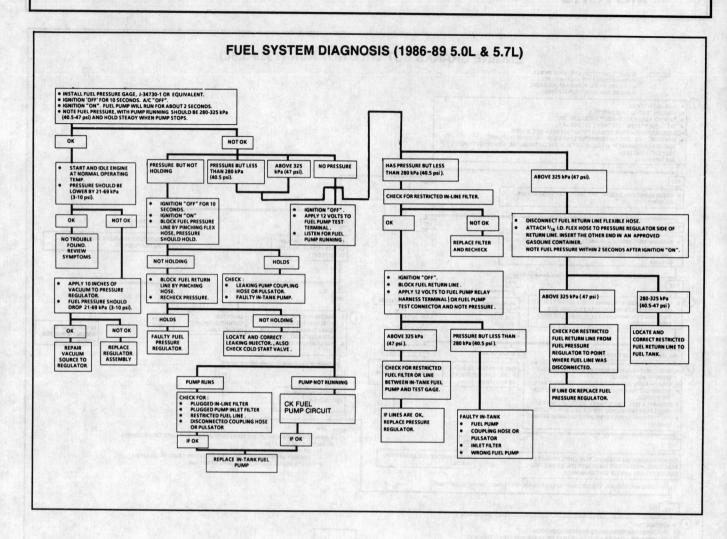

FUEL SYSTEM DIAGNOSIS (1984-85 MODELS)

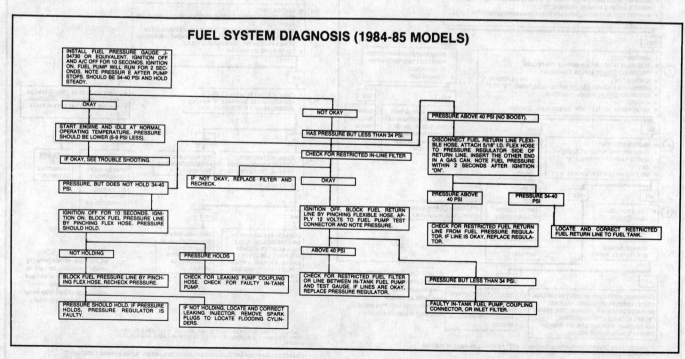

FUEL SYSTEM DIAGNOSIS (1988 3.0L)

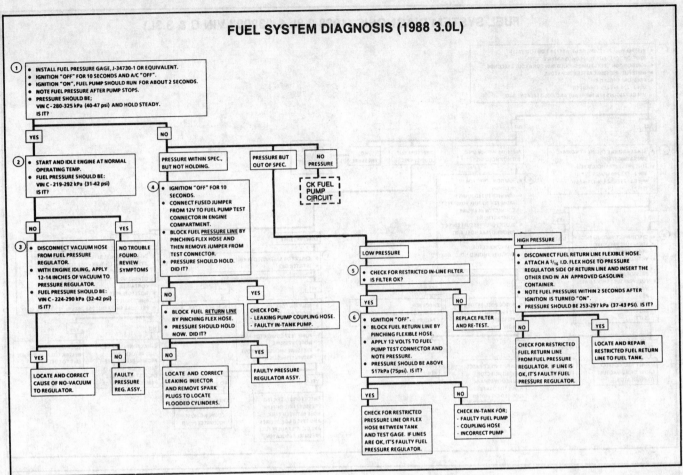

FUEL SYSTEM DIAGNOSIS (1987-89 2.0L)

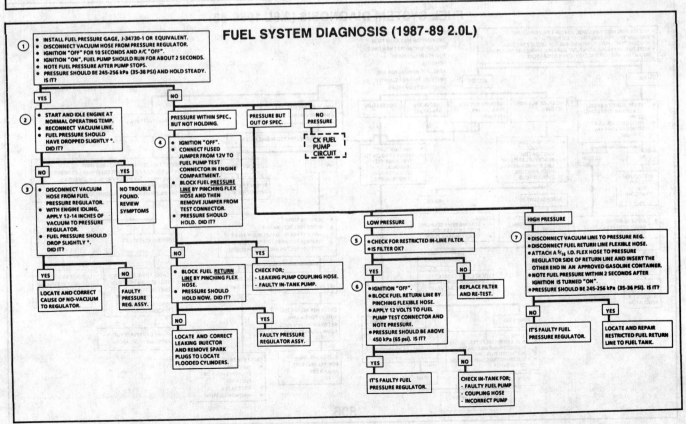

FUEL SYSTEM DIAGNOSIS (1988 3.8L & "3800" VIN C & 3.3L)

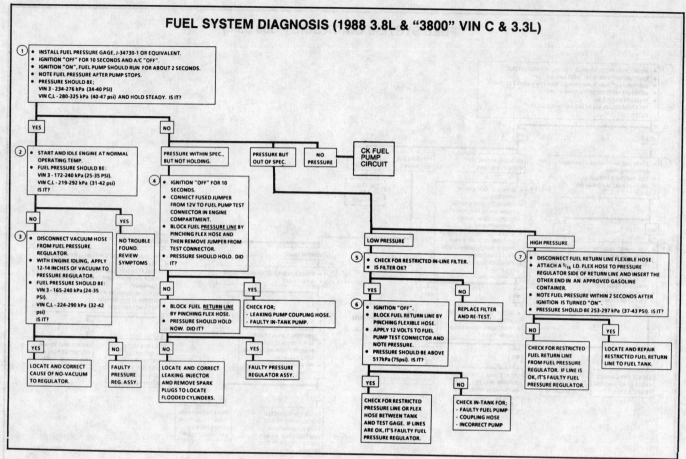

FUEL SYSTEM DIAGNOSIS (2.8L 1986-88)

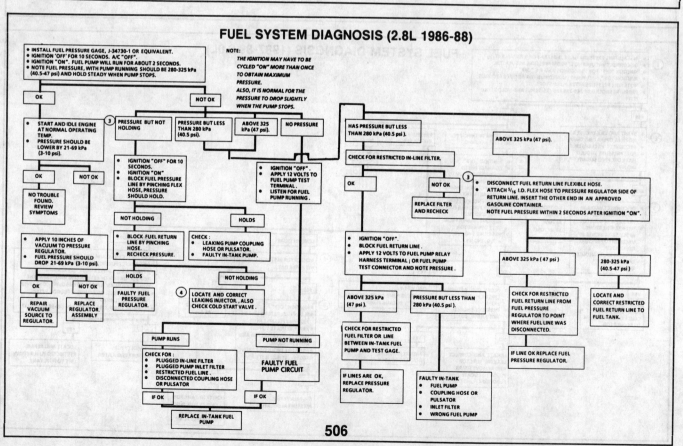

FUEL SYSTEM DIAGNOSIS (1986-87 3.0L)

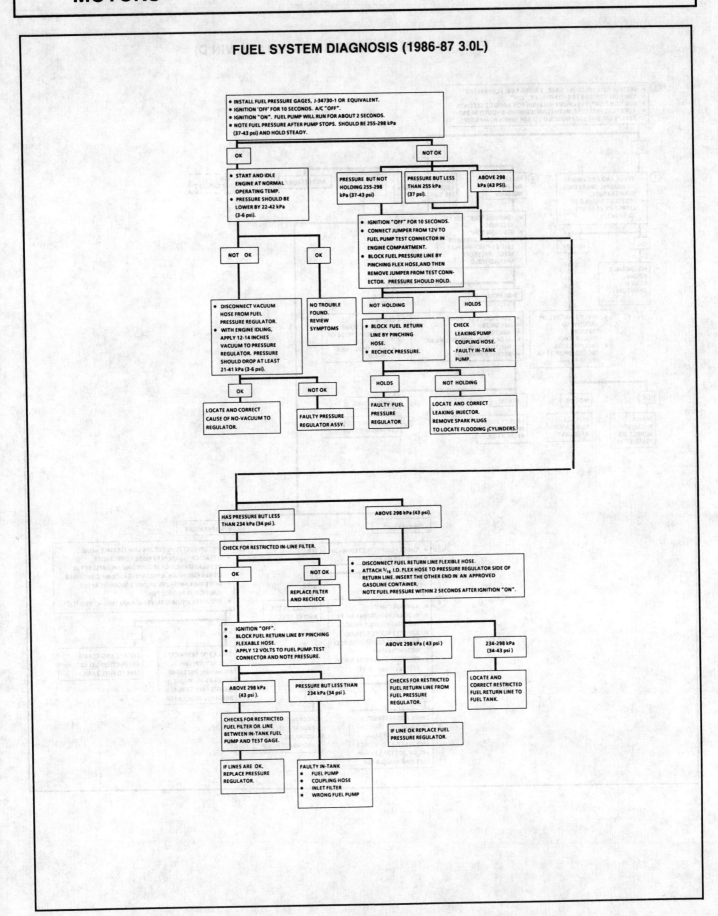

FUEL SYSTEM DIAGNOSIS (1988-89 2.3L VIN D)

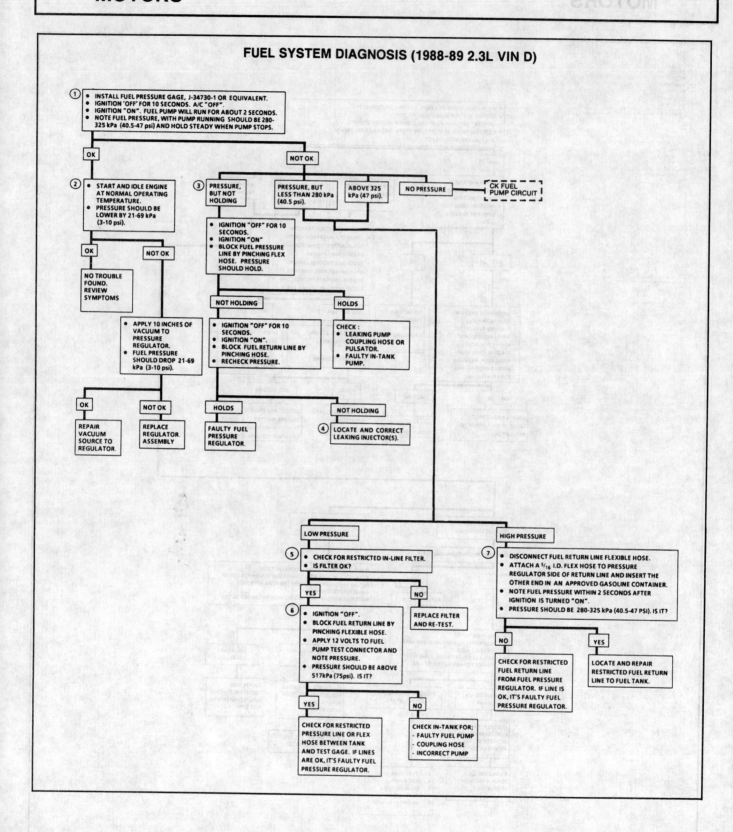

① • INSTALL FUEL PRESSURE GAGE, J-34730-1 OR EQUIVALENT.
• IGNITION 'OFF' FOR 10 SECONDS. A/C "OFF".
• IGNITION "ON". FUEL PUMP WILL RUN FOR ABOUT 2 SECONDS.
• NOTE FUEL PRESSURE, WITH PUMP RUNNING SHOULD BE 280-325 kPa (40.5-47 psi) AND HOLD STEADY WHEN PUMP STOPS.

OK | NOT OK

② • START AND IDLE ENGINE AT NORMAL OPERATING TEMPERATURE.
• PRESSURE SHOULD BE LOWER BY 21-69 kPa (3-10 psi).

③ PRESSURE, BUT NOT HOLDING | PRESSURE, BUT LESS THAN 280 kPa (40.5). | ABOVE 325 kPa (47 psi). | NO PRESSURE | CK FUEL PUMP CIRCUIT

OK | NOT OK

NO TROUBLE FOUND. REVIEW SYMPTOMS

• IGNITION "OFF" FOR 10 SECONDS.
• IGNITION "ON"
• BLOCK FUEL PRESSURE LINE BY PINCHING FLEX HOSE. PRESSURE SHOULD HOLD.

• APPLY 10 INCHES OF VACUUM TO PRESSURE REGULATOR.
• FUEL PRESSURE SHOULD DROP 21-69 kPa (3-10 psi).

NOT HOLDING | HOLDS

• IGNITION "OFF" FOR 10 SECONDS.
• IGNITION "ON".
• BLOCK FUEL RETURN LINE BY PINCHING HOSE.
• RECHECK PRESSURE.

CHECK :
• LEAKING PUMP COUPLING HOSE OR PULSATOR.
• FAULTY IN-TANK PUMP.

OK | NOT OK

REPAIR VACUUM SOURCE TO REGULATOR. | REPLACE REGULATOR. ASSEMBLY

HOLDS | NOT HOLDING

FAULTY FUEL PRESSURE REGULATOR. | ④ LOCATE AND CORRECT LEAKING INJECTOR(S).

LOW PRESSURE | HIGH PRESSURE

⑤ • CHECK FOR RESTRICTED IN-LINE FILTER. IS FILTER OK? | ⑦ • DISCONNECT FUEL RETURN LINE FLEXIBLE HOSE.
• ATTACH 5/16 I.D. FLEX HOSE TO PRESSURE REGULATOR SIDE OF RETURN LINE AND INSERT THE OTHER END IN AN APPROVED GASOLINE CONTAINER.
• NOTE FUEL PRESSURE WITHIN 2 SECONDS AFTER IGNITION IS TURNED "ON".
• PRESSURE SHOULD BE 280-325 kPa (40.5-47 PSI). IS IT?

YES | NO

⑥ • IGNITION "OFF".
• BLOCK FUEL RETURN LINE BY PINCHING FLEXIBLE HOSE.
• APPLY 12 VOLTS TO FUEL PUMP TEST CONNECTOR AND NOTE PRESSURE.
• PRESSURE SHOULD BE ABOVE 517kPa (75psi). IS IT?

REPLACE FILTER AND RE-TEST.

NO | YES

CHECK FOR RESTRICTED FUEL RETURN LINE FROM FUEL PRESSURE REGULATOR. IF LINE IS OK, IT'S FAULTY FUEL PRESSURE REGULATOR. | LOCATE AND REPAIR RESTRICTED FUEL RETURN LINE TO FUEL TANK.

YES | NO

CHECK FOR RESTRICTED PRESSURE LINE OR FLEX HOSE BETWEEN TANK AND TEST GAGE. IF LINES ARE OK, IT'S FAULTY FUEL PRESSURE REGULATOR. | CHECK IN-TANK FOR;
- FAULTY FUEL PUMP
- COUPLING HOSE
- INCORRECT PUMP

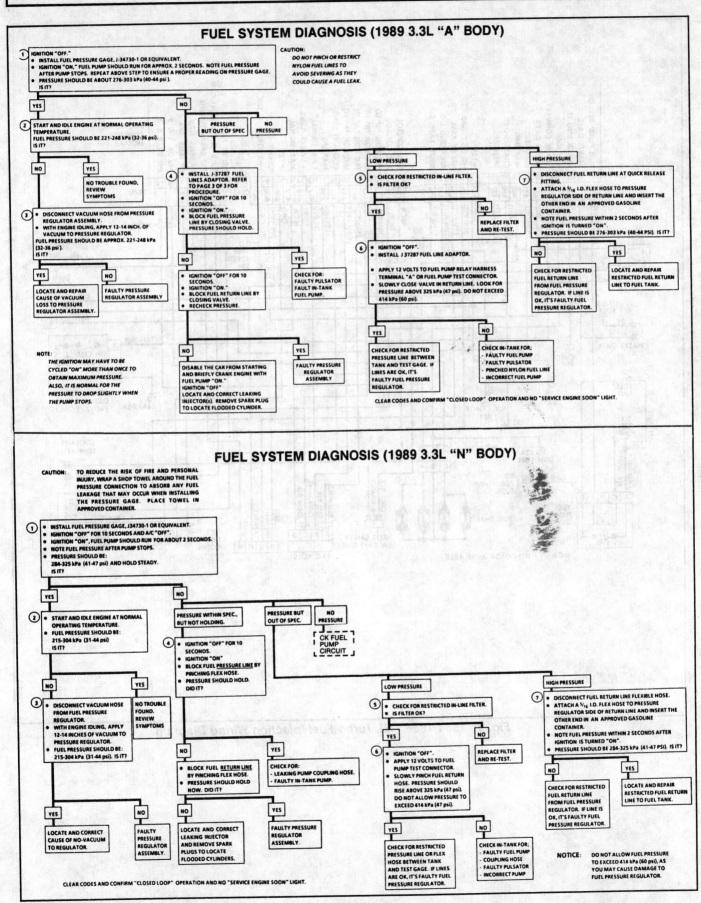

FUEL SYSTEM DIAGNOSIS (1989 3.3L "A" BODY)

FUEL SYSTEM DIAGNOSIS (1989 3.3L "N" BODY)

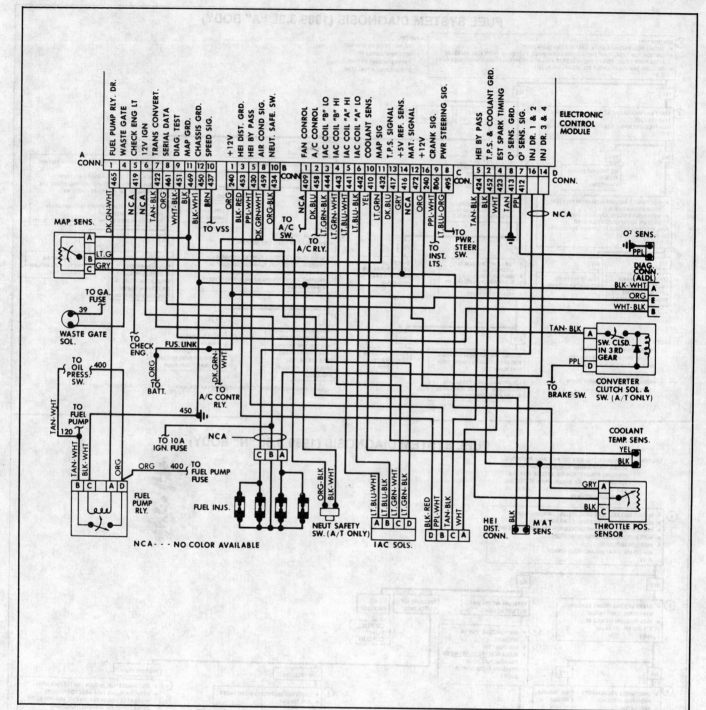

Fig. 9: 1984-1986 1.8L Turbo Fuel Injection Wiring Diagram

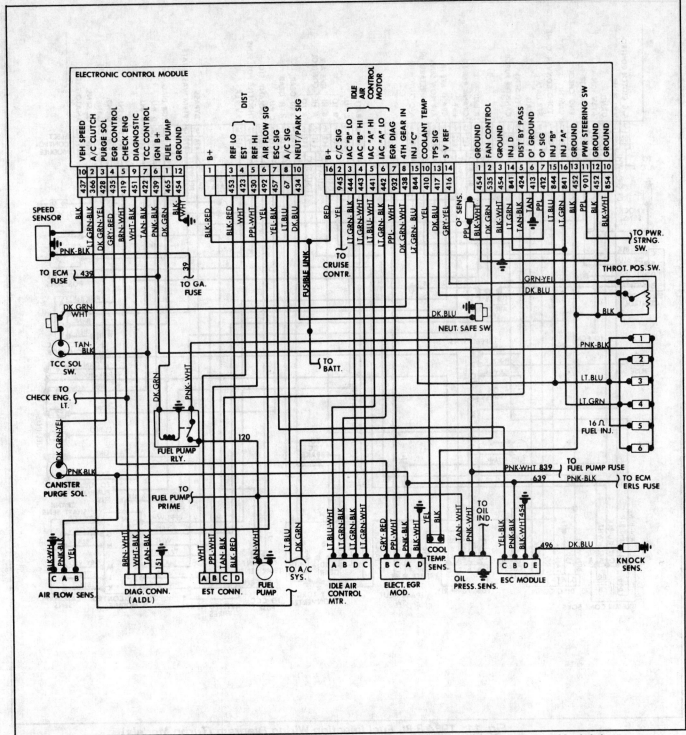

Fig. 10: 1984 3.8L Fuel Injection Wiring Diagram (Non-Turbo Models)

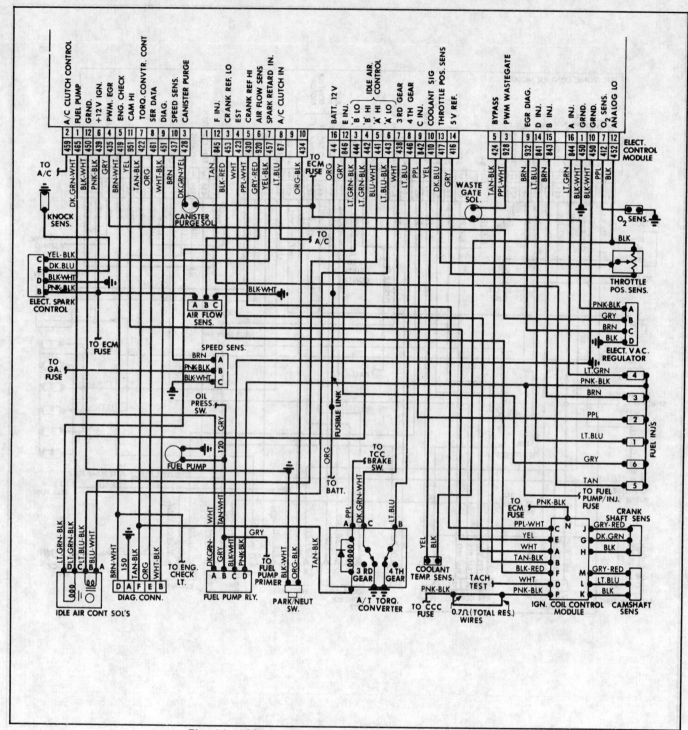

Fig. 11: 1984 3.8L Fuel Injection Wiring Diagram (Turbo Models)

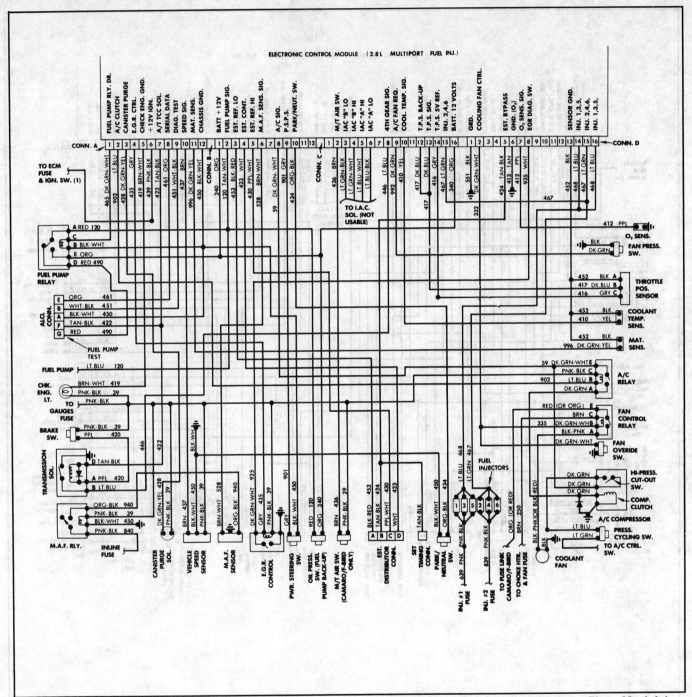

Fig. 12: 1985 2.8L Fuel Injection Wiring Diagram (All RWD Except Fiero Models)

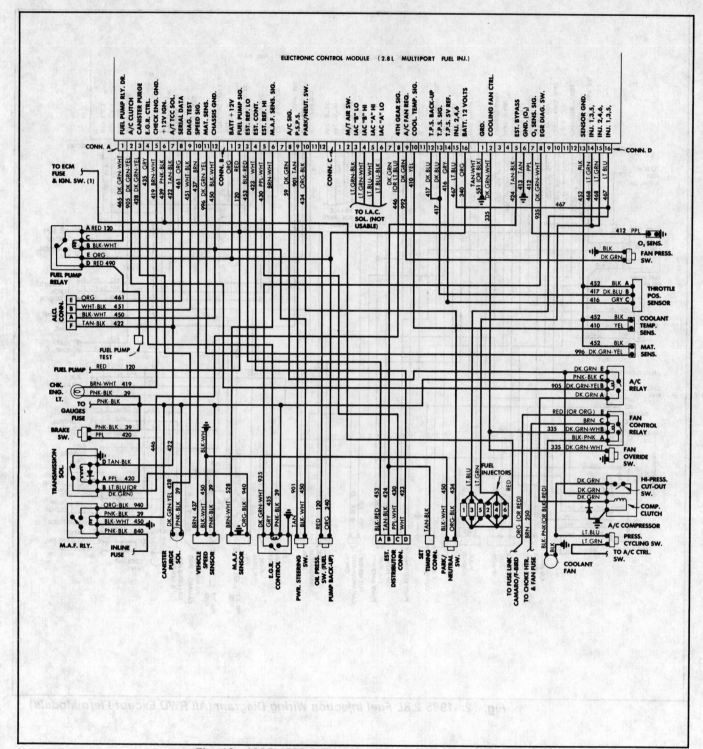

Fig. 13: 1985-1986 2.8L Fuel Injection Wiring Diagram (FWD Models)

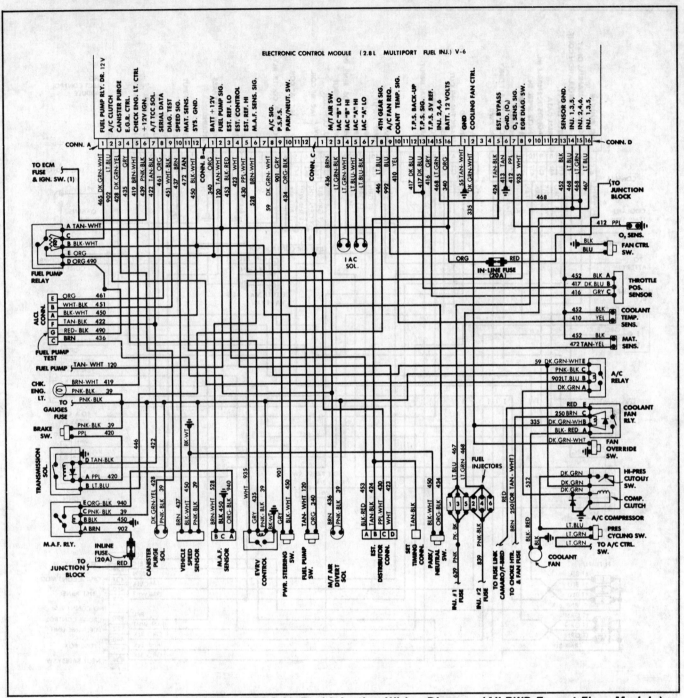

Fig. 14: 1986 2.8L Fuel Injection Wiring Diagram (All RWD Except Fiero Models)

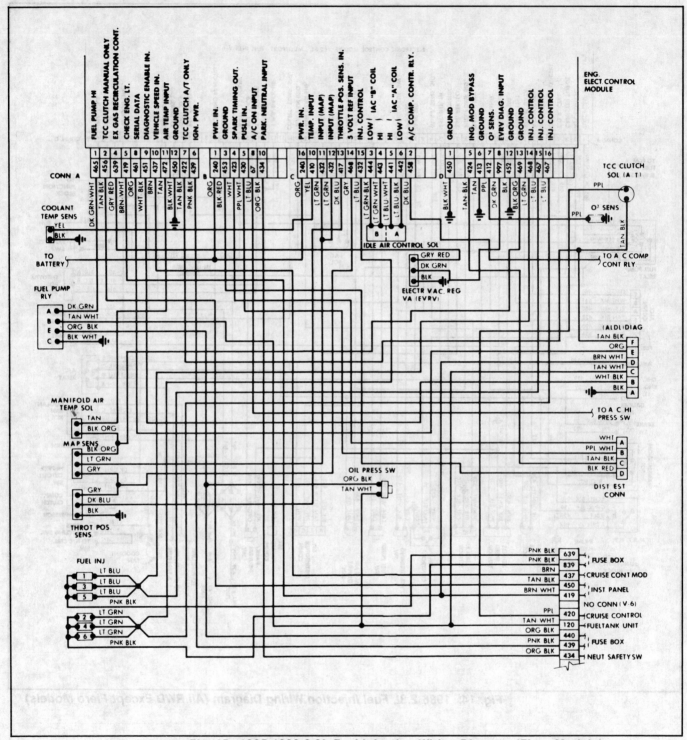

Fig. 15: 1985-1986 2.8L Fuel Injection Wiring Diagram (Fiero Models)

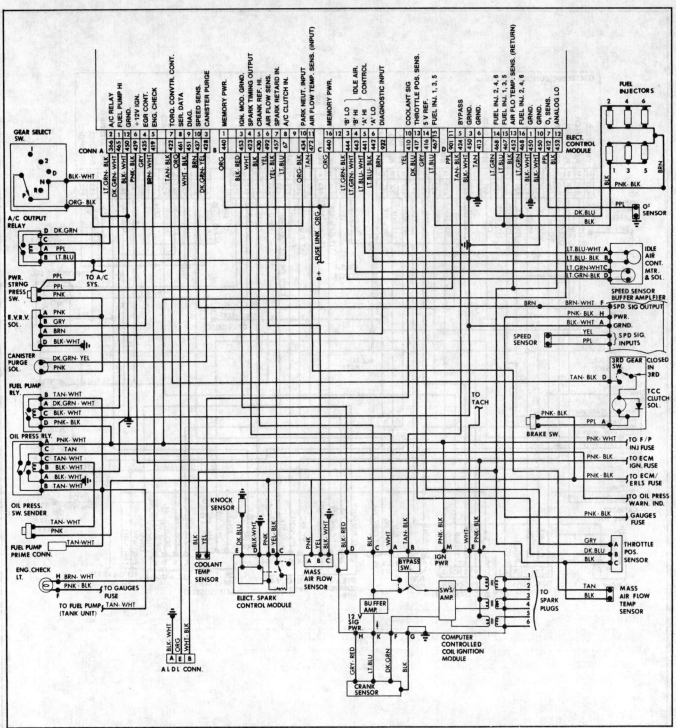

Fig. 16: 1985-1986 3.0L Fuel Injection Wiring Diagram

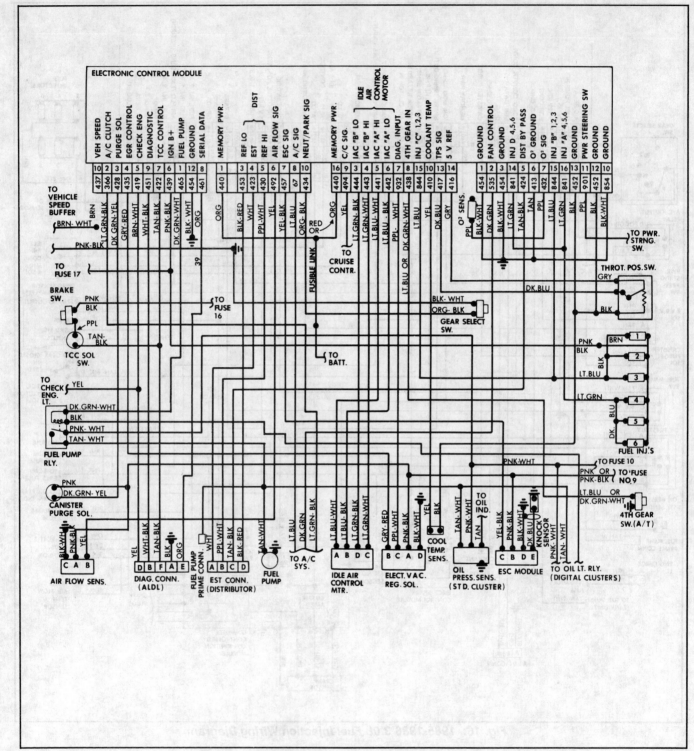

Fig. 17: 1985-1986 3.8L Fuel Injection Wiring Diagram (Non-Turbo)

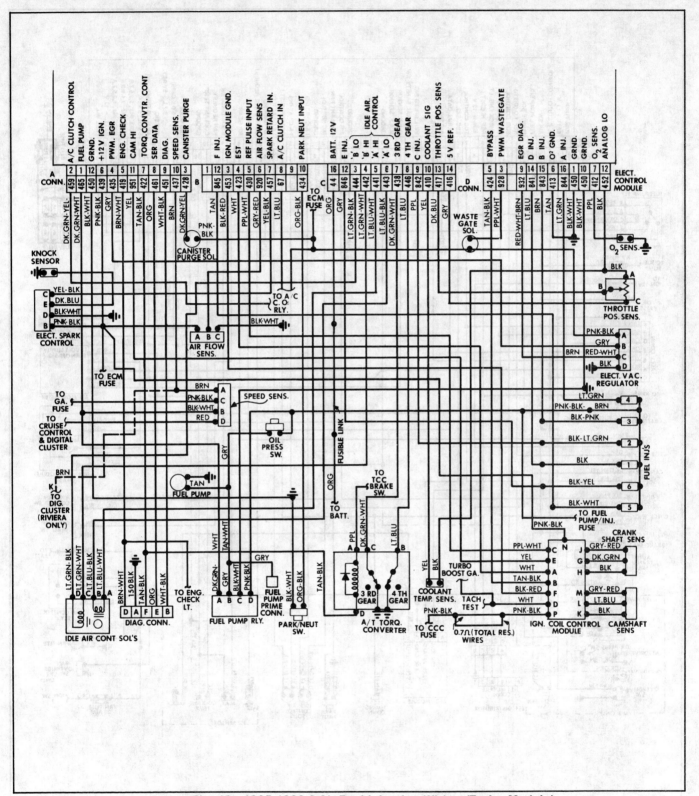

Fig. 18: 1985-1986 3.8L Fuel Injection Wiring (Turbo Models)

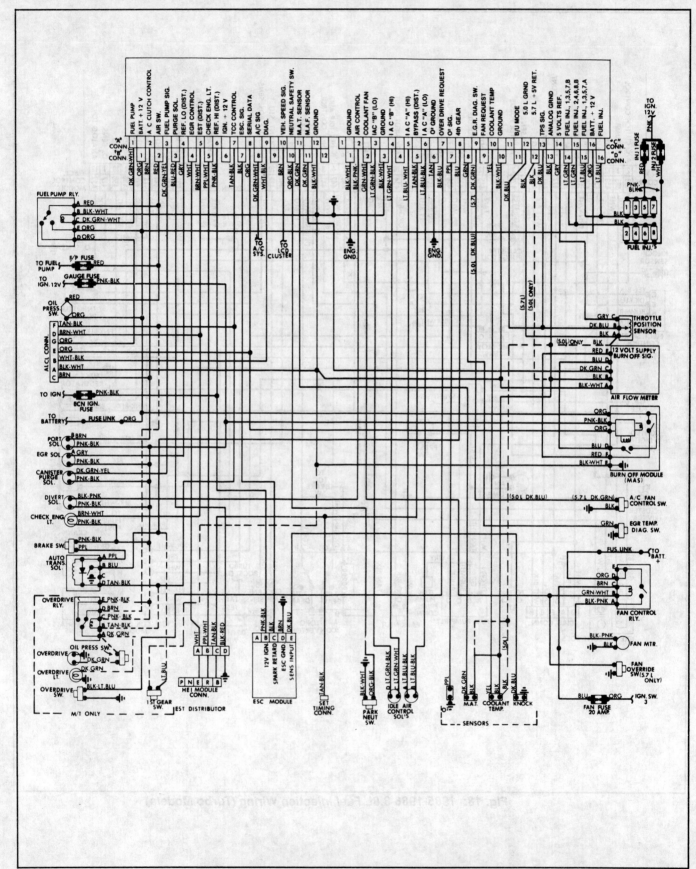

Fig. 19: 1985 5.0L & 5.7L Fuel Injection Wiring Diagram

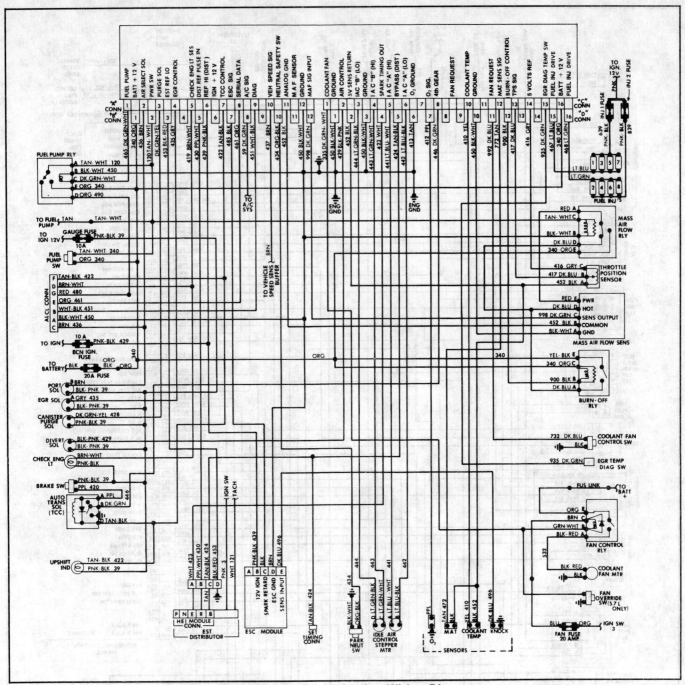

Fig. 20: 1986 5.0L Fuel Injection Wiring Diagram

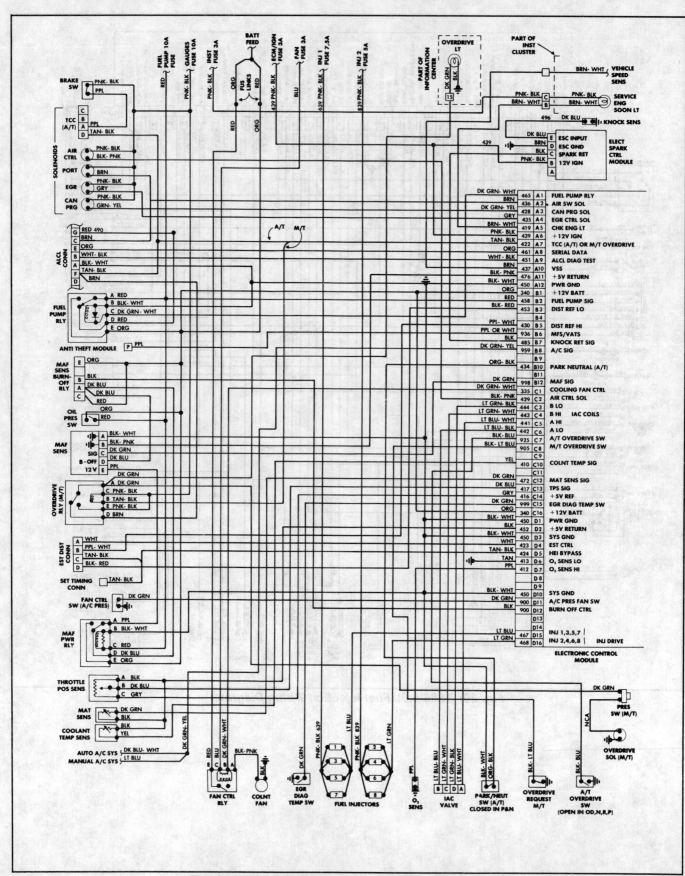

Fig. 21: 1986 5.7L Fuel Injection Wiring Diagram

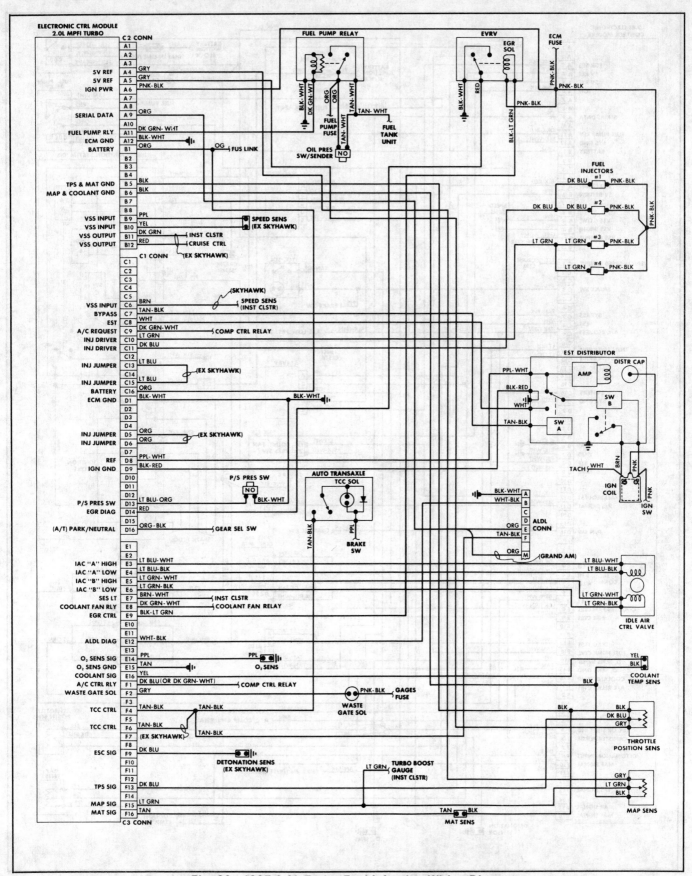

Fig. 22: 1987 2.0L Turbo Fuel Injection Wiring Diagram

GENERAL MOTORS   PORT FUEL INJECTION

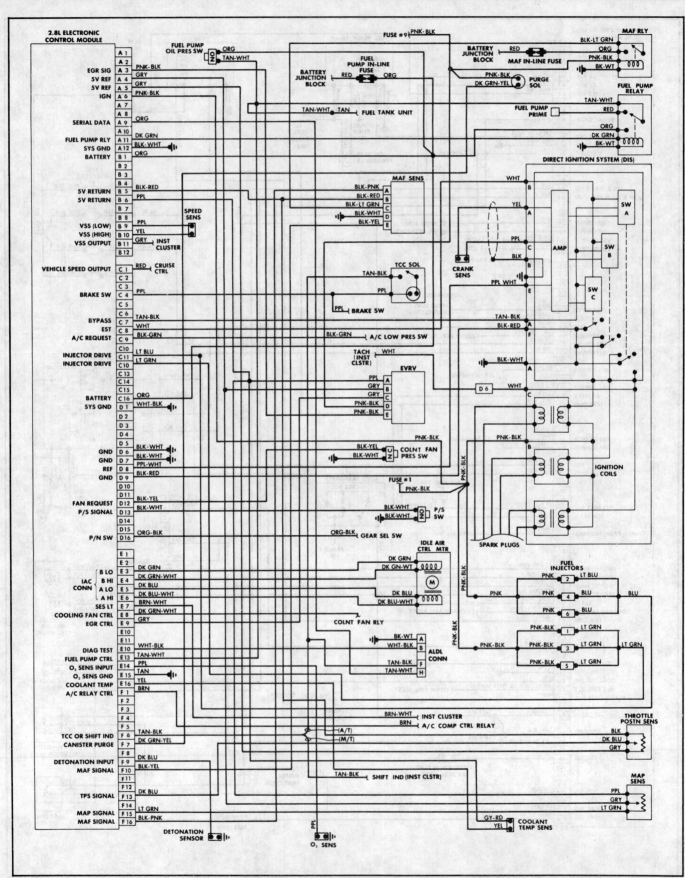

Fig. 23: 1987 Beretta & Corsica 2.8L Fuel Injection Wiring Diagram

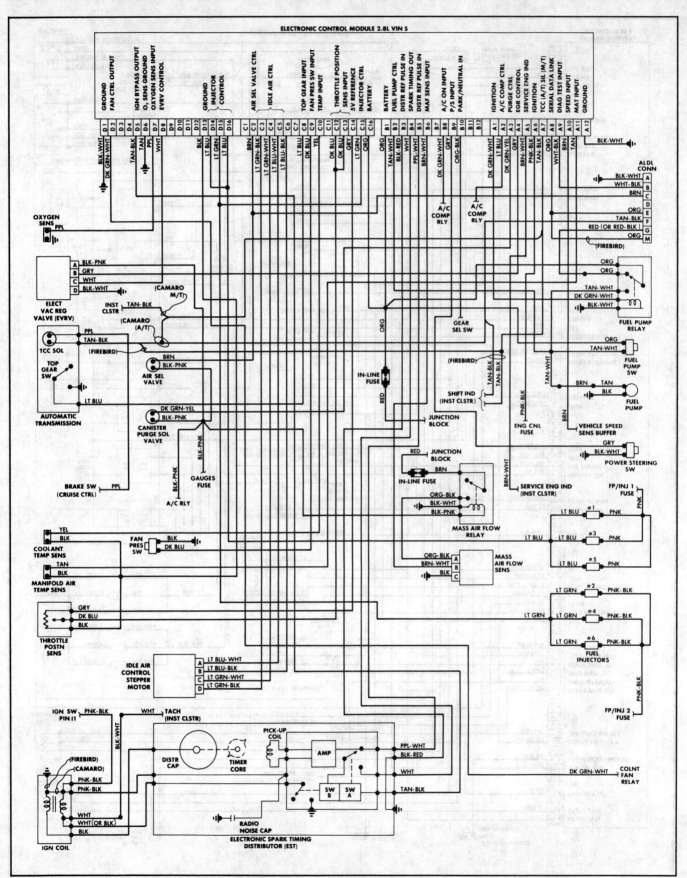

Fig. 24: 1987 Camaro & Firebird 2.8L Fuel Injection Wiring Diagram

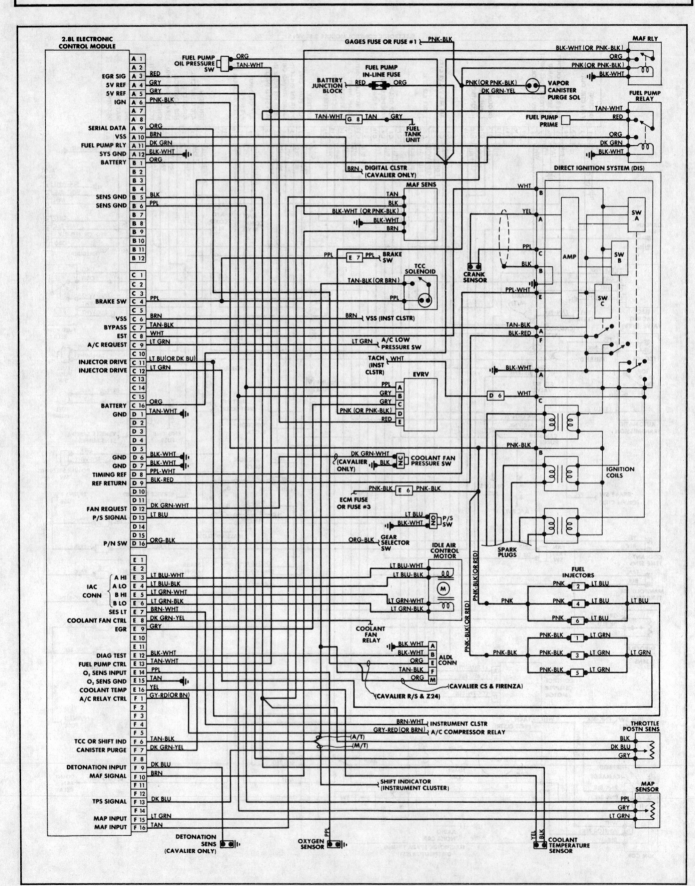

Fig. 25: 1987 Cavalier & Firenza 2.8L Fuel Injection Wiring Diagram

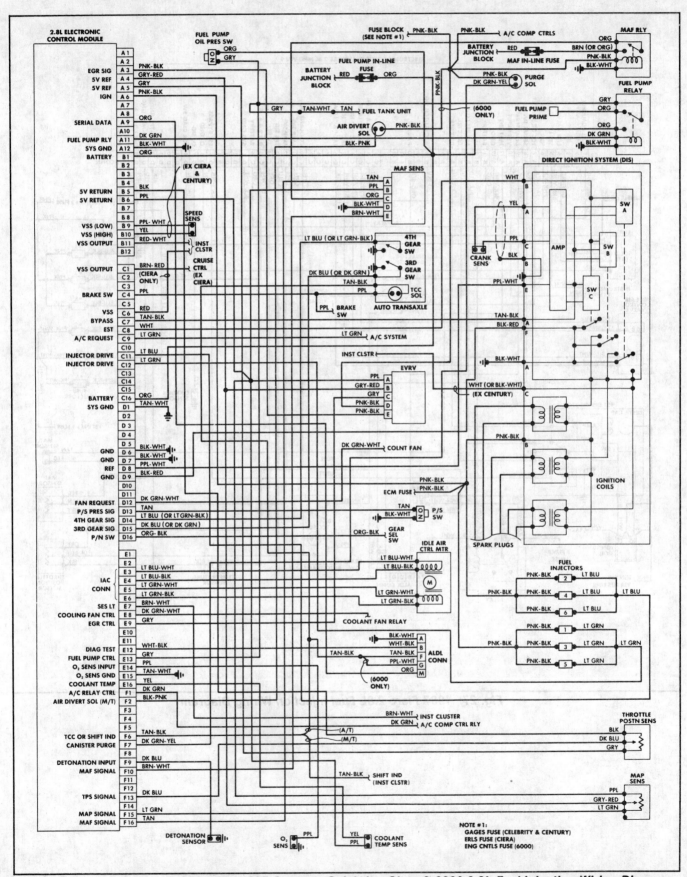

Fig. 26: 1987 Century, Celebrity, Ciera & 6000 2.8L Fuel Injection Wiring Diagram

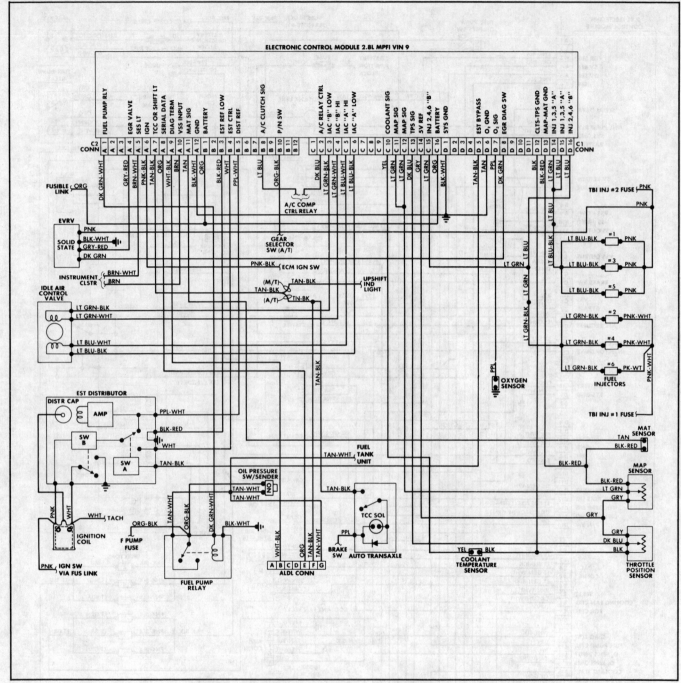

Fig. 27: 1987 Fiero 2.8L Fuel Injection Wiring Diagram

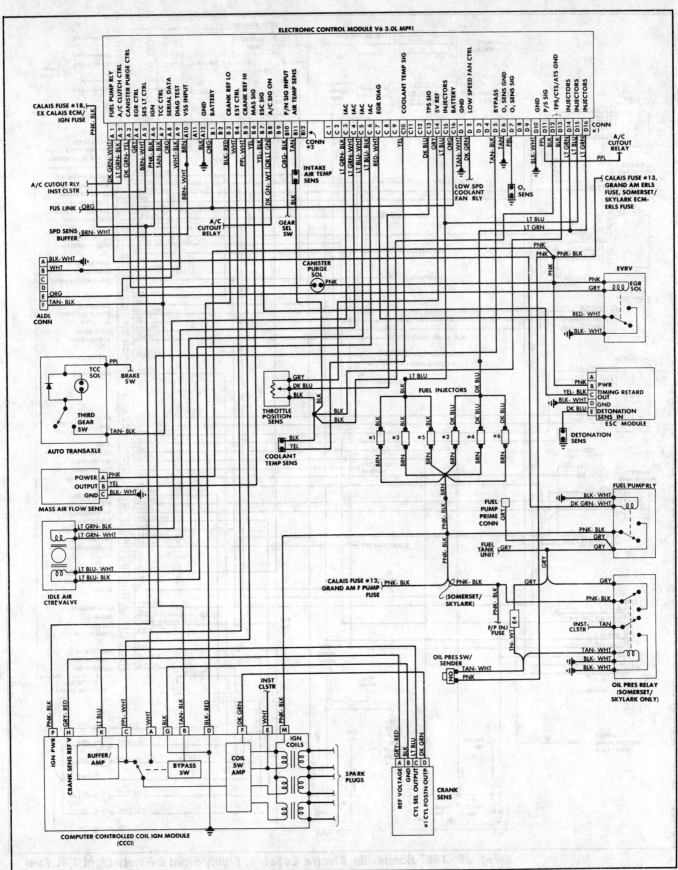

Fig. 28: 1987 Calais, Grand Am, Skylark & Somerset 3.0L Fuel Injection Wiring Diagram

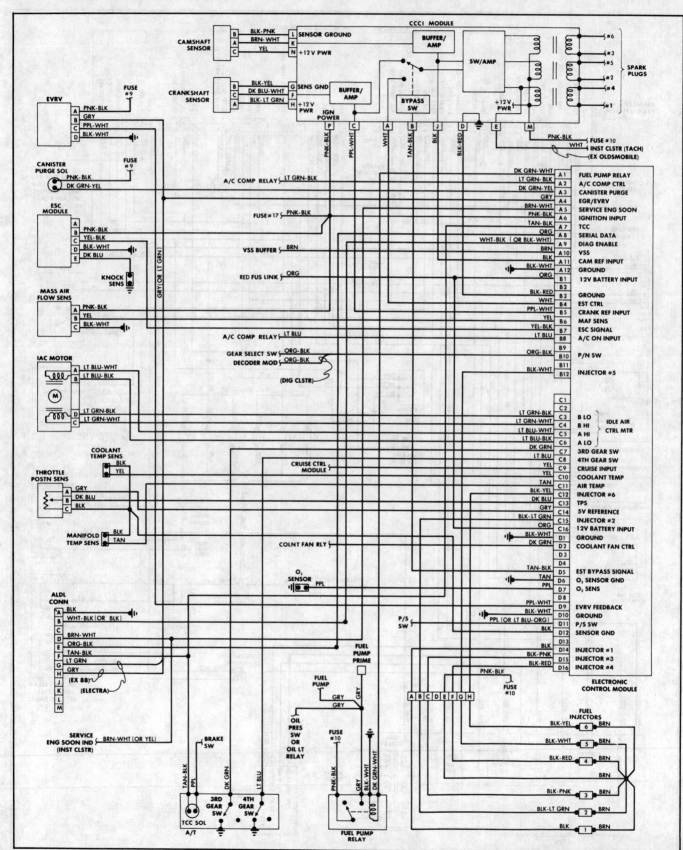

Fig. 29: *1987 Bonneville, Electra, LeSabre, Eighty-Eight & Ninety-Eight 3.8L Fuel Injection Wiring Diagram*

530

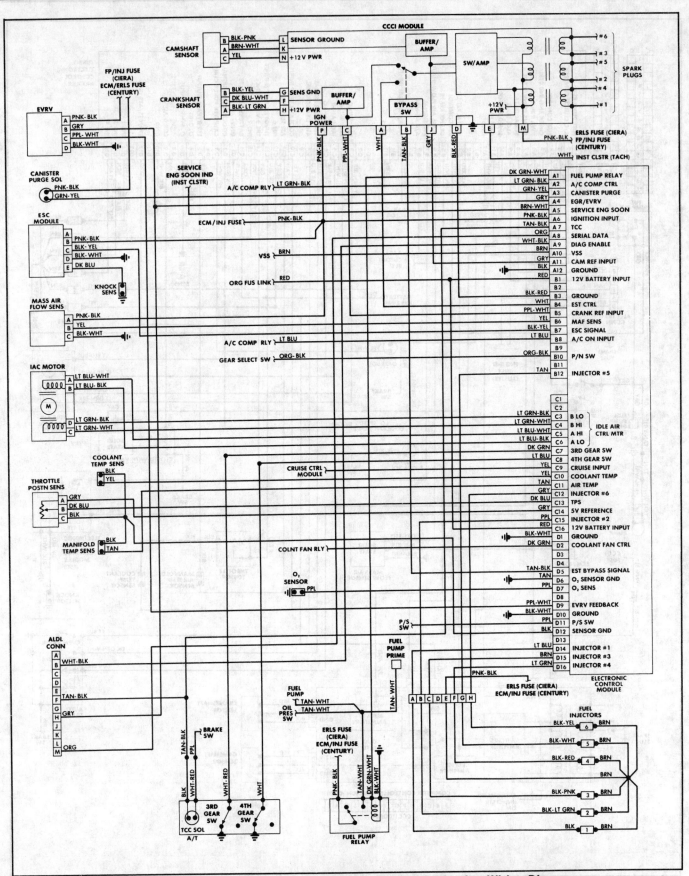

Fig. 30: 1987 Century & Ciera 3.8L Fuel Injection Wiring Diagram

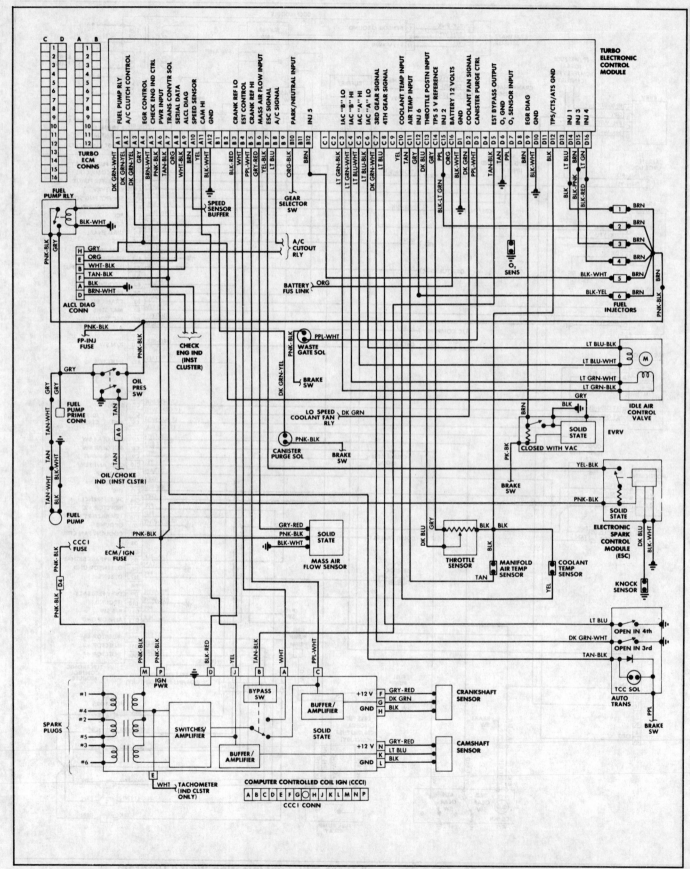

Fig. 31: 1987 Regal Turbo 3.8L Fuel Injection Wiring Diagram

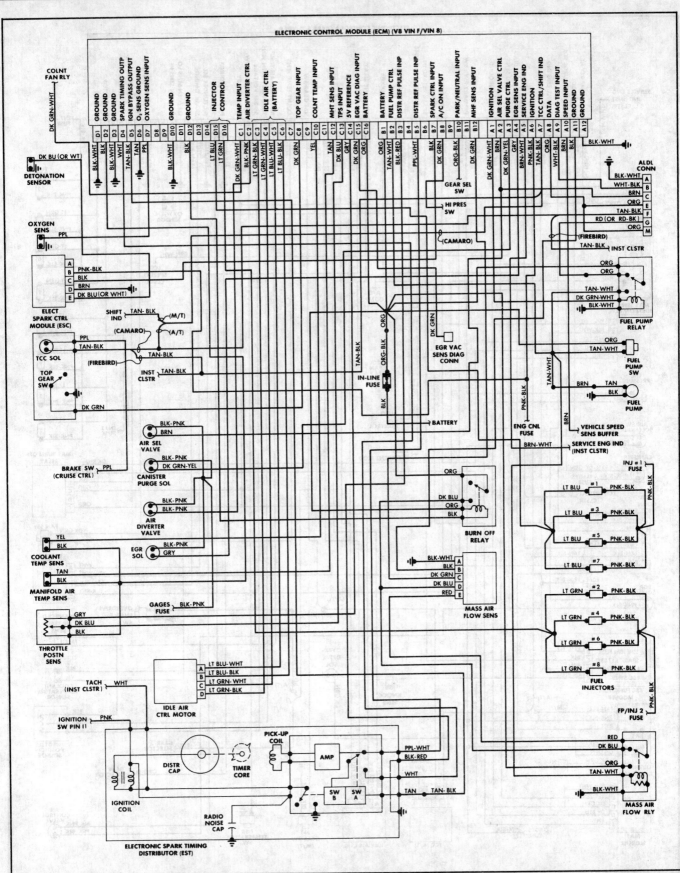

Fig. 32: 1987 Camaro & Firebird 5.0L & 5.7L Fuel Injection Wiring Diagram

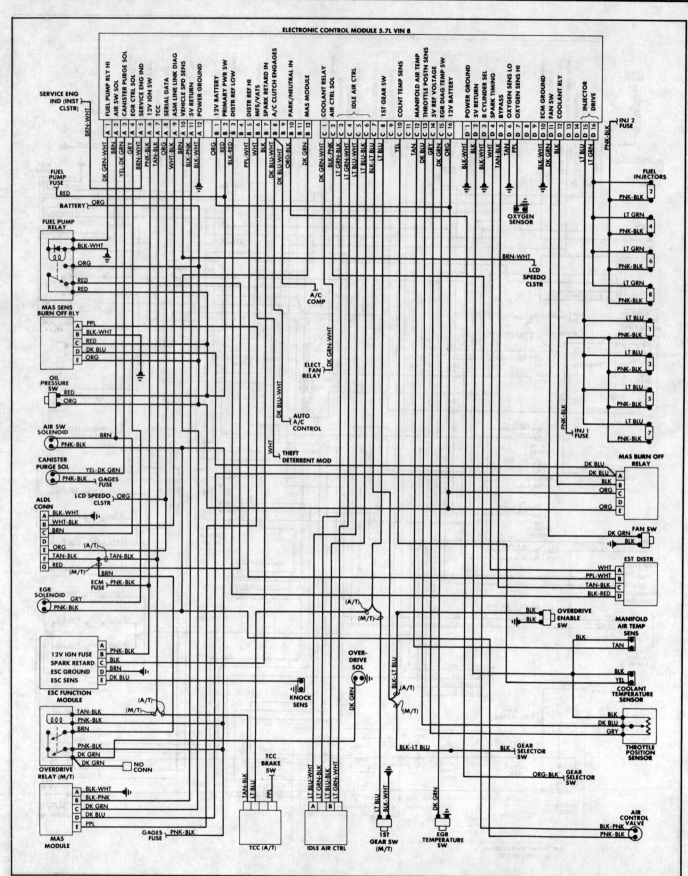

Fig. 33: 1987 Corvette 5.7L Fuel Injection Wiring Diagram

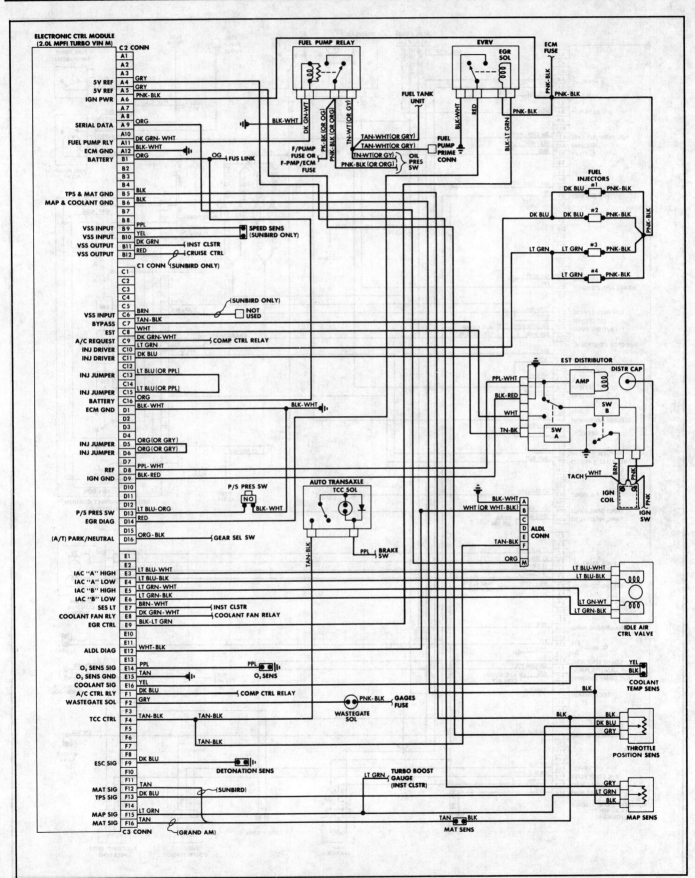

Fig. 34: 1988 Grand Am & Sunbird 2.0L Fuel Injection Wiring Diagram

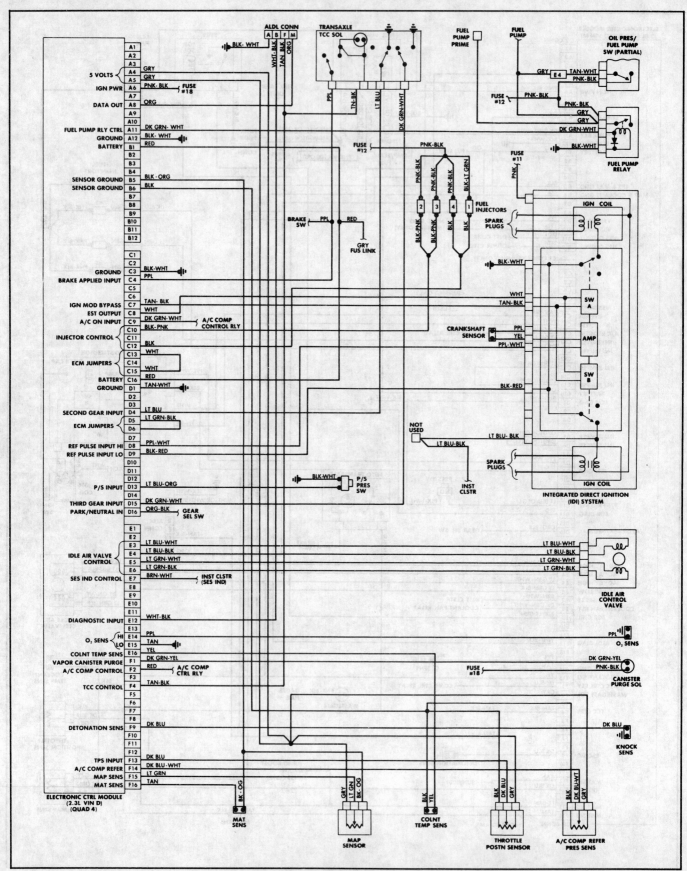

Fig. 35: 1988 Calais & Grand Am 2.3L Fuel Injection Wiring Diagram

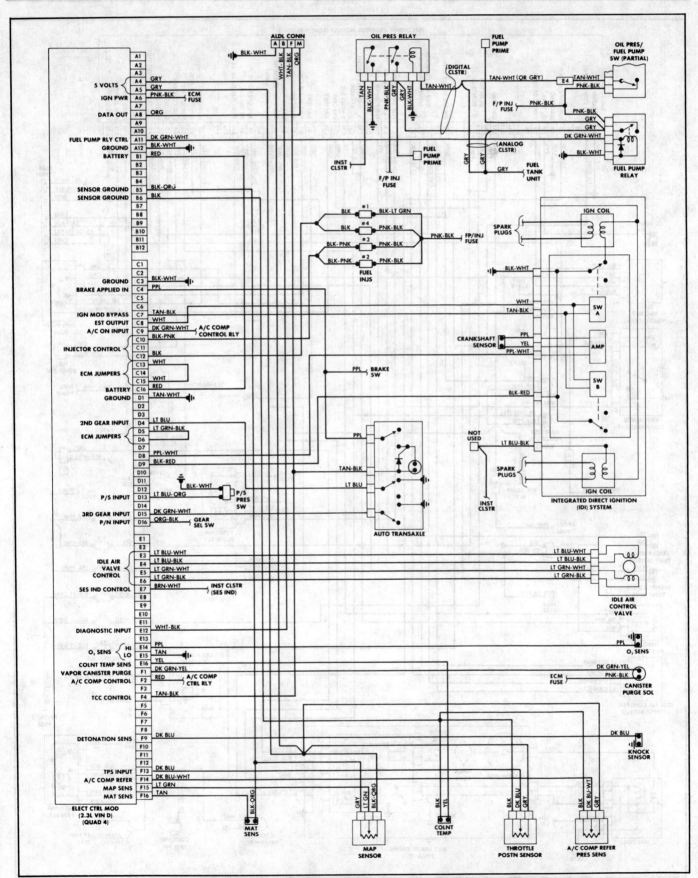

Fig. 36: 1988 Skylark 2.3L Fuel Injection Wiring Diagram

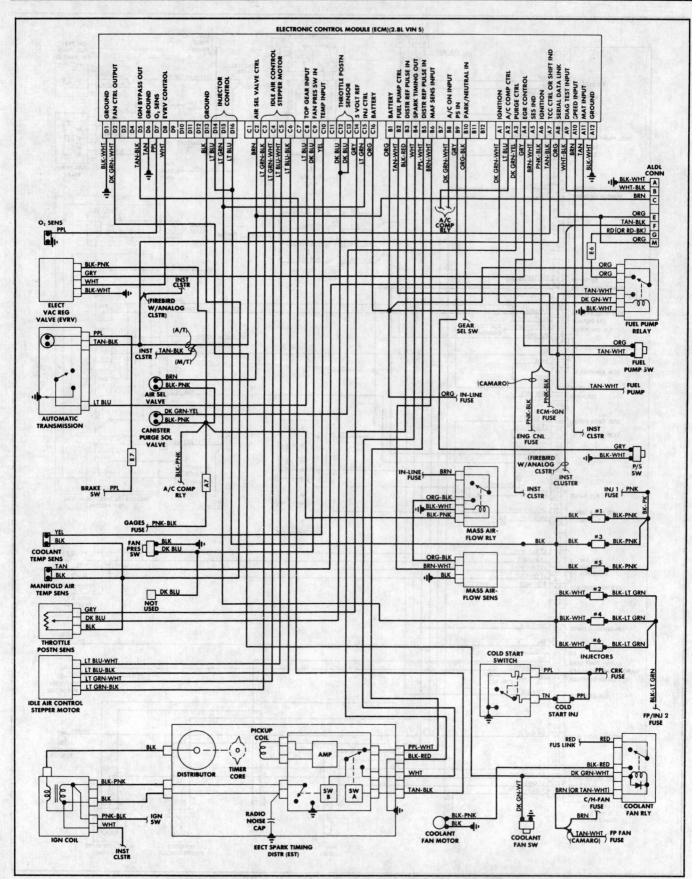

Fig. 37: 1988 Camaro & Firebird 2.8L Fuel Injection Wiring Diagram

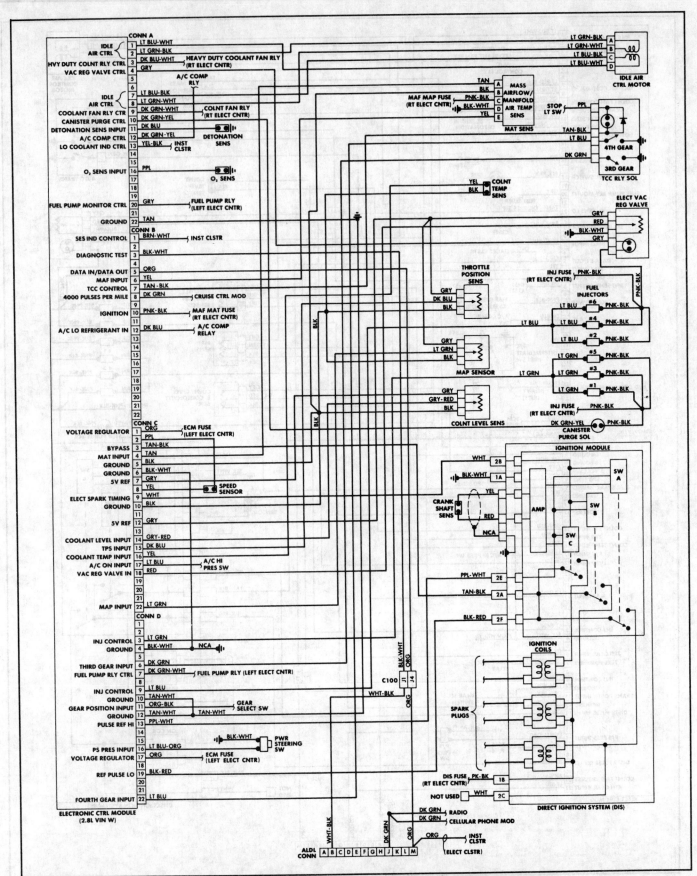

Fig. 38: 1988 Regal FWD 2.8L Fuel Injection Wiring Diagram

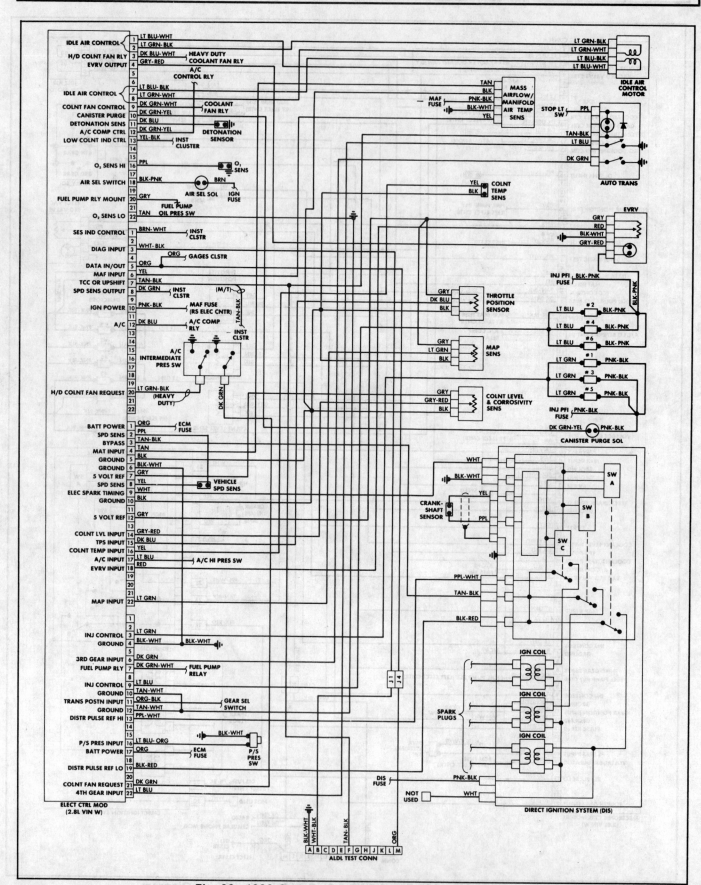

Fig. 39: 1988 Cutlass Supreme FWD 2.8L Fuel Injection Wiring Diagram

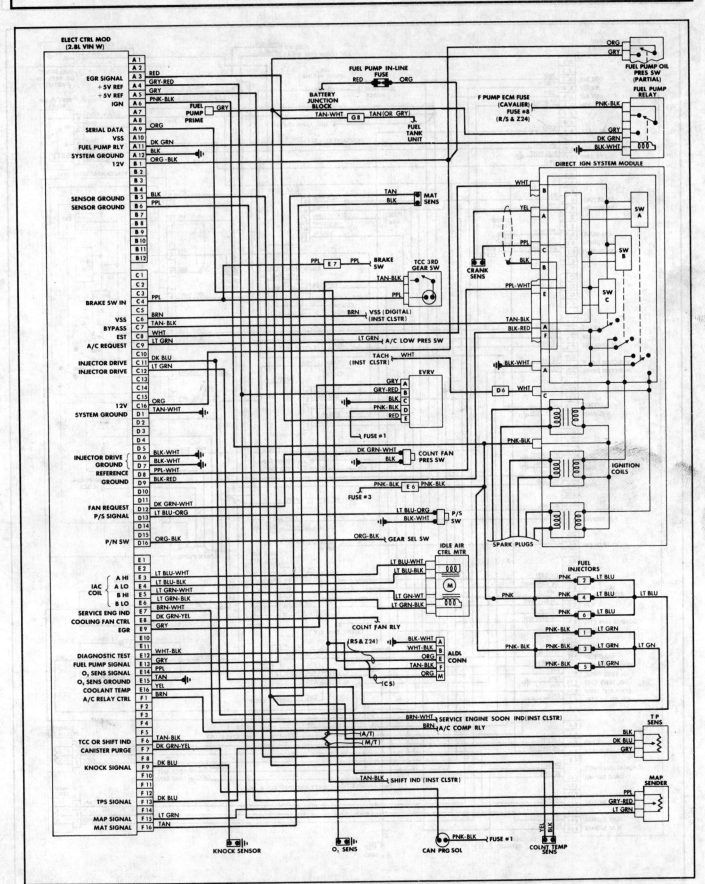

Fig. 40: 1988 Cavalier 2.8L Fuel Injection Wiring Diagram

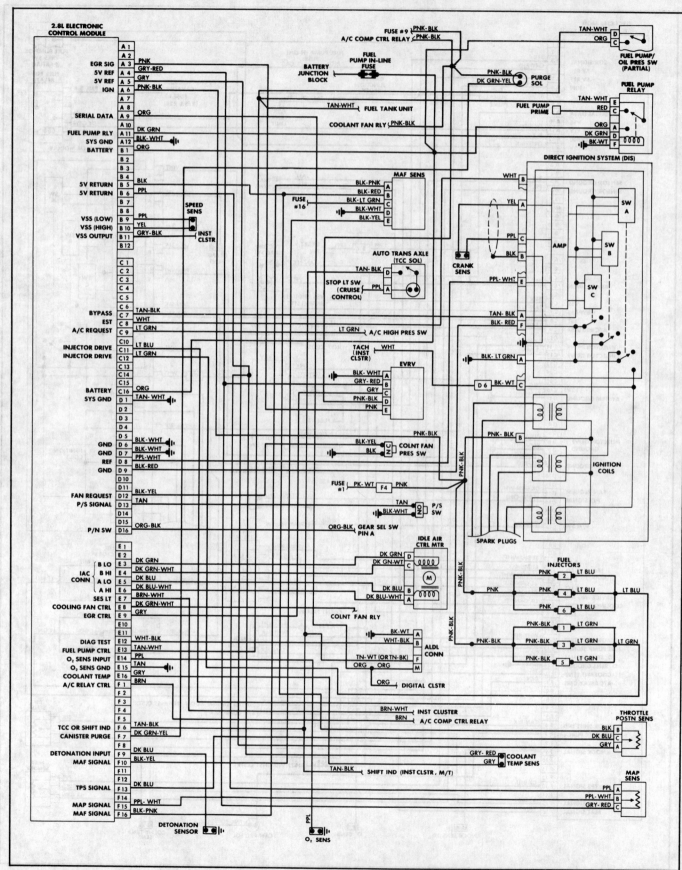

Fig. 41: 1988 Beretta & Corsica 2.8L Fuel Injection Wiring Diagram

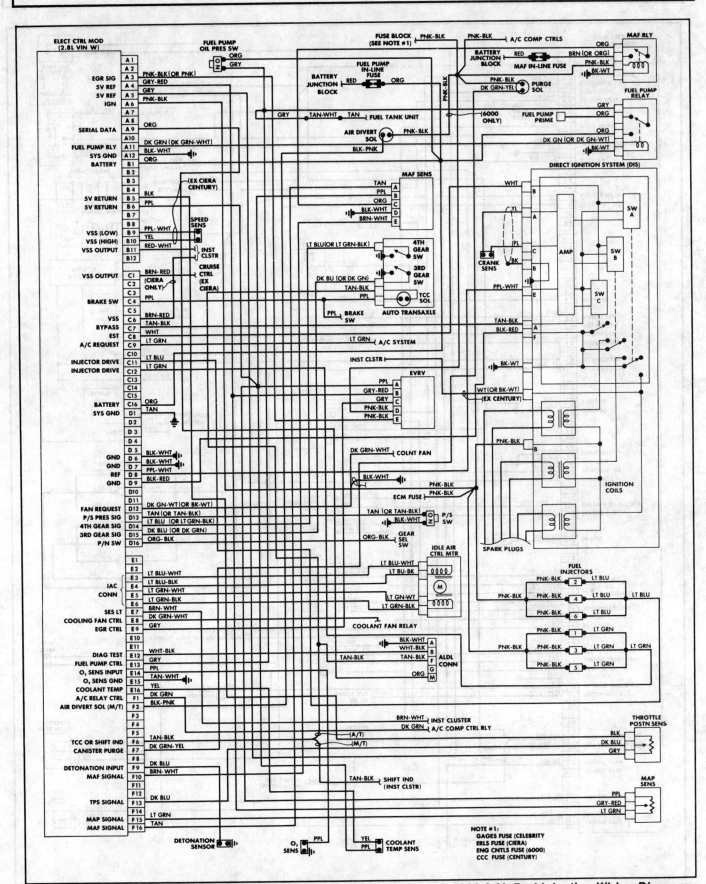

Fig. 42: *1988 Celebrity, Century, Ciera & 6000 2.8L Fuel Injection Wiring Diagram*

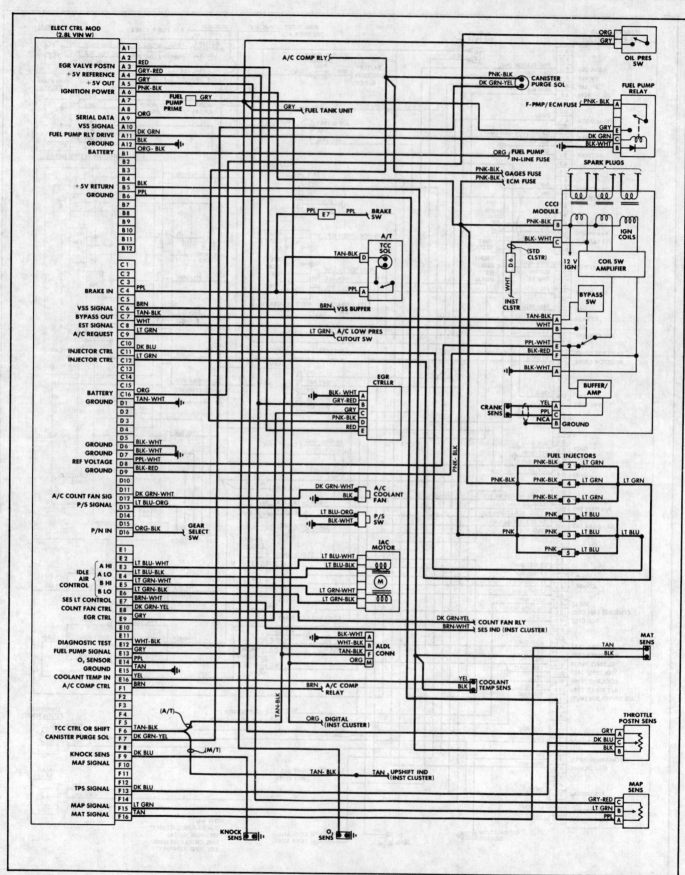

Fig. 43: 1988 Cimarron 2.8L Fuel Injection Wiring Diagram

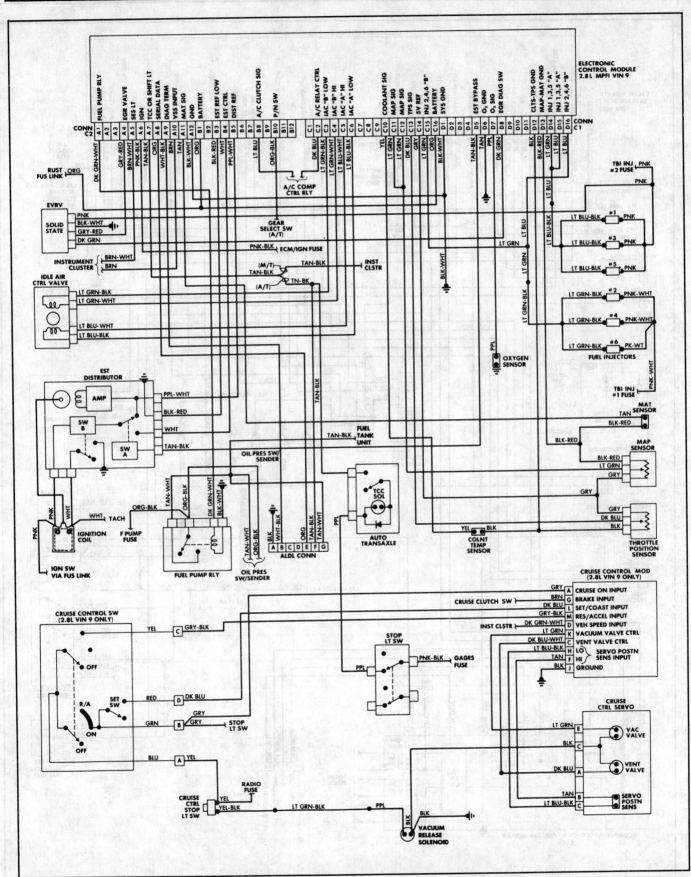

Fig. 44: 1988 Fiero 2.8L Fuel Injection Wiring Diagram

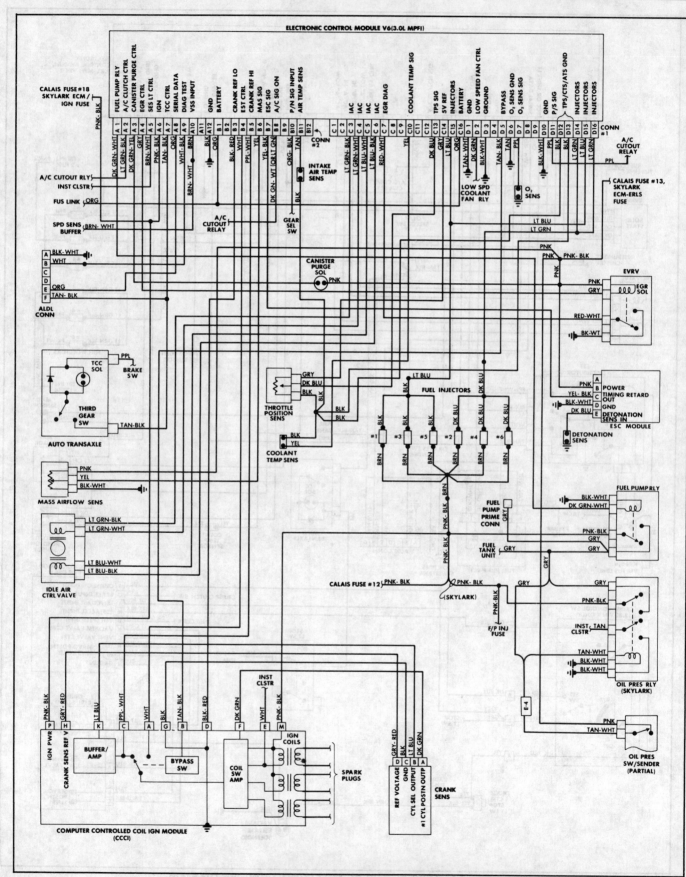

Fig. 45: 1988 Calais & Skylark 3.0L Fuel Injection Wiring Diagram

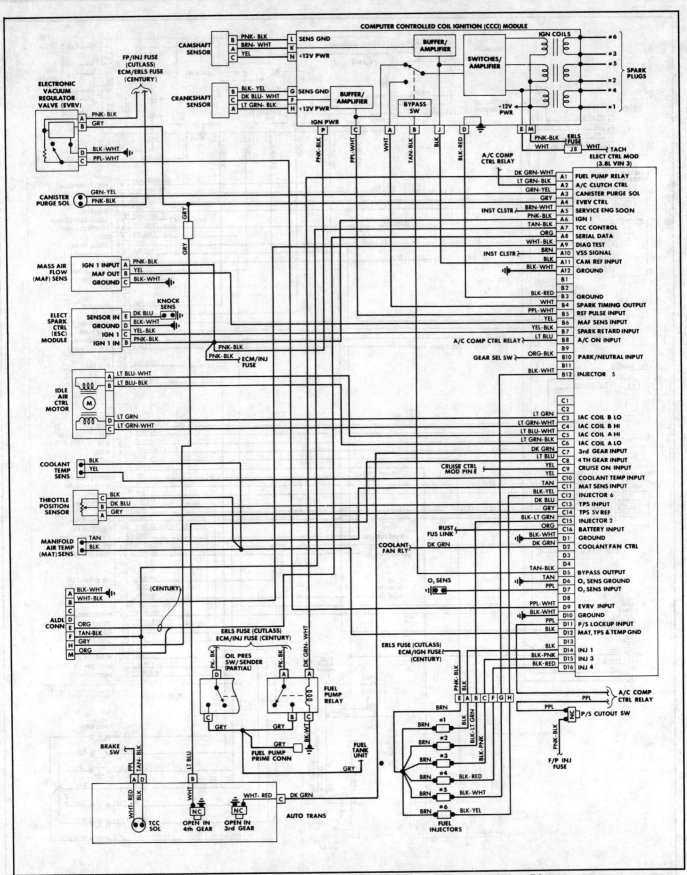

Fig. 46: 1988 Century & Ciera 3.8L Fuel Injection Wiring Diagram

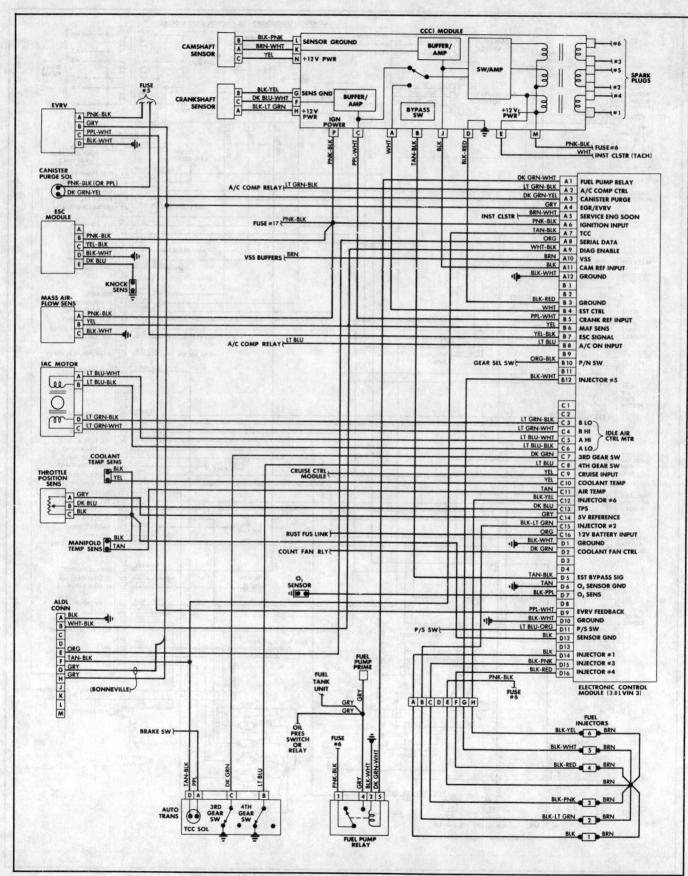

Fig. 47: 1988 Bonneville, Delta 88, LeSabre & Ninety-Eight 3.8L Fuel Injection Wiring Diagram

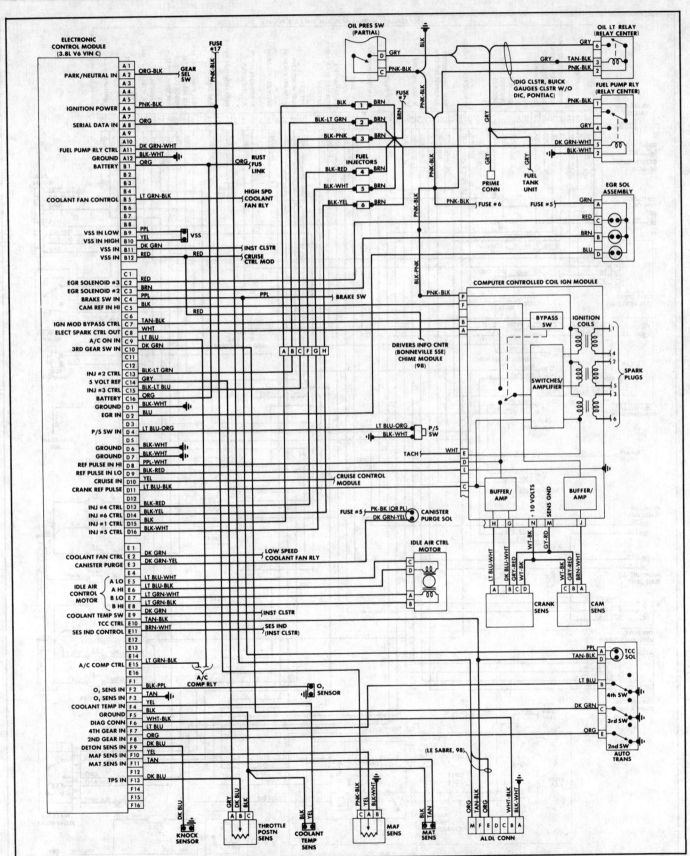

Fig. 48: 1988 Bonneville, Electra, LeSabre & Ninety-Eight 3.8L "3800" Fuel Injection Wiring Diagram

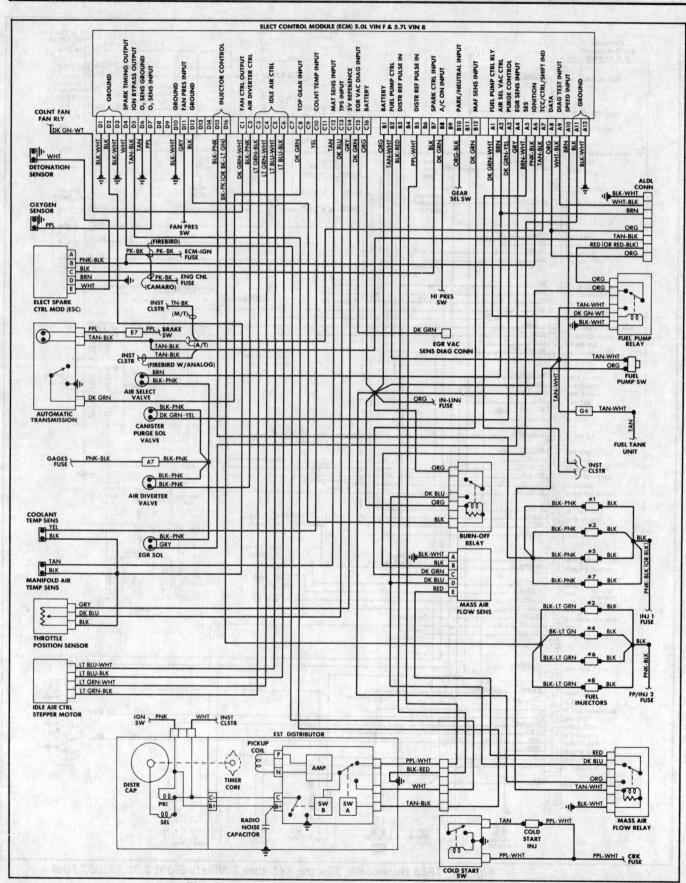

Fig. 49: 1988 Camaro & Firebird 5.0L & 5.7L Fuel Injection Wiring Diagram

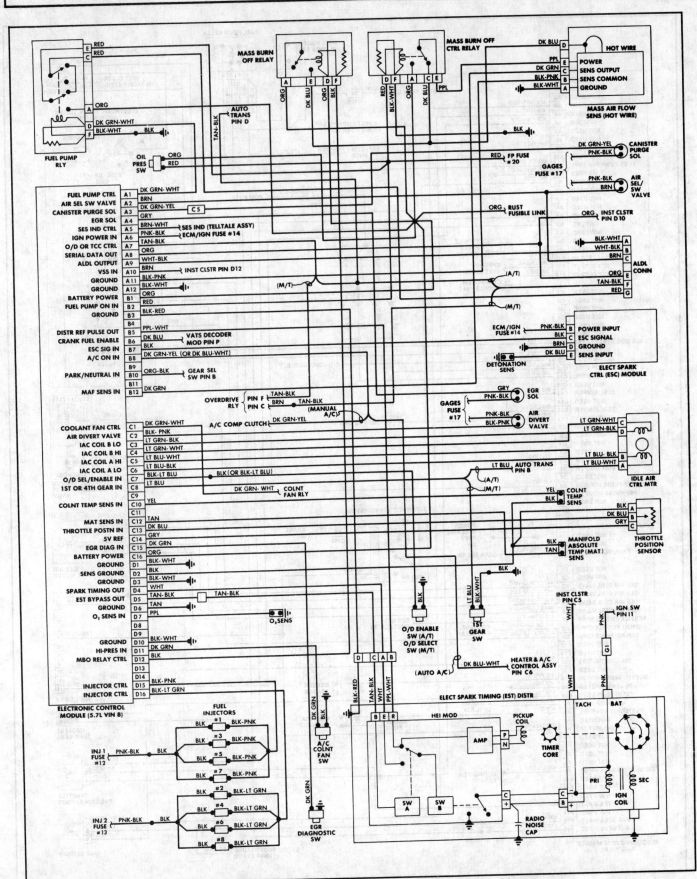

Fig. 50: 1988 Corvette 5.7L Fuel Injection Wiring Diagram

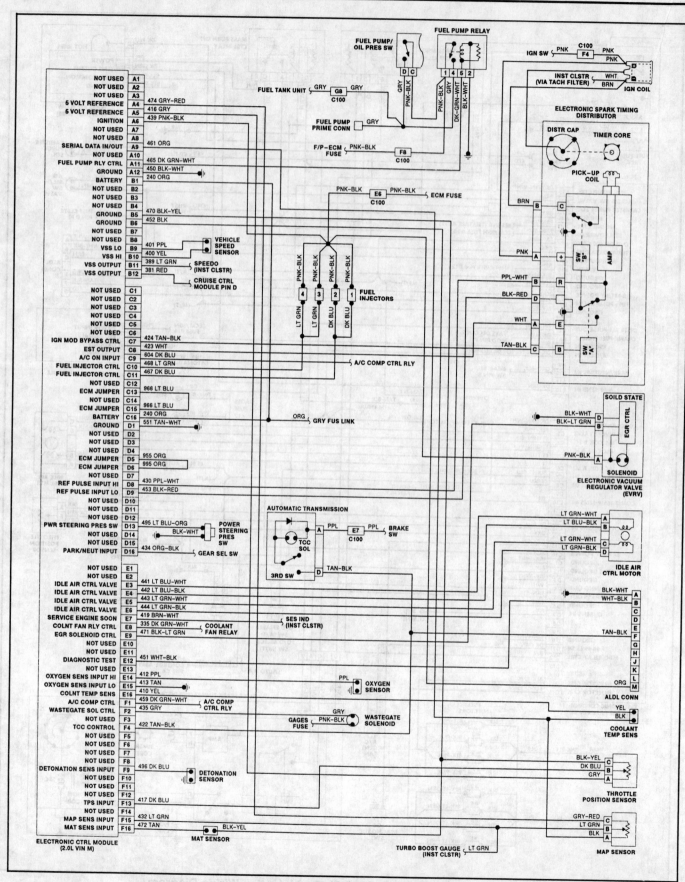

Fig. 51: 1989 2.0L (VIN M) J Body Fuel Injection Wiring Diagram

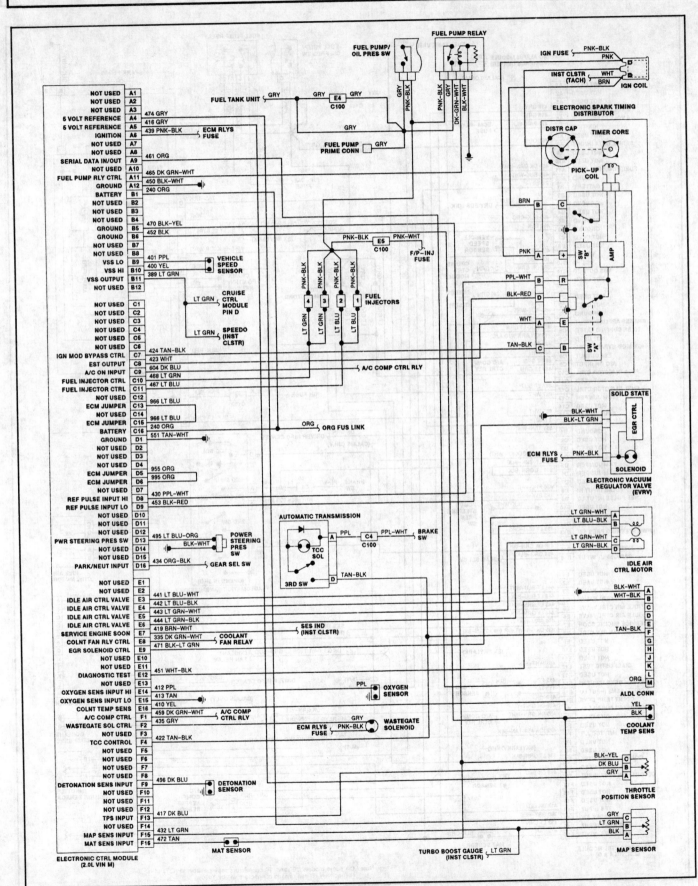

Fig. 52: 1989 2.0L (VIN M) N Body Fuel Injection Wiring Diagram

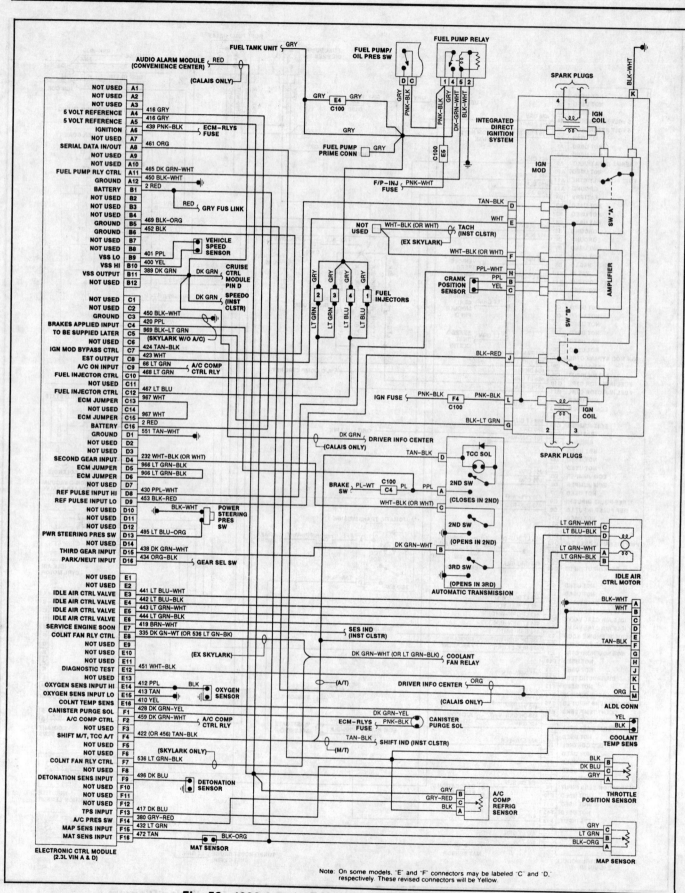

Fig. 53: 1989 2.3L Fuel Injection Wiring Diagram

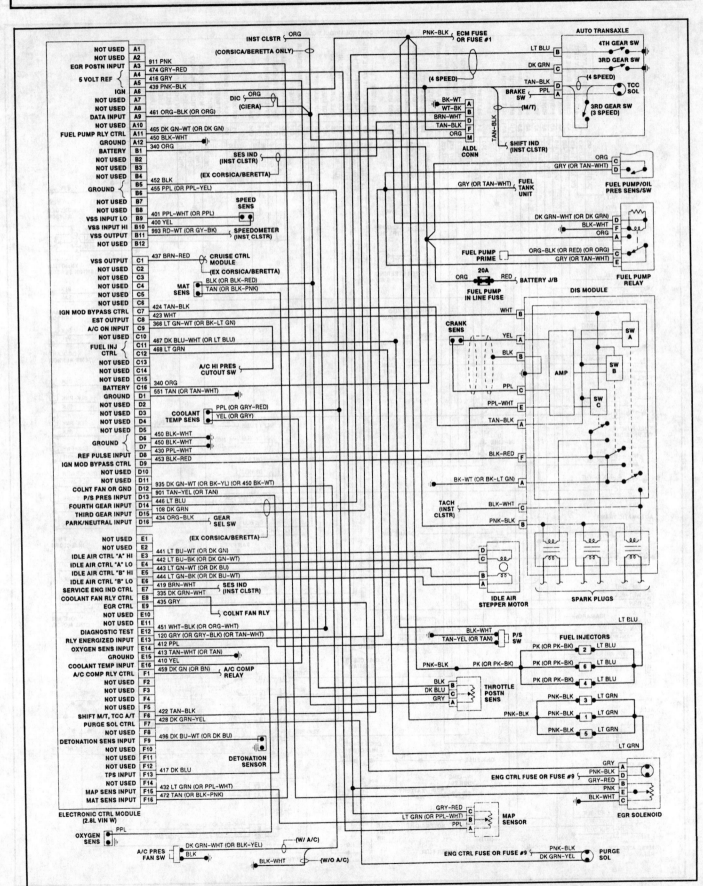

Fig. 54: 1989 2.8L (VIN W) A & L Bodies Fuel Injection Wiring Diagram

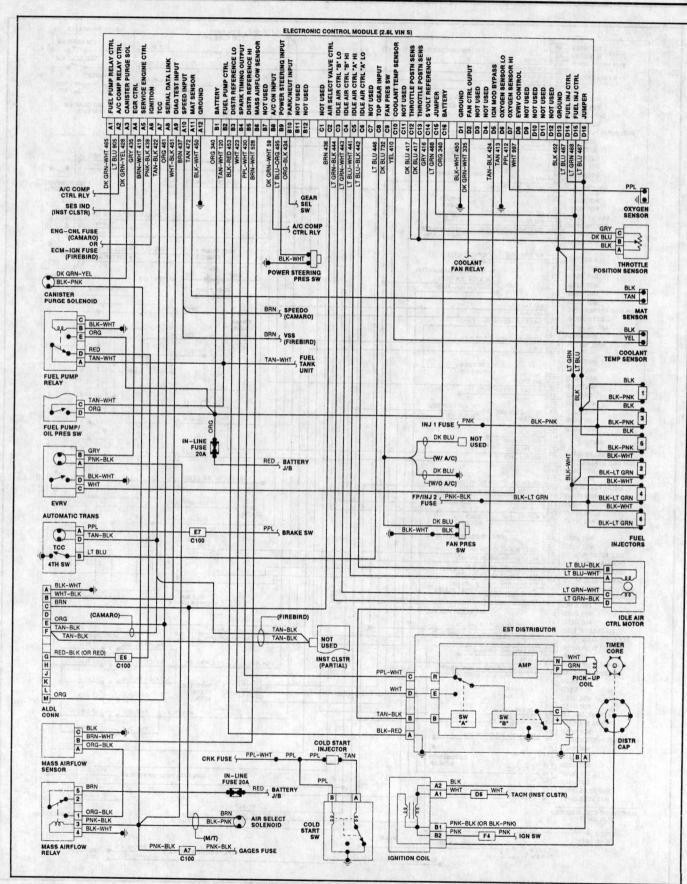

Fig. 55: 1989 2.8L (VIN S) F Body Fuel Injection Wiring Diagram

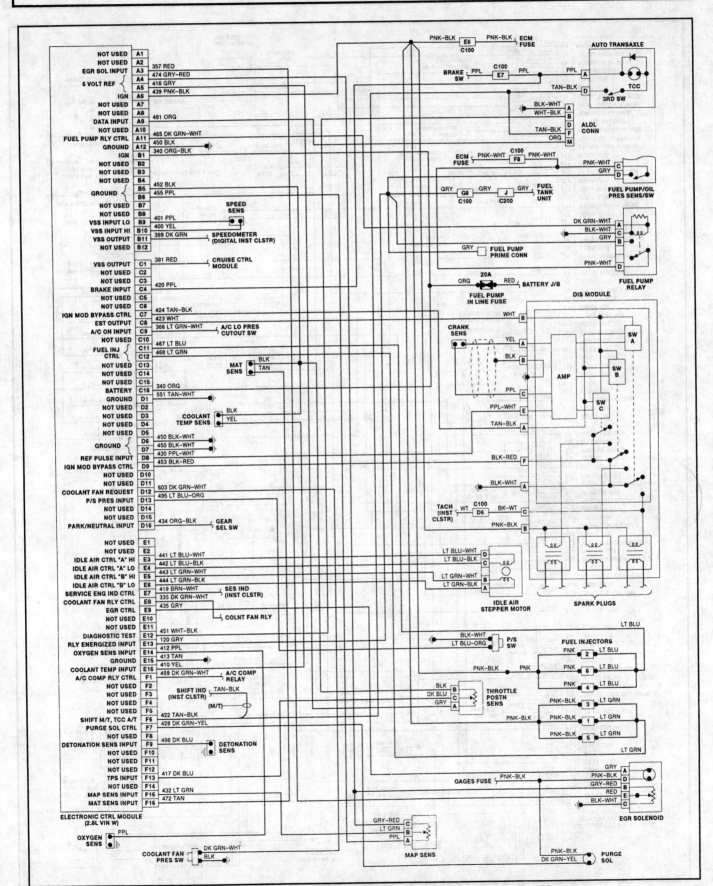

Fig. 56: 1989 2.8L (VIN W) J Body Fuel Injection Wiring Diagram

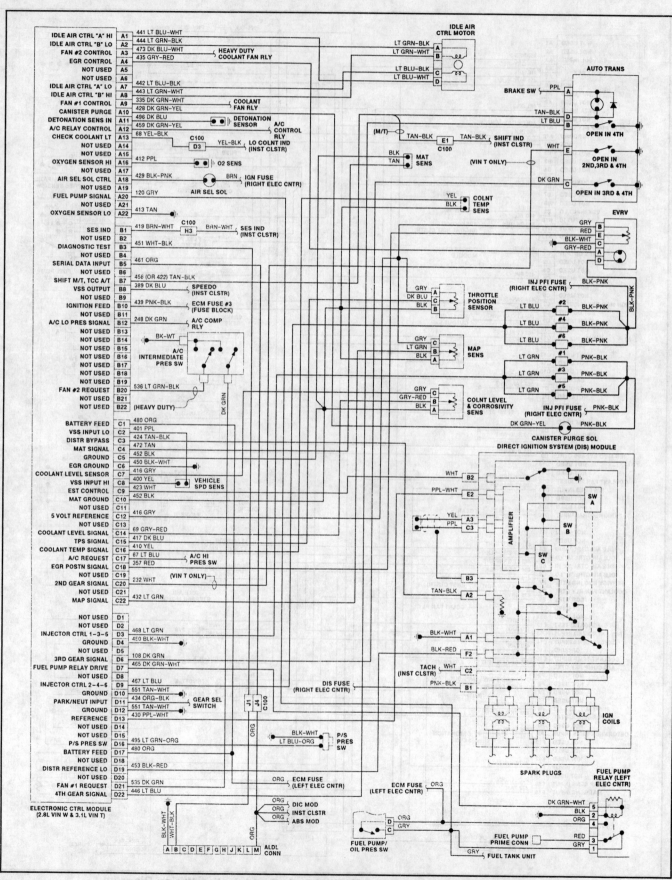

Fig. 57: 1989 2.8L (VIN W) Cutlass Supreme & Regal Fuel Injection Wiring Diagram

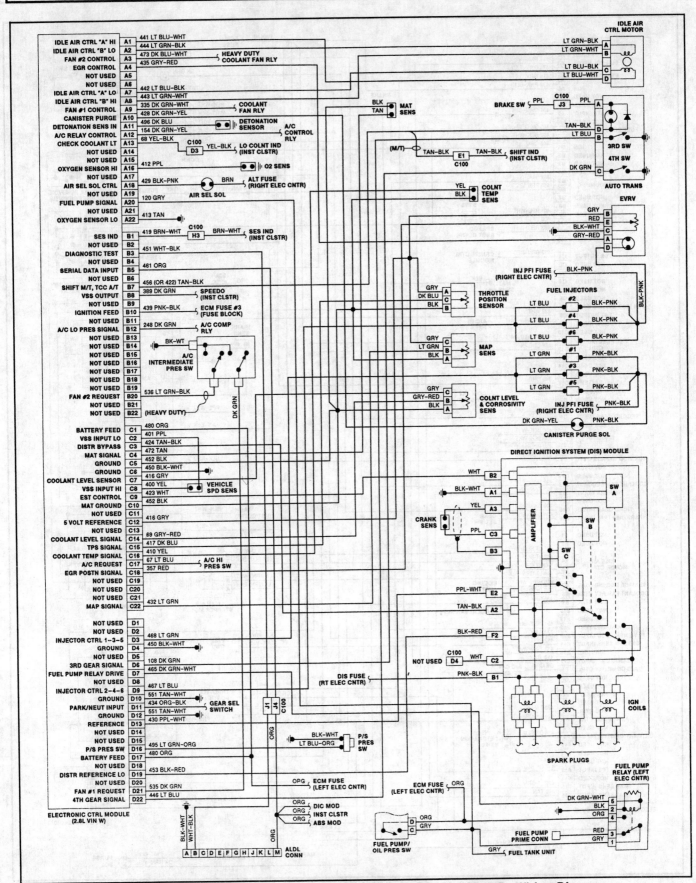

Fig. 58: 1989 2.8L (VIN W) Grand Prix Fuel Injection Wiring Diagram

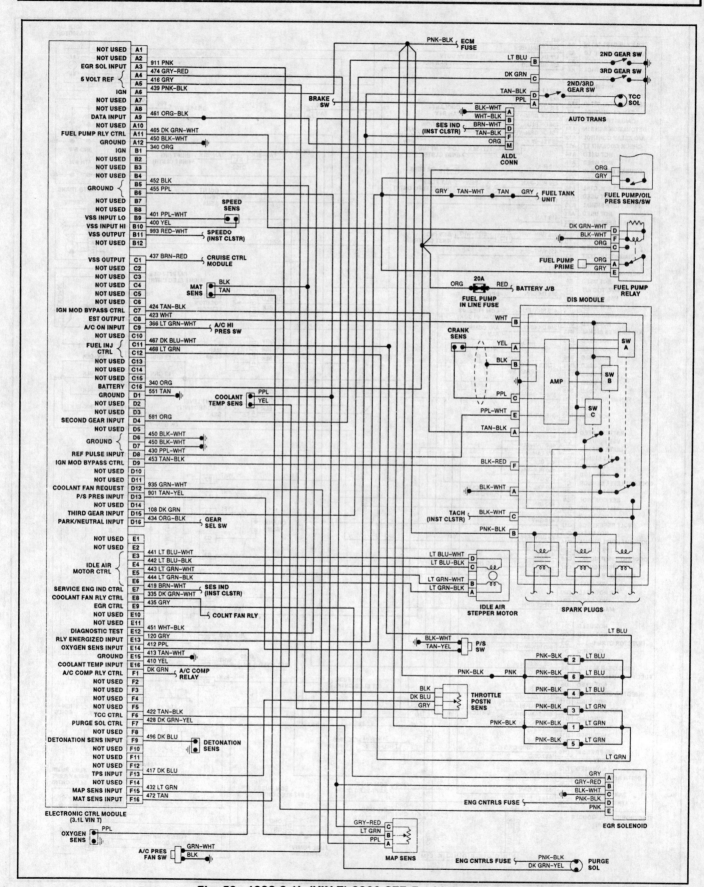

Fig. 59: 1989 3.1L (VIN T) 6000 STE Fuel Injection Wiring Diagram

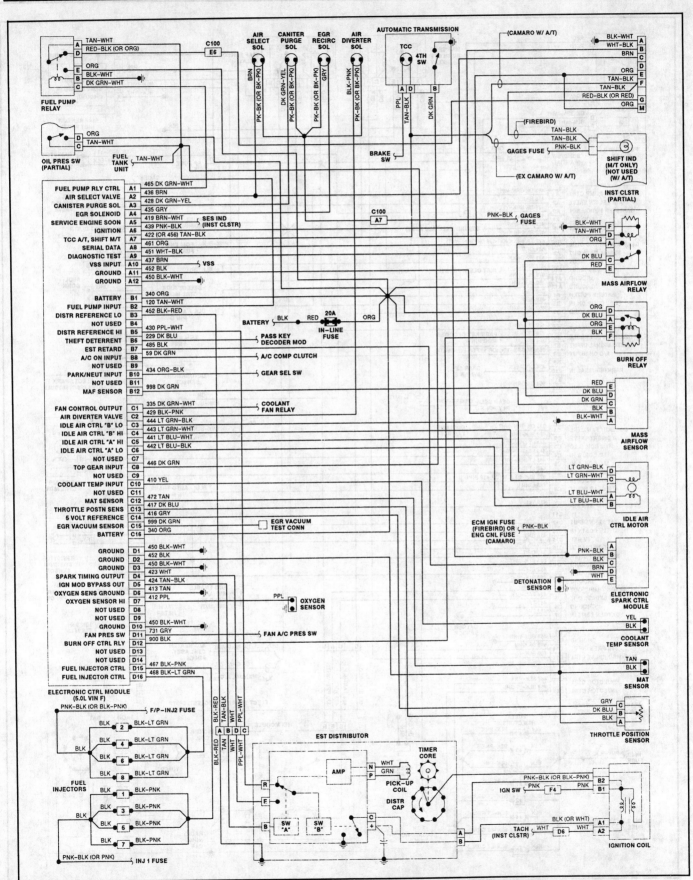

Fig. 60: 1989 5.0L (VIN F) & 5.7L (VIN K) F Body Fuel Injection Wiring Diagram

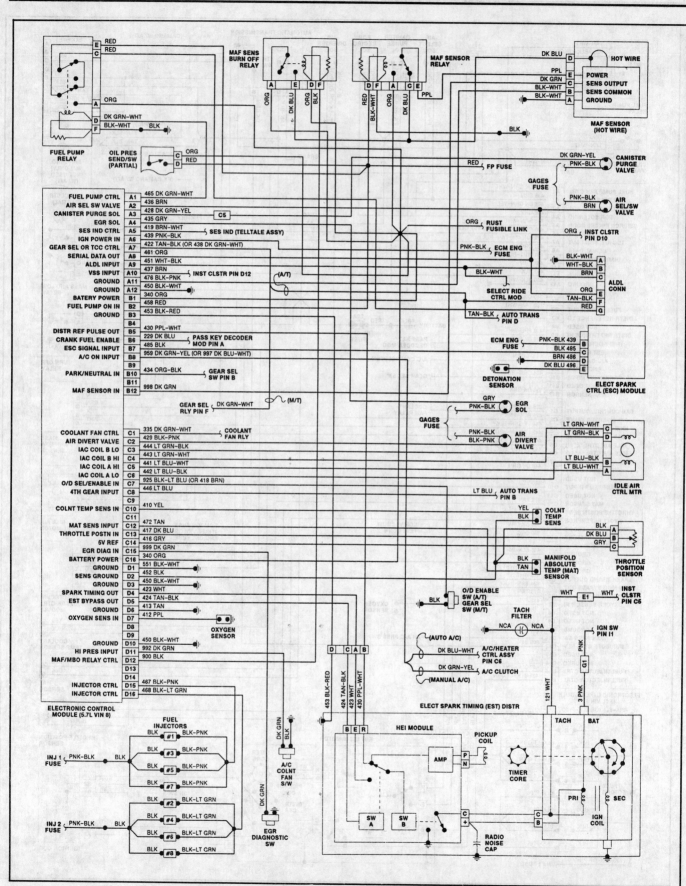

Fig. 61: *1989 5.7L (VIN 8) Corvette Fuel Injection Wiring Diagram*

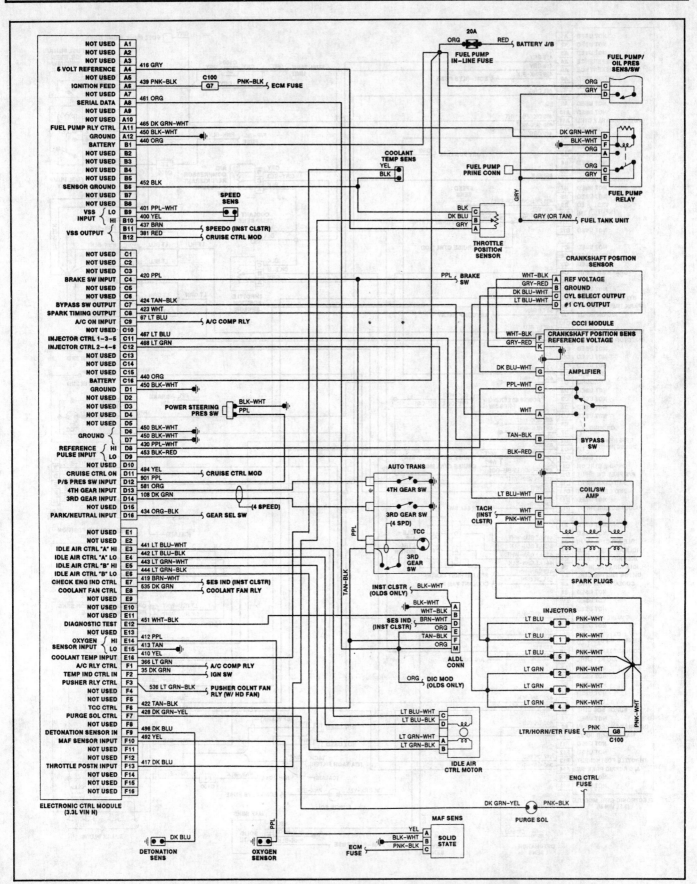

Fig. 62: 1989 3.3L (VIN N) A Body Fuel Injection Wiring Diagram

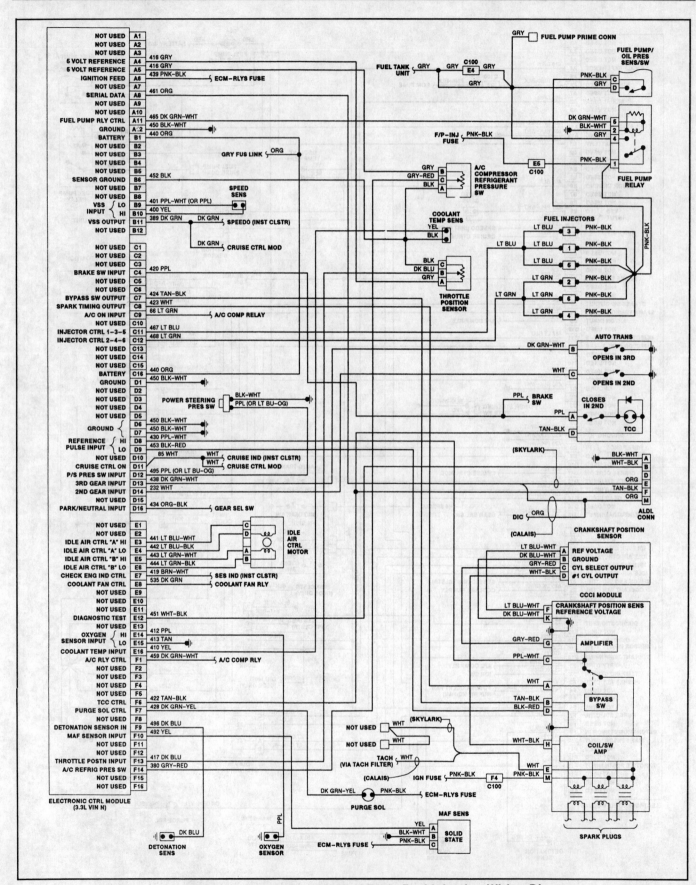

Fig. 63: 1989 3.3L (VIN N) N Body Fuel Injection Wiring Diagram

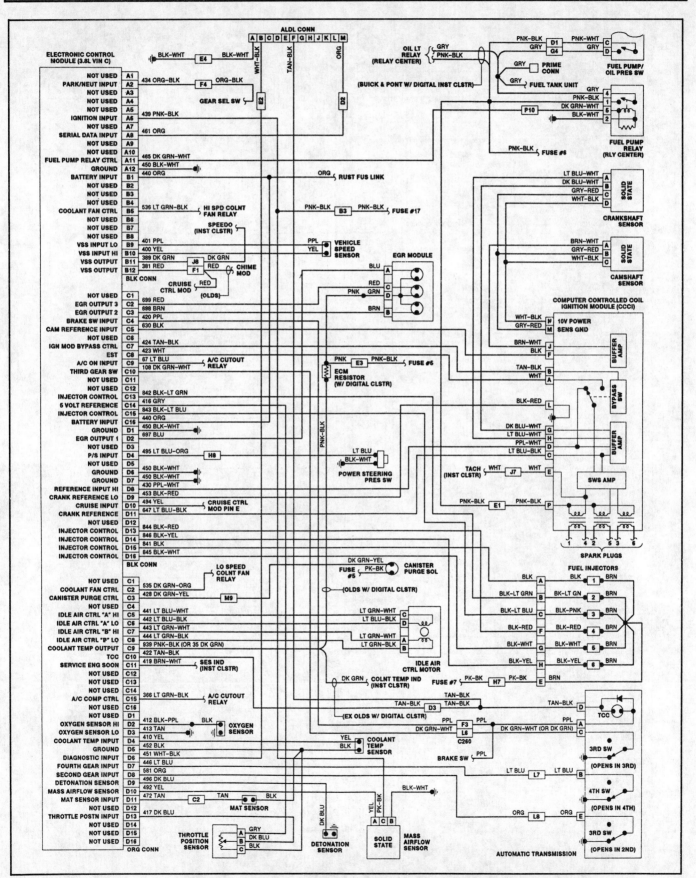

Fig. 64: 1989 3.8L (VIN C) C & H Bodies Fuel Injection Wiring Diagram

GLOSSARY OF TERMS

GLOSSARY

AFC	Airflow Control. A type of fuel injection system that measures the amount of air flowing past a sensor to determine engine fuel requirements.
AIRFLOW METER	Used to measure the volume of air entering the engine on many fuel injection systems.
AIRFLOW SENSOR	See AIRFLOW METER.
AIR MASS METER	See AIRFLOW METER.
AUXILIARY AIR VALVE	A special valve which provides additional air into the intake manifold during cold engine starting and operation.
CIS	Continuous Injection System. A Bosch fuel injection system which injects a steady stream of pressurized fuel into the intake manifold. Technically a mechanical system, albeit a very elaborate one, its wide use throughout the industry justifies its coverage in this manual. With the introduction of the Lambda feedback sensor in 1977 (CIS/Lambda system) and CIS-E in 1984, electronic control of the CIS was established.
CFI	Central Fuel Injection. A Ford Motor Co. fuel injection system that uses an injector-mounted throttle body assembly.
CLOSED LOOP OPERATION	Applies to systems utilizing an oxygen sensor. In this mode of operation, the system uses oxygen sensor information to determine air/fuel ratio. Adjustments are made accordingly and checked by comparing the new oxygen sensor signal to previous signals. No stored information is used. Compare to Open Loop Operation.
COLD START INJECTOR	An auxiliary fuel injector which injects additional fuel into the intake manifold during cold engine starting and operation.
CONTROL MODULE	One of several names for a solid state micro-computer which moniters engine conditions and controls certain engine functions, i.e. air/fuel ratio, injection and ignition timing, etc.
C³I	Computer Controlled Coil Ignition. This is General Motors computerized ignition coil system used on many different engine applications.
"D" JETRONIC	See MPC. "D" Jetronic is the term used by Bosch to describe a fuel injection system controlled by manifold pressure.
DEFI & DFI	A General Motors system, similar to earlier electronic fuel injection systems but with digital microprocessors. Analog inputs from various engine sensors are converted to digital signals before processing. The system is self-monitoring and self-diagnosing. It also has the capabilities of compensating for failed components and remembering intermittent failures.
ECA, ECM & ECU	Electronic Control Assembly, Electronic Control Module, and Electronic Control Unit. See CONTROL MODULE.
EFI	Electronic Fuel Injection. A fuel injection system which uses a micro-computer to determine and control the amount of fuel required by, and injected into, a particular engine.
EGI	Electronic Gasoline Injection. This is Mazda's fuel injection system used on RX7, RX7 Turbo, 323 and 626 models.
FUEL DISTRIBUTOR	Used on the Bosch CIS fuel injection system, the distributor is supplied with fuel from the fuel tank. Fuel leaves the distributor via 1 fuel line for each injector at a constant, pre-determined pressure.
FUEL INJECTOR	In all except CIS, CIS/Lambda and CIS-E systems, a spring-loaded, electro-magnetic valve which delivers fuel into the intake manifold, in response to electrical signals from the control module. In CIS, CIS/Lambda and CIS-E systems, a spring-loaded, pressure sensitive valve which opens at a pre-set value.
IDEAL MIXTURE	The air/fuel ratio at which the best compromise of engine performance to exhaust emissions is obtained. Typically 14.7:1.
"K" JETRONIC	See CIS. "K" Jetronic is the term used by Bosch to describe a fuel injection system which features continuous injection.
"L" JETRONIC	See AFC. "L" Jetronic is the term used by Bosch to describe a fuel injection system controlled by the air flowing through a sensor.

GLOSSARY (Cont.)

FUEL INJECTOR
In all except CIS, CIS/Lambda and CIS-E systems, a spring-loaded, electro-magnetic valve which delivers fuel into the intake manifold, in response to electrical signals from the control module. In CIS, CIS/Lambda and CIS-E systems, a spring-loaded, pressure sensitive valve which opens at a pre-set value.

IDEAL MIXTURE
The air/fuel ratio at which the best compromise of engine performance to exhaust emissions is obtained. Typically 14.7:1.

"K" JETRONIC
See CIS. "K" Jetronic is the term used by Bosch to describe a fuel injection system which features continuous injection.

"L" JETRONIC
See AFC. "L" Jetronic is the term used by Bosch to describe a fuel injection system controlled by the air flowing through a sensor.

LAMBDA SENSOR
A feedback device, usually located in the exhaust manifold, which detects the amount of oxygen present in exhaust gases in relation to the surrounding atmosphere.

"LIMP-IN" OR "LIMP-HOME"
This term is used by many manufacturers to explain the driveability characteristics of a failed computer system. Many computer systems store information that can be used to get the vehicle to a repair facility. In this mode of operation, driveability is greatly reduced.

LOGIC MODULE
See CONTROL MODULE.

MPC
Manifold Pressure Controlled fuel injection system. A fuel injection system which determines engine load based upon intake manifold pressure.

MPFI
Multi-Point Fuel Injection. A fuel injection system that uses 1 injector per cylinder, mounted on the engine to spray fuel near intake valve area of combustion chamber.

OPEN LOOP
An operational mode during which control module memory information is used to determine air/fuel ratio, injection timing, etc., as opposed to "real" sensor input. This occurs during cold engine operation, or when a particular sensor malfunctions. COMPARE To CLOSED LOOP OPERATION.

OXYGEN SENSOR
See LAMBDA SENSOR.

PGM-FI
Programmed Fuel Injection. This is Honda's fuel injection system used on Accord, Civic, Civic CRX and Prelude models.

PRESSURE REGULATOR
A device to control the pressure of fuel delivered to the fuel injector(s).

POWER MODULE
On Chrysler Motors vehicles, this module works in conjunction with the Logic Module. The power module is the primary power supply for the EFI system.

REST PRESSURE
Fuel pressure maintained within the system after engine shut down.

SENSORS
Devices which provide the control module, by electrical signal, with engine information as needed to properly control engine function.

SEFI OR SFI
Sequential Electronic Fuel Injection or Sequential Fuel Injection. A fuel injection system which uses a micro-computer to determine and control the amount of fuel required by, and injected into, a particular engine in the same sequence as engine firing sequence.

SYSTEM PRESSURE
The fuel pressure maintained in the system during normal engine operation.

TBI
Throttle Body Injection. Any of several injection systems which have the fuel injector(s) mounted in a centrally located throttle body, as opposed to positioning the injectors close to the intake ports.

TPI
Tuned Port Injection. A General Motors fuel injection system that uses tuned air intake runners for more precise delivery of airflow.

NOTES

NOTES

NOTES

NOTES